BTEC
Level 3

CONSTRUCTION & THE BUILT ENVIRONMENT

LEVEL 3

BTEC National

Simon Topliss | Mike Hurst

A PEARSON COMPANY

Published by Pearson Education Limited, a company incorporated in England and Wales, having its registered office at Edinburgh Gate, Harlow, Essex, CM20 2JE. Registered company number: 872828

www.pearsonschoolsandfecolleges.co.uk

First published 2010

13 12 11 10
10 9 8 7 6 5 4 3 2 1

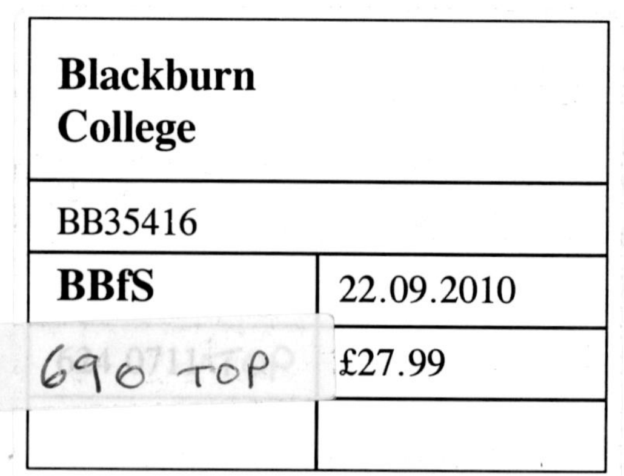

British Library Cataloguing in Publication Data
A catalogue record for this book is available from the British Library.

ISBN 978 1 846906 56 5

Designed by Wooden Ark
Typeset by HL Studios
Original illustrations © Pearson Education Limited 2010
Illustrated by HL Studios
Cover design by Visual Philosophy, created by eMC Design
Cover photo © Corbis: Artiga Photo
Back cover photos © (tl) Shutterstock: OrionTrail; (c) Pearson Education Ltd: Clark Wiseman, Studio 8; (tr) Image Source Ltd
Printed in Spain by Graficas Estella

Websites
The websites used in this book were correct and up to date at the time of publication. It is essential for tutors to preview each website before using it in class so as to ensure that the URL is still accurate, relevant and appropriate. We suggest that tutors bookmark useful websites and consider enabling learners to access them through the school/college intranet.

Acknowledgements
The publisher would like to thank the following for their kind permission to reproduce their photographs:

(Key: b-bottom; c-centre; l-left; r-right; t-top)

akg-images Ltd: Erich Lessing 419; **Alamy Images:** 1best of Photo 415, Adrian Sherratt 151, Alan Oliver 186, Chris Selby 16r, David J Green 126, David Robertson 88, David Young-Wolff 223, DJG Technology 349bl, E.D. Torial 189, Justin Kase 13, 229, Martin Mayer 325, Paul Williams 349cl; **Construction Photography:** Adam Greeman 200, Andrew Benton 117, Andrew Jankunas 16l, Buildpix 33, 236, Chris Henderson 331, ITP Images 245, Mike St Maur Sheil 30, Paul McMullin 357, 404, QA Photos / Jim Byrne 194, SImon Turner 326, Tom Lee 187, Xavier de Canto 148; **Corbis:** Construction Photography 153, Jeremy Horner 14; **Getty Images:** Gallo Images 9, Iconica 51; **Image Source Ltd:** 61, 225, 413; **iStockphoto:** Wojtek Kryczku 279; **Pearson Education Ltd:** Clark Wiseman, Studio 8 3, 119, 185, David Sanderson 181, Gareth Boden 201, 263, 268, 293, 358, Jules Selmes 24, Lord & Leverett 305, Mind Studio 35, 155, 267, 345, 391, 417, Rob Judges 227, 309; **Photolibrary.com:** Stockbyte 59; **Photos.com:** 115; **Science Photo Library Ltd:** Tony McConnell 122, 401, Victor de Schwanberg 402cr; **Shutterstock:** Alex Yeung 231, Andre Helbig 402cl, Andrew Lever 341, Carlos Arranz 147, Chris Pole 343, Christina Richards 321, David W Hughes 424, Dwight Smith 1, Ingvar Bjork 402tl, Joe Gough 307, Karol Kozlowski 40, Mark William Richardson 389, OrionTrail 249, PhotoGL 63, 66, Rob Marmion 387, 433, Stephen Coburn 31, Tiplyashin Anatoly 183, Tracy Whiteside 65, Zlatko Guzmic 265

All other images © Pearson Education

Every effort has been made to contact copyright holders of material reproduced in this book. Any omissions will be rectified in subsequent printings if notice is given to the publishers.

Contents

About the authors

Simon Topliss (BSc(Hons) PGCE MiFL) has worked for Edexcel for over six years as an External Verifier and Examiner. He has taught for over eleven years on a variety of Construction Technical Qualifications at Further and Higher Education levels. Simon has developed a suite of published resources for Levels 2 and 3, written unit specifications for the Firsts, Nationals and Higher Nationals, and is Chief Examiner and Principal Moderator on the CBE Diploma Level 2.

Mike Hurst (MSc. BSc(hons) ICIOB CertEd MifL) has taught a variety Construction and Civil Engineering qualifications in Further and Higher Education for over twenty years. Mike has been an External Verifier for Edexcel for a number of years, and has published a number of study guides in several areas, including health and safety in construction, construction management and civil engineering technology.

About your BTEC Level 3 Construction and the Built Environment

Choosing to study for a BTEC Level 3 National Construction and the Built Environment qualification is a great decision to make for lots of reasons. The Construction industry is a growing sector with many different specialisms within it. Studying for this qualification will give you a clear path of progression towards eventually working in this industry.

Your BTEC Level 3 National in Construction and the Built Environment is a **vocational** or **work-related** qualification. This doesn't mean that it will give you *all* the skills you need to do a job, but it does mean that you'll have the opportunity to gain specific knowledge, understanding and skills that are relevant to your chosen subject or area of work.

What will you be doing?

The qualification is structured into **mandatory units** (ones that you must do) and **optional units** (ones that you can choose to do). How many units you do and which ones you cover depend on the type of qualification you are working towards. For the certificate and subsidiary diploma, you will choose units from two groups of optional units.

Unit number	Credit value	Unit name	Certificate	Subsidiary Diploma	Diploma	Extended Diploma
1	10	Health, safety and welfare in construction and the built environment	OA	OA	M	M
2	10	Sustainable construction	OA	OA	M	M
3	10	Maths in construction and the built environment	OA	OA	M	M
4	10	Science and materials in construction and the built environment	OA	OA	M	M
5	10	Construction technology and design in construction and civil engineering	OB	OB	M	M
6	10	Building technology in construction	OB	OB	M	M
7	10	Project management in construction and the built environment			O	O
8	10	Graphical detailing in construction and the built environment			O	O
9	10	Measuring, estimating and tendering processes in construction and the built environment			O	O
10	10	Surveying in construction and civil engineering	OB	OB	O	O
15	10	Building surveying in construction			O	O
17	10	Building regulations and control for construction			O	O

OA – Optional Units Group A, OB – Optional Units Group B

How to use this book

This book is designed to help you through your BTEC Level 3 National Construction and the Built Environment course.

This book covers all the mandatory units and the most popular optional units.

This book contains many features that will help you use your skills and knowledge in work-related situations and assist you in getting the most from your course.

Introduction

The introduction gives you a snapshot of what to expect from each unit – and what you should be aiming for by the time you finish it!

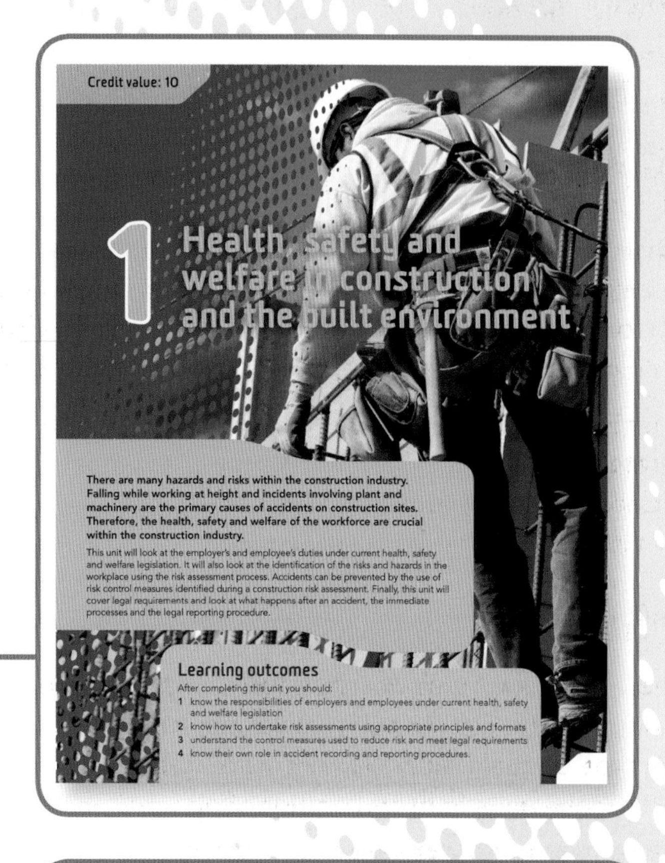

Credit value: 10

1 Health, safety and welfare in construction and the built environment

There are many hazards and risks within the construction industry. Falling while working at height and incidents involving plant and machinery are the primary causes of accidents on construction sites. Therefore, the health, safety and welfare of the workforce are crucial within the construction industry.

This unit will look at the employer's and employee's duties under current health, safety and welfare legislation. It will also look at the identification of the risks and hazards in the workplace using the risk assessment process. Accidents can be prevented by the use of risk control measures identified during a construction risk assessment. Finally, this unit will cover legal requirements and look at what happens after an accident, the immediate processes and the legal reporting procedure.

Learning outcomes

After completing this unit you should:

1. know the responsibilities of employers and employees under current health, safety and welfare legislation
2. know how to undertake risk assessments using appropriate principles and formats
3. understand the control measures used to reduce risk and meet legal requirements
4. know their own role in accident recording and reporting procedures.

1

Assessment and grading criteria

This table explains what you must do to achieve each of the assessment criteria for each unit. For each assessment criterion, shown by the grade button P1, there is an assessment activity.

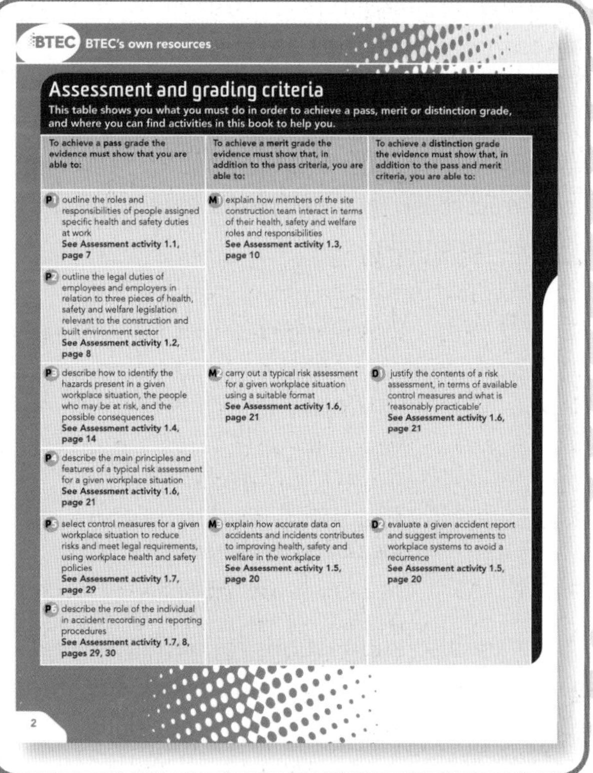

BTEC BTEC's own resources

Assessment and grading criteria

This table shows you what you must do in order to achieve a pass, merit or distinction grade, and where you can find activities in this book to help you.

To achieve a pass grade the evidence must show that you are able to:	To achieve a merit grade the evidence must show that, in addition to the pass criteria, you are able to:	To achieve a distinction grade the evidence must show that, in addition to the pass and merit criteria, you are able to:
P1 outline the roles and responsibilities of people assigned specific health and safety duties at work **See Assessment activity 1.1, page 7**	M1 explain how members of the site construction team interact in terms of their health, safety and welfare roles and responsibilities **See Assessment activity 1.3, page 10**	
P2 outline the legal duties of employees and employers in relation to three pieces of health, safety and welfare legislation relevant to the construction and built environment sector **See Assessment activity 1.2, page 8**		
P3 describe how to identify the hazards present in a given workplace situation, the people who may be at risk, and the possible consequences **See Assessment activity 1.4, page 14**	M2 carry out a typical risk assessment for a given workplace situation using a suitable format **See Assessment activity 1.6, page 21**	D1 justify the contents of a risk assessment, in terms of available control measures and what is 'reasonably practicable' **See Assessment activity 1.6, page 21**
P4 describe the main principles and features of a typical risk assessment for a given workplace situation **See Assessment activity 1.6, page 21**		
P5 select control measures for a given workplace situation to reduce risks and meet legal requirements, using workplace health and safety policies **See Assessment activity 1.7, page 29**	M3 explain how accurate data on accidents and incidents contributes to improving health, safety and welfare in the workplace **See Assessment activity 1.5, page 20**	D2 evaluate a given accident report and suggest improvements to workplace systems to avoid a recurrence **See Assessment activity 1.5, page 20**
P6 describe the role of the individual in accident recording and reporting procedures **See Assessment activity 1.7, 8, pages 29, 30**		

2

Assessment

Your tutor will set **assignments** throughout your course for you to complete. These may take the form of projects where you research, plan, prepare, make and evaluate a piece of work, sketchbooks, case studies and presentations. The important thing is that you evidence your skills and knowledge to date.

Stuck for ideas? Daunted by your first assignment? These learners have all been through it before…

Unit 1 Health, safety and welfare in construction and the built environment

How you will be assessed

The evidence requirements for pass, merit and distinction grades are shown in the grading criteria grid. Evidence for this unit may be gathered from a variety of sources, including well-planned investigative assignments, practical work or reports of practical assignments. You will be given written assessments to complete for the performing element of this unit. These will contain a number of assessment criteria from pass, merit and distinction.
This unit will be assessed by the use of two assignments:

- Assignment one will cover P1, P2, P3, M1, M2 and D1
- Assignment two will cover P4, P5, P6, M3 and D2.

Doug

This unit opened my eyes to the hazards associated with working on a construction site. Many of the unfortunate fatalities that occur are caused by silly mistakes which could be avoided if operatives take time to carefully consider the risks of an activity.

I now know that a risk assessment clearly identifies the hazards associated with a working environment. I understand that you have to reduce the high risks to acceptable levels using control measures.

I also now know what to wear when I visit construction sites and how personal protective equipment prevents me from being harmed by hazards.

Over to you

- What do you already know about health, safety and welfare legislation?
- What are the health and safety roles of the employee, the employer, and the local authorities?
- What are you looking forward to learning about in this unit?

3

Activities

There are different types of activities for you to do: **Assessment activities** are suggestions for tasks that you might do as part of your assignment and will help you develop your knowledge, skills and understanding. **Grading tips** that clearly explain what you need to do in order to achieve a pass, merit or distinction grade.

Assessment activity 1.1 P1

You have just started work as an assistant site manager for a housing refurbishment company. You are not sure of anyone's duties. Outline the health and safety responsibilities of the following roles:

- client
- CDM coordinator
- architect. 

Grading tip

For P1 outline the responsibilities as they would be listed in the CDM regulations on duties of the various parties to a contract.

There are also suggestions for **activities** that will give you a broader grasp of the industry, stretch your imagination and deepen your skills.

Activity: Risk calculation

Construction sites are full of physical hazards. On the site where you work, there is a reinforcing bar sticking out of the ground which has the potential to cause injury to a worker should they fall on it and get impaled.

Divide yourselves into teams and discuss the likelihood that a worker could fall on the reinforcing bar. If there was an injury, what would be its severity?

Personal, learning and thinking skills

Throughout your BTEC Level 3 National Construction and the Built Environment course there are lots of opportunities to develop your personal, learning and thinking skills. Look out for these as you progress.

> **PLTS**
>
> Planning and carrying out research on health and safety legislation, and appreciating the consequences of decisions made in its use, will help develop your skills as an **independent enquirer**.

Functional skills

It's important that you have good English, maths and ICT skills – you never know when you'll need them, and employers will be looking for evidence that you've got these skills too.

> **Functional skills**
>
> By selecting relevant information on health and safety legislation you will develop your **English** skills.

Key terms

Technical words and phrases are easy to spot. You can also use the glossary at the back of the book.

> **Key term**
>
> **Panel fencing** – this is constructed out of mesh squares welded to a framework. The panels sit into feet and are bolted together with clips; they are high enough so they cannot be climbed over.

WorkSpace

Case studies provide snapshots of real workplace issues, and show how the skills and knowledge you develop during your course can help you in your career.

There are also mini-case studies throughout the book to help you focus on your own projects.

WorkSpace

Simon Green

Health and safety officer

Simon has worked on several construction sites for the company he works for. During this time he has seen several accidents and a few near misses that have occurred as a result of poor health and safety control. Simon is often involved in first aid and the completion of risk assessments and is keen to reduce the accident record of the company.

As a result of Simon's interest he was offered the position of trainee health and safety officer three years ago. Simon relished the chance to take this opportunity to influence the health and safety policies of the company and to try to improve its health and safety performance.

Simon reviewed the company's health and safety policy and made several changes to the way in which work is undertaken and activities supervised. Employees are now educated during their induction to each construction site about the health and safety arrangements. This has resulted in a reduction of the accident rate over the past three years.

Simon has managed to accomplish this through thinking about health and safety and evaluating the accidents that had been occurring. He then devised policies to reduce and target the specific accidents that were occurring.

The managing director was extremely impressed with Simon's influence on health and safety, especially since the liability insurance premiums that the company were paying have almost halved. Simon was promoted to Health and Safety Officer last year.

Think about it!

- How would you reduce a high accident rate?
- How would you win employees over?
- What support would you need from management?

31

Just checking

When you see this sort of activity, take stock! These quick activities and questions are there to check your knowledge. You can use them to see how much progress you've made or as a revision tool.

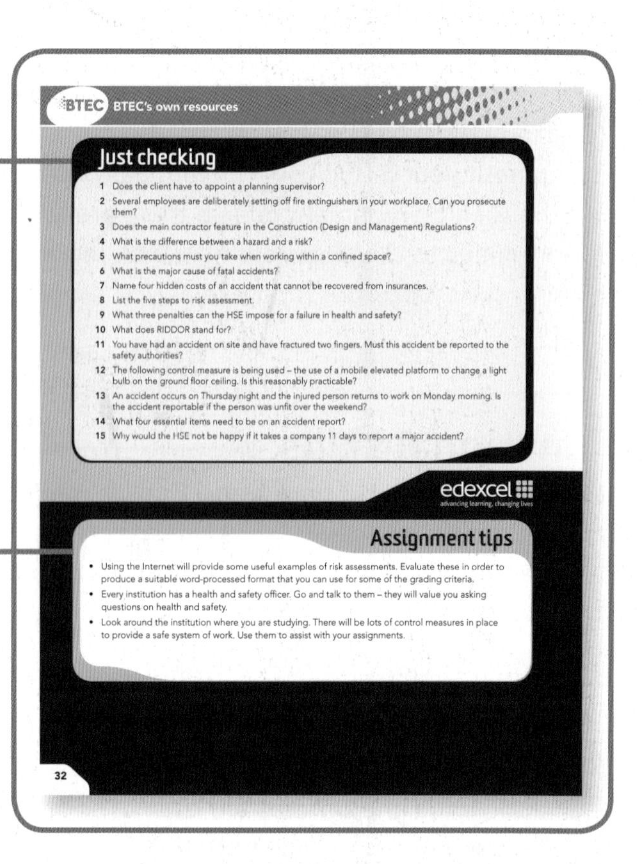

BTEC BTEC's own resources

Just checking

1 Does the client have to appoint a planning supervisor?
2 Several employees are deliberately setting off fire extinguishers in your workplace. Can you prosecute them?
3 Does the main contractor feature in the Construction (Design and Management) Regulations?
4 What is the difference between a hazard and a risk?
5 What precautions must you take when working within a confined space?
6 What is the major cause of fatal accidents?
7 Name four hidden costs of an accident that cannot be recovered from insurances.
8 List the five steps to risk assessment.
9 What three penalties can the HSE impose for a failure in health and safety?
10 What does RIDDOR stand for?
11 You have had an accident on site and have fractured two fingers. Must this accident be reported to the safety authorities?
12 The following control measure is being used – the use of a mobile elevated platform to change a light bulb on the ground floor ceiling. Is this reasonably practicable?
13 An accident occurs on Thursday night and the injured person returns to work on Monday morning. Is the accident reportable if the person was unfit over the weekend?
14 What four essential items need to be on an accident report?
15 Why would the HSE not be happy if it takes a company 11 days to report a major accident?

edexcel advancing learning, changing lives

Assignment tips

- Using the Internet will provide some useful examples of risk assessments. Evaluate these in order to produce a suitable word-processed format that you can use for some of the grading criteria.
- Every institution has a health and safety officer. Go and talk to them – they will value you asking questions on health and safety.
- Look around the institution where you are studying. There will be lots of control measures in place to provide a safe system of work. Use them to assist with your assignments.

32

Edexcel's assignment tips

At the end of each unit you'll find hints and tips to help you get the best mark you can, such as the best websites to go to, checklists to help you remember processes and useful facts and figures.

Have you read your **BTEC Level 3 National Construction and the Built Environment Study Skills Guide**? It's full of advice on study skills, putting your assignments together and making the most of being a BTEC Construction and the Built Environment student.

Ask your tutor about extra materials to help you through your course. **The Tutor Resource Pack** which accompanies this book contains presentations and information about the Construction and the Built Environment sector.

Your book is just part of the exciting resources from Edexcel to help you succeed in your BTEC course.

Visit:

- www.edexcel.com/BTEC or
- www.pearsonfe.co.uk/BTEC 2010

:redit value: 10

1 Health, safety and welfare in construction and the built environment

here are many hazards and risks within the construction industry. alling while working at height and incidents involving plant and ıachinery are the primary causes of accidents on construction sites. herefore, the health, safety and welfare of the workforce are crucial vithin the construction industry.

his unit will look at the employer's and employee's duties under current health, safety nd welfare legislation. It will also look at the identification of the risks and hazards in the orkplace using the risk assessment process. Accidents can be prevented by the use of sk control measures identified during a construction risk assessment. Finally, this unit will over legal requirements and look at what happens after an accident, the immediate rocesses and the legal reporting procedure.

Learning outcomes

After completing this unit you should:

1. know the responsibilities of employers and employees under current health, safety and welfare legislation
2. know how to undertake risk assessments using appropriate principles and formats
3. understand the control measures used to reduce risk and meet legal requirements
4. know their own role in accident recording and reporting procedures.

Assessment and grading criteria

This table shows you what you must do in order to achieve a pass, merit or distinction grade, and where you can find activities in this book to help you.

To achieve a **pass** grade the evidence must show that you are able to:	To achieve a **merit** grade the evidence must show that, in addition to the pass criteria, you are able to:	To achieve a **distinction** grade the evidence must show that, in addition to the pass and merit criteria, you are able to:
P1 outline the roles and responsibilities of people assigned specific health and safety duties at work **See Assessment activity 1.1, page 7**	**M1** explain how members of the site construction team interact in terms of their health, safety and welfare roles and responsibilities **See Assessment activity 1.3, page 10**	
P2 outline the legal duties of employees and employers in relation to three pieces of health, safety and welfare legislation relevant to the construction and built environment sector **See Assessment activity 1.2, page 8**		
P3 describe how to identify the hazards present in a given workplace situation, the people who may be at risk, and the possible consequences **See Assessment activity 1.4, page 14**	**M2** carry out a typical risk assessment for a given workplace situation using a suitable format **See Assessment activity 1.6, page 21**	**D1** justify the contents of a risk assessment, in terms of available control measures and what is 'reasonably practicable' **See Assessment activity 1.6, page 21**
P4 describe the main principles and features of a typical risk assessment for a given workplace situation **See Assessment activity 1.6, page 21**		
P5 select control measures for a given workplace situation to reduce risks and meet legal requirements, using workplace health and safety policies **See Assessment activity 1.7, page 29**	**M3** explain how accurate data on accidents and incidents contributes to improving health, safety and welfare in the workplace **See Assessment activity 1.5, page 20**	**D2** evaluate a given accident report and suggest improvements to workplace systems to avoid a recurrence **See Assessment activity 1.5, page 20**
P6 describe the role of the individual in accident recording and reporting procedures **See Assessment activity 1.7, page 29, and Assessment activity 1.8, page 30**		

How you will be assessed

The evidence requirements for pass, merit and distinction grades are shown in the grading criteria grid. Evidence for this unit may be gathered from a variety of sources, including well-planned investigative assignments, practical work or reports of practical assignments. You will be given written assessments to complete for the performing element of this unit. These will contain a number of assessment criteria from pass, merit and distinction.

This unit will be assessed by the use of two assignments:

- Assignment one will cover P1, P2, P3, M1, M2 and D1
- Assignment two will cover P4, P5, P6, M3 and D2.

Doug

This unit opened my eyes to the hazards associated with working on a construction site. Many of the unfortunate fatalities that occur are caused by silly mistakes which could be avoided if operatives take time to carefully consider the risks of an activity.

I now know that a risk assessment clearly identifies the hazards associated with a working environment. I understand that you have to reduce the high risks to acceptable levels using control measures.

I also now know what to wear when I visit construction sites and how personal protective equipment prevents me from being harmed by hazards.

Over to you

- What do you already know about health, safety and welfare legislation?
- What are the health and safety roles of the employee, the employer, and the local authorities?
- What are you looking forward to learning about in this unit?

1 Know the responsibilities of employers and employees under current health, safety and welfare legislation

Build up

Site hazards

The safety of construction workers is paramount. Employers have a duty to provide a safe working environment for their employees, visitors or anyone else affected by their actions.

In groups discuss why the fatality rate within the construction industry is not falling as much as hoped.

Roles and responsibilities

The client

The **client**, whether a landlord, private individual or a company, must demonstrate an acceptable standard of health and safety. Under the Construction (Design and Management) Regulations 2007 (usually referred to as the CDM Regulations), they have specific health and safety responsibilities:

- they have to appoint the CDM Co-ordinator.
- they must appoint a principal contractor.
- they must ensure that the construction phase health and safety plan has been produced before commencement.
- they have to store the health and safety file on completion.

Designers and architects

Designers have a duty to reduce the effect of or eliminate any hazards associated with the building's design. This would extend to looking at items such as how light bulbs can be changed safely, especially in large halls where bulbs may be located at extreme heights!

The employer – directors and managers

Employers have a general duty under the Health and Safety at Work Act (HASAWA) 1974 to ensure, so far as is reasonably practicable, their employees' health, safety and welfare at work.

Key terms

Client – the person who will ultimately own the constructed building or project and who pays for the work.

Employer – the person who owns the company which is constructing the building or project; the employer may be represented by a managing director or by a multinational organisation with shareholders and a chief executive officer.

Specific responsibilities of employers, as listed in the HASAWA under the general duties to employees, are as follows:

- to ensure the health and safety of all employees
- to provide safe systems of work, including handling, storage and transport as well as information training and supervision
- to provide a health and safety policy if there are five or more employees
- to appoint union safety representatives
- to consult and cooperate with employees on safety measures
- to observe the regulations on safety committees
- not to charge for anything provided for safety (Health and Safety Executive).

This list is quite comprehensive and covers all the vital elements to provide a safe environment for employees.

The employee

Employees, under the HASAWA, have the following general duties:

- to act with due care for themselves and others, e.g. to walk rather than run down a corridor
- to cooperate with the employer, e.g. taking part in **toolbox talks**
- to correctly use anything provided for health and safety in accordance with any instruction or training
- not to misuse or damage equipment provided for health and safety purposes.

The principal contractor

The **main contractor** is often referred to as the 'principal contractor' under the terms of the CDM Regulations 2007. The main contractor has general and specific duties as an employer under the HASAWA 1974, and the CDM Regulations 2007 place specific responsibilities upon them as follows:

- plan, manage and monitor the construction phase of the contract in liaison with contractors employed to do the work
- prepare, develop and implement a written plan and site rules (formally known as the health and safety plan)
- give contractors relevant parts of the plan as information and instructions
- make sure suitable welfare facilities are provided and maintained throughout the construction phase
- check the competence of all people involved in construction activities
- ensure all workers have site inductions and any further information and training needed for their work
- consult with the workers on site
- liaise with the CDM coordinator regarding ongoing designs and changes
- make sure that the site is secure.

Contractors

Contractors generally work under contract to the principal contractor. The principal contractor has to ensure that contractors are competent and are provided with the relevant information including a site safety induction. The contractor has to provide **method statements** and risk assessments so that the principal contractor can safely coordinate the work. There is quite a long list of duties for contractors, which can be found in the HSE Guidance.

Key terms

Employees – workers who receive wages for their skills from the employer.

Toolbox talks – a time when everyone stops work to discuss a safety aspect of a current job.

Main contractor – the company constructing the building or project; this may be a large or small organisation.

Contractors – separate companies who work for the main contractor, e.g. a heating engineering company.

Method statements – documents which identify the methods used to price the work items, the plant and labour required for each activity.

Remember!

Subcontractors are employers in their own right and have to fulfil the duties of employers (see page 4).

Health and Safety Executive

The **Health and Safety Executive (HSE)** was set up to regulate and control health and safety in the UK. The HSE has many divisions that cover industries from nuclear power to agriculture and railways to construction. The HSE plays a wide role in controlling health and safety in construction. Its responsibilities are:

- to advise
- to inspect
- to enforce.

The HSE also offers and promotes health and safety information. If it wishes, the HSE has powers to inspect any construction site notified to it under the notification rules associated with the F10. The F10 is the official form that is completed by the contractor and sent to the HSE to advise it that works are about to commence. The F10 contains a brief description of the work, where it is being carried out, for how long and who will be working on it.

The HSE can enforce health and safety legislation in two ways, either by an improvement notice or by a prohibition notice. An improvement notice is served when a potentially harmful defect has been found during an inspection that requires correcting within a certain time. A prohibition notice is served when there is a serious and imminent danger to persons on the site. Work or activity is stopped immediately and cannot be restarted until the defect is corrected.

Remember!

The HSE also carries out extensive inspections and investigations following an accident, especially if there has been a fatality. In the last 25 years, nearly 3000 people have been killed within the construction sector. These investigations may often lead to a conviction for failure of duties under the HASAWA, with either a fine or imprisonment or both.

The local authority

Local authorities' responsibilities for safety mainly cover environmental health and other general duties such as highways and road safety. The environmental health officer may visit a site on grounds of noise, nuisance or environmental issues. For example an environmental health officer may come out if a contractor is disturbing surrounding residents with noise from a compressor, or even if excessive dust is blowing into people's homes.

Key terms

Health and Safety Executive (HSE) – a body set up by the Health and Safety Commission acting under the Health and Safety at Work Act. The HSE is responsible for inspecting, and enforcing health and safety.

F10 – the official document that informs the HSE that a company is undertaking a project.

Local authority – the elected local council which runs the services within a geographical area.

The construction design and management (CDM) coordinator

The CDM coordinator has the following key responsibilities under the CDM Regulations:

- to ensure that the HSE is notified with the signed F10 of the project
- to ensure that there is coordination and cooperation between designers if there is more than one designer on the contract, e.g. a structural designer, an architect, an interior designer and a landscape architect
- to ensure that there is good communication between client, designers and contractors
- to make sure that the health and safety file is maintained
- to advise and assist the client
- to check a health and safety file is prepared and handed over to the client.

Remember!

The CDM Coordinator is given specific roles and responsibilities on planning the health and safety before and after a contract under the Construction (Design and Management) Regulations 2007.

Consequences for individuals and employers

When an accident occurs the consequences can be devastating for all involved especially if the accident results in a fatality. Individuals and employers who are found to be at fault can be:

- fined up to a maximum of £20,000 for a breach of regulations
- imprisoned
- charged with corporate manslaughter. This carries unlimited fines and imprisonment.

Case study: Factory health and safety

Read the following newspaper article on an accident that occurred when a roofing contractor fell through a fragile roof covering.

When you have fully read the article, answer the following:

- Identify and describe the roles of the people responsible for health, safety and welfare on a construction project. In the case study who was responsible for health and safety in this workplace.
- What were their roles and responsibilities in this case?
- Was there anything that could have been done to prevent Mr Cartwright's death?

Roofer dies in fall

Health and safety officials are considering a criminal prosecution after an experienced roofer fell through a roof panel to his death.

Graham Cartwright was checking for leaks on an asbestos roof at a local factory. Factory workers at the inquest described hearing the roof crack. They saw Mr Cartwright fall through the air then hit his head on some machinery before landing.

Mr Cartwright had worked at the factory as a contractor many times. The inquest heard that the factory had a detailed health and safety manual and inducted subcontractors on health and safety procedures. The factory's managing director told the inquest that repair and maintenance of the roof was carried out regularly by subcontractors.

An inspector from the Health & Safety Executive confirmed that the HSE had investigated Mr Cartwright's death. She explained that asbestos sheeting panels sometimes shattered to the touch and were the cause of several fatalities each year.

Assessment activity 1.1

BTEC

You have just started work as an assistant site manager for a housing refurbishment company. You are not sure of anyone's duties. Outline the health and safety responsibilities of the following roles:

- client
- CDM coordinator
- architect. P1

Grading tip

For P1 outline the responsibilities as they would be listed in the CDM regulations on duties of the various parties to a contract.

PLTS

By analysing and evaluating CDM regulations summary information in order to judge its relevance and value in outlining the roles and responsibilities of specific health and safety duties at work you will develop your skills as an **independent enquirer**.

Functional skills

By selecting and using appropriate sources of ICT-based and other forms of information to research the CDM regulations you can improve your **ICT** skills.

Legislation

Legislation on health and safety is there to protect *everyone* involved in and around the construction workplace. It also covers people who are not employees, such as delivery drivers and visitors, and places that are not necessarily the employer's, for example parts of the site that are let to another employer.

Assessment activity 1.2

Identify three main pieces of health, safety and welfare legislation relevant to the construction and built environment sector and describe the legal duties of employees and employers in terms of such legislation. P2

Grading tip

For P2 you need to describe legal duties of employees and employers as identified in three pieces of legislation. Look for summaries of the three pieces of legislation as they will put the legal language into a more understandable form.

PLTS

Planning and carrying out research on health and safety legislation, and appreciating the consequences of decisions made in its use, will help develop your skills as an **independent enquirer**.

Functional skills

By selecting relevant information on health and safety legislation you will develop your **English** skills.

Remember!

The HSE website (www.hse.gov.uk) has a lot of information that can help you.

The Health and Safety at Work Act (HASAWA) 1974

The HASAWA 1974 is a very important piece of legislation from which many regulations have been developed. There are numerous sections in the Act that cover where the duty of care lies on a construction site. For example, manufacturers of materials used to construct a building must ensure these materials are safe and contractors must provide a safe means of access and egress from the place of work, such as scaffolding.

There has been a considerable amount of health and safety legislation since the HASAWA 1974. What follows are just some of the primary pieces of legislation that are concerned with the construction industry.

Remember!

Health and safety legislation continually evolves and changes. Look for updated laws and regulations.

Construction (Design and Management) Regulations 2007

The Construction (Design and Management) Regulations 2007 were developed from a European directive that had looked at the principal cause of accidents. The research showed that nearly a third of accidents could be traced back to the design stage of a project. Thus, the onus of a risk assessment now has to be undertaken by the designer, so that the contractor and the client are aware of the inherent risks associated with constructing and maintaining a building. The regulations placed duties upon five main parties to the design and construction process:

- the client
- the designer
- the CDM coordinator
- the principal contractor
- contractors.

Specific duties concerning these five are covered later in the chapter.

Work at Height Regulations 2005

The Work at Height Regulations were introduced to try to control the large number of fatalities in construction that result from falls from a height.

To control this, the regulations state that a person must avoid working at height where an alternative method can be used. For example, cleaning first-floor windows to a building can be done using a specialist pole, or if it is a new building, altering the design to install tilt and turn windows which can be cleaned from the inside.

Where there is no alternative and work at height is necessary, the regulations state that: 'Where work is carried out at height, every employer shall take suitable and sufficient measures to prevent, so far as is reasonably practicable, any person falling a distance liable to cause personal injury' (Work at Height Regulations, © Crown Copyright 2005). All the following precautions can be used to help prevent a person falling from height:

- using scaffolding instead of ladders
- mobile elevated platforms
- guard rails, barriers and handrails
- toe boards.

Another major item the regulations cover is to restrict the distance a person can fall. This can be done using:

- **PPE** suspension harnesses to secure a person to a solid structure and restrict the distance they fall, through a lanyard
- netting or airbags to catch a person falling.

The Work at Height Regulations ask employers to check the competency of each individual asked to work at height. To do this, employers may provide suitable training, instruction and supervision and make sure that employees are happy and feel confident at the height at which they are being asked to work. The regulations also require that the work is suitably planned and organised and has sufficient supervision. The schedules at the end of the regulations provide details of requirements for working platforms; ladders for risk-assessed, short-duration work; access and egress, requirements for PPE and **collective means of protection**.

Key terms

PPE – personal protective equipment which is provided for the individual to use to protect themselves against certain hazards where there is no alternative method.

Collective means of protection – a system that protects the whole workforce and not just the individual. For example, a scaffold with guardrails, handrails, toe boards and netting protects everyone working on or using it.

This is definitely not an acceptable method of working at height

Management of Health, Safety and Welfare Regulations 1999 (MHSW)

The Management of Health, Safety and Welfare Regulations were introduced to reinforce the message of risk assessment. The five steps will be covered later in this unit. Some of the regulations that apply to construction work which illustrate the wide range of duties that an employer has under these regulations include:

- health and safety arrangements, e.g. first aid provision
- the surveillance of employees' health, e.g. hearing tests
- informing employees on safety aspects
- judging the capabilities of employees, e.g. can a person work at height?
- risk assessment processes to ensure employee's safety
- the protection of people under 18 years of age

- provision for expectant mothers
- provision for temporary workers.

Employees also have duties under these regulations which are:

- to use any plant or machinery provided in accordance with any training in its use
- to inform their employer of a work situation that poses a serious and imminent danger to employees
- to inform the employer of any safety protection measure or arrangement that may be defective (Management of Health, Safety and Welfare Regulations, © Crown Copyright).

Construction (Health, Safety and Welfare) Regulations 1996

The Construction (Health, Safety and Welfare) Regulations were introduced in order to try to reduce the high level of accidents that were occurring within the construction industry year on year.

The employer's duties under these regulations include the following:

- ensuring that construction workplaces are safe
- putting in place measures to prevent operatives falling
- providing toe boards, netting or a physical barrier to prevent objects from falling
- undertaking measures to prevent people and plant from falling into the excavations
- putting in place preventive measures to ensure employees' safety when working above water, considering the movement of traffic on construction sites, both on and off
- putting in place temporary emergency routes and procedures while the building is incomplete
- providing suitable welfare facilities for workers
- providing suitable lighting and fresh air to workplaces
- undertaking some specific safety inspections, e.g. scaffolds and excavations.

There are no specifically defined duties for employees within these regulations apart from Regulation 4, which states the following.

- It is the duty of every employee to comply with these regulations.
- Employees should report to a supervisor any defect that could cause harm to themselves or colleagues.
- Every employee has to cooperate regarding health and safety.

Assessment activity 1.3

Explain how the members of the site construction team interact in terms of their health, safety and welfare roles and responsibilities.

Take a blank sheet of A4 paper and on each corner write down one member of the construction site team:

- Health and Safety Officer
- Site Manager
- Bricklayer
- General Operative.

Then identify their roles and responsibilities and write them down next to their titles. M1

Grading tip

For M1 you will write down the roles and responsibilities of four members of the construction site team. Think about the typical safety activities that will be carried out on site and who performs them; for example, site inductions.

PLTS

Exploring issues on roles and responsibilities, events or problems from the different construction team perspectives will help you develop your skills as an **independent enquirer**.

Functional skills

Presenting the information on the subject of roles and responsibilities will help your **English** writing skills.

2 Know how to undertake risk assessments using appropriate principles and formats

Identification of hazards

Direct observation of work environment

How can you identify the hazards on a construction site? The most useful method is by direct observation. This technique requires practice spotting hazards that are not obvious. This can be aided by the use of photographs, which enable you to check the image later to identify or confirm a hazard in the workplace.

Remember!

The most useful tool you have is your eyes. Directly observing a site should enable you to spot many potential hazards.

The use of accident data

The statistical analysis of accident data is another tool used to help spot potential hazards. Looking at a pyramid of injuries, with the minor ones at the base and the major ones at the top, can be used to help identify the causes of the injuries and hence a control method. For example, an injury pyramid could help identify the cause of a large volume of eye injuries as cement mortar entering the eye on a windy day; the control method must be to wear eye protection or cease work when the wind speed exceeds a certain number of kilometres per hour.

The analysis of risk assessments is another method of hazard identification. By looking through a large number of risk assessments to locate a common hazard, this can be collectively dealt with by global control measures. In effect, you are using this data as a set of fresh eyes to assess the situation and to point out something you may have missed.

A large number of minor accidents can be a cause for concern as eventually there will be a major accident. By reducing the volume of minor accidents, the statistical probability of a major accident is reduced. For example, if there are many injuries to the hands, then you could make it a site rule that everyone on site has to wear safety gloves while working.

Checklists and method statements

Checklists are a standard sheet produced for a particular workplace environment. They are useful for complex construction sites where a large number of processes and substances are utilised. The hazards are then identified by a safety audit or inspection, which involves walking around the site and ticking off the hazards from the pre-set list. When new hazards are identified, these can be added to the existing checklist to build up a more comprehensive checklist.

Method statements are produced as part of the CDM Regulations. They state the methods to be used to construct a particular item. For example, if you were going to drill through a wall for a toilet connection, the method statement would list how you were going to do this, the equipment you were going to use (hand or machine) and a list of safety precautions. Method statements enable you to analyse the correct and safest way to undertake a task.

Regular safety inspections

Safety inspections are a hands-on approach to safety. They also get you out of the office environment and on to the site! Inspecting plant and machinery to ensure all test certificates are up to date is essential. If operatives know they will be subject to inspection, they will act more appropriately in their work environment. Coupling this with a competent supervisor who can inspect as the work as it proceeds will ensure that all control measures are in place and used.

Inspections by the HSE will, of course, have the most effect on health and safety on site, but sadly there are just not enough inspectors to cover the tremendous workload the UK construction industry generates; only the major accidents appear to get investigated. Company-employed health and safety inspectors appointed in accordance with the Management of Health and Safety at Work (MHSW) Regulations (see p. 9) can be used to visit company construction sites. They provide an excellent resource in experience, training and knowledge on all aspects of health and safety and will know the company's procedures to be adopted on all sites.

The difference between hazards and risks

A hazard is something that has the potential to cause harm, such as electricity, hot water, steam or noise. A risk is the potential of a hazard to actually cause someone injury. For example, steam within an insulated stainless steel pipe 4 metres above floor level has a very low potential to cause someone harm.

People who may be at risk

On a construction site, there are many people who may be at risk from a hazard, including:

- employees
- site visitors
- the general public.

The nature of the hazard will determine how many people will be harmed; for example, a gas escape from a site may affect many people, whereas minor incidents may involve only one person.

Employees

Employees are the people who directly undertake the construction of the project. They are the most vulnerable to potential hazards, as the majority work on site and are not based within an office which may be some distance away from the hazards. Each employee, whether very experienced or new to the project, will require information, instruction and training about the potential site hazards.

Site visitors

Visitors to a construction site should first report to the site manager's office. Here the manager will confirm that the visitor has signed in on a register (in the event of a fire everyone needs accounting for), has been given appropriate personal protective equipment and has been inducted on the site hazards. The visitor may be accompanied on their visit by a supervisor.

General public

The main contractor has a duty to protect the general public. This is normally achieved by fencing off the construction site with **panel fencing** so that people cannot wander onto it. In addition, signs should be put up on the fencing to warn intruders of the inherent dangers of trespass.

Key term

Panel fencing – this is constructed out of mesh squares welded to a framework. The panels sit into feet and are bolted together with clips; they are high enough so they cannot be climbed over.

Risk rating of hazards

The identification of hazards relies on training, knowledge and experience. This is especially so when identifying hidden hazards. For example, asbestos insulation had a hidden hazard – the asbestos fibres. In order to understand the risk of hazards they are classified as high, medium or low.

Part of the risk assessment process assesses the potential harm from the hazard. Risks are rated by the likelihood that the hazard will cause harm and the severity of that harm were it to occur. Severity and likelihood are each rated on a scale of 1 to 4.

Likelihood:	Severity:
1 = slight	1 = no injury
2 = possible	2 = minor injury
3 = very likely	3 = major injury
4 = certainty	4 = fatal

Case study: Site protection

Stephanie was employed as an assistant site manager on an out-of-town supermarket shopping development. The site was on the outskirts of a town next to the local school. One of her responsibilities was to prepare the site before the main construction work commenced. She set up the on-site compound and accommodations and fenced in the site on three sides using 2.4 metre high temporary **panel fencing** bolted together on rubber feet. The fourth side was an existing farmer's fence that was 1.2 metres high and topped with two strands of barbed wire.

Divide yourselves into teams and discuss the following questions:

- Had Stephanie taken all reasonable precautions to protect employees and the general public from the construction site activities?
- What other measures would you have undertaken?

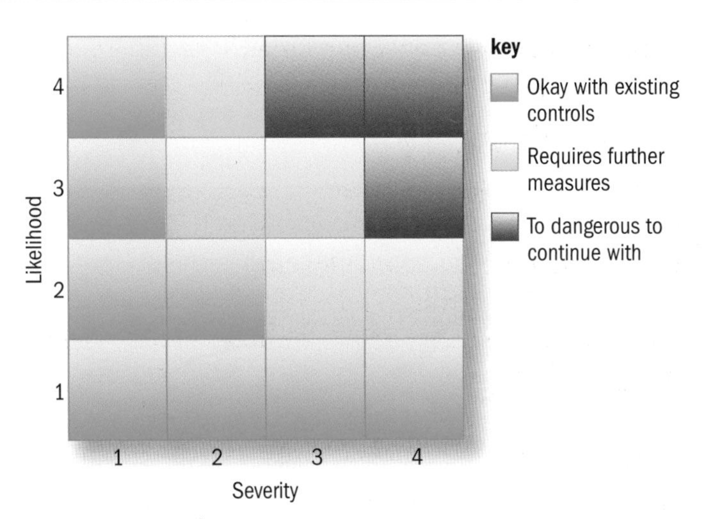

Figure 1.1: A risk matrix

You then multiply likelihood and severity to work out the potential risk.

Risk = Likelihood × Severity

For example, a calculation may place the risk rating into a 4 × 4 = 16 rating, which means the likelihood of it occurring is certain and the severity is fatal. This formula can be used to produce a risk matrix (see Figure 1.1). The table has been coloured:

- green is okay with existing controls
- yellow requires further measures
- red means the work is too dangerous to continue with.

The more complex the work, the bigger the matrix can become. High ratings must have further control measures placed upon them. These are then reassessed to check that the rating has been reduced to an acceptable level. Employers often use numbers on risk assessment forms as they take up less space. They then have a key to reflect what the numbers mean.

Potential to cause harm

Potential to cause harm refers to the possible dangers of the identified hazard and the likelihood that it may cause an injury. For example, gas contained within a gas pipe serving your home has the potential to explode if ignited and cause a fatal injury.

Activity: Risk calculation

Construction sites are full of physical hazards. On the site where you work, there is a reinforcing bar sticking out of the ground which has the potential to cause injury to a worker should they fall on it and get impaled.

Divide yourselves into teams and discuss the likelihood that a worker could fall on the reinforcing bar. If there was an injury, what would be its severity?

Theory into practice

Look closely at the reinforcing rods sticking out of this slab that is ready to be cast. The rods are capped to protect operatives should they fall onto the rods. This is called a 'control measure' and is used to reduce the risk to an acceptable level.

Assessment activity 1.4

P3 

1. Look at the photo above and list at least three hazards that you can see. Describe how you identified these hazards.
2. Look again at the photo and decide who might be at risk from the hazards you can see – don't forget the people you cannot see. What would be the possible consequences for such individuals? P3

Grading tip

For P3 list three hazards you see in the picture. Concentrate on the activities within the photograph and remember that a hazard is something that has the potential to cause harm.

PLTS

Exploring issues, activities or hazards from different perspectives will help develop your skills as an **independent enquirer**.

Functional skills

Discussing your analysis of the photograph will help your speaking skills in **English**.

Environmental hazards

Different workplaces present different potential dangers. Environmental concerns should not be forgotten. Hot and cold environments can present many hazards. Hot environments may affect operatives working within them. They have a drying-out effect and individuals may suffer from dehydration, which can lead to unconsciousness and, ultimately, death. Cold environments cause the body to shiver and can be followed by hypothermia, eventually lowering the body's core temperature, which may result in death. These extremes are usually only found when working outdoors or within cold stores or boiler rooms. Working alone in any of these environments is considered dangerous.

Working over water also presents environmental difficulties, mainly the risk of drowning. Additional control measures must be used when working over water. Contractors must make sure that anyone working above water has a stable platform with guard rails to work on. Operatives working above water must also wear lifejackets (a suitable PPE) and they must be able to swim. Most importantly, there must be means of rescue such as a boat with a trained operative.

Confined spaces

'A confined space is any enclosed space that has restricted natural ventilation and is not intended for continual occupancy by people. By virtue of its enclosed nature there is a reasonably foreseeable risk of injury to workers (Confined Spaces Regulations 1997). In construction, confined spaces might be basements, cellars or under-floor spaces. Under-floor ducts for services would also conform to the above definition. Inspection chambers are also a confined space with the added hazard of gas.

In order to work within a confined space, a risk assessment for the work has to be carried out. It would also be a good idea to put in place a safe system of working. An example of this is **a permit to work**, which lists all the requirements for working in a confined space. Also, when working in confined spaces, the following must always be considered and set up:

- emergency arrangements in the event of an accident in the confined space
- risk assessment for people who will have to enter the confined space should a rescue be required
- available personnel who are trained to remove unconscious operatives from the confined space.

Access and egress

Construction sites are notoriously difficult to enter and exit since the permanent structure has not yet been completed. Temporary **access and egress** can be achieved through:

- mobile elevated platforms
- scissor lifts
- scaffolding
- scaffolding stairs
- ladders for short risk-assessed durations
- tower scaffolds
- lifts.

The most common way is the use of scaffolding, but stair towers must now be incorporated as it is not safe to use ladders with scaffolding. Access platforms with roll-over guards have to be provided for the safe lifting and placing of materials onto the scaffolding. This is done with a rough terrain forklift with telescopic boom.

Key terms

A permit to work – a document issued by the person responsible for a particular work area. Such an area is usually one of high and complex activity; the permits detail who is working in that area and lists any precautions that must be taken and any isolation of services that may be required.

Access and egress – entrance and exit.

Lanyard – an attachment that clips between the safety harness and a secure point to gradually slow the rate of descent.

Working at height

Working at height is inherently dangerous not only because accidents can often be fatal but also because of the danger of falling objects such as tools, plant and materials which can injure workers at ground level. Access and egress issues with working at height can be solved with mobile elevated platforms (MEPs) and stair scaffolds. This enables plant and materials to be moved safely to the workplace.

The photo on page 16 illustrates a typical scaffold that has protective covering to prevent materials from falling and injuring the general public. MEPs provide a safe and secure means of access by allowing you to drive and rotate the guarded cage you work in through 360 degrees. Training to use an MEP is essential.

A harness is the personal protective equipment used for working at height. It is worn by the operative and contains a **lanyard** which is secured to a physical fixing point that will support a fall. A lanyard slows down the rate of descent should a fall occur. A rescue procedure must also be in place because the harness needs to be removed within 10 minutes of a fall as it slows down the blood flow to the brain.

Other prevention measures for working at height involve physical barriers which are horizontal rails set at certain distances as specified in the Working at Height Regulations. There should be no unprotected gap of more than 470 mm. Toe boards must also be provided to stop objects being pushed off the working platform.

As mentioned earlier, the Working at Height Regulations state that you must not work at height unless it is reasonably practical to do so. If not, working at ground level is advised. Examples of this would be external lighting columns where the bulb cluster can be winched down to ground level to maintain the bulbs.

Protected scaffolding

Working safely at height. What safety measures can you see there?

Case study: Working at height

Look at the photograph of the worker on the edge of a building. As you can see, he is working at a considerable height.

- Would you be happy to carry out this work?
- What are the risks in doing so?
- Is there any other way this work could be undertaken without working at height?
- What precautions have to be taken to control the risks involved in undertaking this work?

Accident data

Principal major causes of accidents and fatalities

The construction industry currently employs around 2.2 million people and accounts for over 50 fatalities every year. The most frequent cause of death is through falling. Figure 1.2 shows the four major causes plotted against time which are:

- falls from height
- struck by a moving vehicle
- struck by a falling object
- trapped by collapse or overturning.

As you can see, falls from height have steadily decreased.

Figure 1.3 illustrates the current trend in fatalities over the past nine years.

Demographics

The **demographic** statistics looks at the people involved in the accidents and their classification. The government uses demographics to shape its policy and procedures through the Department of Environment and through safety legislation. By looking at the occupations of those involved in major accidents, correct legislation can be written and statutory bodies, such as the HSE, can provide the correct inspections. For example, if a large number of accidents involve people aged 64 the retirement age might be reduced to 60. The HSE publishes statistics classified by gender, location or age (see the HSE's website at www.hse.gov.uk).

Key term

Demographics – the characteristics of a population such as gender, location or age.

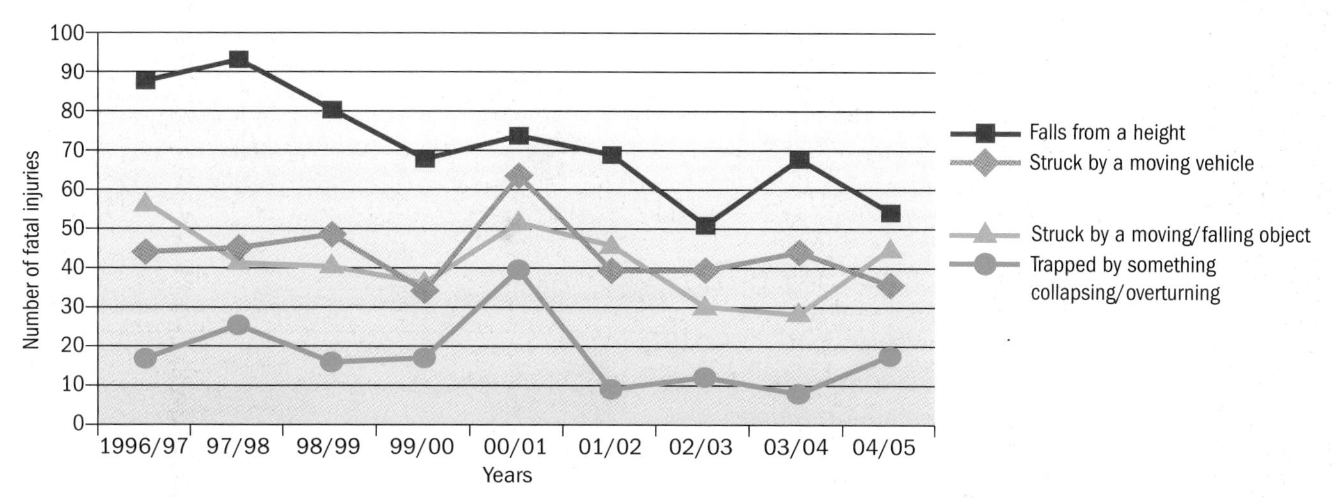

Figure 1.2: Number of fatal injuries to workers by kinds of accident, 1996/97–2004/05
Source: Health & Safety Executive

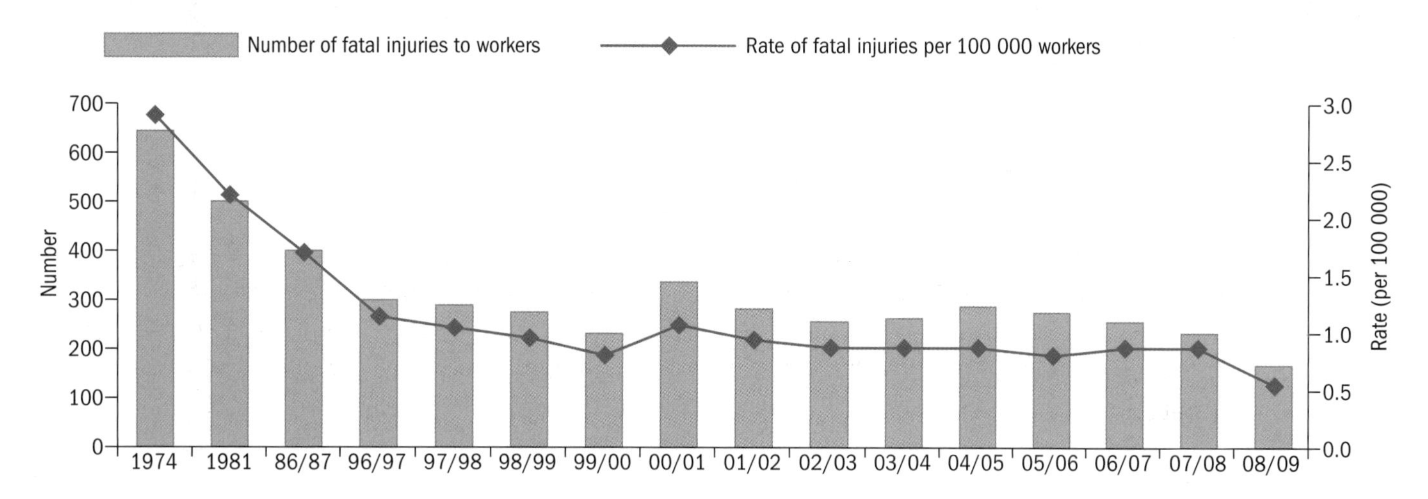

Figure 1.3: Number and rate of fatal injuries to workers, 1974–2008/09
Source: Health & Safety Executive

UK and European safety statistics

Figure 1.4 illustrates how the UK compares with the rest of Europe in terms of safety. As you can see, the UK has the lowest rate of fatal injury. This may be due to several reasons: the UK may report all its accidents (see Observable trends below), it may have more inspectors touring the country, it may introduce more health and safety legislation or it may take the forefront in safety campaigns.

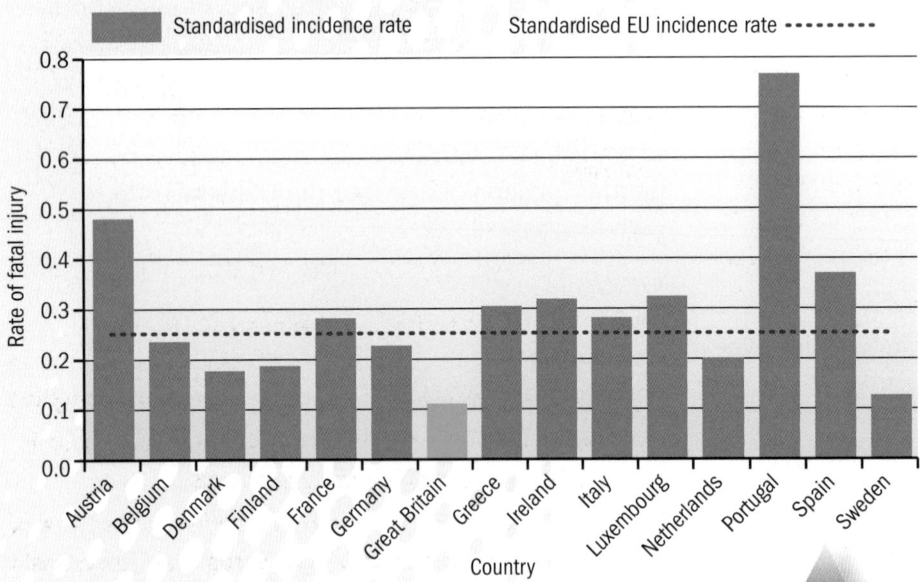

Figure 1.4: Rate of fatal injuries in the UK and the rest of Europe

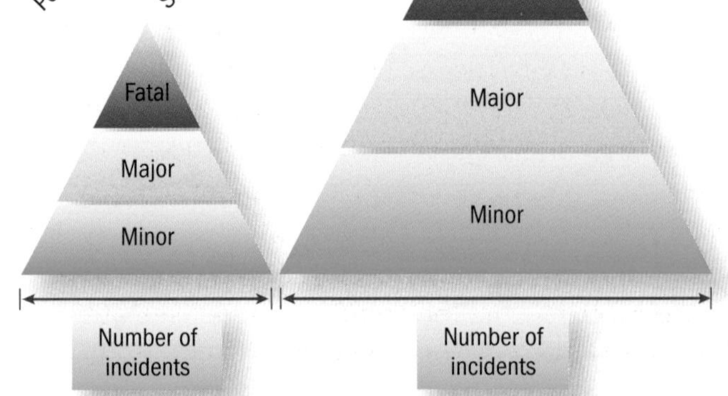

Figure 1.5: The number of incidents that occur is represented by the width of the triangle at the base

Observable trends

The UK keeps records of its accidents, which helps to prevent future accidents from occurring. For example, if the number of reported accidents increases in a certain area or location, then this would indicate the need for an investigation into why these are occurring. The accidents could be due to a change in the process or procedure. Minor-level accidents require acting upon so they do not become major or fatal accidents. The wider the base of the triangle (see Figure 1.5), the more danger that a fatal accident may occur.

Consequences of accidents

Consequences to humans

The consequences of an accident can be devastating to both victims and their families. The disability caused by the fracture, breaking or amputation of a limb may give an employee no option but to leave their employment. Victims of workplace accidents may suffer with poor health for many years. Diseases of the lung caused by asbestos poisoning, for example, are just one long-term consequence of dealing with a product that had a hidden hazard.

Case study: Consequences

John Curtis had been working as a site engineer at a local construction firm. He started this job straight from university. He eventually settled down, married and had two children. His career was fast developing and he was rising quickly through the ranks to senior engineer and, eventually, project manager. One day, John was setting out some road kerb lines for a 250 housing-unit development when he looked up from the surveying instrument and received a hot blast of sparks from a saw being used to cut road kerbs. This permanently damaged his eye sight.

What are the human consequences for John and his family?

Moral consequences

The causes of accidents need to be determined, not pushed aside. Managing directors and chief executives need to lead from the front. Safety before costs should be the ethical ethos of a business. A company that has a high accident rate and does nothing to control this will eventually have a demotivated workforce.

Financial consequences

Employees who have accidents at work may have to stay off work as a consequence. Should the employer have a contractual agreement of only two week's sick pay, then after this period an employee will be paid only statutory sick pay, which is a very small amount, each week after. This may lead to short-term financial hardship for the employee or a legal claim against the company for the injury. The company itself may face a large financial fine as a result of a prosecution by the HSE, along with the unrecoverable costs that are explained below.

The cost of an accident

The cost of an accident can be split into the costs that can be recovered from an insurance company and the uninsured costs that cannot. An analogy of an iceberg can be used. See Figure 1.6. Part of the iceberg can be seen above the water line – this represents the direct costs of the accident. Below the water line is two-thirds of the iceberg which represents the hidden costs, or uninsured costs, that the company cannot see at the time of the accident but eventually will have to pay for.

All businesses have to carry employer's and public liability insurance to cover claims against their business, but there are exclusions within these policies.

The following are the types of insurance cover that may be in place on a construction site and represent insured costs that can be recovered:

- employer's liability insurance on its workers to insure against claims
- public liability insurance to insure against claims from the general public
- motor vehicle insurance to cover company transport
- product liability insurance from manufacturers and suppliers to cover defective products.

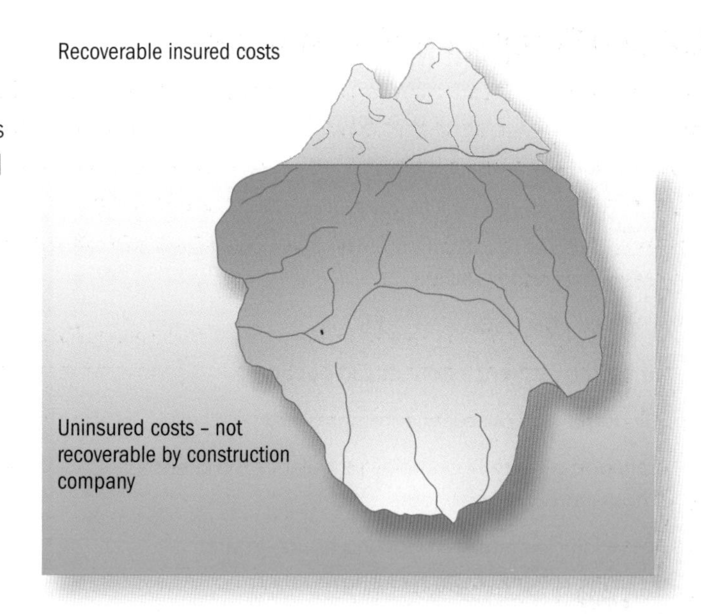

Figure 1.6: Insured and uninsurable costs

Uninsured costs cannot be accounted for at the time of the initial accident; it is only afterwards that these costs become apparent. Several major accidents can have a very detrimental effect on a company's performance and may put it out of business. Uninsured costs, which include non-monetary costs, include the following:

- court fines imposed as a result of a prosecution from the HSE
- costs of clearing up the accident
- costs of investigating the accident
- costs of overtime or additional labour as required
- loss of reputation leading to loss of business and sales
- loss of production time for the damaged product
- demotivating effects on employees.

Activity: Accident costs

The managing director of a housing construction company has asked you to calculate the cost of a recent accident. The accident involved a joiner using a bench saw on site to cut roof timbers to length. Look at the accident cost data that has been analysed from timesheets and invoices.

Accident cost data:

First aid treatment 1 hour at £30 per hour

Bandages and miscellaneous items £20

Transportation to hospital by ambulance: NHS cost 2 hours at £23.55 per hour

Loss of immediate staff working hours on site:

- Site manager 2 hours at £45 per hour
- General foreman 2 hours at £25 per hour
- Joiners (2) 2 hours at £18.50 per operative per hour
- Bricklayers (5) 2 hours at £18.50 per operative per hour
- Labourers (3) 2 hours at £12.50 per operative per hour
- Store person 2 hours at £10.50 per hour

Accident Investigation by C. M. 12 hours at £50 per hour

In-house meeting with three people 3 hours at £55 per person per hour

Agency joiner to replace injured worker 6 weeks at £250 per week

On-site cleaning by contract company £550

Disposal of old saw £120

Purchase of new site bench saw £2500

Hire charges temporary saw 3 weeks at £65.50 per week

Staff retraining on portable electric tools (4) 4 hours at £20 per operative per hour

Sick pay to injured person 12 weeks at £64.66 per week

HSE prosecution and subsequent fine £35,000

Insurance premium increase £2500

- Now calculate the total cost of the accident.
- Which of the costs are insured? Which are uninsured?

Assessment activity 1.5

BTEC

1. The company you work for is very keen to keep accurate accident data records. Explain why accurate data on accidents and incidents or near misses can contribute to reducing accidents and improving health and safety on site. M3
2. Obtain an accident report from a construction press article and evaluate it in terms of how the situation could have been improved in order to prevent a reoccurrence. D2

Grading tips

For M3 you will need to look at and explain why saving accident data is important.

For D2 you will need to find an accident report which is construction-related; look on websites, Construction News and other sources for a construction-related case.

PLTS

Planning and carrying out research on an accident report will help you appreciate the consequences of decisions made within it and will help develop your skills as an **independent enquirer**.

Functional skills

Selecting and using different types of accident reports will help your reading skills in **English**.

Principles of risk assessment

Obligations under the Management of Health and Safety at Work Regulations

The principles of risk assessment are set out under Regulation 3 of the Management of Health and Safety at Work Regulations 1999 are summarised as follows.

- Every employer must make a risk assessment for their workers and non-employees at work.
- The risk assessment that is undertaken must be reviewed in the light of any changes.

- Specific risk assessments must be carried out on young persons employed at work.
- Where an employer employs five or more people, the assessment must be recorded in writing.

The concept of what is 'reasonably practicable'

'Reasonably practicable' is an expression that you may see in some of the legislation discussed at the beginning of the chapter. It means that a company should take all reasonable steps to protect workers against risks and balance these against the cost and time of doing so. For example, the best solution for dust in a worker's environment may be to provide a mask rather than full-extract ventilation systems. This is a judgement based upon knowledge and experience.

The five steps to risk assessment

The risk assessment process can be broken down into five main steps, as outlined by the HSE.

1. Identify the hazards.
2. Decide who might be harmed by these hazards.
3. Evaluate the risks and precautions to be taken.
4. Record your findings.
5. Review your assessment and update the records.

Remember!

Take the first letters of each of the steps – IDERR and use these as a prompt to remember each step.

Evaluation of risks

You need to be fully aware of the primary hazards. Besides these principal hazards, there may be hidden hazards that you cannot directly see or hazards that are brought in temporarily. For example, during a site visit, everything may appear safe. However, a crane could be brought onto site afterwards which may bring with it some extra risks when combined with other hazards such as electricity pylons running over the site. Look back at the risk assessment matrix (Figure 1.1) and you will see that an injury such as electrocution as a result of an accident could mean certain death.

Remember!

A fatal accident is reportable to the HSE even if the person concerned dies at a later date as a result of their injuries.

A high score from the risk assessment matrix requires you to look closely at the nature of the risk. You then need to identify the existing **control measures** that are in place and establish if they are sufficient to reduce the risk from the hazard to an acceptable level. You could do this by looking at the product data and checking the control measures against the manufacturer's recommendations. You would need to identify any further controls that are required to put the risk at an acceptable level.

Key term

Control measure – a method, system or product used to reduce a high risk to an acceptable risk; for example using a fork-lift truck to lift a heavy object rather than trying to lift it by hand.

Assessment activity 1.6

BTEC

1. Look back at the photo on page 14 and the series of hazards that you identified for it. Describe the main principles and features of a typical risk assessment that would be used to control these hazards. One of these principles is to list the existing control measures you would apply to the hazards. **P4**
2. It is your first job on a construction site as the trainee site manager. The site manager has asked you to undertake a risk assessment for excavating a house foundation 1m deep using a mechanical excavator. The soil will be placed in a dumper and moved to a tip on site. Carry out a risk assessment for this activity. What control measures will be required during the work? **M2**
3. State the reasons why you have chosen the control measures in your risk assessment. Are they 'reasonably practicable'? **D1**

Grading tip

For P4 you will need to describe how you can identify hazards, the people who might be at risk, and the possible consequences. Put yourself in the working situation.

For M2 you actually have to undertake a risk assessment. Use the five steps that have been covered in this chapter.

For D1 you should obtain a definition of reasonably practicable from a health and safety source; this will help you to justify the control measures selected.

PLTS

Planning and carrying out research on risk assessments while appreciating the consequences of control measure decisions will help develop your skills as an **independent enquirer**.

Functional skills

By selecting and using appropriate sources of ICT-based information, you can improve your skills in **ICT**.

Formats

A risk assessment form should contain the following:

- the place the work or activity is to be carried out. Undertake the same operation in two different places and the hazards will be different
- the date the assessment took place. This will give an indication of how old the risk assessment is. Does it need reviewing in the light of a change in conditions?
- the nature of the work activity, for example using a pedestal drill to drill holes through some steelwork
- the primary and most important hazards
- the people who might be harmed in this workplace For example, visitors, employees, supervisors and members of the general public
- an evaluation of the risk – what risk is there from the hazards you have identified?
- the existing control measures. Look closely and record what is currently being done to control the risk from the hazards
- any further action/controls required to reduce risks to an acceptable level
- the person who must make sure the additional controls are in place. You may need to identify the person who undertook the risk assessment, especially in an accident investigation
- A date for the risk assessment to be reviewed. Risk assessments will need reviewing periodically with any changes in systems of work, technology and further health and safety legislation.

The advantages of using a standard risk assessment form are as follows.

- All the boxes must be completed, so nothing can be missed during your assessment.
- The risk assessment can be continually reviewed and improved on.
- Health and safety policy can be written into the risk assessment format.
- It can be as simple or as complex as necessary for the type of work being assessed.

Theory into practice

Carry out a risk assessment on a local construction site, your place of work or within the educational establishment you are enrolled at. You can find a suitable risk assessment form either from a place of work, your tutor or the HSE website. Locate an area on the site that contains a great deal of activity so that you have some principal hazards to identify. Fill in your risk assessment form, completing all the stages mentioned above. You may want to also take a photograph which is an advantage as you can study it later in your own time.

3 Understand the control measures used to reduce risk and meet legal requirements

Control measures

Workplace procedures

These are safe systems of work that should be specifically designed for the workplace. Workplace procedure could include separating delivery traffic from on-site operatives, clear signage and pedestrian walkways. Tool box talks are another example of a safe system of work; they identify daily or weekly hazards and keep all employees informed of site dangers.

Workplace procedures take time to develop, implement and maintain and they must be reviewed to check that they are working. Advice and input from the workforce via safety committees act as a useful third set of eyes. Workers can immediately see any problems with these systems. Corrective action can then be put in place.

Hazardous substances

Hazardous substances are controlled through a set of regulations known as Control of Substances Hazardous to Health (COSHH). The regulations mainly covers the chemicals used, such as adhesive glues to stick laminate onto kitchen worktops, or paint thinners in decorating. Every chemical that is used on a construction site has to be risk assessed for its harm potential under the COSHH Regulations. The regulations advise eight steps:

1. Undertake a risk assessment. Use the manufacturer's data sheet on the substance.
2. Decide what precautions are needed. This may involve ventilation, PPE or isolation.
3. Prevent or adequately control exposure through the use of gloves, dust masks or ventilation.
4. Ensure that control measures are used and maintained.
5. Monitor exposure, usually through measurements.
6. Carry out appropriate health surveillance such as blood tests.
7. Prepare plans and procedures to deal with accidents, incidents and emergencies should a spillage of the substance occur in the workplace.
8. Ensure that employees are properly informed, trained and supervised (Health & Safety Executive).

It is worthwhile building up a COSHH library, where manufacturers' data sheets can be kept up to date with the necessary precautions for using that particular product. When there is a choice, it is best to use a less dangerous chemical.

Remember!

It is worthwhile substituting a chemical that harms the environment with a 'greener' version. Green chemicals also require fewer control measures as they have little effect on the environment and the person using them.

Lifting and manual handling

If properly lifting close to the body, a man can normally lift 25 kg; a woman can lift 16 kg. However, this is only a guide and there is no set limit to how much a person can lift. Lifting must be subject to a risk assessment which must take into account the weight to be lifted and how far it has to be moved and where the centre of gravity is.

Working at height

This has already been discussed on page 15.

Physical safeguards

Physical safeguards include secure fencing, barriers and guards that can be fitted to plant and machinery. They are used to reduce and prevent hands, feet, arms, hair and any extremity from becoming entrapped and injured within a moving piece of equipment or plant. Extra protection occurs with interlocking, where the machine has to be switched off before a guard can be removed.

Working in excavations

When working in excavations, there are several hazards that need controls such as falling into the excavation, drowning (should it fill with water), gas and collapse of the sides of the excavation. Control measures should include:

- the addition of secure fencing to the perimeter of an excavation
- a pumping system to remove any water that is building up within the excavation

- a physical barrier to prevent plant falling into the excavation
- a gas test to detect the presence of poisonous gases
- a support system for the sides to prevent collapse.

Site traffic and plant

A large number of accidents are attributable to workers coming into contact with plant and machinery on a noisy construction site. Often the plant operator can not hear or be heard over the noise. Control measures primarily involve segregation of workers and machinery. Clear traffic routes for each must be established on the site. Signage must be displayed to direct one-way traffic on a congested site. Traffic lights and footpaths must be used to good effect. Reversing warning lights and pre-recorded voice warnings are also very effective. Operatives can assist the process by wearing high-visibility clothing that reflects light.

Contaminated ground

Construction regularly takes place on 'brown field sites' which are development sites that tend to be in inner cities and have been cleared of the existing building structures. However, they are usually old and run-down sites with contaminated soil. An example would be old petrol station sites that contain spillages of detergents, petrol and diesel which can ignite and can also cause dermatitis and skin diseases if in contact with the skin.

Remember!

The 'PIGSRISE' system can be employed when implementing control measures.

PIGSRISE is a backwards acronym made up of the first letters of the following and is taken from HSG65 (Health & Safety Executive). You have to start with the last letter and work backwards.

E is for eliminate.

S is for substitute.

I is for isolate.

R is for reduce.

S is for safe systems of work.

G is for good housekeeping.

I is for information instruction and training.

P is for PPE.

This system can easily be applied to COSHH assessments.

Activity: Control measures

A bricklayer's operative is using a diesel cement mixer to mix the ingredients for the bricklaying mortar. This requires that 25 kg bags of cement have to be opened and the correct quantities poured into the mixer by hand using a shovel. It is a windy day and every time the operative places the cement into the mixer, cement dust blows into his eyes.

Divide yourselves into teams. Taking each of the following in turn, identify a control measure from the list that you would use to reduce the risk to an acceptable level.

- Leave the task until a calmer day.
- Replace the cement with an alternative.
- Provide the correct eye protection to the operative.
- Obtain cement in silos with a mortar plant.
- Form a shelter around the mixer.
- Obtain mortar in ready-mixed tubs.

Bricklayer's operative mixing bricklaying mortar in a diesel cement mixer

Soils must be tested to see what type of contamination is present. Then, the contamination can be treated in two ways: by removal to a licensed tip or by ground treatment.

Legal requirements

The duty of everyone

Individuals and employers cannot afford to ignore current health and safety legislation. Workplace policies are there for a reason: to prevent accidents occurring and diseases developing. Everyone has a **duty of care** and no one should ignore unsafe practice. The human, moral and legal consequences should outweigh this situation. Thinking 'It's not my job' is not the answer! Health and safety should be driven by senior management so that it becomes the normal process in working safely.

Remember!

Everyone has a duty of care in law on a construction site and must act if they see something that would be considered dangerous and might harm a person.

The consequences of non-compliance

Improvement and prohibition notices can be issued by the HSE and either state a time limit on when a safety defect should be corrected or a prohibition notice which stops work on site because of a serious and imminent danger to workers.

- Non-compliance can also lead to financial penalties. The severity of an offence will decide which court – crown court or magistrates' court – deals with a health and safety prosecution. The type of penalty depends upon the scale of the breach and can include:
- a £5,000 fine at a magistrates' court
- a £20,000 fine or six months' imprisonment or both at a higher court
- two years' imprisonment or an unlimited fine or both at a crown court.

The charge of corporate manslaughter, where a senior manager or the owner of the company is taken to court and charged with the death of an individual(s), has until recently been very difficult to prove and successfully prosecute, but the HSE will still take this line should it be warranted.

Procedures after an accident or incident

Under the Reporting of Injuries, Diseases, Dangerous Occurrences Regulations 1995 (RIDDOR), you have a legal duty to report the following to the HSE:

- deaths
- major injuries
- accidents resulting in an **over-three-day injury**
- diseases
- dangerous occurrences
- gas incidents.

Key terms

Duty of care – the duty placed upon everyone by the HASAWA to take care of themselves and others about them.

Over-three-day injury – an injury which results in the person being away from work or unable to do the full range of their normal duties for more than three days.

After a serious accident, police also may have to be informed, as there could be suspicious circumstances that may need to be investigated. The coroner also has to be informed should a death occur as they may also wish to hold an inquiry.

Case study: Builders Construction Ltd

When XYZ Ltd decided to refurbish its office building in early 2002, it asked Builders Construction Ltd to take part in initial discussions, with a view to appointing the company as the principal contractor. The refurbishment was extensive, and included the removal of an asbestos ceiling. Before starting work, Builders Construction carried out a risk assessment. However, no control measures were put in place to protect workers and the general public from the effects of asbestos fibres. The ceiling was removed in a few days by operatives who had no training in handling asbestos or the correct PPE. Builders Construction Ltd was taken to court and found guilty of two breaches of the Control of Asbestos Regulations 2006 and one breach of the Health and Safety at Work Act 1974. The company was fined £20,000.

Work with a partner and answer these questions:

- What breaches of the health and safety legislation was the defendant prosecuted on?
- What was the nature of the offence and who was harmed?
- What control measures could have been put in place?

General workplace health and safety policies

The Health and Safety at Work Act 1974 states that:

'Except in such cases as may be prescribed, it shall be the duty of every employer to prepare and as often as may be appropriate revise a written statement of his general policy with respect to the health and safety at work of his employees and the organisation and arrangements for the time being in force for carrying out that policy, and to bring the statement and any revision of it to the notice of all of his employees.'

Employers must write down their health and safety arrangements if they employ five or more people. This is called a health and safety policy and is divided into three main sections:

1. A general statement which gives a key outline as to how the employer will observe their duties under the Act. Specifically, the statement must:
   - demonstrate a company's commitment to health and safety
   - state how the company intends on dealing with this
   - specify who is responsible for health and safety
   - contain a paragraph that all necessary resources will be provided in the pursuance of health and safety
   - inform all employees about the policy's contents
   - be signed and dated by a prominent person in charge, usually the head of the organisation.
2. A listing of specific responsibilities such as health and safety representative and employees' duties.
3. The health and safety administration, including permit to work systems, accident reporting and risk assessment procedures.

Remember!

As an employee, you have responsibilities. You must not ignore any unsafe acts and should bring these to the attention of a manager or supervisor.

Other policies

As well as the health and safety policy produced as a requirement of the HASAWA 1974, companies may also produce their own safety policies. These cover many aspects such as:

- how the company will deal with employees found under the influence of drugs during their employment
- how the company will deal with employees who are clearly under the influence of alcohol
- a driving policy requiring all employees to hold a clean driving licence while using company vehicles
- a smoking policy. Smoking is now banned under recent UK legislation in many open access areas and within buildings.

Remember!

Workplace policies include no smoking, drugs, alcohol and other site rules such as the use of radios and the noise they create.

Thinking point

Look at a typical non-smoking policy below. Do you think that this smoking policy should apply to all construction sites? Give your reasons.

COMPANY SMOKING POLICY

This company operates a no-smoking policy within the company premises. We request that all employees please respect this policy.

No smoking is allowed within the company's buildings or construction sites.

Facilities have been provided to enable you to smoke. These are designated as smoke rooms. Non-smokers use them at their own risk.

Any person found not to be utilising this facility when they smoke will be dealt with under the company's disciplinary procedure.

Arrangements for implementation

Allocation of roles and responsibilities

Different posts hold responsibility for the different aspects of the health and safety policy. This may be broken down into main roles and responsibilities for the following key people in a typical organisation:

- directors
- contracts managers
- site supervisors
- health and safety officer
- employees.

Often the roles are displayed on a chart which illustrates the management structure of the organisation from the top down.

Procedures

Monitoring, review and inspection

Health and safety must be monitored, checked and reviewed regularly. This is because legislation is frequently changing and policies will need updating. There should be regular audits and inspections to ensure that employees are complying with the law, are not taking any unnecessary risks to their health and are acting in accordance with any training given. It is when these controls are relaxed and no one is looking that there might be a tendency to take a risk that results in a serious accident.

Remember!

Monitoring and reviewing means taking time to look carefully through what has been produced in the past and improving it, often in line with new systems, procedures and legislation.

Arranging implementation of health and safety policies

The main part of a company's health and safety policy should state how they are going to carry out and implement it, and may contain any of the following:

- dealing with asbestos
- manual handling
- COSHH
- accident reporting procedure
- health surveillance
- permits to work
- lifting operations
- safety committees
- smoking policy and regulations.

Remember!

This section is where many of the company's policies are placed, so this document must be available for all employees to read.

Use of permits to work

Permits to work are used to safely control any activity on large projects and sites. Large complex sites may have several levels and floors and the permits of work may list all the people and different trades that are working in one particular area, so supervisors can plan safely. For example, if one set of workers are scheduled to weld some steel stairs on one floor, the supervisor will need to check if anyone is scheduled to work at the same time on the floor below, and rearrange if possible. The welders should check nobody is working on the floor below before they start work to avoid accidents. This is also true of workers working on roofs, confined spaces, and doing hot work such as welding, cutting, grinding. Permits make supervisors aware of who is working, where they are working, what isolation may be required, when the work will be finished and what safety coordination is required. Permits may be required for the following work:

- working on roofs
- confined space working
- sewer works
- electrical installation works
- gas installation works.

Some permits may require operatives' signatures so that supervisors know who is working on the site in the event of an emergency. The permit will list the control measures that will be required to be in place before the permit can be authorised. At the end of the working day the permit must be signed off by all who have been working on it so that the supervisor knows that the work is done. On large factory complexes there may be more than one permit station within definable boundaries of the factory. This is done so that adequate supervision can be assigned to each area. Permits often are not issued without visible proof that a risk assessment and a method statement have been carried out and they will highlight the hazards and the control measures that are being used to reduce the risk to an acceptable level.

Remember!

Some factory construction sites are vast, covering many square miles. If a supervisor knows who is working where, then all employees, visitors and workers can be accounted for should an emergency situation occur.

Method statements

Method statements analyse in detail the appropriate method to use in a work activity. They list whether the work will be undertaken by hand or plant, what equipment will be used and the logical sequence of the work. This is good procedure as it makes the person writing the statements think in detail about the hazards and how they will control them.

Induction and training

An induction is essentially a comprehensive introduction to the rules and regulations of the particular site. All employees, visitors and workers on the site must have the induction. Uninformed people on the site may cause a hazard in the event of an emergency. The induction covers several aspects of the site and may include information about:

- hazards
- site rules
- transport movements
- first aid
- fire alarm
- evacuation procedures
- site accommodation facilities
- waste removal
- car parking
- site working hours.

Training is an essential element of health and safety. Highly qualified employees who have the correct training and equipment will result in a workforce with a lower accident rate than a workforce who has not had the correct training. Training can simply take the form of a tool box talk which involves employees stopping work and gathering around a common meeting area to discuss the hazards of the day, control measures and any other high-risk activity on site. These are very useful for passing on safety information to employees.

Good site management procedures

Good site management procedures are enforced by supervisors and managers of the company. They can be as simple as signing in on a register on site that keeps track of how many people are on site, to site meetings that help control health and safety on site. Setting up site rules and procedures is a vital health and safety tool, but to be effective it must be monitored, maintained and obeyed by all without exception.

4 Know their own role in accident recording and reporting procedures

Accident definition

An accident as defined by the HSE is:

'any unplanned event that results in injury or ill-health to people, or damages equipment, property or materials but where there was a risk of harm'.

The key term is 'unplanned' as no one plans for accidents to occur.

Recording and reporting

We have already looked at the reporting of fatal injuries under the RIDDOR Regulations. There are additional conditions required in the reporting of fatal or major injuries. If a member of the public is killed or taken to hospital, the enforcing authority (HSE) must immediately be notified by telephone. Within ten days, a completed accident report form (F2508) needs to be filled out. Over-three-day injuries must also be reported to the HSE.

When reporting injuries, the following needs to be recorded:

- date and time of injury
- a brief description of what happened
- the name and address of the person injured
- the date and method of reporting.

The RIDDOR Regulations give some guidance on the classification for recording injuries. Minor injuries are those that keep an employee off work for more than three days. While major injuries as listed in RIDDOR include:

- fractures other than to fingers, thumbs or toes
- amputation
- dislocation of the shoulder, hip, knee or spine
- loss of sight (temporary or permanent)
- chemical or hot metal burn to the eye or any penetrating injury to the eye

- injury resulting from an electric shock or electrical burn leading to unconsciousness or requiring resuscitation or admittance to hospital for more than 24 hours
- any injury leading to hypothermia, heat-induced illness or unconsciousness; or requiring resuscitation; or requiring admittance to hospital for more than 24 hours
- unconsciousness caused by asphyxia or exposure to harmful substance or biological agent
- acute illness requiring medical treatment, or loss of consciousness arising from absorption of any substance by inhalation, ingestion or through the skin
- acute illness requiring medical treatment where there is reason to believe that this resulted from exposure to a biological agent or its toxins or infected material.

Remember!

Under the regulations, if a person is off work from normal duties for a period of over three days (excluding the day of the injury, but including weekends where they might not be at work), it has to be reported.

Reporting dangerous occurrences

The HSE publishes a list of dangerous occurrences which need to be reported. The following are specific to construction:

- collapse, overturning or failure of load-bearing parts of lifts and lifting equipment
- plant or equipment coming into contact with overhead power lines
- electrical short circuit or overload causing fire or explosion
- collapse or partial collapse of a scaffold over 5 metres high or one erected near water where there could be a risk of drowning after a fall.

The role of the individual in accident recording and reporting procedures

The following procedure outlines what you should do following an accident.

- A trained first aider should administer first aid to the casualty.
- If required, an ambulance should be called to take the casualty to hospital or an on-site facility.
- Rescue teams may assist if it is too dangerous to approach the casualty.
- The accident scene must be left intact if at all possible.
- Inform the casualty's immediate supervisor.
- The company's health and safety department must be informed.

Assessment activity 1.7

BTEC

1. The company you work for has just introduced a set of workplace policies to reduce the amount of accidents to hands. For the following activities, list the control measures that should be used to reduce the risk of harm:
   - replacing a broken window at first floor level
   - replacing an inspection chamber cover in a driveway
   - placing roof tiles on a roof. P5
2. An accident has occurred while replacing the broken window; while removing it the joiner has severely cut his thumb on the glass. Explain the role of the individual supervisor in recording and reporting this accident. P6

Grading tip

For P5 look at each activity, place yourself in that position and write down the control measures you will need to have in place to keep you safe. For P6 look at the role of the supervisor in recording and reporting this accident; write it as a storybook with the full process described.

PLTS

Presenting a persuasive case for action in suggesting control measures will help develop your skills as an **effective participator**.

Functional skills

By describing the control measures for others to pay attention to you will be developing your writing skills in **English**.

Remember!

Only qualified first aiders should administer first aid – it could be dangerous for an untrained person to do so.

The accident investigation

An accident investigation analyses what went wrong. The process must be thorough and methodical. If the accident is a fatality or major, it may be conducted by the HSE. Accident investigations are done:

- so a similar accident does not occur again
- because it is a legal requirement under RIDDOR.
- to provide defensive evidence in a civil claim from the injured party.

Can accidents be completely avoided? Well, no, because we are only human! However, an accident investigation helps reduce the risk of future accidents. If an accident occurs at work, then the correct procedure must be followed to ensure that it is dealt with effectively and efficiently so that any investigation of the accident can ensure that the same occurrence will not happen in the future.

The process may follow these steps:

1. The accident location is photographed to record the area.
2. A statement is taken from the person who had the accident.
3. The accident report form is filled in and sent off if it is reportable under RIDDOR.
4. Witnesses are interviewed to provide further evidence as to the cause of the incident.
5. An analysis of the accident is undertaken to establish the primary cause.
6. New control measures are devised to establish if the existing system of working can be revised.
7. Any changes to the system of working or control measures are then implemented.

Changes that are implemented as a result of an accident investigation should be reviewed periodically to see if they are working. If they are not, the process is reviewed until a successful outcome is established. It is then monitored periodically. When a hazard cannot be adequately controlled, then the potential for it to cause an accident is raised.

Assessment activity 1.8 P6 

You are the assistant health and safety manager. The contracts manager has approached you to report an accident that occurred yesterday on one of the company's construction sites. The accident involved a collision between an employee and a rough terrain forklift: the rear wheel ran over the employee's foot.

What recording and reporting procedure would you undertake? P6

Grading tip

For P6 write an accident report. List everything from the point of the accident just occurring to the employee's return to work.

PLTS

By generating ideas and exploring possibilities on reporting procedures you will develop your skills as a **creative thinker**.

The importance of collecting accurate accident data

A construction company should collect accident data:

- so it can be used to prevent future accidents
- because many clients make it a requirement to see your accident data as part of the tendering procedure; the client checks the contractor for competency. This can be done by looking at the number of accidents that a company has had over the last three years.

Activity: Health and safety

You have been asked to review a company's accident statistics from the last three years to identify if there are any trends. You find that there are a large number of foot injuries to a set of workers who manufacture concrete lintels in the construction yard.

- What could you do to reduce the number of injuries?
- How would you monitor improvements?

Simon Green

Health and safety officer

Simon has worked on several construction sites for the company he works for. During this time he has seen several accidents and a few near misses that have occurred as a result of poor health and safety control. Simon is often involved in first aid and the completion of risk assessments and is keen to reduce the accident record of the company.

As a result of Simon's interest he was offered the position of trainee health and safety officer three years ago. Simon relished the chance to take this opportunity to influence the health and safety policies of the company and to try to improve its health and safety performance.

Simon reviewed the company's health and safety policy and made several changes to the way in which work is undertaken and activities supervised. Employees are now educated during their induction to each construction site about the health and safety arrangements. This has resulted in a reduction of the accident rate over the past three years.

Simon has managed to accomplish this through thinking about health and safety and evaluating the accidents that had been occurring. He then devised policies to reduce and target the specific accidents that were occurring.

The managing director was extremely impressed with Simon's influence on health and safety, especially since the liability insurance premiums that the company were paying have almost halved. Simon was promoted to Health and Safety Officer last year.

Think about it!

- How would you reduce a high accident rate?
- How would you win employees over?
- What support would you need from management?

Just checking

1. Does the client have to appoint a planning supervisor?
2. Several employees are deliberately setting off fire extinguishers in your workplace. Can you prosecute them?
3. Does the main contractor feature in the Construction (Design and Management) Regulations?
4. What is the difference between a hazard and a risk?
5. What precautions must you take when working within a confined space?
6. What is the major cause of fatal accidents?
7. Name four hidden costs of an accident that cannot be recovered from insurances.
8. List the five steps to risk assessment.
9. What three penalties can the HSE impose for a failure in health and safety?
10. What does RIDDOR stand for?
11. You have had an accident on site and have fractured two fingers. Must this accident be reported to the safety authorities?
12. The following control measure is being used – the use of a mobile elevated platform to change a light bulb on the ground floor ceiling. Is this reasonably practicable?
13. An accident occurs on Thursday night and the injured person returns to work on Monday morning. Is the accident reportable if the person was unfit over the weekend?
14. What four essential items need to be on an accident report?
15. Why would the HSE not be happy if it takes a company 11 days to report a major accident?

Assignment tips

- Using the Internet will provide some useful examples of risk assessments. Evaluate these in order to produce a suitable word-processed format that you can use for some of the grading criteria.
- Every institution has a health and safety officer. Go and talk to them – they will value you asking questions on health and safety.
- Look around the institution where you are studying. There will be lots of control measures in place to provide a safe system of work. Use them to assist with your assignments.

Credit value: 10

2 Sustainable construction

For thousands of years, humans have exploited the natural resources of the world, for example, coal, oil, gas, metals, gypsum and aggregate. All of these resources are consumed in the production of construction materials. The burning of fossil fuels has contributed to an annual increase in the Earth's temperature, known as global warming. Many measures – such as legislation to increase sustainable construction, and tighter building regulations – have been put into place to reduce the factors that contribute to global warming.

Many measures have been implemented to protect the natural environment, for example, designating certain areas as National Parks, protecting green belts of land and efficient use of recycling techniques that reduce the need to mine or develop new resources. The government's recent initiative to redevelop brownfield sites (areas of land with existing buildings on them that have exceeded their life span) has reduced the need to develop greenfield sites.

The current emphasis is on the sustainable use of construction materials and processes so homes can be made that have low carbon emissions and are more energy efficient, therefore reducing the reliance on oil-based resources. Sometimes this can be as simple as buying materials manufactured locally, which saves transport costs.

Learning outcomes

After completing this unit you should:

1. know the important features of the natural environment that need to be protected
2. understand how the activities of the construction and built environment sector impact on the natural environment
3. understand how the natural environment can be protected against the activities of the construction and built environment sector
4. understand sustainable construction techniques that are fit for purpose.

Assessment and grading criteria

This table shows you what you must do in order to achieve a pass, merit or distinction grade, and where you can find activities in this book to help you.

To achieve a **pass** grade the evidence must show that you are able to:	To achieve a **merit** grade the evidence must show that, in addition to the pass criteria, you are able to:	To achieve a **distinction** grade the evidence must show that, in addition to the pass and merit criteria, you are able to:
P1 describe six different features of the natural environment that must be considered at the planning stage of a construction project **See Assessment activity 2.1, page 39**	**M1** assess the potential environmental impact of a proposed construction project on the local natural environment **See Assessment activity 2.2, page 42**	**D1** assess the importance of addressing environmental issues for the mutual benefit of the community and individual construction firms **See Assessment activity 2.6, page 55**
P2 explain four different forms of global pollution arising from construction projects **See Assessment activity 2.3, page 47**		
P3 explain how four different forms of local pollution arising from construction projects may harm the local environment **See Assessment activity 2.4, page 51**		
P4 explain four key methods used to protect the natural environment from the impact of the construction and built environment sector **See Assessment activity 2.5, page 53**	**M2** compare the four key methods used to protect the natural environment in terms of cost, effectiveness and public perception **See Assessment activity 2.5, page 53**	
P5 explain three different, fit-for purpose sustainable construction techniques **See Assessment activity 2.7, page 57**	**M3** compare sustainable construction techniques in terms of relative cost and performance **See Assessment activity 2.7, page 57**	**D2** justify the use of appropriate sustainable construction techniques for a specified construction project **See Assessment activity 2.7, page 57**

How you will be assessed

The evidence requirements for pass, merit and distinction grades are shown in the grading criteria grid. Evidence for this unit may be gathered from a variety of sources, including well-planned investigative assignments, practical work or reports of practical assignments. You will be given written assessments briefs to complete for the assessment. These will contain a number of assessment criteria from pass, merit and distinction.
This unit will be assessed by the use of two assignments:

- Assignment one will cover P1, P2, P3, M1 and D1
- Assignment two will cover P4, P5, M2, M3 and D2.

Carter

Before I studied this unit I'd never considered things like protecting our resources for the future. I did not realise that oil and gas cannot be replaced and are fast running out. This is a very serious issue.

I now realise why saving energy and resources is important for the Earth and why we should do everything possible to protect our natural environment.

This unit has made me aware of the factors that have a major influence on the built environment both globally and at a local level and what can be done to prevent any further damage, preserving the future for our children.

Over to you

- What does the phrase 'sustainable construction' mean to you?
- What do you already know about importance of sustainable construction?
- What are you looking forward to learning about in this unit?

1 Know the important features of the natural environment that need to be protected

Build up

The environment

The environment is rapidly changing. Flash floods are becoming much more common. In groups, think about the following questions.

- How is the environment is changing as a result of human actions?
- What can be done to reduce the impact that humans are having on the Earth's climate?
- How can you make more people aware of what is happening?

Features

Air quality

Good air quality is vital for life and is an essential part of a healthy environment. Large factories tend to be sited away from population centres in order to prevent pollution entering the breathable atmosphere over towns and cities; indeed, large chimneys push the pollution further up into the atmosphere avoiding any fallout to local inhabitants.

This was not always the case. During the Industrial Revolution, coal replaced water as the main source of fuel used to power the pulleys and belts that drove industrial machinery. There was no control over the use of this new technique, and pollution became a serious problem as smog developed over the large industrial cities of the UK. Smog brought with it poor visibility as well as breathing difficulties for the cities' inhabitants – smog depleted oxygen levels. Deaths from asthma, bronchitis and other lung diseases became common.

The development of the petrol engine led to another rise in pollution, this time caused by lead-based petrol emissions. The government at the time passed acts to control waste emissions into the atmosphere. The Clean Air Act of 1956 was one example, and smog over large population centres was quickly reduced as a result of this legislation.

Today, the level of carbon dioxide (CO_2) emissions is strictly controlled so the effects of global warming are not increased. Tree and landscape planting improves the quality of air and is considered to be a sustainable part of any housing development.

Air quality obviously differs with geographical location. The centre of London or Birmingham will have a vastly poorer quality of air than sparsely populated areas of Scotland. Generally, the more population and industry in an area, the poorer the air quality.

Ozone quality

Ozone is a gas that occurs naturally in the Earth's upper atmosphere. It shields the Earth from harmful ultraviolet (UV) radiation. Without the ozone layer, there would be no life on Earth.

Chemicals such as chlorofluorocarbons (CFCs) were used as propellants in aerosols, as well as cooling gases in fridges and freezers when they were first developed. However, in the 1980s, scientists made a link between the use of these CFCs, and the hole opening up in the ozone layer above the Antarctic. Since 1987, many of the world's governments have signed up to the Montreal Protocol on Substances that Deplete the Ozone Layer. As a result, CFCs have been replaced with less harmful chemicals. Scientists now believe that the hole in the ozone layer is getting smaller.

Soil quality and natural drainage

The benchmark for soil quality is hard to define. No two soils are the same and therefore there is no British Standard for a soil to be compared against. However, two environmental committees have formulated up to 67 measured variables for soil quality. The **Soil Association** grades soils against a certain standard for the organic growing of fruit and vegetables.

The following are just a few of the categories against which soils may be measured:

- drainage properties
- texture
- acidity
- **pH** balance (see Figure 2.1) – ideally a good soil should have a pH of between 6 and 7.
- use
- level of contamination
- fertility
- mineral content
- organic content
- structural properties.

Extremely acid	4.3 or over
Highly acid	4.3 – 4.8
Moderately acid	4.8 – 5.5
Mildly acid	5.5 – 7.0
Mildly alkaline	7.0 – 7.7
Moderately alkaline	7.7 – 8.5
Highly alkaline	8.5 or higher

Figure 2.1: pH range of soil

A quality soil is one that will sustain life. Soil is used in construction to provide attractive and environmentally landscaped areas for the community to interact with. New housing schemes must include these areas as part of the government's sustainability policy.

The drainage of soils is a vital environmental consideration. Tree roots bind soil together. In areas that have been deforested (as has been the case in many parts of the UK, and in many developing countries) the soil is left unbound. Therefore, when these areas are exposed to high levels of moisture, for example, when it rains, the water runs off and sits on the surface. In small amounts this does not cause a problem, but if the rainfall is very heavy, it can result in flash flooding. Recent government planning policy which has allowed developments within flood plains, has added to this problem. Overburden on a river's **levees** causes them to break and localised flooding to occur.

Soil drainage depends greatly on the structure of the soil, that is, how many pores or open voids are contained within it. The voids allow water to penetrate through the soil, eventually ending up within an **aquifer** below ground. Clay soils tend to resist the passage of moisture, whereas limestone rock allows the percolation of water through it. Therefore, geography and location play an important part in soil quality and drainage, along with the substrata of the rocks below the surface soils.

Key terms

Soil Association – organisation promoting healthy soils via a certification scheme that enables a producer to use the term 'organic' produce.

pH – the measure of acidity or alkalinity of a substance.

Levees – natural banks of silt deposits which are left after a river floods. These are shaped into higher banks to control flood waters. In 2005, some of the levees protecting New Orleans in the USA broke and flooded the city.

Aquifer – an underground storage area created naturally within the Earth's rock strata.

Landscape

During the Ice Age, millions of tons of ice were pushed southwards down half of Britain. This action moulded and shaped the landscape beneath it, often leaving behind boulder clay and rocks, and this provided the basis for the natural landscape that is familiar to us today. However, the UK has undergone many changes since the ice melted away.

There are many varied and diverse areas of natural beauty within easy reach of major towns and cities that are protected, for example the highlands of Scotland, the Cotswolds, Cornwall, the Lake District and the Pennines. Local natural landscapes comprise forests, rivers, streams, hills, topography and the unspoilt countryside.

The landscape is a valuable resource. Tourists travel to the UK to see pieces of history, and the unique landscape. An attractive view with open spaces allows the use of the landscape for recreation purposes such as horse riding, fishing and hill walking.

The coastal landscape is continually changing, as erosion from the action of the waves moves parts of the coastline steadily inward. The different coastal rocks produce different landscapes, for example, the clay boulder slopes of the Yorkshire coastline are dramatically different from the chalk-white cliffs of Dover.

Urban landscapes carry a unique signature. For example, in the Lake District, most houses are built of slate, whereas sandstone is the predominant building material in Edinburgh. A lot of inner cities have a vertical landscape; London is becoming a rival to the buildings of New York with many multi-storey constructions such as the Gherkin (Swiss Re Tower). Leeds city centre is also developing vertically.

Natural amenities

The facilities and services afforded by the natural environment cover a wide range of activities:

- Rivers – used for a variety of water-based sports such as fishing, angling and canoeing. Fishing is regulated under licence by the Waterways Environment Agency.
- Lakes – the Lake District, for example, is enjoyed by boating enthusiasts; a ferry service and pleasure craft attract a healthy tourist industry.
- Fenlands – the naturally occurring seasonal flooding areas of the fenlands and the Norfolk Broads produce large areas for migrating birds to feed upon during their annual migration from other climates. This enables the development of protected areas for wildlife conservation and the hobby of bird watching to flourish.
- Moors – the natural moors of the Yorkshire and Pennine regions are unique, although it may be said that humans have developed this environment by burning large areas for the promotion of grouse shooting as a sport. Large areas of the moors tend to be managed estates.
- Mountains – the Scottish and Welsh mountain areas offer hill walking and mountain climbing. Scotland also offers skiing facilities in winter.
- Natural forests – forests used to cover most of Britain during the Middle Ages, but they have been gradually cut down and used for fuel and ship building. There are small areas of natural woodland that are now protected from felling. Natural woodlands provide a landscape that can be used for mountain biking, dog walking and exercise.
- The sea and beaches – Britain's coastal areas are unique and diverse. From the pebble beaches of the south coast to the sandy beaches of Norfolk and the rugged granite cliffs of parts of Cornwall and Scotland, beaches attract tourists during the summer months. Many are protected under an SSSI (site of special scientific interest). The sea is a valuable source of food, although stocks of fish are protected by European Union quotas.

Land use

The use of the land has been defined over hundreds of years and has been largely governed by the local population and the resources that are available at a given time. As villages developed into towns and then major cities; expansion took place into the surrounding land. In order to feed the population, farming and agriculture developed in the fields surrounding the villages.

During the Industrial Revolution, resources such as coal, oil, water and wood were increasingly used, and the industrial use of plots of land was born. The waterway network developed with a canal system that was later superseded by the railway network.

More recently, planning control has given the community more say in the choice of land use and has limited development where it is considered inappropriate. For example, the expansion of out-of-town shopping developments has been reversed through government policy on strategic planning.

Land use can therefore be broadly categorised into:

- agricultural
- heavy industrial
- housing
- commercial
- natural landscape.

Green belts

Green belts are the areas of green land that surround communities and provide open parkland for the community to enjoy, away from the industrial use of the land. Green belt land is protected; no development is permitted on it. Green belts provide an attractive and aesthetic area that breaks up the large conurbations; they also provide a buffer zone between different land uses and help to maintain a clean, fresh and natural land that all of the community can experience and enjoy.

Thinking point

In 2006, over 300 applications were made to develop on areas of green belt. The Office of the Deputy Prime Minister (now Communities and Local Government) approved 150 of these!

Agriculture

Land that is used to produce food is known as agricultural land. It can be classified into two broad areas:

- Arable – the growing of crops.
- Livestock – the raising of animals.

Arable farming in the UK includes cereal crops, such as wheat and barley, and horticulture, which is the growing of fruit and vegetables. There has recently been a drive for organically grown crops, i.e. crops produced without the use of chemicals. Livestock includes the production of meat from cattle, sheep and pigs, and the use of cattle for producing milk.

Agriculture plays a significant part in developing the fertile regions of the landscape by ploughing the land and adding fertilisers to grow crops. This alters the natural landscape from its raw state into a condition that can be used for food production. Large areas each side of a major river are the most fertile for this use, as the river floods depositing minerals and materials that feed nutrients into the soils.

Forestry

The use of land for forestry can be divided into:

- naturally occurring, established woodlands that are hundreds of years old and are carefully managed
- plantations – areas of land that have been deliberately planted to grow and harvest timber resources.

In Britain, natural woodlands are limited to small pockets. There is a national forest at Nottingham, but a lot of woodland was cleared for industrial development, ship building and to use as fuel. Forests also provide an opportunity for recreation, offering an ideal environment for walking, bird watching, horse riding and biking.

The UK government's Forestry Commission looks after many of the forests and protects them as well as developing and expanding the use of timber as a resource. Timber, unlike oil, gas and coal, is a renewable resource that is not finite.

Countryside

The British countryside is not developed in strong concentrations. Small villages and settlements, such as farms, are surrounded by green areas of farming or natural landscape which is known as the countryside.

In the UK, these areas are green due to high concentrations of rainfall; in hotter climates they would be brown.

Assessment activity 2.1 — P1 BTEC

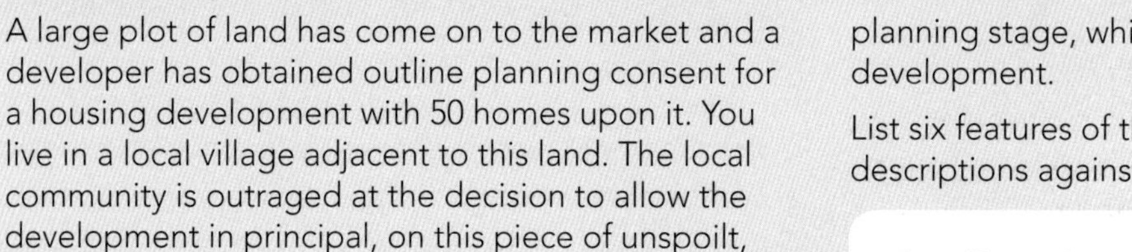

A large plot of land has come on to the market and a developer has obtained outline planning consent for a housing development with 50 homes upon it. You live in a local village adjacent to this land. The local community is outraged at the decision to allow the development in principal, on this piece of unspoilt, green landscape.

In order to help fight an objection to full planning consent, identify and describe the features of the natural environment that must be considered at the planning stage, which could be used to contest the development.

List six features of the natural environment with descriptions against each. **P1**

Grading tip

For **P1** place yourself on the land and list six features that will want protection from development.

PLTS

By exploring the environmental issues from different perspectives, you could develop your skills as an **independent enquirer**.

Functional skills

Reading and summarising information and ideas from different sources on the natural environment will help your reading skills in **English**.

There are very distinctive areas of countryside, for example the Wolds of Lincolnshire, the Lowlands of Scotland and the South Downs of Dorset.

Remember!

The countryside is an area of green scenery that is clean, attractive and unique within easy reach of major living areas.

Heritage

Britain has been inhabited by humans for thousands of years. Over this period, we have developed a unique and enviable heritage that is now protected. Heritage encompasses not just the land but the structures created upon it and could include the following:

- battlefields
- ancient monuments and icons
- castles
- manor houses
- streets
- archaeological sites
- bridges
- parks and gardens.

It is vital that these are taken care of for future generations to enjoy. They are part of the culture of the UK and stand out as distinctive structures. A prime example is the stone circle of Stonehenge. A system of **listing** important buildings has protected many of them, as has the National Trust which owns and runs some of the historic locations in the UK.

Key term

Listed building – a building of special architectural or historic interest in the UK. Alterations to these buildings cannot be made without consent or careful consideration.

Remember!

English heritage attracts a large tourist industry, especially from the USA, Japan and other countries, which substantially adds to the economic well-being of the UK. It is vital that it is protected.

Castles are an important part of British heritage

Theory into practice

Find out more about listed buildings in England at English Heritage's website, www.english-heritage.org.uk; for information on listing in Scotland, see Historic Scotland's website, www.historic-scotland.gov.uk.

Water resources

Over the past 100 years, the use of water as a resource has increased. It is used not just for drinking but also for washing, flushing toilets, cleaning and bathing. A dishwasher and washing machine are now considered essential items in many homes. Indeed, the growth in hot tubs now puts even more strain on the distribution systems that supply the water.

Water resources include the extraction of water above and below ground. Above ground, water is captured in reservoirs or by damming a river supply, extracting from rivers, or in hot climates by desalination plants from seawater. Below ground, boreholes are sunk within permeable rocks to extract the water by pumping. Water is a valuable commodity – a licence is needed to extract it and it must be protected from contamination through pollution leaking into the soils and rocks.

Remember!

The Earth has the same body of water as the day it was created; nothing has been taken away or added. The water cycle describes how water evaporates from the sea into the atmosphere, moves as clouds, then falls as rain water, and runs into rivers, and back to the sea.

Water quality

Water for human consumption is extracted from reservoirs, rivers and boreholes. It then has to be treated to enable it to be classified as drinking water. Water is distributed around the UK using a system of pipework, which is then rated or metered as the consumer uses it.

Water cannot contain any harmful elements such as bacteria, as this would affect people's health. Pure, fresh and clean water can be extracted at source and bottled as a mineral water. Non-drinking water supplies tend to be used in European countries where water is in short supply; we, for example, use drinking water to flush toilets in the UK.

Water quality is often determined by the material through which it percolates during its journey to the aquifer, which is an underground storage area created naturally within the Earth's rock strata, for example limestone.

Marine environment

The marine environment covers many aspects, including:

- harbours
- the sea
- estuaries
- marshes
- beaches
- cliffs.

The marine environment differs from the river environment in that it is salt-water based and not fresh water, but it is worth noting that several rivers are tidal. The tides are a gravitational effect caused by the moon's influence on the Earth. Along the coast there will be high-tide and low-tide marks. Global warming has been blamed for leading to the steady rise in sea levels as a result of the melting of polar ice. This rise could threaten many cities throughout the world, for example, London and Venice would be in danger if the sea levels rise much more. Hurricanes and the storm surges that arrive with rising sea levels are a great threat too, as the flooding and devastation of New Orleans in 2005 shows.

Marine wildlife is extensive in its range and biodiversity. The surface of the Earth is covered with a great deal of salt water, much of which has not been explored, so there is still a great deal to discover about this environment. Britain's coastline differs depending on the surrounding rock structures that meet the sea – soft rocks and clays are easily washed away by the action of the waves, whereas igneous rocks take longer to be broken down into fine sands.

Theory into practice

Look up 'biodiversity' using an Internet search engine. Write your own definition in no more than 20 words.

Wildlife

Wildlife is the native life that exists within a geographical location. Native wildlife tends to be unique to its area, for example the colonies of birds that congregate in certain areas, like the Dartford warbler, or ospreys in Scotland. However, some species migrate with the seasons, for example, the swallows of North Africa visit Britain each year.

Wildlife, as the name suggests, refers to life that is truly wild, i.e. not animals that have been tamed by human interference, for example, horses, pigs and sheep. There is a wide range of wildlife in Britain, for example, birds, seals, whales, fish, snakes, otters, which are all left alone to develop and establish themselves within the landscape.

It is important to remember that the agriculture industry has an effect on wildlife. The crops that are grown and the land that is cleared for cattle have an impact on the diversity and location of wildlife. Humans have to be very careful to avoid disturbing wildlife or this may tragically mean extinction for some species.

Biodiversity

Biodiversity is a term used to describe the amount of living matter and range of biological species present in a particular environment (or geographic area). It covers everything from micro-organisms to wildlife such as ducks and swans. Biodiversity is essential to ensure the continued success of an ecosystem. If an imbalance develops, then one species can take over; this can be seen most obviously with rat or mice infestations.

Any activity humans undertake has an effect on biodiversity. For example, if a building is constructed on a plot of land in a green belt, an area the size of the footprint of the building has been lost, and all the biodiversity contained within it. However, the loss can be compensated by external landscaping to encourage more wildlife in the area.

Remember!

Biodiversity evolves; there are many species and micro-organisms that have yet to be discovered.

Natural habitat

Natural habitat refers to the place where any living thing naturally lives. Most of the British Isles was originally covered in woodlands, but a large proportion of this has been removed for the construction of buildings and to be used as a fuel source.

Natural habitats include the heaths, meadows, limestone pavements and moorlands that can be found in the countryside. Humans have had a considerable impact on British natural habitats, which now require more protection than ever, so those that remain can be enjoyed in the future. Even grazing sheep on land destroys the natural habitat.

Nearly half of the ancient woodlands, and almost three-quarters of the ponds in the UK have been lost. This has had a detrimental effect on biodiversity and some species are already extinct. Environmentalists have responded to the loss of so much of the natural habitat by attempting to restore many of the natural habitats in Britain.

Assessment activity 2.2

M1 BTEC

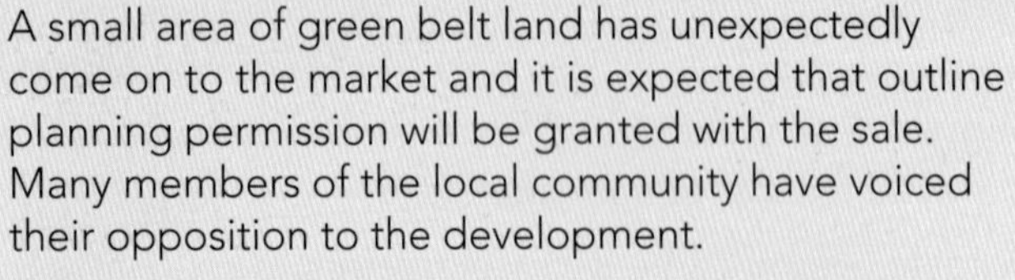

A small area of green belt land has unexpectedly come on to the market and it is expected that outline planning permission will be granted with the sale. Many members of the local community have voiced their opposition to the development.

You work for the developer. To help defend the development, undertake an assessment of the potential environmental impact a proposed construction project will have on such a piece of land and on the local natural environment. M1

Grading tips

For M1 you will need to look at what impact a development would have, what potential harm could occur and how the development could minimise this impact.

PLTS

By presenting a persuasive case for action on this development you will develop your skills as an **effective participator**.

Functional skills

Presenting information and your ideas on the important environmental issues concisely, logically and persuasively will help your writing skills in **English**.

2 Understand how the activities of the construction and built environment sector impact on the natural environment

Globally

Greenhouse gases and global warming

Everyone in the world needs a roof over their head to protect them from the environment they inhabit. In the UK, the winter is colder than in the warmer climates and so more substantial buildings are required.

The construction industry uses raw materials, such as limestone and clays for cement and brick manufacturing. Both these and many other processes require energy, which can be in the form of gas or electricity supplies.

The process of turning the raw materials into a construction product releases carbon, a greenhouse gas, into the atmosphere. Furthermore, many trees have been cut down in order to process the timber for construction. Trees absorb carbon and produce oxygen as a by-product of photosynthesis.

The construction industry cannot solely be blamed for global warming. The production of cars and the fuels they burn, along with air conditioning and several other sources add to the problem.

Remember!

A natural layer of greenhouse gases – water vapour, carbon dioxide, methane, nitrous oxide and ozone – surrounds the Earth and keeps it warm. Without this layer, temperatures on our planet would be much cooler and life on Earth, as we know it, would not exist.

In recent times, man-made greenhouse gases – carbon dioxide, nitrous oxide, sulphur hexafluoride, hydrofluorocarbons, perfluorocarbons and chlorofluorocarbons (CFCs) – have been released into the atmosphere and been added to the natural greenhouse gas layer. The result has been global warming – heat which would normally escape through the greenhouse gas layer into space remains trapped, causing the temperature on Earth to rise. Scientists predict that the increase in temperature will have serious consequences on the environment, including a change in rainfall patterns and a rise in sea levels.

Theory into practice

The consequences of global warming are now a reality that is affecting all of us. Undertake a web-based search on the effects of global warming on the Earth and list three of the major causes and three of the consequences of this.

There are two sides to any discussion on global warming. Some scientists disagree that the Earth is entering a warmer phase and point to the fact that the Earth was warmer in medieval times than it is now. Other factors known to influence the Earth's climate are:

- The sun's activity goes in cycles. During the last ice age, the sun was less active than it is now.
- The tilt of the Earth is off-centre. As the Earth rotates around its axis the Northern hemisphere is nearer the sun at certain times of the year than the Southern hemisphere (and visa versa), which affects temperatures.
- The Earth's orbit around the sun is elliptical, which means that it is sometimes closer to the sun than at others.
- Volcanic eruptions on Earth throw dust into the atmosphere, which blocks the sunlight. This can cause the Earth's climate to cool, sometimes for several years.

Acid rain

Acid rain has an unusually low pH value. When the sulphur dioxide, nitrogen oxide and hydrogen chloride gases combine with water droplets in the atmosphere, weak solutions of sulphuric, nitric and hydrochloric acids are formed (see Figure 2.2). The gases come from two main sources: hot magma ejected into the atmosphere during volcanic activity and the burning of fossil fuels, namely oil, gas and coal. Sulphur dioxide and nitrogen oxides can be carried by the prevailing winds and may not combine with water droplets until they are many miles from the initial source of pollution.

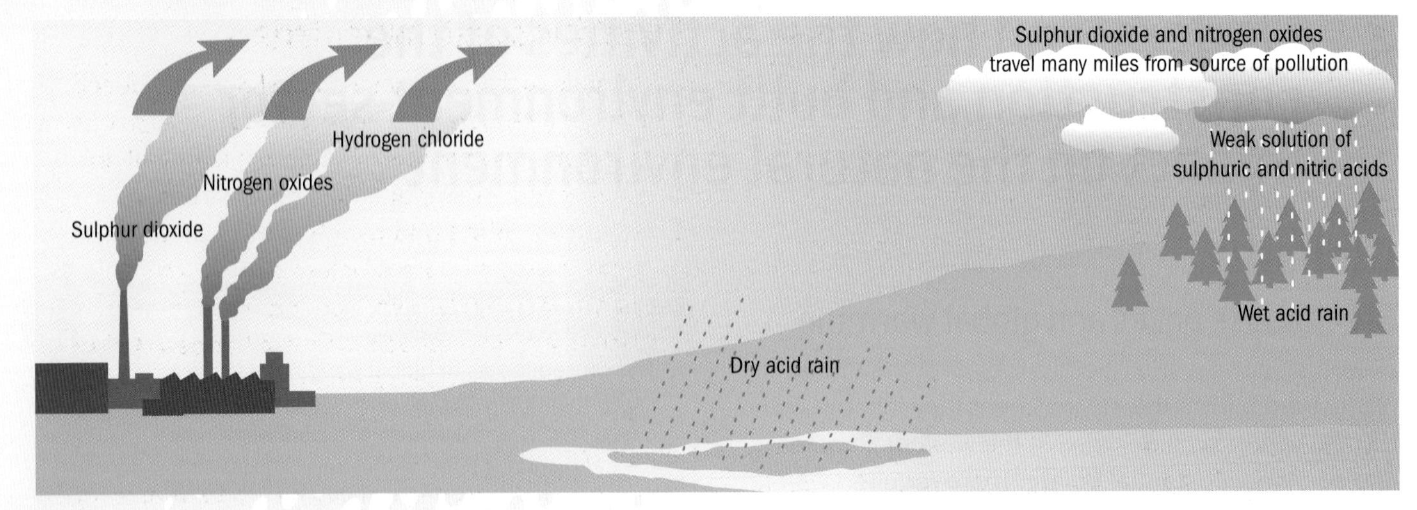

Figure 2.2: The acid rain cycle

Consequently, the pollution of one country can seriously affect the environment of another.

The acid pollution can fall to the ground in two forms:

- Wet acid rain – precipitation, such as rain, fog and snow. It can affect the area's biodiversity by upsetting the chemical balance of soils and the environment.
- Dry acid rain – falls as a dust or smoke that coats the ground, the buildings, the plants and the land. This pollution combines with the surface water to produce an acid water that has serious effects where it collects, especially in lakes and ponds as well as when it is taken up by tree roots.

Effects of acid rain

Acid rain can concentrate in thin soils, causing slow growth and eventually death of trees and plants. It affects the environment of lakes and rivers, killing fish and other aquatic life. Acid rain also causes damage to buildings where it reacts with some types of stone gradually removing and destroying the surface over time.

Remember!

Because acid pollution is carried by the wind, there is nothing the receiving country can do about the acid rain that falls upon its lands. Only a global agreement can help to prevent future contamination.

Ozone depletion due to CFCs

Chlorofluorocarbons (CFCs) are man-made chemicals that, until recently, were used in the following appliances:

- air conditioning units
- refrigerators
- cold stores
- freezers
- aerosols
- cleaning solvents
- foam products.

As mentioned on page 36, CFCs released into the atmosphere have damaged part of the protective ozone layer surrounding the Earth, leading to increased amounts of UV radiation reaching the Earth. UV radiation can:

- cause sunburn and lead to skin cancer
- affect the biodiversity and ecosystems of marine environments by killing certain micro-organisms
- lead to eye disorders such as blindness and cataracts.

Remember!

According to scientists, a single CFC molecule can destroy over 100,000 ozone molecules, so it does not take much of this product to have an effect. Despite the reduction in use of CFCs, they are still affecting the atmosphere as they gradually decay.

Over-extraction of water

The population of the UK has grown over hundreds of years, and in order to house this growth villages have turned into towns and towns have turned into cities. This expansion is reflected in the increased demand for water which is extracted from several sources by water companies. There are several ways to increase supply:

- create more reservoirs
- increase pumping rates from boreholes
- increase extraction from rivers
- reduce the level of wastage from distribution supply pipes
- install water meters.

The first three methods have detrimental effects on the environment. Creating more reservoirs means turning suitable river valleys into lakes by damming. This floods valuable fertile agricultural land and any buildings in the area are also lost.

Increasing pumping rates from boreholes drops the level of water contained within the aquifer. Although this has no visible effect on the surface, the borehole could run dry unless the water level is replaced by rainfall. Increased extraction of water from rivers can lead to environmental damage in the areas of wetlands. With the reduction of water levels, the wetlands dry out and their biodiversity and ecosystems change. In fact, during hot summers, over-extraction of rivers causes them to run dry, which kills off all aquatic life.

Reducing the level of wastage from the distribution supply pipes is a sensible alternative. Water is wasted through leakage due to broken pipes and connections which have outlived their service life. Reduce this leakage and you increase the supply to consumers. Installing water meters is another positive alternative and can have an immediate effect on consumption. This occurs as end users realise that every drop not used efficiently increases costs not only for the supply but also for the disposal.

Fossil fuels and raw materials

Fossil fuels such as oil, gas and coal were produced by the compression of the detritus materials (plants) from forests. This compression occurs over millions of years as tonnes of plants pile on top of each other and are inter-layered with rocks. Fossil fuels have to be extracted from below ground which has had a detrimental effect on the environment. Further environmental pollution can happen during transportation of these materials, for example the Torrey Canyon oil disaster of 1967. The supertanker leaked thousands of gallons of oil onto the coast of Cornwall. All of these resources are *finite*, that is, when they run out they cannot be replaced. Of all the raw materials, only timber can be regrown and used again.

Oil extraction in the UK only affects small areas on land. It is much more prevalent in the North Sea, for example the Brent oilfields off the coast of Scotland. Off-shore oil extraction still has a huge impact on the environment. Distribution and storage facilities are required for transporting the oil from the oilfields to processing plants. Heavier oils are imported using cargo ships that require docks and harbours for berthing and unloading.

It is a similar situation with natural gas, which is extracted from the North Sea gas fields and, more recently, also delivered by pipeline from Russia and Norway. Large areas are required to store and pressurise gas, preparing it for distribution through the underground pipelines.

There has been a recent decline in the coal industry. This is due to industrial disputes as well as the fact that burning coal emits a large amount of pollution. There are two methods of coal extraction: underground

Case study: UK coal industry

The UK coal industry in its era of maximum production extracted many thousands of tonnes of coal each year, mostly by underground mining. Part of the extraction process involved filling up large areas of land with the waste product removed with the coal. Large areas of land were used for these slag heaps. Coal was mainly produced for the UK electricity industry where it was burnt to produce steam that drove the turbines which generated electric currents.

These processes also had consequences on the environment.

Divide yourselves into teams and undertake research into the two areas identified above which are:

- the land use associated with coal extraction
- the environmental effects of burning coal.

(Hint: There are many coalfields within the South Yorkshire region.)

mines and open cast. Underground mines produce a substantial amount of waste that is landscaped into heaps which alter the surrounding environment. Open cast extraction, which strips the surface areas of the land to excavate the coal, involves replacing the removed spoil so the land can be restored to a new use.

Increased energy consumption and electricity generation

As the use of fossil fuels, such as coal, to produce electricity has declined, the burning of gas, a cleaner fuel, to produce electrical energy has increased to keep up with the modern consumers' demands. Electricity is required to power the variety of electrical items bought, from flat screen televisions to microwave ovens.

The problems with increasing energy consumption to supply this demand are not only the limited supply of fossil fuels but also the increased amounts of greenhouse gases. In order to start to address this issue, energy is now being generated from renewable sources. The current sources used in the UK are:

- wind power developed from wind turbines sited on land or at sea
- wave power produced from the action of the waves forcing air to drive a turbine
- hydro-electric power produced by damming rivers and directing water into turbines
- landfill gas that is tapped and burnt to produce heat, steam and electric turbine energy
- combustion of waste and sewage sludge to produce gas which is converted into steam
- the growing of crops to produce biofuels
- solar heating in which the sun's rays warm water in coils of pipes
- geothermal aquifers, where the Earth's magma energy near the crust is used to produce steam
- solar voltaic, which is the production of electricity using photo cell panels that convert the sun's energy to electricity.

Another energy option, nuclear power, still exists and produces energy in the UK, but public opinion on its use is influenced by green issues.

More and more of these renewable energy sources will have to be used to produce electricity if we are to reverse the effects of global warming. Using energy-saving technology, for example low voltage light bulbs, also contributes to saving valuable energy and extends the life of the fossil fuels that we have.

Deforestation

The hot and humid tropical areas of the world provide ideal conditions for natural rain forests, which have grown over thousands of years. Unfortunately, the location of most rain forests are within countries whose infrastructure and economic wealth are only just developing. The rain forests are disappearing for a number of reasons:

- expansion of towns and cities
- slash and burn agriculture where trees are cleared and the land used for the growing of crops
- trees being used for firewood
- clearing for cattle production
- hardwood timber is important to economic wealth and therefore the forest is seen as a resource to help the government
- illegal logging
- forest fires.

The UK was also once covered in deciduous forests whose timber was used for fuel, house building and ship building. Now these forests have all been nearly cleared.

So what impact will deforestation have on the natural environment?

- The water cycle may be affected with the loss of trees as they absorb a large volume of moisture and water in the tropical areas.
- The amount of carbon within the atmosphere will increase since there will be fewer trees to absorb these harmful gases and thus fewer trees to give out oxygen via photosynthesis.
- As the trees and ground covering are stripped, rainfall will erode the surface of the soil.
- This eroded soil may end up deposited in, and eventually silting up, rivers, lakes and ponds.
- The changes in environment may result in deserts being formed.
- With the loss of this habitat, species contained within it may become extinct.

Loss of natural habitat

Some people say that Britain has lost all of its natural habitats with the building of roads, railways, canals, airports, buildings and fields for agriculture, to name a few. As our population has expanded, so has industry and agriculture. The balance has favoured humans until recently.

The loss of natural habitat also leads to the non-recoverable loss of biodiversity. The loss of natural habitat in one area may also have an effect on another; animals that migrate between habitats no longer have places to migrate to, thus the natural cycle between environments is being lost.

Reduction in biodiversity

Biodiversity provides many natural resources, from clean air to clean water, allowing crops to propagate and fruit. Damaging the environment and thus affecting its biodiversity will always have an effect somewhere in the food chain.

The process of natural selection, the way different species rely on one another, the predator and the prey, are all examples of how biodiversity is linked between species. Take out or change a process and the results could lead to a domino effect on other species. For example, rats are now becoming resistant to the poisons used to control them; this will affect other species populations.

Assessment activity 2.3

BTEC

You work at a UK construction company. The managing director wants the company to have a green policy on all its activities. You have been tasked with researching this topic and have to initially identify four forms of global pollution arising from the company's activities and describe how each may harm the local environment where the company operates. P2

Grading tip

For P2 you need to first identify a form of global pollution then open this up with a description of how it harms the environment.

PLTS

Analysing and evaluating information you find on global pollution, judging whether it is relevant and has any value will develop your skills as an **independent enquirer**.

Remember!

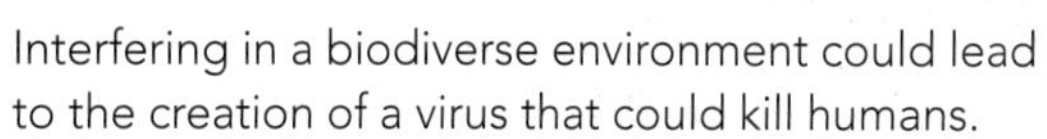

Interfering in a biodiverse environment could lead to the creation of a virus that could kill humans.

Locally

Air pollution by combustion products and volatile organic compounds

Combustible pollutants are compounds produced by burning a fuel. For example, petrol or diesel engines produce exhaust gases that pollute and affect the quality of air. These pollutants include carbon monoxide, nitrogen dioxide and carbon dioxide. Smoking cigarettes also gives off pollutants such as nicotine and tar.

Combustion products can be dangerous to our health and in concentrated doses can damage respiratory tissue which lead to lung diseases and, in some cases, death through asphyxiation. Local combustion products can enter your home and affect you via several sources:

- gas boilers
- gas fires
- car engines running beside an open window
- traffic on busy roads
- smoking indoors.

Volatile organic compounds (VOCs) are given off by certain solids or liquids. VOCs are more concentrated indoors due to the lack of ventilation. In the home VOCs are found in such items as spirit-based paints; these are now gradually being replaced by water-based paints which do not contain the harmful compound.

At work, VOCs are still present in photocopiers, inks, cleaning materials, and many other products. Local exhaust ventilation (LEV) in offices and workshops removes these to the atmosphere.

Polluting discharges to water by communities

The sewer system of the UK, from both home and industry, is made of separate and combined systems. With a separate system, the surface rainwater and foul water are two separate drains; thus, the rainwater does not have to be treated. Foul drainage requires treatment to remove solids and bring the water up to a quality standard before its discharge into rivers or sea outlets. The alternative to connecting to the main sewer is a septic tank which filters out the solids, keeping them in the tank which has to be emptied, and the water from the tank treated using reed bed techology.

Thinking point

Some of London's water supply has passed through the human body three times!

The UK's sewer system was designed in the nineteenth century. There are many problems with this ageing system. Capacity is the main problem, as the system was not designed for today's large towns and cities. When there is heavy rainfall the sewer system is unable to cope. The water then backs up and floods areas, leaving environmental health consequences associated with raw sewage.

Another modern problem is off-road parking. Many driveways used to be grassed and unpaved areas. Since the boom of the motor industry, many of these areas have been paved over. Any rainfall, therefore, does not soak into the ground. Instead it enters the sewer system by running off the driveway and into the road gullies. Further contamination and pollution of the sewers occurs from car-washing detergents, petrol and coolant, which are washed into the drainage system.

Modern industry also adds to the water pollution problem. Industrial pollution takes several forms. Heavy metals that have been used in the past such as cadmium and mercury collect in lakes, sometimes affecting drinking water supplies. Also, micro-organisms, such as cryptosporidiosis, enter the system through water leaks; they can cause sickness and upset stomachs. The only remedy is to boil the drinking water supply. In developing countries, where water is supplied from wells, diseases such as cholera and typhoid can easily spread.

Industry and agriculture

Industry and agriculture have a great impact, both locally and globally, on the environment. Both require land for their processes; for example, factories are needed for such processes as the conversion of sugar beet into sugar granules or the conversion of crude oil into petrol. These factories also have other environmental links to global warming, energy usage and pollution.

Agriculture takes up vast areas of land and has changed our natural landscape through ploughing, the use of fertilisers and so on. Industry has tended to concentrate within certain areas and regions, from the Sheffield's steel industry to the Humber estuary, where a number of factories use the river for the export and import of goods. Industry tends to be located where there is easy access to raw materials.

Contaminated land

Human activity has often been unregulated and uncontrolled. The rapid expansion of the Industrial Revolution of the eighteenth and nineteenth centuries needed natural resources for power and transportation links. Factories were set up and used whatever method was needed to produce a finished product. This often involved using chemicals to process the raw material. At the time no one realised that chemicals such as arsenic, cyanide, acids and alkalis used in the petro-chemical, oil, paint and iron and steel industries would contaminate the soil and environment.

Old industrial areas have been classified as brownfield sites by the government and developers are encouraged to use these rather than take large areas of **greenfield land**. The soil below brownfield sites has to be tested for contamination before being built on and suitable action taken to prevent any contamination which could eventually cause ill health.

Key term

Greenfield land – area which is undisturbed by previous construction – in effect, a green field.

This may involve removing the pollutants to licensed tips or treating the contamination in the ground by adding another chemical to it to balance the pH of the soil. Soil barriers can also be inserted to prevent the spread of contamination.

Waste disposal

As the population of the UK has grown and expanded, so has the demand for consumables. These generate food waste as well as packaging waste in the form of paper, cardboard and plastics. There are only a few methods of dealing with this waste:

- disposal to landfill sites – this is now subject to a landfill tax levied to prevent waste and encourage recycling
- incineration – waste is managed as a fuel and burnt to produce electricity
- recycling – materials such as cardboard, newspaper, metals, plastics and glass
- composting into mulch – this involves composting green waste from gardens into a useful compost.

Suitable sites for waste disposal are becoming increasingly difficult to find. Recycling needs to be increased to reduce the amount of wastage produced. Current local authority levels of recycling are at 40 per cent; European legislation will further increase this percentage. New technology has helped; there are now plants capable of recycling fridges and freezers.

Remember!

The more we recycle and conserve, the fewer finite resources we will need to use which is more beneficial for the construction and built environment and global warming.

Existing site dereliction

Many old buildings may become derelict because:

- the cost of refurbishing the building is too high
- the building has exceeded its life expectancy
- the building has been damaged by fire and/or vandalism
- the occupier has gone bankrupt or ceased trading
- the economy of the local area is in decline.

Often these buildings are listed and have a recognised English Heritage status. This means they have to be preserved in the condition in which they were built and any change of use and alteration to the exteriors is prohibited.

Empty buildings are targets for vandalism and destruction; therefore, any valuable item is removed and the exteriors are boarded up to prevent entry. Local authority building control officers may get involved to make safer a derelict building which has become a dangerous structure so that no member of public is injured by it.

Comfort disturbance

Comfort within a home or working environment can be disturbed by:

- too little or too much heat
- poor ventilation
- noise
- unpleasant smells
- lack of cleanliness.

While many people like to live in cities, city life has its downside. Uncollected refuse on the streets is unsightly and can cause unpleasant smells. Sewer systems may not be able to cope with the large demand. Noise is another disturbance associated with a densely populated area, from people playing music, or arguing as well as from the large volume of traffic that the city has to cope with during rush hours.

Traffic can also cause cracking and structural damage to buildings from vibrations. Furthermore, the exhaust from traffic congestion has an environmental impact when it mixes with the midday heat plus other combustion products to form a smog over large city areas. Smog can affect the respiratory system causing breathing problems such as asthma. Industry can also often cause dust and dirt problems in cities. This does, to a certain degree, depend on the local climate; for example a hot climate may be dustier than a wetter climate where rainwater washes away the dust and dirt.

Thinking point

Litter is a common sight in our towns and cities. Think of ways to keep our streets and environment cleaner.

Increased pressure upon existing services and infrastructure

With 60 million people now living in the UK, many of our towns and cities have expanded into huge conurbations. With this expansion has come increased pressure on the following types of infrastructure:

- the road network – roads were not built for today's amounts of traffic; for example, the M25 is now full to capacity at most times
- the railway network – increasingly passengers have to stand for their journey due to the sheer volume of people using the network services
- air traffic numbers – these have increased dramatically with the rise of low-cost airlines offering budget fares
- the water supply network – some areas of the UK struggle with inadequate supplies and water pressure
- gas supply – more demand has meant obtaining supplies from continental Europe and Russia.

Specification of hazardous materials

Lead and asbestos were historically used in construction before their possible health risks were determined. Lead is present in old water pipes, some gas pipes and in paint and it can be absorbed into the blood stream through the drinking water supply. Lead can also be inhaled when old paint is burnt during redecoration. Outside of construction, lead was also used as a lubricant in petrol and has entered the atmosphere as a result of combustion.

Asbestos is present in old vinyl tiles, pipe lagging, roof sheeting, latex and many other materials. Asbestos is known to cause lung disorders and cancers. As a result of this, asbestos is removed during alterations and refurbishments and is taken to licensed tips and disposed of safely in controlled conditions.

Extraction of raw materials

There are many raw materials that are mined, quarried or extracted by drilling in the UK, including:

- coal
- oil and gas
- gypsum
- various rocks for crushing into hardcore
- roofing materials such as slates
- gold
- various minerals.

The effect on the environment depends on the extraction method used to obtain the raw product. Coal mining where seams of coal are taken out of the ground can cause long-term settlement to the surface. Disposal of the unwanted material that is brought to the surface and tipped can also lead to detrimental environmental problems. Open cast mining requires the **overburden** to be removed so the raw material can be mass excavated and removed; when the mine is exhausted the overburden is replaced. Ground levels are subsequently lower and the environment has to recover from this change.

Key term

Overburden – the layer of material that has to be removed to get at the minerals beneath.

Electromagnetic radiation from overhead power lines

Transporting electricity using pylons is cheaper than burying cables across miles of countryside. Electromagnetic radiation forms from the conduction of electricity along the power line. There have been several studies published that seem to prove a link between an increased risk of childhood leukaemia and the proximity to power lines. However, this needs to be confirmed from other long-term studies as there may be other factors involved.

The electricity companies try to build pylons away from high centres of population. They also tend to bury cables when nearing a town or city, as pylons are also an eyesore.

Sick building syndrome

Sick building syndrome refers to a building that is causing its occupants to feel ill. The illness can take the form of headaches, flu-like symptoms, ear, nose and throat problems, nausea and tiredness. When the occupants leave the building their health improves. There could be several factors causing sick building syndrome, including:

- build-up of dust particles due to lack of cleaning
- chemicals that have been released into the atmosphere.

- outside pollution that has entered the building
- damp, dark conditions that cause the growth of mould and bacteria which release spores into the atmosphere
- lack of natural light needed for normal active health
- lack of fresh air from inadequate ventilation
- poor, dirty conditions
- high humidity which can cause breathing problems
- high temperatures that cause discomfort.

Good building design needs to also take in the importance of the health of the occupants. Illness caused by sick building syndrome has a detrimental effect on the financial side of the business, through sickness benefit and potential claims for ill health.

Assessment activity 2.4

BTEC

Residents of a local housing association have complained about some local pollution issues. You are asked to visit the site and establish what their concerns are.

Identify some forms of local pollution that could be affecting them and describe how each may harm the local environment around the housing development. P3

Grading tip

For P3 identify some forms of local pollution, for example traffic exhausts, and describe how this local pollution would harm the environment.

PLTS

Questioning your own views on local pollution issues and others assumptions on this will help your development as a **creative thinker**.

Functional skills

Ensuring that your written work has accurate grammar, punctuation and spelling and that its meaning is clear will help your skills in writing **English**.

Pylons transport electricity across the countryside

3 Understand how the natural environment can be protected against the activities of the construction and built environment sector

Legislation

The Environmental Protection Act 1990 was a powerful piece of legislation introduced to protect the UK's environment. In 1995, the Environment Agency was set up (see below). These are the main Acts of Parliament introduced to protect the environment:

- Water Act 1989
- Control of Pollution (Amendment) Act 1989
- Environmental Protection Act (EPA) 1990
- Land Drainage Act 1991
- Water Resources Act 1991
- Environment Act 1995.

Theory into practice

Find out more about legislation that protects the environment at the Office of Public Sector Information website (www.opsi.gov.uk).

The Environment Agency has passed legislation to cover the following environmental areas:

- air
- chemicals
- conservation
- energy
- land
- noise and statutory nuisance
- pollution prevention and control (PPC)
- plant protection
- radioactive substances
- waste
- water.

Environmental regulations developed from Acts of Parliament include the End of Life Vehicle Regulations 2003 and Agricultural Waste Regulations 2006. There are hundreds of European directives concerning the environment created by the European Parliament for which each member country develops environmental regulations.

Thinking point

Legislation has been passed to protect the environment; people and individuals can now be brought to court and prosecuted when they have caused damage to the environment. Is bringing them to court after the damage too late?

Control

Health and Safety Executive

The Health and Safety Executive (HSE) has powers to enforce government legislation. The HSE has very little input into the environmental side of the law apart from the nuclear directive governing the releases of radiation. The HSE primarily investigates safety in environmental issues, for example they make sure the correct personal protective equipment (PPE) is worn when cleaning up a spillage. Should a breach of a safety regulation occur, then the HSE has the power of prosecution.

Environment Agency

The Environment Agency (EA) is a public body whose role is to protect the environment. According to the EA's chairman the EA is 'the leading public body for protecting and improving the environment in England and Wales. It's our job to make sure that air, land and water are looked after by everyone in today's society, so that tomorrow's generations inherit a cleaner, healthier world' (Environment Agency website, 2007). The EA undertakes many checks including air and water quality monitoring. It also prosecutes offenders and gathers evidence in support of cases going to court.

Local authorities

Local authorities have many powers including:

- planning
- environmental services
- building control.

The planning department establishes the **local plan** under planning legislation. This enables control over the erection and alteration of buildings, the removal of trees and hedges and the construction of roads and other hard-landscaped areas. The local plan helps protect and maintain the environment by listing areas which are earmarked for developments in such categories as industrial, residential and commercial.

Key term

Local plan – the document that sets out and controls the planning policy for the local authority's area. It sets out where the authority wants industry and housing to grow.

Theory into practice

Visit your local authority's website to find out the local plan.

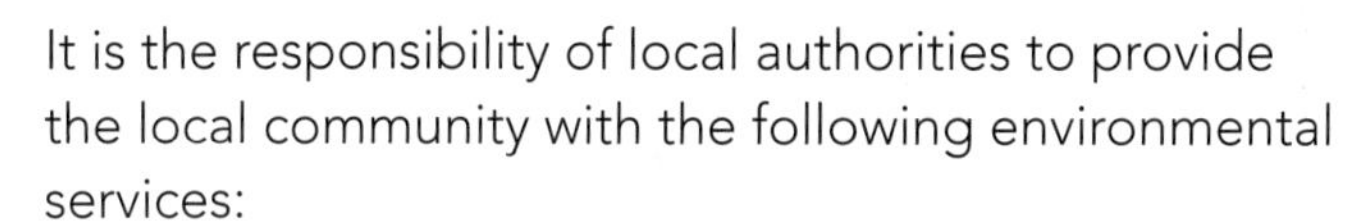

It is the responsibility of local authorities to provide the local community with the following environmental services:

- municipal waste collection and disposal
- recycling
- pollution control
- noise control
- clean air
- public health
- commercial waste
- food safety.

All of the above can be legally enforced through prosecution and fines.

The building control department regulates the demolition, alteration and erection of both commercial and domestic buildings. Environmental aspects covered by building control include drainage, heating, power and lighting and ventilation. They also control the thermal properties of a dwelling and hence lead to a reduction of carbon emissions. Building control will often enforce regulations through the checking of plans on conformity and by visual site inspections. This department can also legally enforce the regulations.

Assessment activity 2.5

BTEC

1. A new housing developer has purchased the field next to your row of houses. Development work has started on site and there have been several complaints by you and your neighbours. Complaints have been about the size of the properties being constructed, how near they are to your boundary and the noise levels during construction. Describe four key methods that can be used to protect the natural environment from this proposed development. P4
2. When you have described the four key methods, undertake a comparison of these in terms of:

- cost
- effectiveness
- public perception. M2

Grading tips

For P4 you will need to describe in detail four key methods that are used to protect the natural environment.

For M2 you will make a comparison of these four key methods in terms of their costs, how effective they are and what the public perception is of the method.

PLTS

Working within a team discussing environmental protection will help with your development as a **team worker**.

Functional skills

Selecting and using appropriate sources of ICT-based and other forms of information will help develop your skills in **ICT**.

Design and specification

Good building design is essential to protect the local and global environment. Good-quality design to a high standard, using sustainable construction techniques, will pay dividends for the environment. The orientation of buildings, for example, can harness the sun's energy, providing natural light instead of artificial, thereby saving energy. Sourcing local materials helps reduce the transport carbon footprint and using renewable resources are just some of the ways in which designers can help to protect the environment.

The use of better-quality materials plus more time spent on the thermal design will, in the long term, not only save on building maintenance but will also save energy and hence reduce the effects of global warming. This has been further backed up with changes to Part L of the **Building Regulations** which deals with the thermal efficiency of buildings.

Reduction in energy usage

We need to learn how to extend the life of our finite resources. To accomplish this, we need to save energy. This can be achieved through:

- energy saving measures
- renewable energy measures.

There are many ways to save energy in a building including switching off appliances on standby, using grade A appliances, using low energy light bulbs and only filling a kettle with the amount of water needed – but these are just the tip of the iceberg.

Renewable energy sources are a rapidly developing technology. They harness aspects of the environment for use to make energy. However, they can still leave a mark on the environment. For example, both hydroelectric reservoirs and wind farms require large areas of land. Other examples of renewable energy sources in the UK include wave power, geothermal power and solar power.

Activity: Building wind turbines

The village where you have lived for the past 20 years is an area of natural beauty and has unspoilt views of the Lincolnshire Wolds. It is a countryside region of the east coast, in an area that receives a high and constant wind from the North Sea. You are a member of the local parish council and have just received an application for outline planning permission for ten wind turbines to be built in a field just outside the village. The council is outraged and plans to object strongly.

Divide yourselves into teams and discuss on what grounds the council could object to this electricity production development.

Both energy saving and renewable energy measures rely on the education of the UK population and their willingness to accomplish these measures against the cost involved. For example, solar panels on a new house have a heavy initial cost on the selling price and a long payback period. However, the positive effect on the environment will outlast those costs.

Minimisation of pollution

Pollution has a marked effect upon the environment as well as on our health and well-being. There are several ways to minimise pollution including:.

- fitting fume scrubbers to chimney outputs to remove solids and any pollutant chemicals that would be released into the air
- treating water of foul drainage before it is discharged into sea outfalls to reduce the effects on the marine environment
- minimising wastage by recycling construction products such as timber, metal and bricks and thus reducing pollution dumped into landfill sites
- using developments in technology such as biofuels and low sulphur fuels to provide cleaner emissions released during combustion in engines.

Reduction in embedded energy

There is a close relationship between the energy we use and global warming. We can therefore try to reduce the amount of energy we use in construction by specifying materials that have lower **embedded energy**. Natural timber products are excellent examples as they absorb carbon during growth and cost little to convert into a usable construction product. Cement-based products, on the other hand, contain high levels of embedded energy; thus, careful consideration should be given before using these products or alternative products should be sought.

Key terms

Building Regulations – these are produced under the Building Act 1974 and control many aspects of construction in order to ensure that energy saving measures are built into new and existing designs.

Embedded energy – the amount of energy used to produce the material. It is often expressed in terms of how much carbon has been released into the atmosphere during its manufacture and transport.

Environmentally friendly, renewable materials

Architects, designers and clients are now considering the effect that their development will have on the environment. Green issues such as sustainability are an essential environmental consideration in selecting the materials to incorporate into a building. Products need to be environmentally friendly, and cause lower amounts of environmental pollution.

An example of an environmentally friendly material would be timber cedar boarding which is long-lasting and requires no treatment. Natural insulation products developed from newspaper and sheep wool are examples of green products used in a unique way to reduce energy consumption.

Reuse of existing buildings and sites

Earlier in this unit, we mentioned the term brownfield site. This is a site that has had a previous commercial use and where the buildings may have been left standing or demolished. The question that must be asked before considering a development opportunity would be, is it better to reuse a plot of land than to develop on virgin land? Often the ground is contaminated with pollutants that require cleaning up to provide a safe environment for any future use.

Assessment activity 2.6

BTEC

You work at a construction company and need to look into the environmental issues. These are very important considerations when developing the built environment of the UK. Assess the importance of addressing these environmental issues for the mutual benefit of the community and individual construction firms. D1

Grading tip

D1 requires you to undertake an assessment, for example advantages and disadvantages of addressing the environmental issues, for the contractor and the wider community.

PLTS

Exploring the environmental issues from another angle will help develop your skills as an **independent enquirer**.

Functional skills

Using appropriate search techniques to locate and select relevant information on environmental issues will help develop your skills in **ICT**.

Reuse of buildings is often called refurbishment or adaptation. This is a useful technique and will save energy and materials in the long run. Buildings of any merit are conserved under English Heritage's listing scheme.

Management

Simple environmental impact assessments

A simple environmental impact assessment (EIA) will enable the management of an organisation to have a clearer picture of the effect the organisation is having on the environment. An EIA is normally a checklist of items which would need careful assessment, such as:

- waste disposal
- thermal efficiency
- water discharge
- heat discharge
- water vapour discharge.

Remember!

Once completed, environmental impact assessments need to become part of an action plan. Left in a file, there will be no benefit to the environment.

There are now a large number of companies that can undertake EIAs on more complicated projects, such as wind farm locations, where the paperwork may stretch into many volumes.

Improved management of construction sites

The site manager is instrumental in protecting the environment. Double bunded tanks for fuel oils, spillage kits and silt tanks can be deployed by an effective site manager to help prevent illegal spillages into soil or silts pumped into drainage systems.

Waste should be sorted during the construction process to ensure materials can be recycled. A good site manager would provide, for example, the following:

- a metal skip so all waste metals can be melted down and reused in metal products
- a wood skip so waste timber can be manufactured into mdf or chipboard products
- plasterboard skips to recycle off-cuts that can be sent back to the manufacturer
- cardboard and paper skips.

Site managers can also avoid the double handling of a material, by moving it once; thus reducing fuel costs. They can also purchase local raw materials such as gravels for drainage to help reduce the effects of transport on the environment.

As you can see, there are many ways for an able, environmentally aware, site manager to provide added value on a construction project to the local and global environment.

Clear policies and objectives

A construction company that has clear environmental policies and objectives on wastage, recycling, noise management and dust and dirt control makes a statement to all its employees and customers that it cares about the environment.

Local authorities' environmental services departments have a clear mandate from both the government and the European Union (EU) to increase the percentage of municipal waste that is recycled. This should reduce the amount of waste going to landfill sites. The EU has also taken steps to encourage car manufacturers to recycle 70 per cent of a car at the end of its life.

As with any policy, continual monitoring and review is necessary to ensure that there is a marked positive effect on the environment.

Sharing of good practice

Good practice must be shared among sites and organisations for an even greater impact to be felt on the environment. No employee should be ignorant when it comes to taking environmental considerations into account while undertaking the duties of the company. A good idea should be rewarded, recognised and shared and these innovative ideas must be shared freely and not sold for profit, for it is only through the efforts of everyone that the possible effects of global climate change can be reduced or halted.

Raising of awareness

Education of the construction workforce is the first thing that must be accomplished in order to put into place the policies described earlier. If people are unaware of the environmental effect they are having locally on a construction site, they are unlikely to participate in any policy or activity that is put into place. Companies can extend this further and involve local communities in the process through school visits, community activities and advertising green issues related to the company.

Communication of information

Communicating environmental information is an important process and can be verbal and/or written. On-site inductions for all employees can make them aware of any environmental considerations related to the project they are working on. Tool box talks can reinforce this message during the construction phase of any project. Communication must be top-driven, that is, from the managing director downwards, and must be two-way so that all levels of operatives and management are aware.

Assessment activity 2.7

BTEC

The construction company that you work for is considering moving all its traditional methods of constructing houses over to sustainable methods of construction.

1. Explain three different sustainable construction methods which are fit for purpose. P5
2. Form a comparison of these three sustainable construction methods in terms of relative cost and performance. M3
3. The company director is considering using sustainable construction techniques for a prefabricated home. Justify this use. D2

Grading tips

For P5 you must explain three different methods of using sustainability in construction which are fit for purpose.

For M3 you form a comparison of these; do so in table format.

For D2 you must justify these sustainable techniques in terms of advantages and disadvantages against each use.

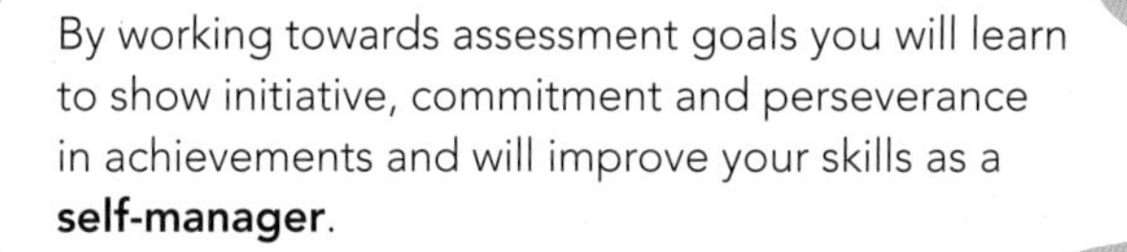

PLTS

By working towards assessment goals you will learn to show initiative, commitment and perseverance in achievements and will improve your skills as a **self-manager**.

Functional skills

Listening to others speak on sustainable construction and replying to discussions will develop your skills in **English**.

4 Understand sustainable construction techniques that are fit for purpose

Fit for purpose

The definition of fit for purpose is that the design, specification and materials used are placed in the environmental location where they can stand up to the elements such as:

- the weather
- flooding
- occupants
- plant and machinery.

All of these factors will need to be carefully considered so a building is initially designed in accordance with the client's detailed brief without a heavy maintenance budget to consider. All materials that are specified must therefore be tested to ensure that they will meet a 'fit for purpose' criteria, this can be achieved by:

- specify materials to a British Standard
- specifying materials to an European standard
- specifying to a manufacturer's recognised standard.

Techniques

Energy-based techniques

Reduction in energy consumption is something that can be easily tackled. Employing more efficient technology and utilising alternative energy sources will help reduce waste and make finite resources last a lot longer. Electrical technology is now producing grade A appliances that require less electricity to run. The cost of manufacturing low energy light bulbs has fallen and they are now available in a wide range of fittings. New buildings are subject to air leakage tests which reduces the amount of energy lost.

Combined plants that produce electricity and heat are a growing technology that reduces our energy consumption. They are a third more efficient than separate plants. Energy efficiency can be improved through construction too. For example, efficient boilers that take the heat out of exhaust gases can be installed, as can improved thermal insulation in housing.

Heat pumps that extract waste heat from systems and harness it to heat water are also a steadily developing technology.

Alternative sources of energy and renewable energy are a newer breed of energy sources that are becoming popular. Examples include:

- wind turbines for home use – a leading DIY store is now providing a fixing service
- solar panels built into roof tiles format so they look attractive and fit into the roof's profile
- geothermal energy using heat within the soil – a trench is dug in the garden to capture latent heat
- biofuels – crops can now be grown and then distilled to produce fuels
- wave turbines
- hydroelectric schemes
- hydrogen fuel cells.

All these energy sources save the Earth's finite resources and with maintenance and repair, they never run out.

Materials-based techniques

Renewable materials in construction is a technology that is slowly developing. Specific materials include:

- sheep's wool can be efficiently used as a thermal insulation
- recycled paper products can be used to form cellulose loose fill insulation for lofts.
- cedar timber cladding is a renewable material that can be used on the exterior of houses
- green roofs that are manufactured from selected plants can now be used as a natural and renewable weatherproof covering.

Timber engineering is taking strides in maximising the use of timber products to develop joists of the same depth, but using smaller sections of timber built up, that can cover large spans so replacing the use of concrete beams (see Figure 2.3). Using high-performance, softwood window frames instead of uPVC saves energy and oil in the long term.

Designers must be the principal driver of low embodied energy materials by specifying these materials within their designs and convincing clients of the benefit to the Earth's environment. Any material that can be recycled halves the amount of embodied energy used in its manufacture. For example, concrete and brickwork from demolitions can be crushed on site and used as a hardcore base for buildings; this also saves transport costs. Steelwork can also be recycled and used to produce new structures. This reduces the amount of energy consumed in the production of the raw materials. Timber beams using recycled timber boarding cores can now be used instead of solid timber joists. These use less timber in their manufacture and are considerably stronger.

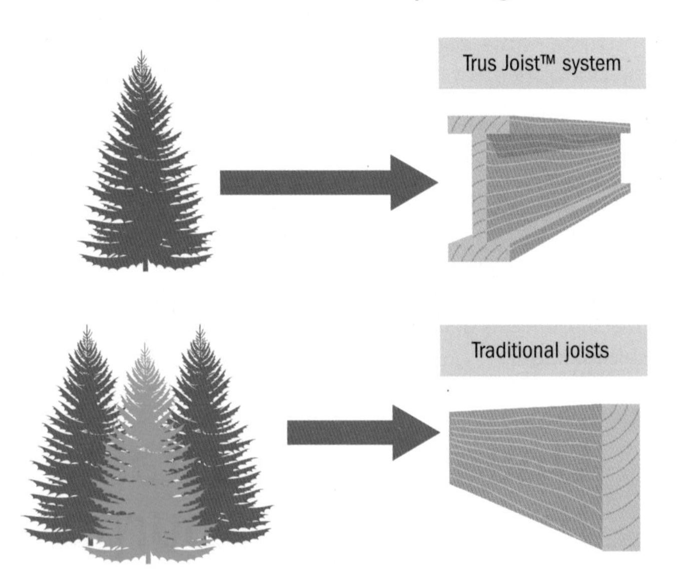

Figure 2.3: Timber engineering replaces the use of concrete beam

Any construction materials manufacturer that can reduce the energy usage in the manufacturing process will lower the embedded energy of the product. This could be achieved in several ways. Waste heat can be recycled and taken out of exhaust gases and used to produce electrical energy which can be recycled back into the product. In fact, many producers are starting up combined heat and power plants to gain the maximum benefit from this type of energy usage on site.

Wind turbines offer an alternative source of renewable energy

Waste-based techniques

Waste costs money, as it has to be disposed of through incineration or landfill. There are several methods that can be used to reduce waste, including:

- recycling materials back into the process, e.g. steelwork
- finding a processor to use the waste produce in the manufacturing process
- ordering the correct amounts of material and managing the use of raw materials efficiently
- using the waste as fuel for energy production
- packaging carefully to prevent breakages
- training in the use of a material
- educating that waste costs the environment
- coordinating modular dimensions to standard lengths throughout the industry.

A reduction in wastage can often be brought about by educating the people using the materials about the environmental costs of wasting these materials. Incentive schemes could help along with a top-down management approach to reducing the effects of construction on the environment. This could even include the manager using a smaller company car.

Recycling

In construction, recycling has to be done cleverly since clients want new builds, not a building that has been rebuilt. There are materials, however, that can be reused and recycled, such as:

- crushed concrete can be used as a fill material to raise levels
- facing bricks can be cleaned and reused to give a house an aged appearance
- slates can be redressed or crushed and combined with an adhesive to form a reconstituted slate tile
- crushed bricks can be used as a fill material
- glass can be melted and reformed
- recycled steel is used in the production of structural steel
- plastics can be used as timber-like products
- timber can be engineered and recycled into structural products.

The possibilities are fast developing with new technologies and new products continually coming into the marketplace; recycling waste into materials is starting to take hold. For example, gabions are steel cages used as earth retainers which are filled with crushed concrete or brick.

Off-site fabrication

Off-site fabrication, a recent development by UK manufacturers, is, for example, where a timber-framed house kit can be assembled off site, delivered by lorries and then site assembled using a crane. The only additional elements needed are the cladding external finish, which can be in a brick skin or a timber-cladded product, and the roof tiles. This type of construction is very thermally efficient, saves time and energy, uses renewable timber products and has relatively low carbon emissions in manufacture.

Off-site fabrication of structural elements of commercial and domestic construction is the new approach to the production of houses and offices. Modules that simply bolt together are fast and efficient methods of producing a structure that is factory produced with minimal resulting wastage. This method is very efficient and is aimed at producing affordable social homes in the current economic climate.

Just in time (JIT) construction takes offsite fabrication one step further. With JIT, all the materials for incorporation into the structure are delivered only when needed in the construction process. This avoids the need to store materials on site for long periods. Furthermore, pre-cast concrete is returning as a modern method of construction enabling benefits from factory-produced concrete that is site assembled with no waste or need for secondary support systems like formwork.

Sue White

Environment Research and Development

Sue works for the government within the environment department. Her role is to research and develop new methods of constructing buildings that reduce our reliance on the use of fossil fuels. This is an important role as the government has signed up to an international agreement on the zero carbon emissions home. This means that the carbon locked up in the home is equal to the amount of carbon it has taken to construct which therefore is carbon neutral.

Sue obtained her job through a Graduate Recruitment Programme. She had attended college and completed a National Diploma and then went on to take a degree at university in Sustainability and the Built Environment.

Sue interacts with many national house builders, which gives her access to construction sites where she can research and develop ideas. She also has control of a budget which is used to build and test homes using research materials. The project she is currently working on involves using straw blocks – a waste product from the harvesting of food stuffs – as a building material to construct homes. Sue loves experimenting with different materials to see what energy efficiency can be obtained. There are several new ideas on the market, for example, ground-source heat pumps, small-scale wind farms, methane gas fuels, and waterless toilets.

Sue works with a dedicated team who have all come from construction and design backgrounds, either through A-levels or Diplomas and a University education. She has been rewarded with the amount of effort that she put into her studies to succeed in a job she really loves doing. The test results she has produced are now being considered by the government and may be introduced into the Building Regulations when they are next revised to become law.

Sue's research methods are now being developed commercially and trialled and included in some test homes that are being built this year. Many of the house developers she has been working with are keen to try out her new ideas on carbon efficiency.

Think about it!

- What ideas could you research on energy efficiency in homes?
- What would be your ideal job in this field?
- Would you enjoy working in a team?

Just checking

1 Name three sources of drinking water.
2 What are the consequences of burning fossil fuels?
3 Identify three forms of air pollution.
4 What does 'EPA' stand for (legislation)?
5 What is embedded energy in a material?
6 How can the amount of construction waste produced during a project be reduced?
7 Identify five techniques that can be used within domestic housing to reduce energy consumption.
8 What is an EIA?
9 Name three products that can be recycled back into building materials.
10 What is sustainability?
11 Give two environmentally friendly techniques used to generate electricity.

Assignment tips

- Every council has an environmental officer. You could contact them for some local knowledge on pollution issues. Make sure you prepare some questions to ask them.
- Look at the websites of some protest groups, for example, Greenpeace. They might have some useful information on how the environment can be protected.
- Have a close look at the websites for timber-frame manufacturers for a sustainable construction technique.

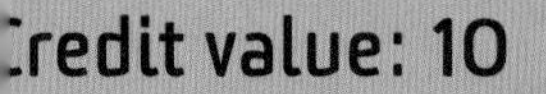
Credit value: 10

3 Mathematics in construction and the built environment

Builders and contractors use a variety of manual and powered tools to construct buildings. Mathematics is simply another tool to assist construction craftspeople and professionals in doing their job. Mathematical knowledge and skills will help solve numerical problems and answer the queries that construction workers come across.

This unit will take you through the basic mathematics to help solve a range of problems faced by designers, surveyors, cost controllers and contractors in the construction industry. It will show you how to work with numbers and formulae using standard techniques and methods. You will learn how to apply theory to practical examples involving perimeters, area and volumes, as well as understand the properties of different shapes, and how to handle calculations involving angles. You will also explore the use of graphs and statistics and see how they can be used to help solve and understand construction-related problems.

Learning outcomes

After completing this unit you should:

1. be able to use basic underpinning mathematical techniques and methods used to manipulate and/or solve formulae, equations and algebraic expressions
2. be able to select and correctly apply mathematical techniques to solve practical construction problems involving perimeters, areas and volumes
3. be able to select and correctly apply a variety of geometric and trigonometric techniques to solve practical construction problems
4. be able to select and correctly apply a variety of graphical and statistical techniques to solve practical construction problems.

Assessment and grading criteria

This table shows you what you must do in order to achieve a pass, merit or distinction grade, and where you can find activities in this book to help you.

To achieve a **pass** grade the evidence must show that you are able to:	To achieve a **merit** grade the evidence must show that, in addition to the pass criteria, you are able to:	To achieve a **distinction** grade the evidence must show that, in addition to the pass and merit criteria, you are able to:
P1 use the main functions of a scientific calculator to perform calculations and apply manual checks to results **See Assessment activity 3.1, page 79**	**M1** use algebraic methods to solve linear, quadratic simultaneous linear and quadratic equations **See Assessment activity 3.1, page 79**	**D1** independently undertake checks on calculations using relevant alternative mathematical methods and make appropriate judgements on the outcome **See Assessment activity 3.1, page 79**
P2 use standard mathematical techniques to simplify expressions and solve a variety of problems using linear formulae **See Assessment activity 3.1, page 79**		
P3 use graphical methods to solve linear and quadratic equations **See Assessment activity 3.4, page 114**		
P4 use mathematical techniques to solve construction problems associated with simple perimeters, areas and volumes **See Assessment activity 3.2, page 91**	**M2** select and apply appropriate algebraic methods to find lengths, angles, areas and volumes for one 2D and one 3D complex construction industry related problem **See Assessment activity 3.2, page 91, and Assessment activity 3.3, page 102**	**D2** independently demonstrate an understanding of the limitations of certain solutions in terms of accuracy, approximations and rounding errors **See Assessment activity 3.1, page 79, and Assessment activity 3.2, page 91**
P5 use trigonometric techniques to solve simple 2D construction problems **See Assessment activity 3.3, page 102**		
P6 use geometric techniques to solve simple construction problems **See Assessment activity 3.3, page 102**		
P7 use graphical techniques to solve practical construction problems **See Assessment activity 3.4, page 114**	**M3** use standard deviation techniques to compare the quality of manufactured products used in the construction industry **See Assessment activity 3.4, page 114**	
P8 use statistical techniques to solve practical construction problems **See Assessment activity 3.4, page 114**		

How you will be assessed

The evidence requirements for pass, merit and distinction grades are shown in the grading criteria grid. Evidence for this unit may be gathered from a variety of sources, including well-planned investigative assignments, practical work or reports of practical assignments. You will be given written assessment briefs to complete for the assessment. These will contain a number of assessment criteria from pass, merit and distinction.

This unit will be assessed by the use of three assignments:

- Assignment one will cover P1, P2, P3, M1 and D1
- Assignment two will cover P4, P5, P6, M2 and D2
- Assignment three will cover P7, P8 and M3.

Rebecca

I like maths and until I did this unit I never realised how important it is to all parts of the construction industry. For example, site surveyors, who peg out the positions of walls and columns so that they can be built on site, could not do their job without a thorough understanding of how to calculate distances and angles from the architect's drawings.

I've learned that maths helps people in the industry to collect and use information in the design and construction process; it helps to keep track of the finances; to calculate the strength of the structure and to plan out the timing of the work. In fact, I've learned that without maths nothing would get built correctly or within budget.

Over to you

- Do you enjoy using maths to sort out practical problems?
- Can you think of any other examples of using maths in construction?
- What are you looking forward to learning about in this unit?

1 Be able to use basic underpinning mathematical techniques and methods used to manipulate and/or solve formulae, equations and algebraic expressions

Build up

Solving problems

The London Eye is a famous iconic structure. If the radius of the wheel is 67.5 m and the individual passenger capsules move at 0.9 km/hour, could you work out how long it would take to do a complete revolution in minutes?

What sort of maths techniques will help you solve this problem?

Mathematical techniques and methods

Using your calculator

It is important that you have a good scientific calculator while you are studying this unit – a basic scientific model is all you will need. Your tutor may advise you on which one to purchase. As well as the standard functions of **add**, **subtract**, **divide** and **multiply** your calculator will also need to perform a range of statistical functions which will be very useful for checking your work in the final section of this unit. It will also need to be able to switch **modes** so that it can calculate angles in decimals or degrees/minutes/seconds or in a special measurement known as radians. Finally, your calculator must have at least one storage area or **memory**; most modern calculators provide more than one. The memory function will be very useful when working out complicated sums where you need to hold part of the calculation safe while working on other parts.

Practise the use of the calculator's keys on simple calculations, and always keep the calculator's instruction book handy so that you can follow the examples it provides!

Mental checks

Mental checks are a good way to make sure the answer that the calculator gives is what you were expecting! To do this, you need to know your multiplication tables – reciting and remembering these basic number relationships will help you with your mental arithmetic.

Basic calculations

The order in which you carry out the separate parts of calculations is important. Look at the following two examples:

$46 - 3 \times 11 = 46 - 33 = 13$

$(46 - 3) \times 11 = 43 \times 11 = 473$

In the first example, we multiplied 3 by 11 and then subtracted the answer from 46 because the multiplication part took precedence over the subtraction part, but in the second example, the subtraction took precedence because it was within brackets.

There is a simple acronym – BODMAS – that can help you remember the order in which to do the separate parts of the calculation:

Bracketed calculations are done first.

Order or powers are calculated next (for example 4^2, where 2 is the power; $4^2 = 4 \times 4 = 16$).

Division and **M**ultiplication take equal priority and are done next.

Addition and **S**ubtraction are done last and are of equal priority.

Remember!

With a scientific calculator, the sequence in which you press the keys will be the same as the sequence in which a calculation is written down.

Worked example

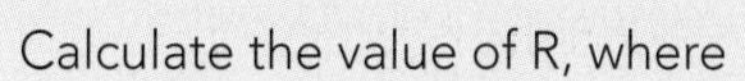

Calculate the value of R, where

$$R = 5000 - \left(\frac{25 \times 3}{2} + \frac{3}{2}\right)^2$$

The multiplication within the bracket is done first:

$$R = 5000 - \left(\frac{75}{2} + \frac{3}{2}\right)^2$$

Next, the division within the brackets is done:

$R = 5000 - (37.50 + 1.50)^2$

Then, carry out the addition within the brackets:

$R = 5000 - (39)^2$

Next, square the value in the brackets:

$R = 5000 - 1521$

Finally, we do the subtraction to give the answer required

$R = 3479$

Worked examples

1. Calculate:

$$52.3 \times \frac{27.4}{91.8}$$

The solution on your calculator (rounded to 2 decimal places) would be:

5 2 . 3 × 2 7 . 4 ÷ 9 1 . 8 = 15.61

and your mental check could be:

$$\frac{50 \times 30}{100} \approx 15$$

The anwers are similar so you can be sure that your calculated answer is correct.

2. Calculate:

$$\frac{(3.2 \times 5.6) + (9.8 \times 2.7)}{5.4}$$

The solution on your calculator (rounded to 2 decimal places) would be:

(3 . 2 × 5 . 6) + (9 . 8 × 2 . 7) = 44.38 ÷ 5 . 4 = 8.22

and your mental check could be:

$$\frac{(3 \times 6) + (10 \times 3)}{5} = \frac{18 + 30}{5} = \frac{50}{5} = 10$$

Note: You must completely work out the top line before dividing the top line by the bottom.

3. Calculate:

$$\frac{78.2 \times 67.3}{31.7 \times 42.5}$$

The solution on your calculator (rounded to 2 decimal places) would be:

7 8 . 2 × 6 7 . 3 ÷ (3 1 . 7 × 4 2 . 5) = 3.91

Note, brackets have been used to keep the bottom line separate from the top. It is important to remember this.

Your mental check could be:

$$\frac{80 \times 70}{30 \times 40} \approx \frac{56}{12} \approx 4.5$$

Activity: Calculate square root

Find the answer to the following:

$$\frac{\sqrt{55 + 31.7 - 7.8}}{2}$$

(*Hint:* As the square root applies to the entire top row, remember to put it in brackets.)

Carry out a mental check to confirm your answer.

Calculation units

Metric

Within the construction industry we use units that are derived from the **International System of Units**, known as **SI**.

The most common units used in our calculations are shown in Table 3.1.

Key term

International System of Units (SI) – from the French *Système International d'unités.* The modern form of the metric system developed in 1960 to promote a worldwide measurement system based on the standard properties of metres, kilograms and seconds.

Imperial

In the UK, you will need to be aware that imperial units of measurement are sometimes quoted. The ones that you might still come across in construction are shown in Table 3.2, with their metric conversions.

Remember!

It is important that you are consistent in the units that you use within a calculation. This means if you are using metres cubed (m^3), all the data must be in metres (m), not in millimetres (mm) or centimetres (cm).

Table 3.1: Common SI units of measurement

Physical property	Unit name	Unit symbol	Conversion
Length	metre millimetre	m mm	1 m = 1000 mm
Area	square metres square millimetres hectare	m^2 mm^2 ha	1 m^2 = 1000 mm x 1000 mm = 1,000,000 mm^2 1 ha = 100 m x 100 m = 10,000 m^2
Volume	cubic metres cubic millimetres	m^3 mm^3	1 m^3 = 1000 mm x 1000 mm x 1000 mm = 1,000,000,000 m^3
Capacity	litre centilitre millilitre	l cl ml	1 l = 100 cl = 1000 ml (also 1 m^3 = 1000 l)
Mass	kilogram	kg	1 kg = 1000g
Force	Newton kiloNewton tonne stress	N kN t N/mm^2 or kN/m^2	1 kN = 1000 N 1 t = 10 kN 1 t = 1000 kg
Pressure	Pascal	Pa	where 1 Pa = 1 N/mm^2
Time	seconds	s	
Temperature	degrees Celsius degrees Kelvin	°C °K	
Luminous intensity	candela	cd	

Table 3.2: Common imperial units of measurement and conversion to metric

Unit	Imperial name	Conversion to metric
Length	inch	1 inch = 25.4 mm
Area	1 foot = 12 inches 1 yard = 36 inches 1 mile = 1760 yards 1 square yard = 9 square feet 1 acre = 4840 square yards 1 square mile = 640 acres	1 mile =1609.344 m 1 square foot = 0.0929 m^2 1 acre = 0.4047 hectares
Mass (weight)	ounce 1 pound = 16 ounces 1 stone = 14 pounds	1 ounce = 28 g
Capacity	pint 1 gallon = 8 pints	1 pint = 0.5683 litre 1 gallon = 4.5 litre

Worked example

An old floor plan is being used to find the width of a new roller shutter door. The plan shows that the existing opening is 18 feet 5 inches wide. Convert this dimension to millimetres.

Step 1: Find the conversion factor: 1 inch = 25.4 mm (see Table 3.2).

Step 2: Find how many inches are in the total distances needed.

18 feet 5 inches = 18 × 12 inches + 5 inches = 221 inches

Step 3:

Therefore the distance in millimetres is:

221 × 25.4 mm = 5613.4 mm

Rounding and estimation

Significant figures

There may be times when quoting the exact answer may be unnecessary. For example, if the tolerance of the new door to be supplied is to the nearest 10 mm, then the above worked example answer could be rounded to a more convenient value, which would be 5610 mm. To help you to round numbers to convenient and useable values we use the rules for significant figures (s.f.) which are as follows:

- figures ending in 5 and above are rounded up and the next figure to the left increases by 1
- figures ending in 4 and below are rounded down and the next figure to the left remains the same.

Using significant figures is good for doing rough mental checks and estimations in all types of calculations. The number to round to will depend upon the required accuracy of the calculation For example:

14.6539 = 14.654 (rounded up to 5 s.f.)
= 14.65 (rounded down to 4 s.f.)
= 14.7 (rounded up to 3 s.f.)
= 15 (rounded up to 2 s.f)

In some situations it is more sensible to round down, even when the answer has a remainder of 0.5. For example let's look at how many 2-metre lengths can be cut from a 6.5 metre long steel tube. The calculation is 6.5 m ÷ 2 m = 3.25. However, only three 2-metre lengths can be cut; the rest (0.25 of a 2-metre length) is wasted. So in this case you should round down.

Decimal places

Decimal places (d.p.) are the numbers appearing to the right of the decimal point. So, in the above example:

14.6539 = 14.654 (rounded to 3 d.p.)
= 14.65 (rounded to 2 d.p.)
= 14.7 (rounded to 1 d.p.)

Decimal places are useful when dealing with distance. We are often required to quote measurements in metres to 3 decimal places so that our accuracy is to the nearest 1 mm. For example, 23647 mm is equivalent to 23.647 m.

Activity: Significant figures, decimal places, conversions and basic calculations

1. Round the following numbers to 5 and then 3 and then 2 significant figures:
   - a 983.5246
   - b 652881.0
   - c 1519.72
   - d 55.6187
   - e 55.4515
2. Write the following numbers to 1 and then 3 decimal places:
   - a 12.87654
   - b 93.6131
   - c 1.59691
   - d 0.0717
   - e 0.1057
3. Convert the following from imperial units to metric units:
   - a a distance of 8 feet 4 inches to millimetres
   - b a length of 12 feet $7\frac{1}{2}$ inches to metres to 3 d.p.
   - c an area of 2.5 square yards to square metres to 1 d.p.
   - s an area of 4 acres to hectares to 2 s.f.
   - e a capacity of $5\frac{1}{2}$ gallons to litres to 3 s.f.
   - f a capacity of 45 gallons to m^3 to 1 s.f.
4. For the following questions do a mental check first using pencil and paper by rounding the numbers to 1 or 2 significant figures, then work out your answer using your calculator:
   - a $\frac{72.5 - 4.5}{3 \times 12 + 32}$
   - b $\frac{\sqrt{46.6 + 17.4}}{\frac{1}{2}}$
   - c $\frac{(9 - 3)^2 + 3}{6}$
   - d $\sqrt{6.8} \times 3 - 5$
   - e $1.4\,(23.5 \times 2.8 - 12.6)^2$

   Did your mental check answers tie up with your calculated results?

Standard form

Standard form is a very convenient way of writing and using very large or very small numbers in technical calculations. Some scientific calculators only have 12 or so digits in their displays, so any number bigger than this could not be shown without using standard form. In standard form a number is split into two parts: a decimal number (N) multiplied by the number 10 raised to a power (n). Such that $N \times 10^n$. For example:

$78531 = 7.8531 \times 10^4$
where $10^4 = 10 \times 10 \times 10 \times 10 = 10{,}000$
and $7.8531 \times 10{,}000 = 78531$ (the original number)

A simple way to remember this is that the decimal point for big numbers moves to the left is the same number that the power to which 10 is raised.
For example:

$235000000000.0 = 2.35 \times 12^{11}$ Positive power of 11 and the decimal point moves 11 places to the left.

Standard form also works with small numbers which are less than 1 by using negative powers, where the powers of 10 mean dividing by 10.

For example:
0.0000000541 is written in standard form as 5.41×10^{-8}.

$$10^{-8} = \frac{1}{10^8} = \frac{1}{10 \times 10 \times 10 \times 10 \times 10 \times 10 \times 10 \times 10}$$

and $5.41 \times \frac{1}{10 \times 10 \times 10 \times 10 \times 10 \times 10 \times 10 \times 10}$
$= 0.0000000541$

The decimal point for small numbers less than 1 moves to the right and is the same negative power to which 10 is raised.

$0.0000000541 = 5.41 \times 10^{-8}$ Negative power of $^{-8}$ means the decimal point moves 8 places to the right.

Remember!

Most scientific calculators allow you to toggle through standard form in powers of multiples of 3. This is often called engineering standard form because we often quote physical properties in kilos, where kilo is 1000 or 10^3.

Take it further!

Powers or indices can be dealt with according to the 'Laws of Indices':

For example:

$a^r \times a^s = a^{r+s}$ and $\frac{a^r}{a^s} = a^r \div a^s = a^{r-s}$ and finally, $(a^r)^s = a^{rs}$.

For example:

$2^3 \times 2^2 = 2^{3+2} = 2^5 = 32$

$\frac{2^3}{2^2} = 2^3 \div 2^2 = 2^{3-2} = 2^1 = 2$

$(2^3)^2 = 2^{3\times 2} = 2^6 = 64$

Powers which are fractions refer to roots, e.g. 3 is the 'square' root of 9, written as $9^{1/2}$. These can be combined. For example:

$2^{4/3}$ $(2^4)^{1/3} = 16^{1/3} = \sqrt[3]{16} = 2.519$ (3d.p.)

Explore the use of the 'power' key on your calculator – repeat the examples above and check that you get the same answers!

Loading calculations

The unit that we use to measure load is the Newton (N) and for large numbers we often work in multiples of 1000 Newtons, called kilo-Newtons (kN). When loads get really huge we use mega-Newtons (MN), where 1 MN is equal to 1,000,000 Newtons. Thus:

500,000 Newtons = 500×10^3 Newtons or 500 kilo-Newtons
= 0.5×10^6 Newtons or 0.5 mega-Newtons

Activity: Standard form

1. Write the following numbers in standard form:
   - a 875
   - b 175113
   - c 0.00221
   - d 0.0000079
2. Write the following standard forms as ordinary numbers:
   - a 8.642×10^3
   - b 5.8×10^{-3}
   - c 2.5875×10^2
   - d 9.11×10^{-1}
3. Write the following loads in kilo-Newtons to 2 decimal places:
   - a 860000 N
   - b 167500 N
   - c 50021.5 N
   - d 0.00467 MN

Formulae, equations and algebraic expressions

Algebraic expressions and equations

In algebra, letters are used together with numbers, with letters representing such things as volume, load, time or temperature. Letters or algebraic terms are dealt with using straightforward rules similar to those used for ordinary numbers and fractions. Algebraic equations can be used to solve many construction maths problems.

Addition and subtraction

Only terms that are alike can be added or subtracted. For example, the expression 30d + 20w – 10d + 30w relates to the number of doors (d) and windows (w) on a particular housing development. Because the doors and windows are different, they have to be treated separately, thus:

$$30d + 20w - 10d + 30w = 30d - 10d + 20w + 30w = 20d + 50w$$

Therefore, on the housing development there are 20 doors and 50 windows.

Multiplication and division

As with addition and subtraction, only like terms can be multiplied or divided. For example:

$6 \times 6 \times \frac{6}{6} = 6^2$ and $d \times d \times \frac{d}{d} = d^2$

Brackets and factors

Brackets are generally used when you are multiplying two terms together, and also avoid having to write an '×', meaning multiply, which could be confused with an algebraic term x representing a physical number or property.

Brackets are used in algebra to simplify expressions, or to change an expression which has terms added together to terms multiplied together. For example, the number 24 can be written as 3 × 2 × 4. Therefore, 24 can be expressed as a sum of numbers such as 14 + 10 or as a product of numbers 3 × 2 × 4.

14 + 10 = 3 × 2 × 4

3, 2 and 4 are known as 'factors' of 24

The same can be done with algebraic expressions as well. For example, the expression $6x + 2y$ can be written as two factors $2(3x + y)$ because 2 is common to both $6x$ and $2y$ and it can be divided out of the expression, leaving $3x + y$ in brackets. Thus:

$6x + 2y = 2(3x + y)$

2 and $3x + y$ are 'factors' of $6x + 2y$

Sometimes both factors can be bracketed.

To multiply them out we use the FOIL rule:

First terms multiplied

Outside terms multiplied

Inside terms multiplied

Last terms multiplied

Worked example

Multiply out the following factors using the FOIL rule:

$(x + 2)(2x + 3)$

First = $x(2x) = 2x^2$

Outside = $3(x) = 3x$

Inside = $2(2x) = 4x$

Last = $2(3) = 6$

Therefore, putting them all together:

$$(x + 2)(2x + 3) = 2x^2 + 3x + 4x + 6$$
$$= 2x^2 + 7x + 6$$

This expression is a quadratic expression because it has a term which is squared (see page 75). We shall be covering this in more detail later.

If you have an expression with three factors such as $x(5x + 2)(2x - 3)$,

then a cubic equation will be formed when the brackets are multiplied out. The way to tackle this expression is to multiply out the two brackets to get a quadratic equation, then multiply each of the quadratic terms by x.

$$x(5x + 2)(2x - 3) = x(10x^2 + 4x - 15x - 6)$$
$$= x(10x^2 - 11x - 6)$$
$$= 10x^3 - 11x^2 - 6x$$

Formulae and equations

Formulae and equations are used in construction to calculate physical properties. Later in this unit we will go over several very important formulae used in construction calculations. One of the most common is the formula for the area of a triangle:

Area = $\frac{1}{2}$ × base × height, abbreviated to $A = \frac{1}{2}bh$

where the letters A, b and h can represent the area, base length and height of *any* triangle.

Activity: Brackets and factors

1. Multiply out the brackets and then write down the following expressions in their simplest form:
   - **a** $5x(3x - 8) - 2x^2$
   - **b** $2(x + 3y - z) - 3(x + y - z) + 4(x - y)$
   - **c** $(2x - 1)(3x^2 - x - 3)$
2. Multiply out the following factors
   - **a** $3(2 + 5b)$
   - **b** $(3x + 2)(x + 1)$
   - **c** $(x - 2)(5 + x)$
3. Find the factors for the following expressions
   - **a** $7y + 49$
   - **b** $12x^2 - 6x$
   - **c** $5c^2 - 25c$

With the equation in the form above the actual values of the base length and height must be known and entered into the equation to find the triangle's actual area. It is also important that the units of both b and h are the same, for example both metres (m). This will then give the value of the Area (A) in square metres (m^2). Both sides of an equation must always balance.

Transposition of formulae

We learned above that the area (A) of any triangle can be found from the following formula:

$$A = \frac{1}{2}bh$$

This formula can also be used to work out height (h) if you already know the Area (A) and the base length (b). This can be done by rearranging the formula using simple rules of transposition to make h the subject, such that

$$h = \frac{2\,(Area)}{base} = \frac{2A}{b}$$

To transpose or rearrange a formula you need to change the subject so that the item you want to find is by itself on the left-hand side (LHS) of the formula. The equals sign in an equation makes sure that the left-hand side (LHS) of the equation balances the right-hand side (RHS).

Remember!

If you have a fraction like $\frac{4}{5}$, the top of the fraction is the numerator and the bottom of the fraction is the denominator.

Worked example

The decking area around a swimming pool is made of a series of rectangles and squares. At the corners of the decking there are 4 squares, s by s m^2. Next to the longest sides of the pool, the decking has dimensions t metres long by s metres wide; and next to the shortest sides the decking has dimensions of s metres long by s metres wide.

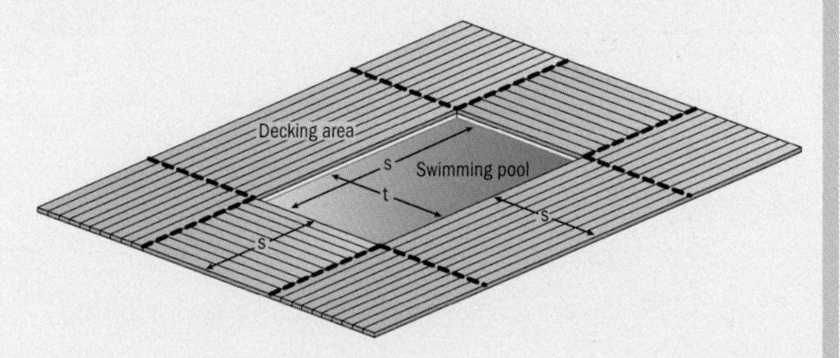

What is the equation used to work out the total area of the decking?

Total area of decking $= 4(s \times s) + 2(s \times t) + 2(s \times s)$

$= 4s^2 + 2st + 2s^2$

$= 6s^2 + 2st$

If s = 5.5 m and t = 12.5 m what would the total area of the decking be?

Total area of decking:

$= 6s^2 + 2st$

$= 6(5.5\text{ m})^2 + 2 \times 5.5\text{ m} \times 12.5\text{ m}$

$= 181.5\text{ m}^2 + 137.5\text{ m}^2$

$= 319\text{ m}^2$

Addition and subtraction

With addition or subtraction, when a number or a term is moved to the other side of the equals sign, it changes its sign to make sure that the equation remains balanced. For example:

$$25 + 10 = 41 - 6$$
$$35 = 35$$

Moving the 6 from the RHS to the LHS changes its sign to '+' and the equation still balances:

$$25 + 10 + 6 = 41$$
$$41 = 41$$

The same is also true for terms or letters, for example:

$$R + S = T + U$$
$$R = T + U - S$$

The equation is still balanced because we changed the sign of '+S' to '–S' as it moved from one side to the other.

Multiplication, division and powers

When dealing with equations which include multiplication, division, powers or roots, whatever is done on one side must be done to the other side. Furthermore, if you want to move a term from one side of the equation to the other, you must do the opposite of the operation that it is currently performing.

Let's look more closely at how we arrived at the equation for the height (h) of a triangle given its base (b) and area (A) above.

$$A = \frac{1}{2} \times b \times h$$

We want to get h on its own on the LHS of the equation. First, switch *both sides* over – the equation remains balanced and the *signs do not change.*

$$\frac{1}{2} \times b \times h = A$$

We now want to remove the b which is at the moment multiplying the h value. So we divide *both* sides by b.

$$\frac{1 \times b \times h}{2 \times b} = \frac{A}{b}$$

Next, we can cancel out the 'b's on the LHS because b into b goes once. Therefore, we have:

$$\frac{h}{2} = \frac{A}{b}$$

Now, we want to move the 2 to the other side of the equation and as it is *dividing* into h we need to *multiply both* sides by 2 to balance out the equation.

$$\frac{2 \times h}{2} = \frac{2 \times A}{b}$$

Finally, 2 goes into 2 once and so it cancels out, leaving the equation that we want:

$$h = \frac{2 \times A}{b} = \frac{2A}{b}$$

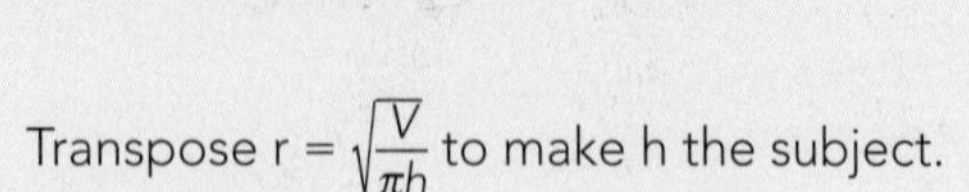

Worked example

Transpose $r = \sqrt{\frac{V}{\pi h}}$ to make h the subject.

Step 1: Square both sides to get rid of the square root sign on the RHS.

$r^2 = \frac{V}{\pi h}$

Step 2: Multiply both sides by h so that h appears in the top line on the LHS.

$hr^2 = \frac{V}{\pi}$

Step 3: Move r^2 to the RHS by dividing both sides by r^2, leaving h as the subject on the LHS.

$h = \frac{V}{\pi r^2}$

Activity: Transposition of formulae and equations

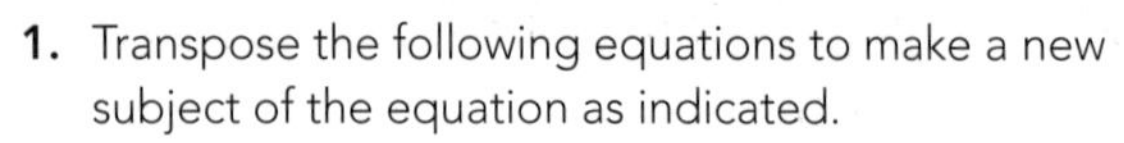

1. Transpose the following equations to make a new subject of the equation as indicated.

a	$v = u + at$	Make *a* the subject
b	$a^2 = b^2 + c^2$	Make *b* the subject
c	$T = 2\pi\sqrt{\frac{l}{g}}$	Make *g* the subject
d	$a = b + \sqrt{b^2 + c^2}$	Make *c* the subject

2. A rectangular concrete block has edge dimensions of x, y and z measured in metres. Write down an expression for the total surface area (A) of the block in terms of x, y and z. (*Hint:* Draw a rough three-dimensional sketch of the block showing dimensions x, y and z.)

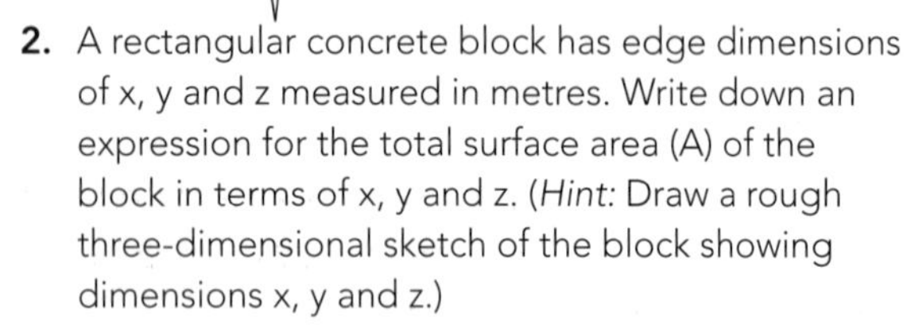

The total surface area of the block above is 7.5 m^2 and x is 1 m and y is 1.2 m. Transpose the formula for its area to make z the subject, and so calculate the length of dimension z in metres to 2 d.p.

Simultaneous equations

So far we have learned about simple linear equations, where we substituted values for all the given data to find the unknown subject. We also need to deal with equations where we have two unknowns. To solve this we need to have two equations known as **simultaneous equations**.

The worked example below shows how to find the values of x and y that make *both* of these simultaneous equations correct:

$x + 2y = 8$

$5x - 3y = 1$

Worked example

Find the value of x and y to solve the following simultaneous equations.

$x + 2y = 8$ [equation 1]

$5x - 3y = 1$ [equation 2]

Step 1: Re-arrange both equations to get x on its own.

$x = 8 - 2y$ [equation 1]

$x = \frac{1 + 3y}{5}$ [equation 1a]

Step 2: As both equations equal x, they must equal each other:

$8 - 2y = \frac{1 + 3y}{5}$

Step 3: We now have one equation and one unknown called y, and we can solve this by transposing and simplifying its terms:

$5(8 - 2y) = 1 + 3y$

$40 - 10y = 1 + 3y$

$40 - 1 = 3y + 10y$

$39 = 13y$

$y = \frac{39}{13} = 3$

Step 4: Put the value that you have found for y back into either equation 1 or 2 to find the other 'unknown' x.

Using equation 1:

$x + 2y = 8$

$x + 2(3) = 8$

$x + 6 = 8$

$x = 8 - 6$

$x = 2$

Step 5: Finally, put the values into equation 2 to check your answers!

$5x - 3y = 1$

$5(2) - 3(3) = 10 - 9 = 1$

Activity: Simultaneous equations

Solve the following equations for x and y.

- $3x + 4y + 6 = 0$ and $x + 6y + 16 = 0$
- $-4y = 7$ and $4x + y = 11$

Quadratic equations

Quadratic equations are used in more complex calculations, often where maximum or minimum values are needed. These equations take the form of:

$ax^2 + bx + c = 0$

where a, b and c are called coefficients, and x can usually have two values. In quadratic equations x can be solved in four ways, by using:

- factors
- completing the squares
- formula
- drawing graphs.

Factors

Given a quadratic equation, we can find its factors in an operation known as **factorisation**.

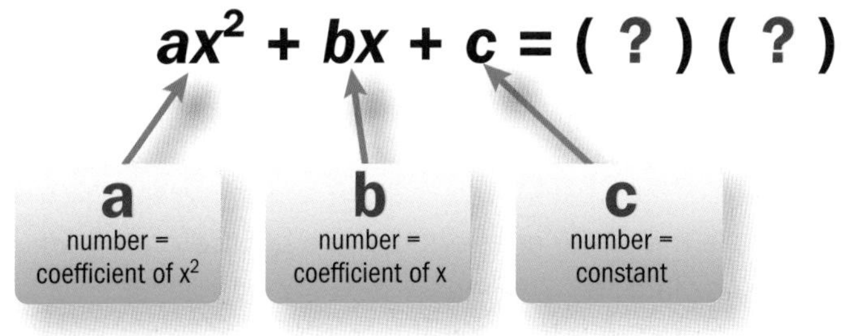

We use factorisation to help us solve quadratic equations for x provided we always rearrange them to equal zero. Consider the following quadratic equation that needs to be factorised:

$x^2 - 2x - 3 = 0$

Start by identifying the coefficients. In this example, they are a = 1, b = –2 and c = –3. The method for factorising them is as follows:

$$x^2 - 2x - 3 = 0$$

When a = 1

Think of two numbers that when **added** together equal -2

Also, the same two numbers when **multiplied** together must equal -3

$x^2 - 2x - 3 = (x + ?)(x + ?)$

Find by trial and error which values are needed; –3 and +1 seem to fit because:

$-3 + 1 = -2$ and $-3 \times 1 = -3$

So let's try –3 and +1 to see if we get what we want when we multiply out the factors:

$$(x - 3)(x + 1) = x^2 - 3x + x - 3$$
$$= x^2 - 2x - 3$$

which is the quadratic expression that we need! Therefore, we can write:

$$x^2 - 2x - 3 = (x - 3)(x + 1) = 0$$
$$(x - 3)(x + 1) = 0$$

where the factors are $(x - 3)$ and $(x + 1)$.

Now, if the expression $(x - 3)(x + 1) = 0$, it must be true that either:

$(x - 3) = 0$ or $(x + 1) = 0$.

which means that:

$x = 3$ or $x = -1$.

To check if this is right, substitute these answers for x in the original quadratic equation. Let's first try $x = 3$ to see if the answer we get is zero.

$$x^2 - 2x - 3 = (3)^2 - 2(3) - 3$$
$$= 9 - 6 - 3$$
$$= 0$$

Let's now try $x = -1$.

$$x^2 - 2x - 3 = (-1)^2 - 2(-1) - 3$$
$$= 1 + 2 - 3$$
$$= 0$$

Both of these make the quadratic equation balance, so the solution to the quadratic equation $x^2 - 2x - 3 = 0$ is $x = 3$, $x = -1$.

Completing the squares

The method of completing the squares converts a quadratic equation to two equal factors or sides of a 'square'. We've learned that the standard format for a quadratic equation is $ax^2 + bx + c = 0$. To find the values of x, we first move the numerical constant c to the RHS of the equation. For example, to find the values of x that solve the equation $2x^2 + 8x - 10 = 0$, first move 10 to the RHS of the equation.

$2x^2 + 8x = 10$

Make a = 1 by dividing through both sides by a. In this example, divide by 2:

$x^2 + 4x = 5$

Now, add $\left(\frac{b}{2}\right)^2$ to both sides of the equation, where b in this example is 4.

$$x^2 + 4x + \left(\frac{4}{2}\right)^2 = 5 + \left(\frac{4}{2}\right)^2 = 5 + \frac{16}{4} = 9$$

Therefore:

$$x^2 + 4x + \left(\frac{4}{2}\right)^2 = 9$$

$x^2 + 4x + 2^2 = 9$

Next, 'complete the square' by factorising the LHS:

$x^2 + 4x + 2^2 = 9$

$(x + 2)(x + 2) = 9$

$(x + 2)^2 = 9$

This can be pictured as a square with sides x + 2, as illustrated.

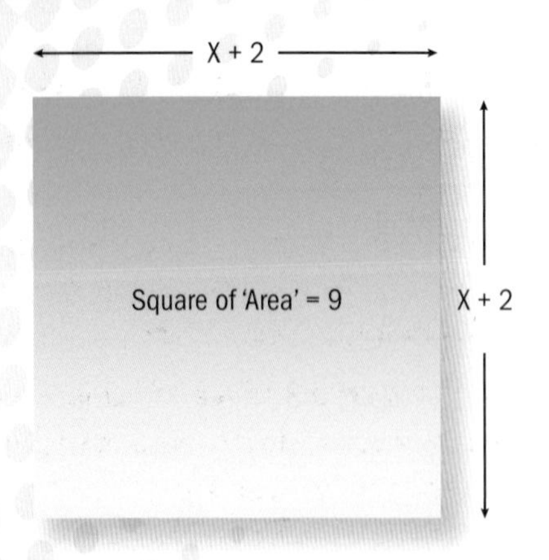

Finally, rearrange and solve the linear equation for x by finding the square root of both sides of the equation.

$(x + 2)^2 = 9$

$x + 2 = \sqrt{9}$

$x + 2 = \pm 3$

Therefore, $x = 3 - 2 = 1$ or

$x = -3 - 2 = -5$.

These values can be substituted into the original equation to check that your answers are correct.

$2x^2 + 8x - 10 = 0$

For $x = 1$

$2(1)^2 + 8(1) - 10 = 2 + 8 - 10 = 0$

And for $x = -5$

$2(-5)^2 + 8(-5) - 10 = 2(25) - 40 - 10 = 50 - 40 - 10 = 0$

Formula

The formula for working out the solution to the quadratic equation $ax^2 + bx + c = 0$ is given by substituting the values into the following equation:

$$x = \frac{-b \pm \sqrt{(b^2 - 4ac)}}{2a}$$

So using the example that we used above:

$x^2 - 2x - 3 = 0$

we see that a = 1, b = –2, c = –3.

Substituting these into the formula, we get:

$$x = \frac{-(-2) \pm \sqrt{((-2)^2 - 4(1)(-3))}}{2(1)}$$

$$x = \frac{2 \pm \sqrt{4 + 12)}}{2} = \frac{2 \pm \sqrt{16}}{2} = \frac{2 \pm 4}{2}$$

Therefore, $x = \frac{6}{2} = 3$ or

$$x = \frac{-2}{2} = -1$$

which are the same answers that we got before when solving by factors.

Drawing graphs

This method is explained on page 103.

Simultaneous linear and quadratic equations

We can now take linear simultaneous equations a step further and solve for unknowns when one or both of the equations are in a quadratic form, that is they have an x^2 term included in them. We must remember that a quadratic equation always has two solutions for x. Therefore, in this situation there will also be two solutions for y. This can be solved graphically, as can be seen on page 107, or algebraically as shown in the worked example opposite.

Worked example

Find the values of x and y that satisfy the following two equations:

$y = 3x + 1$

$y = x^2 + 2x - 1$

Step 1: As both equations equal y, they must both equal each other:

$\therefore 3x + 1 = x^2 + 2x - 1$

Step 2: Rearrange into the standard quadratic form of $ax^2 + bx + c = 0$.

$0 = x^2 + 2x - 1 - 3x - 1$

$\therefore x^2 - x - 2 = 0$

Step 3: Solve the quadratic equation by any of the three methods previously studied. This particular one can be easily factorised thus:

$x^2 - x - 2 = ?$

$= (x + 1)(x - 2) = 0$

Therefore, either $(x + 1) = 0$ or $(x - 2) = 0$

which means $x = -1$ or $x = 2$

Step 4: Now find the corresponding values of y to complete the solution. Substitute the values of x back into one of the original equations to find its y value, and it will be simpler to use the linear equation

$y = 3x + 1$. Hence:

For $x = -1$, $y = 3(-1) + 1 = -2$ and

For $x = 2$, $y = 3(2) + 1 = 7$

Therefore, the solutions to the equations are:

$x = -1$, $y = -2$ and $x = 2$, $y = 7$

Activity: Quadratic equations

1 Use the factors method to solve these quadratic equations:

a $x^2 - x - 12 = 0$ **b** $x^2 - 7x + 10 = 0$

2 Solve the following quadratic equations for x by the completing the squares method, giving your answers as fractions:

a $4x^2 - 3x - 1 = 0$ **b** $6x^2 + x - 35 = 0$

3 Solve the following quadratic equations for x by the formula method. Give your answers to 2 dp:

a $x^2 - 6x - 10 = 0$ **b** $2x^2 + 6x + 3 = 0$

The binominal theorem

The binominal theorem is a mathematical method which allows two terms held within a bracket to be expanded to any power. So far you have learned that $(a + b)^2$ can be expanded to become:

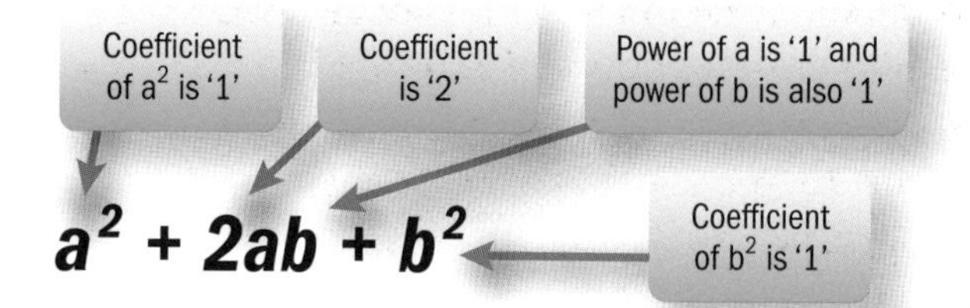

By using the binomial series, brackets can be expanded to any power giving the following coefficients in what is commonly known as Pascal's triangle.

Binomial expression:	Pascal's triangle (coefficients in the expansion)
$(a + b)^0$	1
$(a + b)^1$	1 1
$(a + b)^2$	1 2 1
$(a + b)^3$	1 3 3 1
$(a + b)^4$	1 4 6 4 1
$(a + b)^5$	1 5 10 10 5 1
$(a + b)^6$	1 6 15 20 15 6 1
$(a + b)^7$	1 7 21 35 35 21 7 1

We can see that:

- the number of terms in each expansion is one more than the index, e.g. the expansion of $(a + b)^9$ will have ten terms
- the arrangement of coefficients is symmetrical
- the coefficients of the first and last terms are both always one
- each coefficient in the triangle is obtained by adding together the two coefficients in the line above that lie on either side of it
- that the individual powers of a and b increase or decrease, but their total always equals the value of the original expansion power, e.g.

Power of each term adds up to '3'

$$a + b)^3 = a^3 + 3a^2b + 3ab^2 + b^3$$

- any number or letter by itself is raised to the power of 1, but we do not usually write this, e.g. $3^1 = 3$, $b^1 = b$.

Using Pascal's triangle, we can expand higher powers. For example:

Expand $(2x + 3y)^4$.

Substituting $(2x + 3y)^4$ into Pascal's triangle and following the rules above, we get:

$$[(2x) + (3y)]^4$$
$$= (2x)^4 + 4(2x)^3 (3y) + 6(2x)^2 (3y)^2 + 4(2x) (3y)^3 + (3y)^4$$
$$= 16x^4 + 4(8x^3) (3y) + 6(4x^2) (9y^2) + 4(2x) (27y^3) + 81y^4$$
$$= 16x^4 + 96x^3 y + 216x^2y^2 + 216xy^3 + 81y^4$$

This cannot be simplified any further as none of the terms in x and y have identical powers.

Uses of the binomial theorem

In industry, the binomial expansion is useful when estimating the errors by a certain percentage as seen in the following Worked examples.

Key term

Right cone – cone where the top vertex or point of the cone is vertically above the centre of the circular base.

Worked example

In industry, the binominal expansion is useful when dealing with small errors. Consider $(l + x)^n$, when x is very small compared to l, just like measuring a distance of say 10 m with an error of 5 mm. The binomial expansion for $(l + x)^n$ where n is a power can be approximated to:

$$(l + x)^n \approx l + (n)(x)$$

1. In measuring the area of a square glass cladding panel, the measurement was 1% too large. If this measurement is used to calculate the area of the glass, find what the resulting error in the area would be.

   Let the length of the side of the square be L and the area of the square glass panel be A. Let the error in the measured length be δL. Also let δA be the error in the area, then:

   $$A = L^2 \text{ and } \delta L = \frac{L}{100}$$
   $$A + \delta A = (L + \delta L)^2$$

   Substituting in for δL from above

   $$= \left(L + \frac{L}{100}\right)^2$$

   Now take the L out of the brackets and remember that it must remain squared,

   $$A + \delta A = L^2\left(1 + \frac{1}{100}\right)^2$$

   Since the error is small when compared to the length measured, we can approximate using binomial theory that:

   $$A + \delta A \approx L^2\left(1 + \frac{2}{100}\right)$$

   But $A = L^2$

   $$A + \delta A \approx A\left(1 + \frac{2}{100}\right)$$
   $$A + \delta A \approx A + A\frac{2}{100}$$
   $$\delta A \approx A\frac{2}{100}$$
   $$\delta A \approx 2\% \text{ of } A$$

   Therefore, we can see that if the measurement of the length of the square glass cladding panel is 1% too large, the resulting area calculation will be approximately 2% too large.

2. Find the approximate percentage error in the calculated volume of a **right cone** if the radius is taken as 2% too small and the height as 3% too large.

   Volume (V) of a right cone $= \frac{1}{3}(\pi r^2 h)$

The error in height (δh) = $h\frac{3}{100}$ and

the error in the radius (δr) = $-r\frac{2}{100}$

$$V + \delta V \approx \frac{1}{3}\left(\pi\left(r - r\frac{2}{100}\right)^2\left(h + h\frac{3}{100}\right)\right)$$

$$V + \delta V \approx \frac{1}{3}\left(\pi r^2 h\left(1 - \frac{2}{100}\right)^2\left(1 - \frac{3}{100}\right)\right)$$

Since the errors in height and radius are small when compared to the original lengths, we can approximate using binomial theory that:

$$V + \delta V \approx \frac{1}{3}\pi r^2 h\left(1 - 2\frac{2}{100}\right)\left(1 + \frac{3}{100}\right)$$

Multiplying out the brackets:

$$V + \delta V \approx \frac{1}{3}\pi r^2 h\left(1 - \frac{1}{100}\right)$$ approximately, ignoring the last small term

$$V + \delta V \approx V\left(1 - \frac{1}{100}\right)$$

$$\therefore -\delta V \approx V\left(\frac{1}{100}\right)$$

Therefore, for a right circular cone with a radius taken as 2% too small and a height taken as 3% too large, the calculated volume will be 1% too small.

Activity: Pascal's triangle and binomial theorem

1. Using Pascal's triangle, expand the following bracketed expression:

   $(2 - y)^5$

2. In measuring the internal radius of a circular sewer the measurement is 3% too large. If this measurement is then used to calculate the circular cross-sectional area of the pipe:

   a. Calculate the percentage error that will occur compared to the true area.

   b. Calculate the true area of the pipe when the incorrect measured radius was recorded as 958 mm. Give your answer in m^2 to 6 decimal places.

   Take the cross-sectional area (A) of the pipe to be $A = 3.14r^2$

Assessment activity 3.1

1. A waste management company Easidump has been contracted by Bestend Properties plc to remove construction waste from three of its local construction sites. The charges for each site include a standard charge (S) of £300 for removing the waste, additional labour costs (L) of £80 per operative (for sorting the waste ready for recycling) and a hire cost for the number of site bins (B) they supply at £10 per bin.

   Write down an algebraic formula for the total cost (T) incurred for all 3 sites taking into account the standard charge (S), number of labourers (L) and number of bins (B).

   Use your scientific calculator to calculate the total cost for all 3 Bestend Properties sites if 3 Easidump operators supply 12 site bins for each site. Now apply manual checks to the results. **P1**

   If the standing charge is increased to £390 and the labour cost increases to £90 per operator, rewrite the formula that you wrote above to show these changes. Using this revised formula, find out the new total cost that Bestend Properties plc will have to pay assuming all other information stays the same.

   Bestend Properties plc cannot afford these increased charges – the most its waste budget will stretch to is £2160. Therefore, rearrange the revised formula that you wrote making the cost of the bins (B) the subject of the formula. Use this new formula to calculate the total number of site bins that can be supplied to each site by Easidump for £2160 assuming that they will still be using 3 operatives.

   Comment on the results and advise Bestend Properties plc on what it could do next. **P2**

   The area of the roof-top garden for one of Bestend Properties' new penthouse apartments is shown below. The local planning department has specified that the minimum area for the roof garden should be 126.75 m^2. As can be seen from the diagram, the design team has already allocated an area of 8 m × 5 m for lifts, service ducts and stairwell. (cont.)

2. You are required to work out the dimension 'x', as shown on the diagram, to ensure that the planners' minimum requirements for the roof-top garden are met.

   Write down a quadratic equation for the area of the roof garden in terms of x and the current dimensions of the lift shaft, service ducts and stairwell.

   Solve the quadratic equation giving your answer in metres to 1 decimal place. **M1**

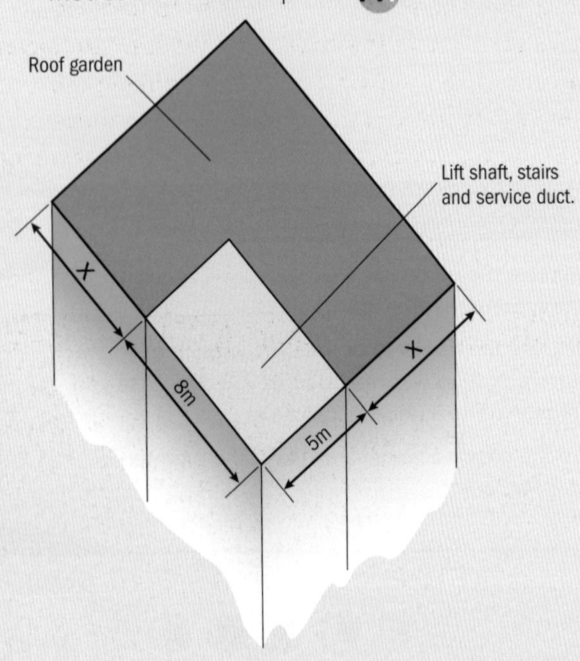

   Check your answer above by carrying out two alternative methods to solve the equation. **D1**

3. Bestend Properties is refurbishing an old hospital that was built in the early 1950s. The building's main structure is a steel frame. The company's structural engineers are trying to determine the strength of the existing floor beams by loading them up and measuring the deflection, which is given by the formula:

   $$E = \frac{5wL^4}{384Iy}$$

   where:

   E is Young's Modulus of elasticity (strength of the material)

   w is the measured uniformly distributed load on the beam

   I is the moment of inertia of the beam (a constant due to the cross-sectional shape)

   L is the measured span of the beam

   y is the deflection of the loaded beam.

   After the on-site testing was done, it was discovered that some of the measuring equipment was wrongly calibrated. The errors that were discovered were as follows:

   The span L was 5% too long.

   The load w was 3% too low.

   Using your knowledge of binomial theory, produce a formula to show the net percentage effect of these calibration errors on the value of E (Young's Modulus). **D2**

Grading tips

For **P1** clearly show in writing how you used your calculator and state any mental checks that you did. Show clearly the progression, line by line, as you work each part of the answer out.

For **P2** you first need to identify the two simultaneous equations from the data given. Then, use either the elimination or substitution method for finding the value of one of the variable.

For **M1** you will need to develop the quadratic equation for the area required and then solve it using your preferred method of solution. Factorisations is usually the quickest way to solve it if you know your factors well.

For **D1** independently undertake checks on your calculations in M1 using relevant alternative mathematical methods and make appropriate judgements on the outcome.

For **D2** you need to demonstrate an analysis of the accuracy and rounding of data and its effect on the calculated outcome, and make suitable conclusions that relate to this situation.

PLTS

Comparing two methods to solve the same problem to check your results will help develop your skills as an **independent enquirer**.

Functional skills

By considering the solution techniques involved in simultaneous equations you have an opportunity to demonstrate your ability to identify the problem and the **mathematical** method needed to tackle it.

2 Be able to select and correctly apply mathematical techniques to solve practical construction problems involving perimeters, areas and volumes

Perimeters and areas

To find the perimeter, area and volumes in a construction activity you need to know some basic formulae; the most common formulae for finding the area and perimeter are shown in Table 3.3.

It is good to note that in some situations a perimeter or area can be broken down into a combination of shapes. See the Worked examples below.

Sometimes assumptions will need to be made, for example, that sides are at right angles to each other or are parallel or that the object can be approximated to a circle. You are making the area formula fit the given situation, but you must be aware how these approximations affect the accurancy of your final result.

Table 3.3: Area and perimeter formulae

Shape	Area (A)	Perimeter (P)
Square (sides L, L)	$A = L \times L$ $= L^2$	$P = 4L$
Rectangle (sides a, b)	$A = a \times b$ $= ab$	$P = 2a + 2b = 2(a + b)$
Triangle 1 – Standard formula (sides a, b, c; height h) Note: h is the perpendicular height	$A = \frac{1}{2} \times b \times h$ $= \frac{1}{2}bh$	$P = a + b + c$
Triangle 2 – Half perimeter formula (sides a, b, c)	$A = \sqrt{s(s-a)(s-b)(s-c)}$ Where $s = \frac{a+b+c}{2}$	$P = a + b + c$

(cont.)

Shape	Area (A)	Perimeter (P)
Trapezium 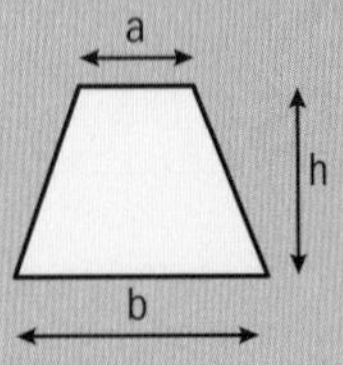	$A = h\frac{(a + b)}{2}$	
Parallelogram 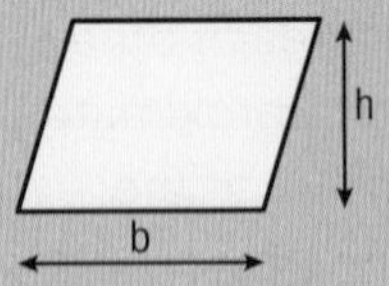	$A = b \times h$ $= bh$	
Circle 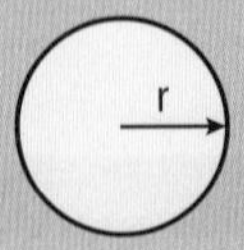 r = radius d = diameter	$A = \pi r^2$ or $A = \frac{\pi d^2}{4}$	Circumference $= 2\pi r$ $= \pi d$ Where d = diameter $= 2r$
Ellipse 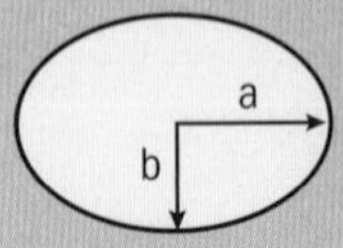 a = half-major axis length b = half-minor axis length	$A = \pi ab$	$P = 2\pi \frac{(a^2 + b^2)}{2}$

Worked example

1. When constructing domestic cavity walls bricklayers need to estimate how many bricks and concrete blocks they will need to build a given area of wall. The coordinating sizes of a brick and blocks are shown below which includes a 10 mm bedding and perpendicular mortar joints.

a To assist the bricklayers in finding how many bricks and blocks they will need, determine the approximate number of bricks and blocks required to build a 1 m by 1 m area of cavity wall.

Remember!

For practical purposes, blocks and bricks have to be cut to fit the given dimensions of the required area and often there is wastage, which has to be allowed for in the estimation calculations. Typically, up to 5% is allowed.

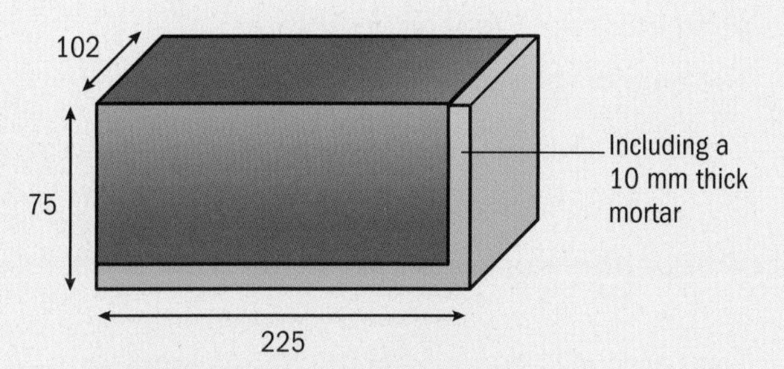

Coordinating dimensions for a brick

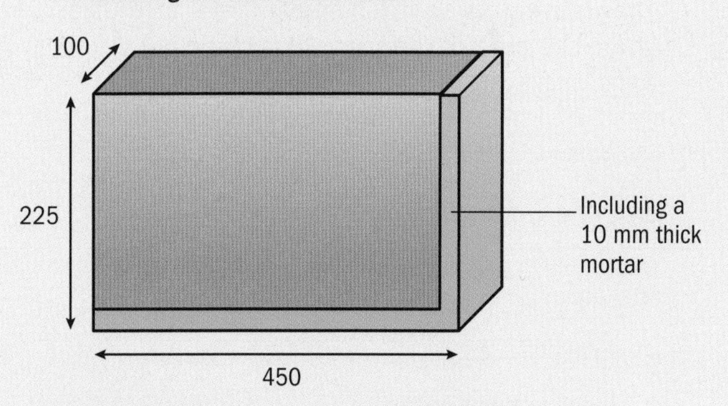

Coordinating dimensions for a concrete block

The area of the face of one coordinated brick

$= 225 \times 75 = 16{,}875\ \text{mm}^2$

and the area of the face of one coordinated

block $= 450 \times 225 = 101{,}250\ \text{mm}^2$

Therefore, the number of bricks that could fit into

$$1\ \text{m}^2 = \frac{1000 \times 1000\ \text{mm}^2}{16{,}875\ \text{mm}^2}$$

$= 59.259$ bricks, say, 60 bricks

The number of blocks that could fit into

$$1\text{m}^2 = \frac{1000 \times 1000\ \text{mm}^2}{101{,}250\ \text{mm}^2}$$

$= 9.877$ blocks, say, 10 blocks

These are very useful approximations for estimating the number of bricks and block required per metre squared for any cavity wall.

b Calculate the number of bricks and blocks to build a wall measuring 6.3 m long by 2.7 m high allowing for wastage of 5%.

Area of wall $= 6.3\ \text{m} \times 2.7\ \text{m} = 17.01\ \text{m}^2$
Number of blocks $= 10 \times 17.01 = 170.1$

$$\text{Wastage at 5\%} = \frac{170.1 \times 5}{100} = 8.505$$

Total number of blocks $= 170.1 + 8.505 = 178.606$, say, 179 blocks.

Number of bricks $= 60 \times 17.01 = 1020.6$

$$\text{Wastage at 5\%} = \frac{1020.6 \times 5}{100} = 51.03$$

Total number of bricks $= 1020.6 + 51.03 = 1071.63$, say, 1072 bricks.

Remember!

The extra allowance for wastage could be included within the overall calculation in one line using brackets in the formula. For example:

Total no. of bricks $= 60 \times 17.01\left(1 + \frac{5}{100}\right)$

1071.63, or 1072 bricks

Worked example

The layout of a proposed modern stained-glass window is shown opposite. It is made up from a semi-circle and a triangle. In order to price the job the glazier needs to calculate both the length of the lead **cames** and the area of glass. Calculate this information. Give the area of glazing in mm^2, and allow 5% wastage for the length of lead came.

Key term

Cames – slender, grooved lead bars used to hold together the panes in stained-glass or latticework windows.

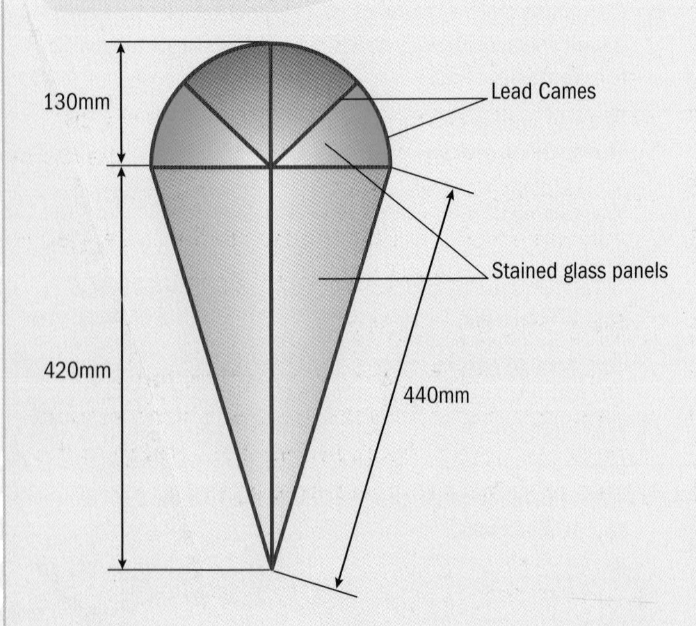

First, calculate the length of cames that is required. The top part of the panel is a semi-circle and the radius of the circle is 130 mm. This allows us to add up all the separate cames that radiate out from the centre of the semi-circle as the length of each of these is the same as the radius.

Length of internal cames in the semi-circle

$= 5 \times 130$ mm $= 650$ mm

Length of internal came in the lower triangle

$= 420$ mm

Length of external cames for lower triangle

$= 2 \times 440$ mm $= 880$ mm

Length of curved came for semi-circle

$= (\frac{1}{2})2\pi r$

$= (\frac{1}{2}) \times 2 \times \pi \times 130 = 408$

Adding up length of lead cames = 2358 mm

Allowing for 5% wastage

$= 2358 \times \frac{5}{100}$

$= 118$ mm

Total length of lead came

$= 2358 + 118$

$= 2476$ mm

Now calculate the whole area by splitting up the window into a semi-circle and a triangle. The area of a semi-circle is half the area of a circle, and the lower triangle can be found by either the standard formula of $\frac{1}{2}$ base multiplied by height or the half perimeter rule. We shall do both.

The area of the semi-circle

$= \pi r^2 \times \frac{1}{2} = \pi \times 130^2 \times \frac{1}{2} = 26{,}546 \text{ mm}^2$

The area of the triangle by standard formula

$= \frac{1}{2} bh = \frac{1}{2} \times 260 \times 420 = 54{,}600 \text{ mm}^2$

Total area $= 26{,}546 + 54{,}600 = 81{,}146 \text{ mm}^2$

Alternatively, using the half-perimeter formula where the area of the triangle is given by:

$\sqrt{s(s-a)(s-b)(s-c)}$

where $s = \frac{a+b+c}{2}$

Hence, $s = \frac{440 + 440 + 260}{2} = 570$ mm

and area, A =

$= \sqrt{570(570-440)(570-440)(570-260)}$

$= \sqrt{570(130)(130)(310)}$

$= \sqrt{2.986 \times 10^9}$

$= 54{,}646.4 \text{ mm}^2 \approx 54{,}600 \text{ mm}^2$

Note: This small rounding error occurs because of the tiny error within the original triangle lengths – the true length of the external came is actually 439.659 mm, not 440 mm!

Activity: Area

1. Calculate the areas of the triangle in m^2 to 3 d.p. where the lengths of the side are 21.30 m, 14.95 m and 18.45 m.
2. Calculate the cross-sectional area of an elliptical steel duct that fits snugly through a rectangular service hole measuring 0.6 m x 0.4 m. Give your answer to the nearest 10 mm^2.
3. Calculate the cross-sectional areas of the gravity retaining walls shown below in m^2 to 2 s.f.

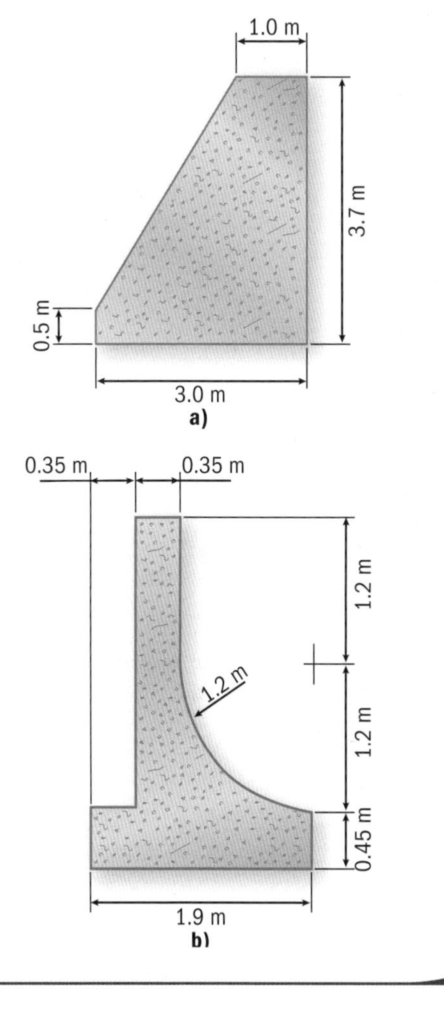

Volumes and surface area

Volume is the space inside a three-dimensional (3D) object. In order to calculate the volume, we need to know the dimensions of the object in three directions. The common name for a 3D object that has been created by extruding a shape from a flat surface area is a prism, where:

Volume of prism = Cross-sectional area × Length.

This can be applied to a whole range of rectilinear objects where a cross-sectional area can be calculated and multiplied by a perpendicular length, as can be seen from Figure 3.1.

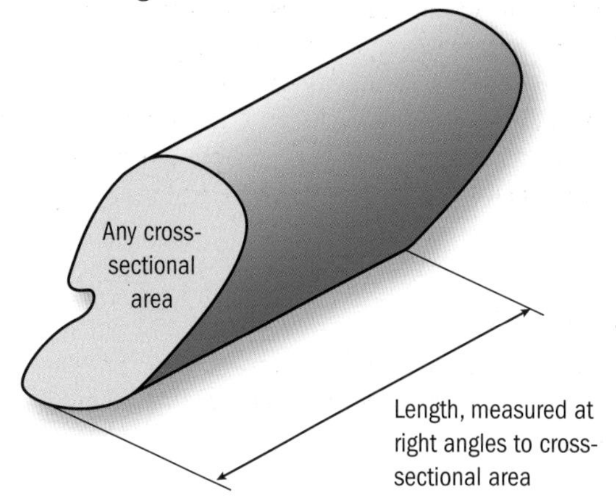

Figure 3.1: A prism

Similarly, the surface area of such an object can be found from knowing the perimeter length of the cross-sectional area and its length.

Surface area = (Perimeter of cross-section × Length) + (Area of the two ends)

Cones and pyramids

Cones and pyramids are common forms found in construction, particularly in roofs and glazed atriums. They can be called **oblique** or right according to their symmetry above a vertical centre line.

Key term

Oblique cone – cone where the top vertex or point is not aligned above the centre of the base.

The calculation of the volume of pyramids and cones is similar to that of prisms, but because their cross-sectional areas taper down with their height, the volume is reduced to a third of the equivalent prism volume.

$$\text{Volume} = \frac{1}{3} \times (\text{area of base}) \times (\text{perpendicular height})$$

Other standard volume formulae are given in Table 3.4.

As with area, the volume may need to be calculated by breaking the object down into one or more standard shapes.

Table 3.4: Volumes and surface area formulae

Object	Cross-sectional area (CA)	Volume	Surface area (SA)
Rectangular prism	$CA = wh$	$V = whL$	$SA = 2(hL + wL + hw)$
Triangular prism	$CA = \frac{1}{2}bh$	$V = \frac{1}{2}bhL$	$SA = L(a + b + c) + 2(CA)$
Cylinder r = radius	$CA = \pi r^2$	$V = \pi r^2 L$	$SA = 2\pi rL + 2(CA)$
Oblique or right cone	$CA = \pi r^2$ r varies according to height of horizontal cross-sections	$V = \frac{1}{3}\pi r^2 h$	For cone only: $SA = 2\pi r \sqrt{(r^2 + h^2)}$
Square oblique or right pyramid The base can be any rectilinear shape	Base area rectangular pyramid $= a^2$	$V = \frac{1}{3}a^2 h$	For sloping sides only: $SA = 2a\sqrt{\left(\left(\frac{a}{2}\right)^2 + h^2\right)}$
Sphere	$CA = \pi r^2$ r varies according to height of horizontal cross-section	$V = \frac{4}{3}\pi r^3$	$SA = 4\pi r^2$

Worked examples

A copper, hot-water tank is in the shape of a cylinder with a hemispherical top. The height of the tank is 1200 mm in total and it has a diameter of 450 mm. Calculate:

- the capacity of the hot-water cylinder in litres
- the surface area of the whole tank in square metres to 3 d.p.

Make a sketch of the hot-water tank. Write in all of the dimensions, being consistent with units.

Note we have split the tank into two parts: the bottom cylinder and the top hemisphere.

Now find the height of the cylinder in order to calculate the volume. The height is:

1200 – 225 = 975 mm

Calculate the volume by working out the volumes of the cylinder and hemisphere separately and adding together for the total volume.

Volume of cylinder

$$V_{cylinder} = \pi r^2 L$$
$$= \pi(0.225\text{ m})^2 \times 0.975\text{ m}$$
$$= 0.155\text{ m}^3 \text{ to 3 d.p.}$$

Volume of hemisphere (half of a sphere)

$$V_{hemisphere} = \frac{1}{2} \times \frac{4}{3}\pi(0.225\text{ m})^3$$
$$= 0.024\text{ m}^3 \text{ to 3 d.p.}$$

Therefore, the total volume of tank

$$= V_{cylinder} + V_{hemisphere}$$
$$= 0.155\text{ m}^3 + 0.024\text{ m}^3$$
$$= 0.179\text{ m}^3$$

When dealing with liquids such as water that would fill this tank, we need to quote the answer in litres.

Therefore,

$$0.179\text{ m}^3 = 1000\frac{\text{litres}}{\text{metres}^3} \times 0.179\text{ m}^3$$
$$= 179 \text{ litres}$$

Now find the surface areas for the cylinder and hemisphere separately.

Surface area of curved side of cylinder

$$= 2\pi rh$$
$$= 2\pi(0.225)(0.975)\text{ m}^2$$
$$= 1.378\text{ m}^2$$

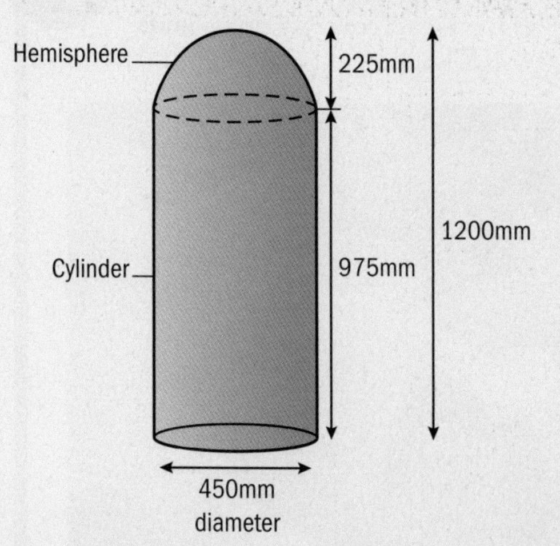

Surface area of bottom of cylinder

$$= \pi r^2$$
$$= \pi(0.225\text{ m})^2$$
$$= 0.159\text{ m}^2$$

Surface area of hemispherical top

$$= \frac{1}{2} \text{ surface area of sphere}$$
$$= \frac{1}{2} \times 4\pi r^2$$
$$= 2\pi r^2$$
$$= 2\pi(0.225\text{ m})^2$$
$$= 0.318\text{ m}^2$$

Therefore, the total surface area of tank can be found by adding up these three separate areas.

Total surface area

$$= 1.378\text{ m}^2 + 0.159\text{ m}^2 + 0.318\text{ m}^2$$
$$= 1.855\text{ m}^2$$

Remember!

It is always good practice when substituting known values into a formula that you also include the symbols that indicate what unit they are measured in, such as 'm' or 'm^2'. This will help you to remember the units your final answer should be stated in. It is also handy in complicated formulae because you can divide or multiply out these 'embedded units' to give you the right units for your final answer.

Look closely at this example converting metres cubed into litres. The metres^3 on the bottom cancel out the metres^3 on top to leave the correct units of litres!

Activity: Volumes

Sketch the following pre-cast concrete **dolos units** described in parts A–C below.

Calculate the volumes of each dolos unit in m^3 and hence find out the weight of each one in kilo-Newtons to 1 d.p. Assume that the weight of 1 m^3 of concrete is 24 kN.

- Dolos A has a triangular cross-sectional base of 250 mm, a height of 350 mm and a length of 2.5 m.
- Dolos B has a parallelogram-shaped cross-section with parallel sides of 325 mm separated by a distance of 200 mm. It is 3 m long.
- Dolos C has dumb-bell shape comprising two different diameter spheres joined by a cylindrical connector. The radii of the largest and smallest spheres are 450 mm and 300 mm respectively. The length and diameter of the cylindrical connector is 1200 mm and 300 mm respectively.

Remember!

Always set out your calculations line by line, just like the examples on the previous page. It is important that you use a clear presentation style so that the work can be checked and people can see what assumptions you have made, and how you have approached solving the problem.

Always write down the formula you are using and explain what you are doing. Then substitute in the given values together with their proper units – identify clearly the units that you are using and be consistent – check what the answer requires and the accuracy needed.

When you have got an answer always try to think of an alternative way of checking your result. Just because the calculator gives you an answer does not mean it's correct. Since you put the values in, if you key in the figures wrongly or miss out a bracket, the calculator won't know.

Key term

Dolos units – man-made sea defence blocks, which are piled against quaysides and jetties to protect them from being eroded by wave action.

Numerical integration methods

Areas

When shapes and objects have irregular boundaries we have to use another technique known as numerical integration to find areas. This involves splitting the object into thin, equal-width strips where the lengths on either side of the strip are called ordinates (see Figure 3.2).

Three useful techniques for carrying out numerical integration are:

- trapezoidal rule
- mid-ordinate rule
- Simpson's rule.

The trapezoidal rule is simple to use and can deal with any number of ordinates.

Area = Width of strip × {$\frac{1}{2}$ (sum of the 1st and last ordinates) + (sum of the remaining ordinates)}

$= w\{\frac{1}{2}(y_1 + y_7) + (y_2 + y_3 + y_4 + y_5 + y_6)\}$

The mid-ordinate rule is also simple to use and can deal with any number of ordinates, provided all the mid-ordinate lengths are supplied.

Area = Width of strip × (sum of all mid-ordinates)

Simpson's rule is the most accurate rule, but can only be used when the total ordinates are an odd number.

Area = $\frac{1}{3}$ × Width of strip × {(1st + last ordinate) + 4(sum of even ordinates) + 2(sum of remaining odd ordinates)}

$= \frac{w}{3}\{(y_1 + y_7) + 4(y_2 + y_4 + y_6) + 2(y_3 + y_5)\}$

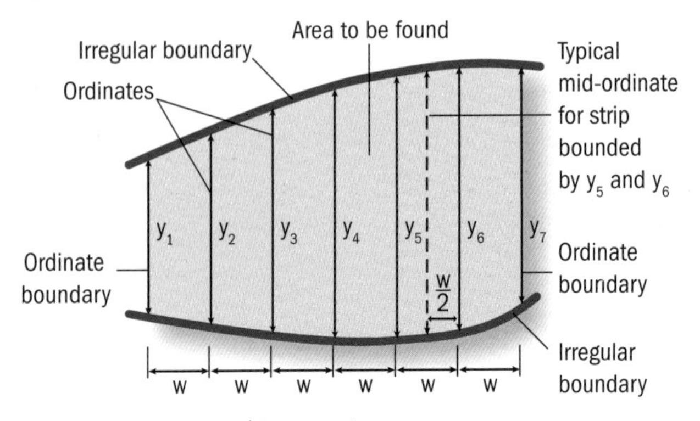

Figure 3.2: Numerical integration terms

Worked example

The cross-section through a river is shown below. Regular depths across the river were recorded at 2.5 m intervals. Determine the cross-sectional area of water in the river in metres squared to 2 d.p. using:

- trapezoidal rule
- Simpson's rule.

Draw a good diagram roughly to scale and mark up the ordinates in sequential order starting at y_1 and working through to the last ordinate. In this example, there are 9 ordinates separated by equal distances of 2.5 m.

Using the trapezoidal rule:

Substituting the ordinate values shown in the diagram into the formula, you get

$= 2.5\text{ m} \times \{\frac{1}{2}(6.65\text{ m} + 0) + (8.25\text{ m} + 7.52\text{ m} + 5.63\text{ m} + 4.87\text{ m} + 3.30\text{ m} + 2.11\text{ m} + 1.91\text{ m})\}$

Working out the ordinate totals within the brackets gives:

$= 2.5\text{ m} \times \{3.325 + 33.590\}\text{m}^2$

$= 2.5\text{ m} \times 36.915\text{ m}^2$

$= 92.2875\text{ m}^2$

$= 92.29\text{ m}^2$ (to 2 d.p.)

Using Simpson's rule:

Cross-sectional area

$= \frac{2.5\text{ m}}{3}\{(6.65 + 0\text{ m}) + 4(8.25\text{ m} + 5.36\text{ m} + 3.30\text{ m} + 1.91\text{m}) + 2(7.52\text{ m} + 4.87\text{ m} + 2.11\text{ m})\}$

Work out all the figures in the brackets:

$= \frac{2.5\text{ m}}{3}\{(6.65\text{ m}) + 4(19.09\text{ m}) + 2(14.5\text{ m})\}$

$= \frac{2.5\text{ m}}{3}(112.01\text{ m})$

$= 93.3416\text{ m}^2$

$= 93.34\text{ m}^2$ to 2 d.p.

Both results are of the same order of size, which is a useful check on our calculations. The Simpson's rule result of 93.34 m^2 is the more accurate answer because it assumes a curved line for the profile of the bed, while the trapezoidal rule always assumes a straight line between each of the ordinates, which is not the case as can be seen from the diagram below. So Simpson's rule is the best one to use when the boundaries are curving. However, if there are straight lines joining the ordinates, then the trapezoidal rule will give just as good a result – and the formula is quicker and easier to use!

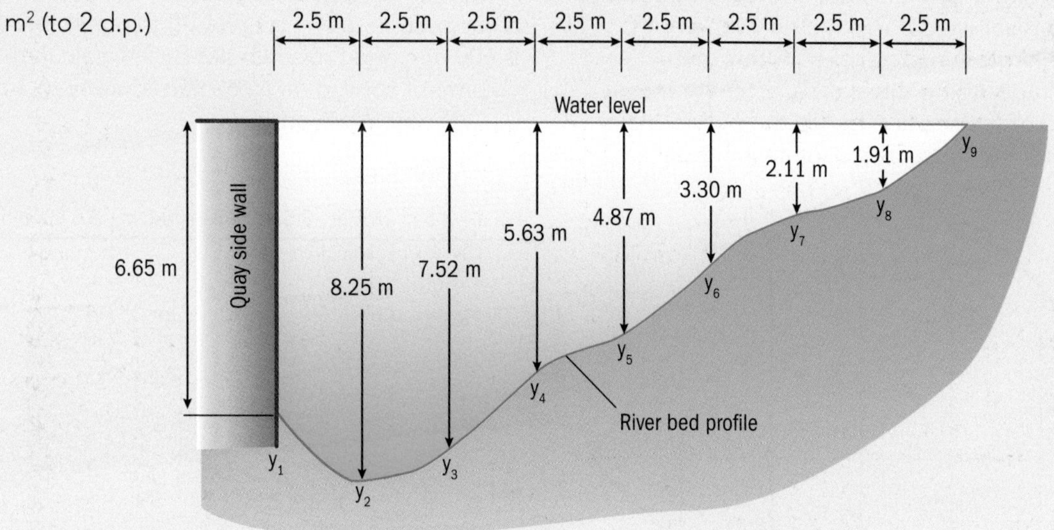

Remember!

If you are going to use Simpson's rule, check to see that there is an odd number of ordinates – in this example, there are 9, so this is fine.

Volumes

We can use the trapezoidal rule or Simpson's rule to find volumes. The only difference is that rather than using ordinates which are distances, we use ordinates which themselves are cross-sectional areas.

Worked example

Look at the diagram below of a large spoil heap. Calculate the volume of the spoil heap to the nearest m^3 by using numerical integration techniques. We have already determined the cross-sections at regular positions along its length.

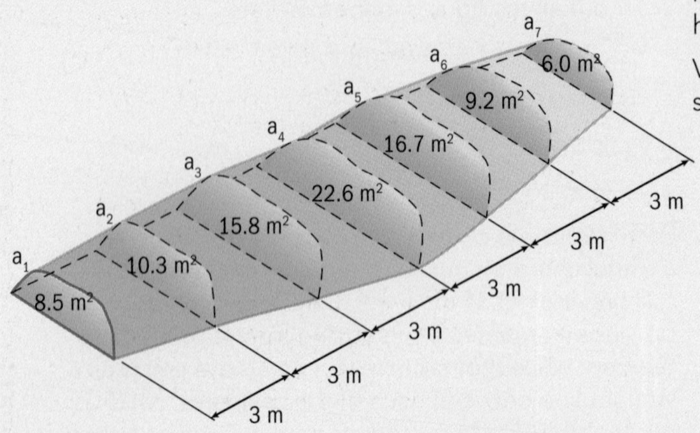

We shall use Simpson's rule because the spoil heap has an irregular curved surface. Replace the 'y' symbol for length of the ordinate to 'a' to represent the 'area' of each 'area ordinate' then note up these values from a_1 to a_7 on the diagram and we have a Simpson's rule for the volume of the spoil heap as:

Volume of spoil heap $= \frac{W}{3}\{(a_1 + a_7) + 4(a_2 + a_4 + a_6) + 2(a_a + a_5)\}$

$= \frac{3\text{ m}}{3}\{(8.5\text{ m}^2 + 6.0\text{ m}^2) + 4(10.3\text{ m}^2 + 22.6\text{ m}^2 + 9.2\text{ m}^2) + 2(15.8\text{ m}^2 + 16.7\text{ m}^2)\}$

$= \frac{3\text{ m}}{3}\{(14.5\text{ m}^2) + 4(42.1\text{ m}^2) + 2(32.5\text{ m}^2)\}$

$= \frac{3\text{ m}}{3}\{247.9\text{ m}^2\}$

$= 247.9\text{ m}^3$

$= 248\text{ m}^3$

Activity: Numerical integration methods

1. A plan view of a greenfield site for a development of 4 houses is shown below together with a set of measurements in metres taken with a metric measuring tape. Using this sketch, estimate the area of land of the site and hence the possible plot size in square metres to 1 d.p. for each house.

Plan view of greenfield site

2. A main sewer is to be excavated as shown in the cross-section drawing below. If the trench width is 600 mm and the soil bulks by 7%, calculate the volume of spoil in m^3 to be carted away to 1 d.p. between points A and B.

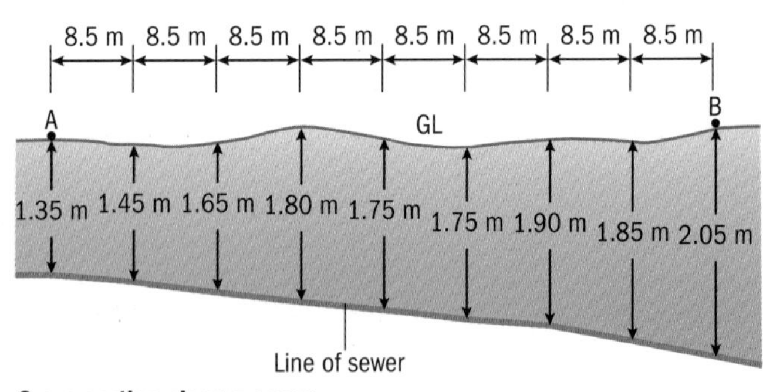

Cross-section along a sewer

Assessment activity 3.2

BTEC

1. A proposed new garden layout is shown below. The new design is for a landscaped lawn and elliptical pond, secured by new vertical boarded timber fencing. Currently, the entire triangular area is used as a car park and has concrete/hardcore surfacing to an approximate depth of 300 mm over the whole site. From the dimensions of the initial site survey shown, estimate the following:
   - The length of new timber fencing required.
   - The amount of concrete/hardcore to be broken up and carted off site.
   - The area of new turf required to cover the garden allowing for the new pond. **P4**

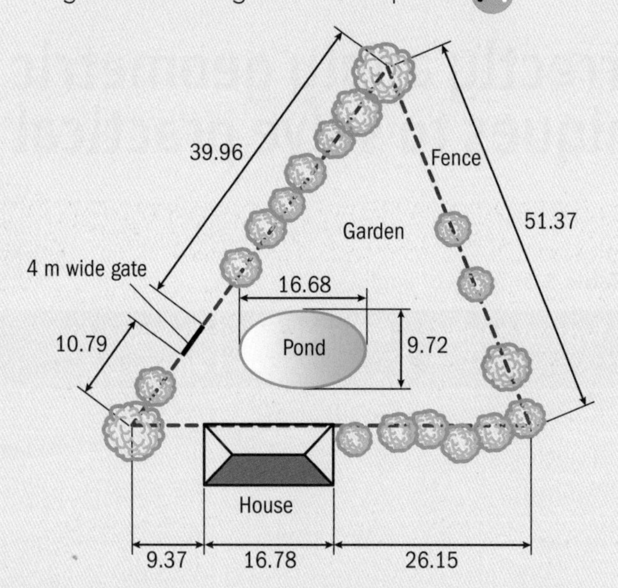

Plan of proposed new garden (All dimensions in metres)

2. A swimming pool with a cross-section as shown below is to be built. A preliminary costing is to be undertaken and you are to provide the information as noted below. Take the width of the pool to be 5 m.

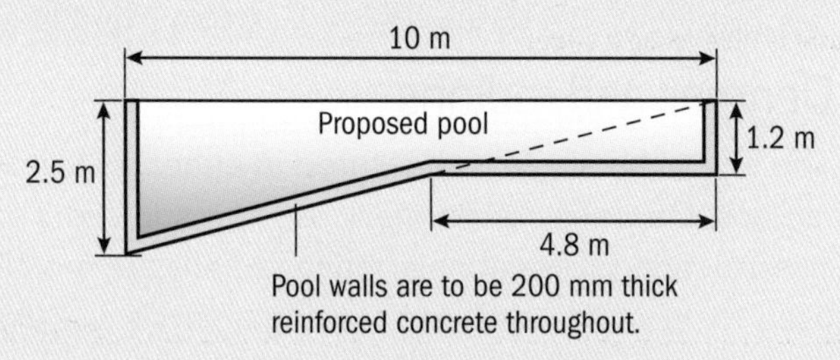

   - Find the amount of excavated soil to be removed from the site to the nearest 0.1 of a m^3 assuming that the soil removed bulks by 20%.
   - Calculate the volume of concrete and reinforcing steel required, given that the thickness of the pool walls is 200 mm and assuming 10% of the volume of the pool walls is reinforcing steel. State your answer for the concrete to the nearest m^3 and for the steel to the nearest 100 mm^3. **M2**

   Discuss how accurate your answer is above when allowing for these rough estimations. **D2**

3. A road embankment is currently under construction. Part of the monitoring process as the embankment is built up is measurement of the cross-sectional areas at regular chainages along the length of the road. A diagram of a typical cross-section of the embankment is shown below.

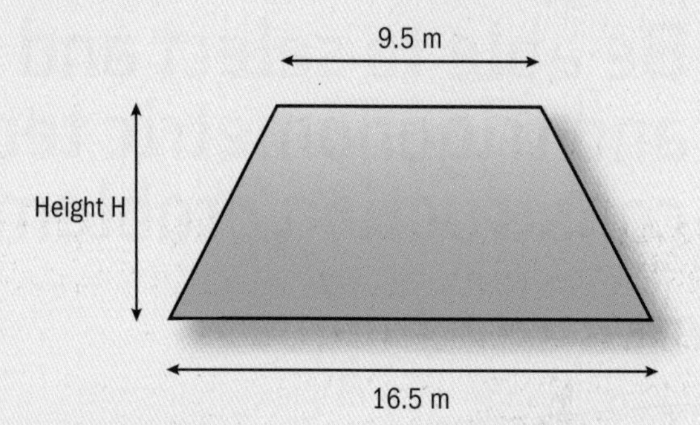

The results of the most recent survey are as follows:

Chainage (m) (distance along road)	0	12	24	36	48	60
Recorded value of height of embankment H (m)	8.6	9.4	10.7	8.9	9.9	11.2
Area of cross-section (m^2)						

Complete the table by working out the cross-sectional areas at each chainage.

Calculate the volume of the embankment using a suitable numerical method. Give your answer in m^3 to the nearest 0.5 m^3.

How might future improvements to the method of carrying out the survey improve the accuracy of the results? **M2** **D2**

For grading tips, see p.92

Grading tips

For **P4** use mathematical techniques to solve construction problems associated with simple perimeters, areas and volumes. A good way to approach this question is to draw out a large A4 sketch of the area showing on it all the key dimensions.

For **M2** you will need to select and apply appropriate algebraic methods to find lengths, angles, areas and volumes for 2D and 3D complex construction industry related problems. In more complicated three-dimensional objects it is useful to sketch out a series of views of the object such as plans and elevations.

For **D2** you need to demonstrate an analysis of the accuracy and rounding of data and its effect on the calculated outcome, and make suitable conclusions that relate to this situation.

The accuracy of the solution as to the amount of volume in the embankment is based upon the number of cross-sections taken – depending on what method you originally chose is there a more accurate one? Which is better Simpson's Rule or the Trapezoidal Rule?

3 Be able to select and correctly apply geometric and trigonometric techniques to solve practical construction problems

Geometric techniques

Geometry involves the study of lines, angles and curves. This section will introduce you to a number of useful techniques to enable you to deal with a variety of practical calculations.

Properties of angles

An angle forms where two lines intersect. The typical types of different angles can be seen in Figure 3.3. Table 3.5 lists the different categories of angles.

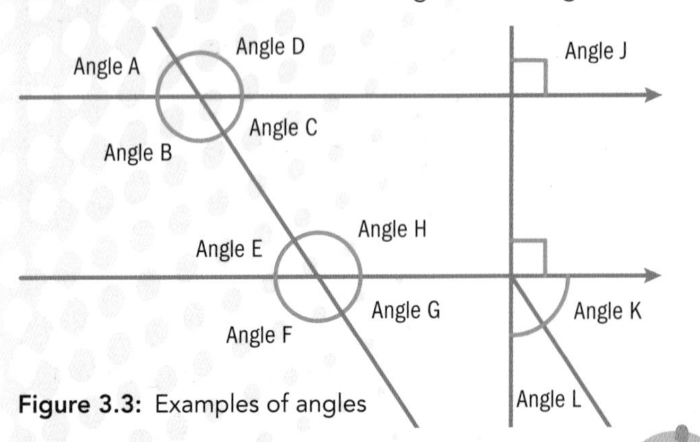

Figure 3.3: Examples of angles

Table 3.5: Types of angles

Type of angles	Degrees	Example from Figure 3.3
Acute	less than 90°	Angle A
Right	90°	Angle J
Obtuse	90°–180°	Angle D
Reflex	180°–360°	Angle B + Angle C + Angle D
Complementary	when added equal 90°	Angle K and Angle L
Supplementary	when added equal 180°	Angle A and Angle B
Alternate or 'Z'	equal	Angle C and Angle E
Opposite	equal	Angle E and Angle G

Degrees and radians

The size of an angle is measured in either degrees or radians. Degrees are normally used by surveyors and construction professionals, while radians are mostly used for complex civil engineering analysis.

Theory into practice

Look at Figure 3.3 again. Find other angles that can be similarly categorised besides the examples already given in Table 3.5. There are many, so look carefully!

Take it further!

Do some research on the internet and find the different measuring systems that are in use around the world.

Angles in the UK are measured using the sexagesimal system where there are 360 degrees (360°) in a whole circle. The sexagesimal system splits up parts of a degree into fractions of a degree as follows:

- 1 whole circle = 360°
- 1° = 60 minutes, written as 60′
- 1′ = 60 seconds written as 60″

So, for example, an angle could be written as 46° 30′ 45″.

In simple calculations, where high accuracy is not required, angles can be quoted in decimal degrees to one or two decimal points for example 72.5° is equivalent to 72° 30′ because 30′ is half a degree.

Theory into practice

Most scientific calculators allow you to input angles in degrees, minutes and seconds and to convert from one to another. Look at your calculator's instruction booklet to see how it is done.

For more complex trigonometry, a radian is the angle made at the centre of a circle when a length, equal to the radius, is traced out around its circumference (see Figure 3.4).

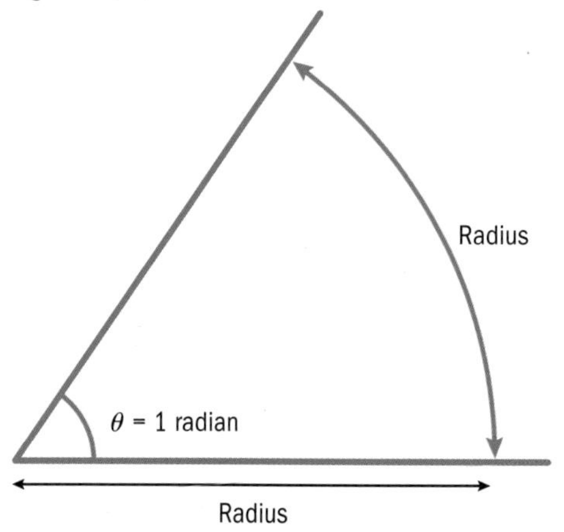

Figure 3.4: Radian measure

The circumference of a circle = $2\pi r$. The number of radians contained within one whole circle of 360° = $\frac{2\pi r}{r}$.

Therefore, 360° = 2π radians, or 180° = π radians which means that:

- 1 radian = $\frac{180°}{\pi} \approx 57°$
- 1 degree = $\frac{\pi}{180°} \approx 0.01745$ radians

Internal angles in polygons

We have already examined a range of regular polygons in the previous section on areas, but in terms of angles, it is useful to know something about the size of angles that define a polygon's shape. The formula that relates the sum of the internal angles of polygon to the number of its sides is:

Sum of internal angles of a polygon = $(2n - 4) \times 90°$ where n is the number of sides of the polygon.

Worked example

In a closed traverse with seven survey stations, the following internal angles were measured with a surveyor's **theodolite**. If the results are to be satisfactory, they must be within +/–1° of what the sum of internal angles should be. Check to see if the results are within that tolerance.

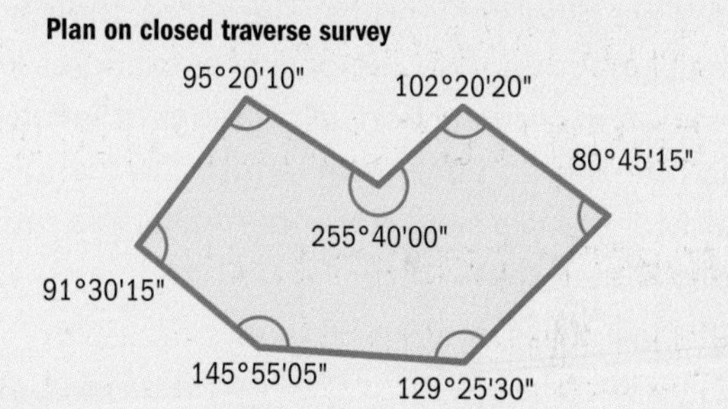

The true value of internal angles

$= (2n - 4) \times 90°$

$= (2 \times 7 - 4) \times 90°$

$= 10 \times 90°$

$= 900°$

The sum of the measured values on site = 91° 30′ 15″ + 95° 20′ 10″ + 255° 40′+ 102° 20′ 20″ + 80° 45′ 15″ + 129° 25′ 30″ + 145° 55′ 05″

= 900° 56′ 35″

which is within the allowed angular tolerance of +/–1° for this survey.

Key terms

Closed traverse – an irregular ring of survey stations linked by measured distances and internal angles which can provide 'easting' and 'northing' coordinates to help set out large structures.

Theodolite – a highly accurate instrument for measuring horizontal and vertical angles for undertaking land surveying and setting out surveys (see Unit 10 Surveying for Construction, page 343).

For example, we know that all the angles in a triangle add up to 180°. Let's try using the internal angle formula.

For a triangle, n = 3, as there are three sides to every triangle.

Therefore, the sum of the internal angles of a triangle

= (2 × 3 – 4) × 90°

= (6 – 4) × 90°

= 2 × 90°

= 180° (which is correct!)

This formula is very when setting up a control network of survey stations as part of a **closed traverse** survey.

Properties of triangles

Similar triangles

Similar triangles contain the same internal angles, but their sides are different lengths. Therefore, their sides are in the same ratio. The use of similar triangles is very common in construction calculations. It involves applying ratios to find unknown lengths. In triangle ABC below there is a similar triangle ADE. The ratio of their sides is such that

$$\frac{AB}{BC} = \frac{AD}{DE}$$

Where ABC is one triangle and ADE is another (see figure in Theory into practice box).

Theory into practice

Draw two similar triangles just like the ones below to a size that conveniently fits on your paper. Measure the four lengths as indicated.

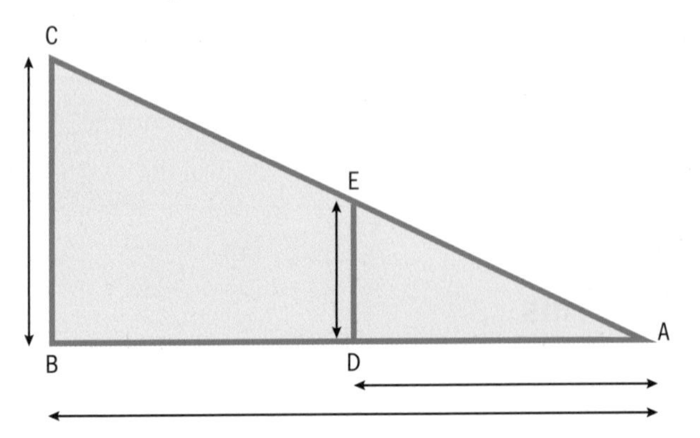

Apply the above formula, to see if they have the same internal angles.

Worked example

A monopitch roof is like the ABC triangle shown in the Theory into practice box. The height of the roof BC = 2.8 m and the span of the roof AB is 5 m. Calculate the height of the vertical post DE, if the distance from the eaves AD is 2 m.

$$\frac{AB}{BC} = \frac{AD}{DE}$$

Use the similar triangle equation

$$\frac{5.0\ m}{2.8\ m} = \frac{2.0\ m}{DE}$$

Re-arranging the formula by cross multiplying we get:

$$DE = \frac{2.0 \times 2.8}{5.0}$$

DE = 1.12 m

Pythagoras' rule

Pythagoras' rule relates the three sides of a right-angled triangle such that

$c^2 = a^2 + b^2$

As shown in Figure 3.5, the longest side of the triangle – the hypotenuse – is opposite the right angle and is labelled 'c'. The two other sides forming the triangle are 'b' and 'a'. This formula is used regularly by surveyors when setting out walls and other features that need to be built at right angles to one another.

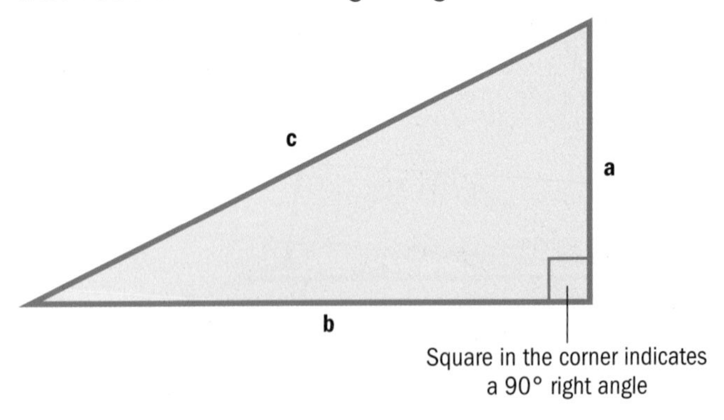

Figure 3.5: Pythagoras' rule – the sides of a right-angle triangle

Pythagoras' rule is shown graphically in Figure 3.6 where:

$c \times c = (a \times a) + (b \times b)$

Area 3 = Area 1 + Area 2

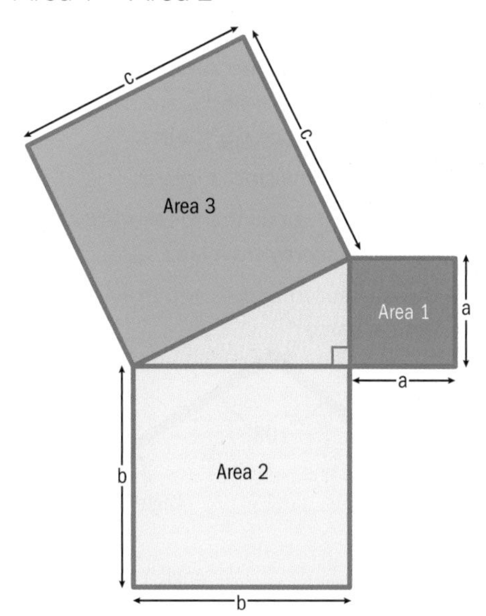

Figure 3.6: Pythagoras' rule – the squares of the sides of the triangle

When given any two of the three sides of a triangle, the third unknown can always be found by transposing Pythagoras' rule. For example:

$a^2 = c^2 - b^2$

Then take the square root of both sides to get a by itself:

$a = \sqrt{(c^2 - b^2)}$

Special triangles

There are two other triangles commonly used in construction calculations: the equilateral triangle and the isosceles triangle (See Table 3.6). Both of these triangles, when cut in half by drawing a vertical line down from the top point to the base, form two identical right-angled triangles.

Table 3.6: Properties of equilateral and isosceles triangles

Property	Equilateral triangle	Isosceles triangle
Basic shape	A, 60°, B 60°, 60° C	A, C, B
Sides	All sides are equal (AB = AC = BC)	Length AC = Length AB
Internal angles	All angles 60°	Angle C = Angle B

Worked example

The cross-section through a timber scissor truss is shown below. Estimate the length of both the rafter and ceiling tie in metres using Pythagoras' rule.

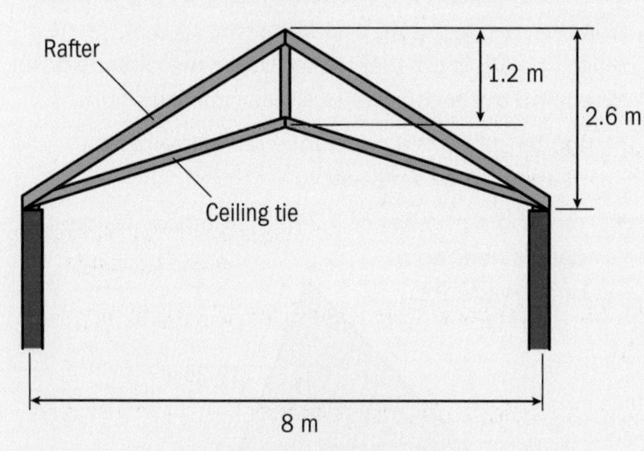

If you study the diagram closely, you can identify two separate right-angled triangles that feature the lengths that we need to find.

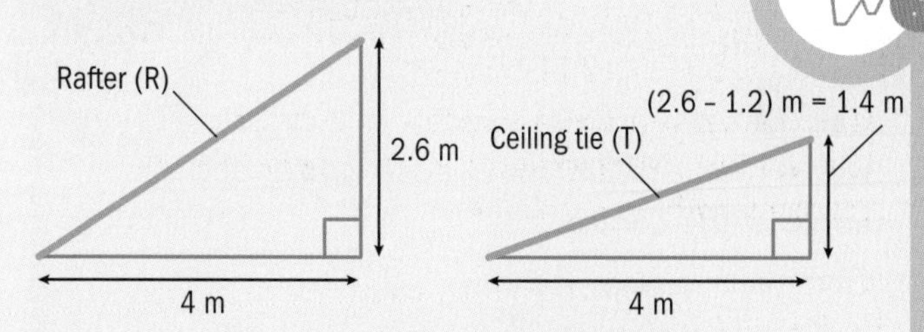

For the length of the rafter (R), we have:

$R^2 = 2.6^2 + 4^2$

$R^2 = 6.76 + 16$

$R = \sqrt{22.76}$

$R = 4.77$ m

And for the length of the ceiling tie (T) we also have

$T^2 = 1.4^2 + 4^2$

$T^2 = 1.96 + 16$

$T = \sqrt{17.96}$

$T = 4.24$ m

Properties of circles

Circles have a number of important geometric properties which occur regularly in construction problems. The basic ones are:

- Circumference of circle = $2\pi r$
- Perimeter of sector = Circumference of circle × $\frac{\theta^\circ}{360^\circ}$
- Area of circle = πr^2
- Area of sector = Area of circle × $\frac{\theta^\circ}{360^\circ}$

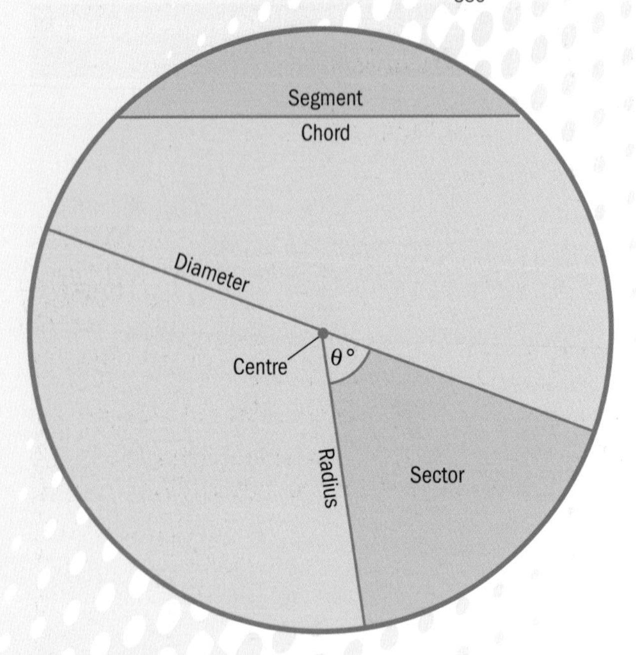

Figure 3.7: Parts of a circle

Worked example

Calculate the area of a sector of a circle with a radius of 350 mm whose subtended angle at the centre is 34°.

Give your answer in mm².

Area of sector = $\pi r^2 \frac{\theta^\circ}{360^\circ}$

$= \pi(350)^2 \times \frac{34^\circ}{360^\circ}$

$= 36{,}346 \text{ mm}^2$

Activity: Geometric techniques

1. In the following five-station survey traverse, the measured angles W, X, Y and Z are missing in the assistant surveyor's booking sheets.
   - What are these missing angles?
   - What is the sum of all the measured internal angles in the survey traverse?
   - Are the measured angles within +/–2° of the theoretical value?

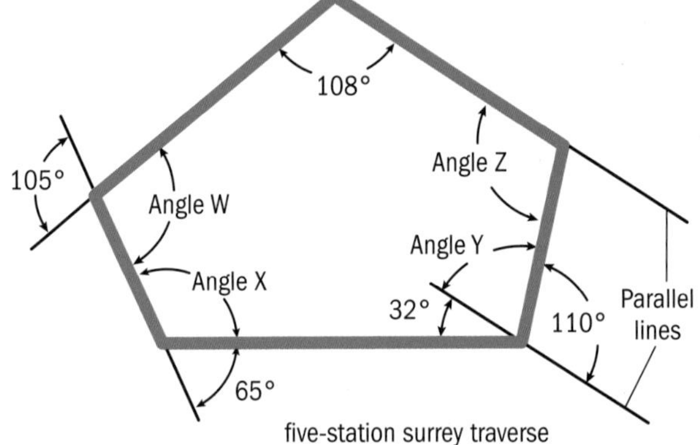

five-station surrey traverse

2. A cross-section through a church spire is an isosceles triangle as shown below. Using Pythagoras' rule, calculate the vertical height of the spire in metres to 2 d.p.

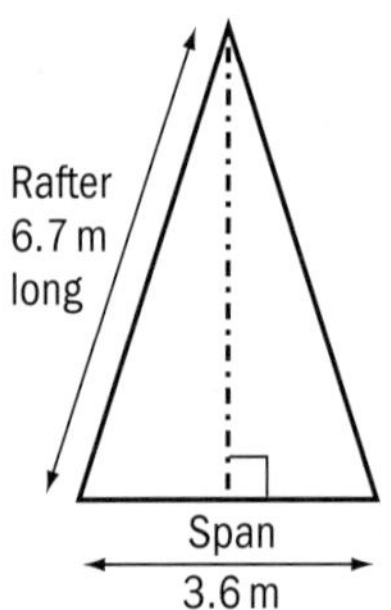

Cross-section through a church spire

3. An area of land in the shape of triangle ABC is to be split into two separate building plots so that the new fence line DE is parallel to AB. Using the dimensions given and the technique of similar triangles find:
   - the length of the new fence DE in metres to nearest 0.1 of a metre.
   - the whole number of 1.2 m long timber fence panels needed to build the fence.

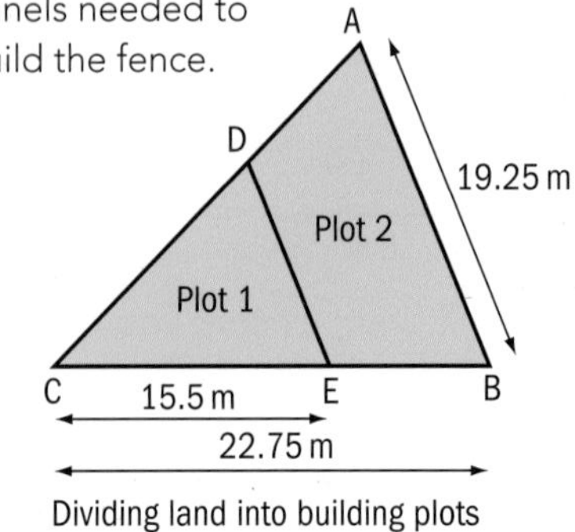

Dividing land into building plots

Trigonometric techniques

Right-angle trigonometry

The study of trigonometry looks more closely at how angles in triangles affect the lengths of their sides. In a right-angled triangle (Figure 3.8) the sine (sin), cosine (cos) and tangent (tan) of angle θ are defined by the lengths of the sides of the triangle such that:

$$\sin \theta° = \frac{\text{Opposite}}{\text{Hypotenuse}}$$

$$\cos \theta° = \frac{\text{Adjacent}}{\text{Hypotenuse}}$$

$$\tan \theta° = \frac{\text{Opposite}}{\text{Adjacent}}$$

These are all known as trigonometric ratios or 'trig ratios' for short.

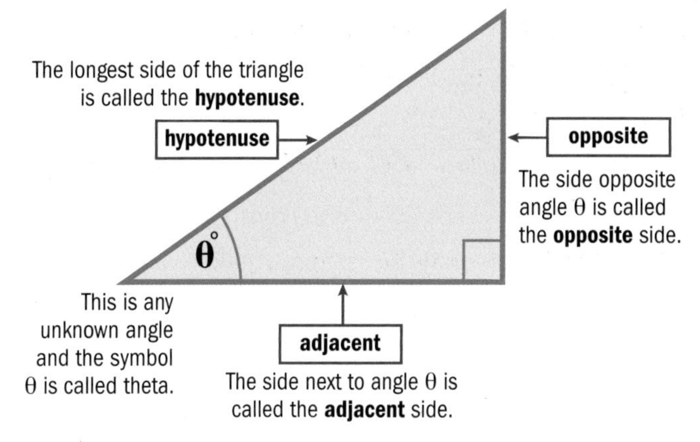

Figure 3.8: A right-angled triangle

Worked example

The cross-section through a triangular, cantilevered canopy for a proposed warehouse is shown below. Using trigonometric ratios, work out:

- the height of the canopy strut BC
- the width of the cantilevered over-hang AC
- the angle that the rear stay makes with the warehouse roof at D
- the length of the rear stay BD.

Undertake a suitable alternative check calculation for BD.

In the diagram, there are two right-angled triangles and in the left-hand side one, labelled ABC, we are given two pieces of information, so we should be able to work out the rest of the dimensions and angles for that triangle. Once this is done, we should be able to move on to triangle BCD using the height BC.

Cross-sectional elevation
(All dimensions in mm)

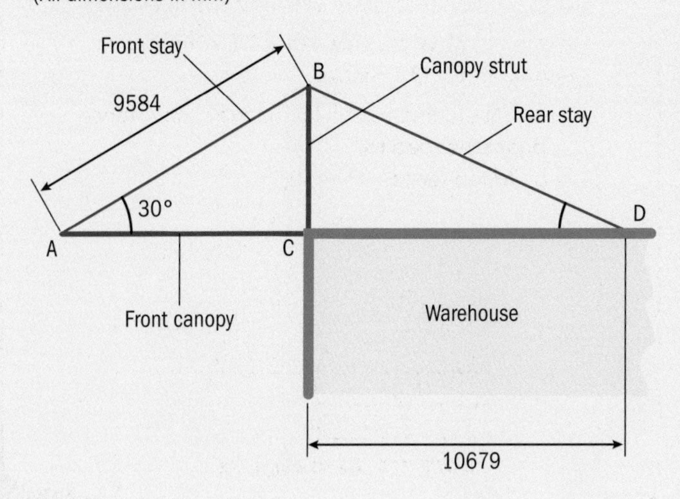

To find the height of the canopy strut BC, look closely at triangle ABC and identify the hypotenuse, opposite and adjacent sides to the angle.

Step 1: It is always a good idea to label up on a separate sketch collating all the information thus:

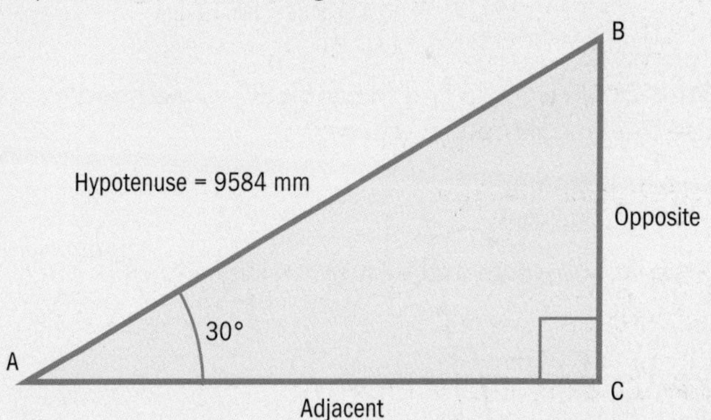

Step 2: Apply the trig ratio. If we want to find BC we need to use the sine ratio because:

$$\sin \theta° = \frac{\text{opposite}}{\text{hypotenuse}}$$

Step 3: Substitute in the known values:

$$\sin 30° = \frac{\text{opposite}}{9584 \text{ mm}} = \frac{\text{BC}}{9584 \text{ mm}}$$

Step 4: Rearrange the equation to make BC the subject of the formula:

$$\text{BC} = 9584 \text{ mm} \times \sin 30°$$

$$= 4792 \text{ mm length of strut BC}$$

The answer will be in the units of mm because the distance we put into the formula was in mm.

(cont.)

To find the width of the cantilevered over-hang AC we need to use the cosine ratio:

$$\cos\theta° = \frac{\text{adjacent}}{\text{hypotenuse}}$$

$$\cos 30° = \frac{\text{adjacent}}{9584\text{ mm}} = \frac{\text{AC}}{9584\text{ mm}}$$

$$\text{AC} = 9584\text{ mm} \times \cos 30° = 8300\text{ mm}$$

To find the angle that the rear stay makes with the warehouse roof at D, let's look at the other triangle that can be found as part of the canopy structure, the right triangle BCD.

Step 1: Draw out the triangle and jot down all the values that we know as well as the unknown angle that we are trying to find.

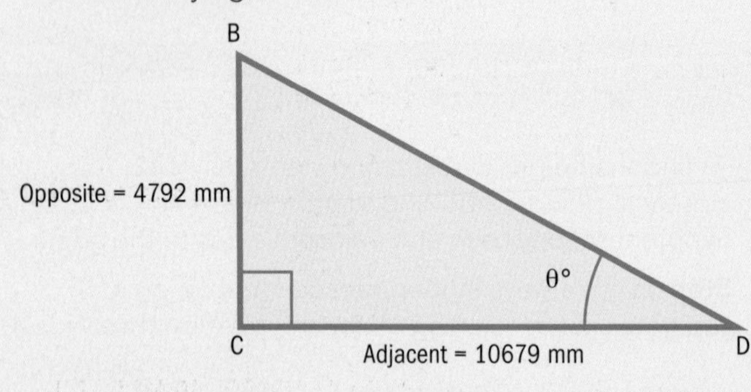

Step 2: We want to find the angle θ°, so we need to use the tangent ratio because:

$$\tan\theta° = \frac{\text{opposite}}{\text{adjacent}}$$

Step 3: Substitute in the known values and work out the ratio.

$$\tan\theta° = \frac{4792\text{ mm}}{10679\text{ mm}} = 0.44873$$

Step 4: Convert the tangent ratio calculated into an angle value that we can recognise. To do this, we need to identify the 'inverse trigonometric ratios' on our scientific calculators. Find the button marked with the symbol [tan⁻¹] (inverse tangent ratio).

$$\tan^{-1}(0.44873) = 24.1672 \text{ decimal degrees}$$

$$= 24° \, 10'$$

To find the length of the rear stay BD, use trigonometric ratios or Pythagoras' rule. Let's use the first one.

Step 1: Collate all the information in a sketch.

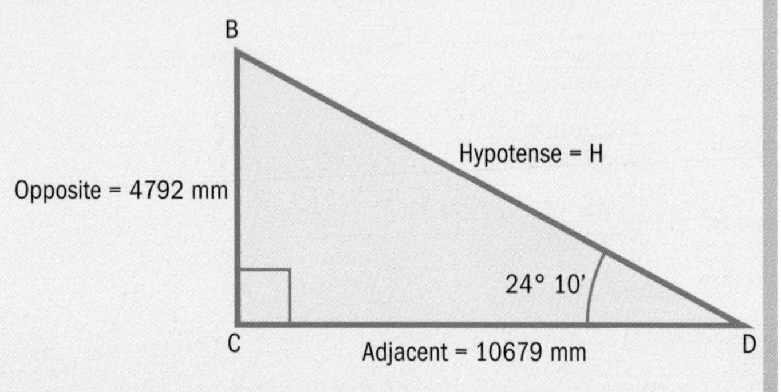

Step 2: Apply the trig ratio. We can use either the sine ratio or the cosine ratio to find the hypotenuse H. Let's do it using the sine ratio, where:

$$\sin\theta° = \frac{\text{opposite}}{\text{hypotenuse}}$$

Step 3: Substitute in the known values and re-arrange to get the unknown on the LHS on the top.

$$\sin 24° \, 10' = \frac{4792\text{ mm}}{\text{H}}$$

$$\therefore \text{H} = \frac{4792\text{ mm}}{\sin 24°10'}$$

$$\text{H} = 11{,}705\text{ mm}$$

An alternative method would be to use Pythagoras' rule

$$\text{BD} = \sqrt{(10679\text{ mm})^2 + (4792\text{ mm})^2}$$

$$= 11705\text{ mm}$$

Non-right-angle trigonometry

A non-right-angled triangle has no hypotenuse. All angles are given capital letters when labelled, e.g. A, B or C, and the sides that are *opposite* these angles are given lower-case letters, e.g. a, b and c (see Figure 3.9).

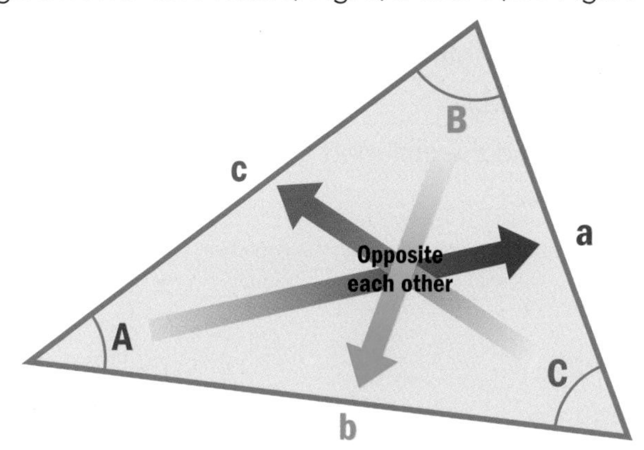

Figure 3.9: A non-right-angled triangle

The two trigonometry techniques for solving problems where there are non-right-angled triangles present are called the sine rule and the cosine rule.

Remember!

Don't confuse the sine and cosine rules that we use in non right-angle trigonometry with the sine and cosine ratios used for right-angled triangles!

The sine rule

The formula for the sine rule looks like this:

$$\frac{a}{\sin A} = \frac{b}{\sin B} = \frac{c}{\sin C}$$

It represents the ratio of the sides of the triangle, divided by the sines of the opposite angles. The ratios for a given triangle are all equal. You can take any part of the expression to make it work as a normal equation. For example:

$$\frac{a}{\sin A} = \frac{b}{\sin B} \text{ or } \frac{b}{\sin B} = \frac{c}{\sin C}$$

or even $\frac{a}{\sin A} = \frac{c}{\sin C}$

Provided you know at least three values from the triangle, you will be able to find the fourth value. In fact, you can solve for all the lengths and angles in a triangle using the sine rule provided you are given either:

- one side and any two angles
- two sides and one angle (but not the angle between the two sides).

Worked example

A reinforced concrete bridge pier supports a sloping bridge deck, as shown below. The top of the pier has been splayed out to increase the bearing area. Calculate the length of bearing b with the information given.

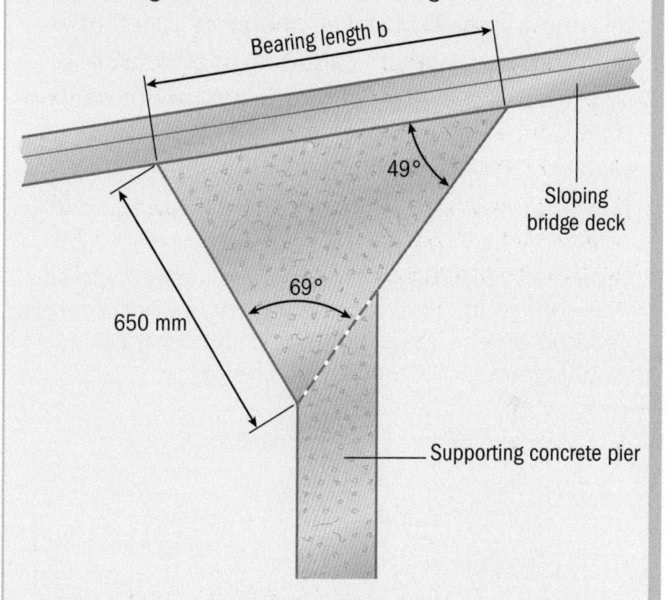

We can solve this problem by using the sine rule.

Step 1: Draw a diagram and label up the sides and angles according to the notation given earlier. Collate the information. Let the bearing length be b.

Step 2: Apply the sine rule:

$$\frac{a}{\sin A} = \frac{b}{\sin B}$$

Step 3: Substitute the values into the formula:

$$\frac{650 \text{ mm}}{\sin 49°} = \frac{b}{\sin 69°}$$

Step 4: Rearrange and carry out the calculations:

$$b = \frac{650 \text{ mm}}{\sin 49°} \times \sin 69°$$

$$b = 804 \text{ mm}$$

The cosine rule

In a triangle with internal angles A, B and C, and opposite sides a, b and c, the cosine rule states:

$a^2 = b^2 + c^2 - 2bc \cos A$, or

$b^2 = a^2 + c^2 - 2ac \cos B$, or

$c^2 = a^2 + b^2 - 2ab \cos C$.

This formula is similar to the Pythagoras' rule that we have used before, with the addition of the cosine of the included angle. This is there as we are not dealing with a right-angled triangle. In fact, if we substitute cos 90° into this 'extra part', all we would get is zero as cos 90° = 0, and the equation reverts back to $a^2 = b^2 + c^2$ which is Pythagoras' rule.

The cosine rule can be used in non right-angled triangles to solve for all lengths and angles, provided the triangle has either:

- three given sides
- two sides and the included angle between those sides.

Worked example

A land survey was carried out where a greenfield site was split up into two triangles, Triangle 1 and Triangle 2 based on one base line AC – all the distances and angles were recorded as shown below.

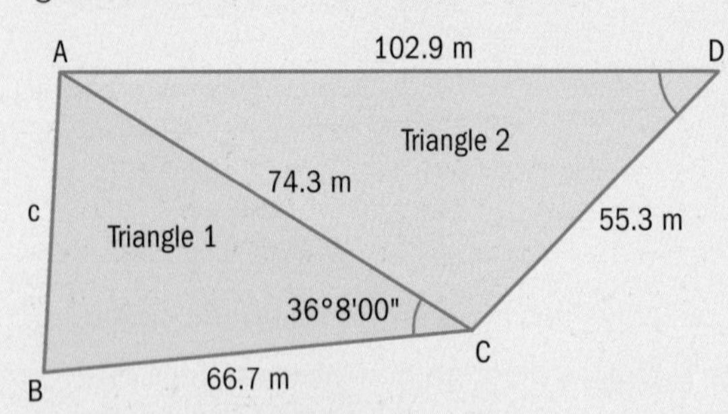

Using the cosine rule, find:

- the length of side c.
- the size of the internal angle at D.

Step 1: Starting with Triangle 1, draw a diagram to collate all the known information and add the correct notation.

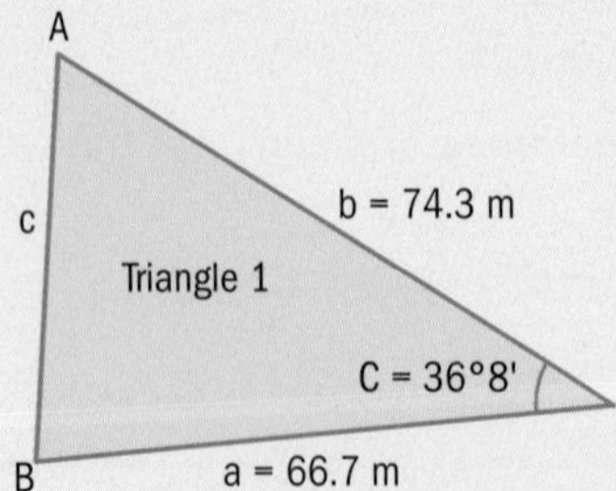

Step 2: Apply the cosine rule. We know sides a and b and the included angle at C, and we want to find the side length c. Studying the cosine rule, we see that we need:

$c^2 = a^2 + b^2 - 2ab \cos C°$

Step 3: Substitute in the values and work out the length of side c.

$c^2 = 66.7^2 + 74.3^2 - 2 \times 66.7 \times 74.3 \times \cos 36°8'$

$c^2 = 1964.29$

$c = \sqrt{1964.29}$

$c = 44.32$ m

Step 4: Move on to Triangle 2, draw a diagram to collate all the known information and add the correct notation.

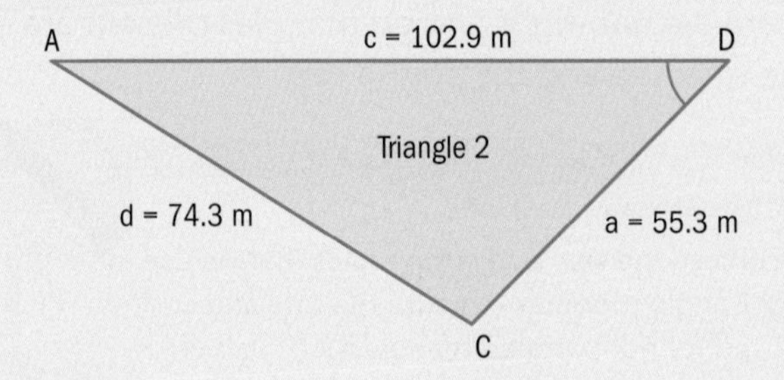

Step 5: Apply the cosine rule. We know all the sides but none of the angles. The angle we want, D, is adjacent to two known lengths. We can now construct a cosine rule for Triangle 2 by looking at the pattern:

$a^2 = b^2 + c^2 - 2bc \cos A°$

Therefore, we have:

$d^2 = a^2 + c^2 - 2ac \cos D°$

Step 6: Rearrange to make the angle D the subject of the formula:

$d^2 = a^2 + c^2 - 2ac \cos D°$

$d^2 - a^2 - c^2 = -2ac \cos D°$

$\cos D° = \dfrac{d^2 - a^2 - c^2}{-2ac}$

Step 7: Substitute in the known values and calculate the required angle:

$\cos D° = \dfrac{74.3^2 - 55.3^2 - 102.9^2}{-2 \times 55.3 \times 102.9}$

$\cos D° = 0.71401$

$D = \cos^{-1} (0.71401)$

$D = 44°26'13''$

Area of a triangle using angles

The general area of any triangle is determined by half the base multiplied by the perpendicular height:

Area = $\frac{1}{2}$ × base × perpendicular height.

With the aid of trigonometry, we can develop this into a formula which ties in the angular measurement techniques used in surveying.

In Figure 3.10, where:

a = side a of the triangle

b = side b of the triangle

C = the given angle of the triangle between sides a and b.

The height (h) can be expressed using the angle and the sine ratio so that:

h = a × sin C

If we substitute that into '$\frac{1}{2}$ × base × perpendicular height', we arrive at:

Area = $\frac{1}{2}$ ab sin C

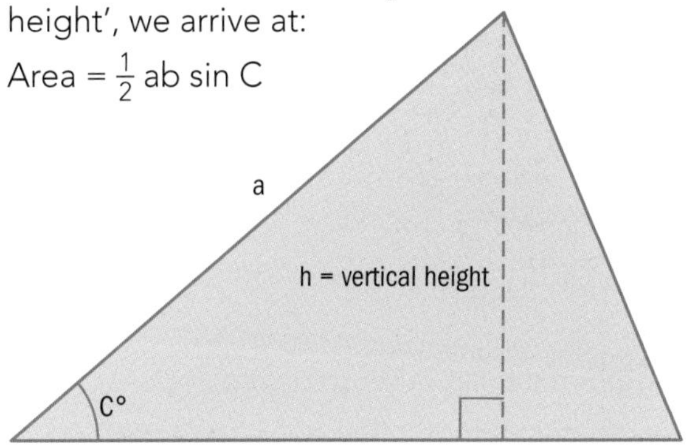

Figure 3.10: Sine area rule

Worked example

An isosceles triangle has two sides, both 2500 mm long with an included angle of 38°.

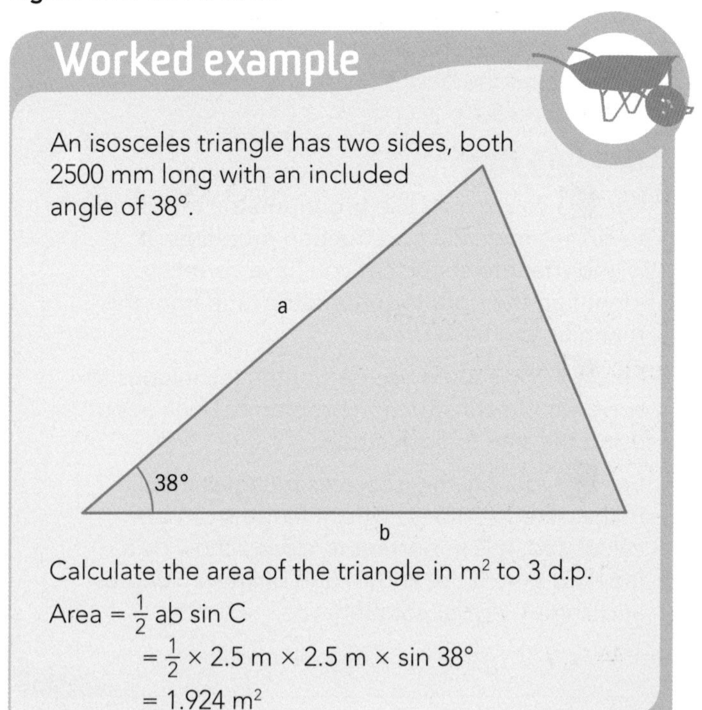

Calculate the area of the triangle in m^2 to 3 d.p.

Area = $\frac{1}{2}$ ab sin C

= $\frac{1}{2}$ × 2.5 m × 2.5 m × sin 38°

= 1.924 m^2

Activity: Trigonometric techniques

1. Sketch and label the following non right-angled triangles using the standard notation. Then find all the side lengths and angles.

   Use the sine rule for the first two, and the cosine rule the last two.

   - A = 37°, B = 73°, b = 4.30 m
   - A = 71°, B = 36°, a = 23.7 mm
   - Y = 11 cm, Z = 15 cm, X = 55°
   - X = 62.8 mm, Y = 41.2 mm, Z = 62°

2. A lean-to conservatory has a base which is 7.5 m long by 4 m wide. The glazed roof makes an angle of 36° to the horizontal, across the 4 m width. Calculate:

   - the length of the sloping roof in m to 1 d.p.
   - the area of glazed roof in m^2 to 1 d.p.

3. A vertical aerial mast RS is 21.8 m high and stands on ground that is inclined 13° to the horizontal. A steel cable connects the top of the aerial R to a point T on the ground 8 m downhill from S at the foot of the aerial mast. Using right-angle trig ratios, calculate the:

   - length of the stay RT to the nearest 10 mm
   - angle that the cable makes with the horizontal ground in degrees, minutes and seconds.

   (*Hint:* Take length TS to be the hypotenuse of a right-angled triangle for the first part of the calculation.)

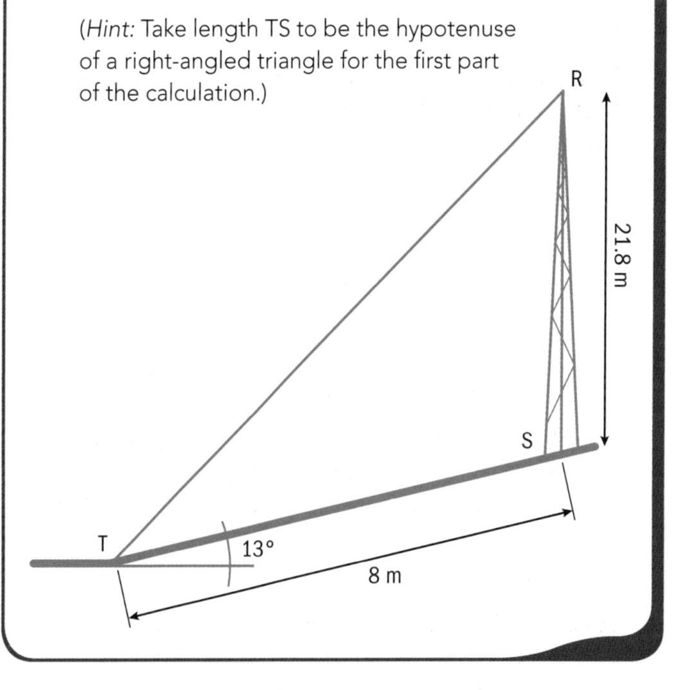

Assessment activity 3.3

BTEC

1. A symmetrical, timber roof truss is to be fabricated as shown below. All the dimensions shown relate to the centre lines of the timber members.

Elevation on flat roof truss

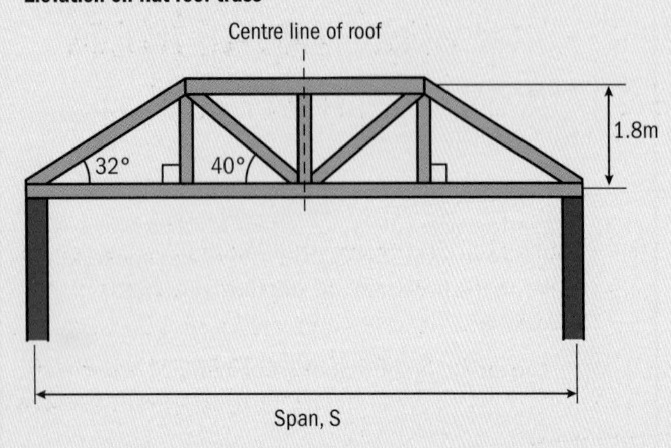

Calculate:

- the lengths of all the separate diagonal timber members, i.e. struts and rafters
- the span distances of the bottom member of the truss. **P5**

2. A stainless steel luminaire, or light fitting, is to be designed for a new shopping centre. The elevation of the new fitting is shown below. The top of the fitting is a dome in the shape of a segment of a circle.

Cross-sectional elevation of luminaire

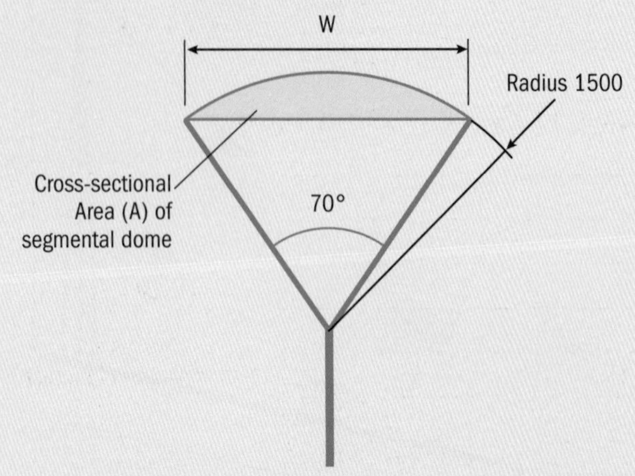

For costing purposes and fabrication processes, the building services engineer needs some information about the proposed fitting. Provide calculations to find:

- the width (W) of the segmental dome
- the cross-sectional area (A) of the segmental dome. **P6**

3. A church spire is being renovated. The timber structure is shown below. It is the shape of an eight-sided (octagonal) pyramid. In plan, the edges of the octagon are 3.2 m long and the vertical height of the spire is 8.4 m.

View on octagonal church spire

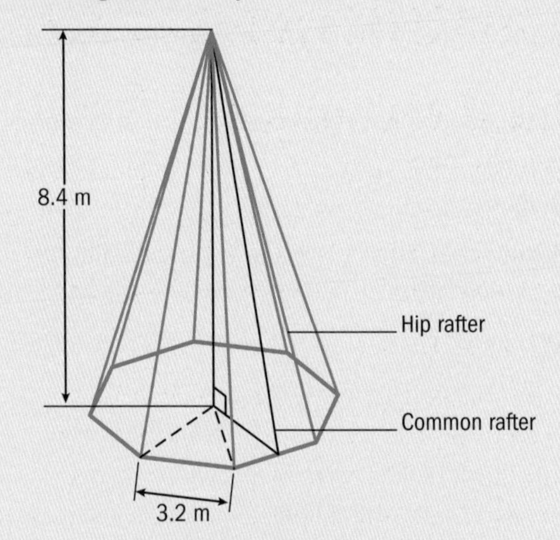

Calculate:

- the length of the hip rafter to be replaced
- the length of common rafter to be replaced
- the total area of the roof to be re-covered with copper sheet. **M2**

(*Hint:* Consider the angles around the centre of the pyramid, at the base.)

Grading tips

For **P5** you should use trigonometric techniques to solve simple 2D construction problems. It is important to show that you have correctly identified the right trigonometric ratio from the triangles you have drawn.

For **P6** you should use geometric techniques to solve simple construction problems. Think about using the sine or cosine rule.

For **M2** identify the isosceles triangles that lie in the octagonal base whose lengths can be calculated. It is important to clearly show by a labelled sketch the known and unknown lengths and angles in your solution.

PLTS

By analysing and evaluating information and judging its relevance and value, you will develop your skills as an **independent enquirer**.

Functional skills

By carrying out this type of approach you have the opportunity to demonstrate **mathematical** ability to identify the problem and the mathematical methods to tackle it.

4 Be able to select and correctly apply graphical and statistical techniques to solve practical construction problems

Graphical techniques

Graphs are useful for showing how one measurable quality varies against another. For example, in Figure 3.11 the graph shows how the output of construction work varies over time in the East of England region.

Cartesian coordinates

All graphs are based on the Cartesian coordinate system which is used to fix a point by two numbers within a square grid. The numbers are called the *x-coordinate* and the *y-coordinate* of the point and are written (x,y). To define the coordinates, two perpendicular directed lines (the *x-axis* and the *y-axis*) are set up on the grid and equal scales are marked off on the two axes. The point where the axes cross is called the origin, O.

The coordinates that fall on the lines in the Cartesian coordinate system will fit the algebraic equations of that line. For example, the circle of radius 3 units, and centre the origin, may be described by the equation $x^2 + y^2 = 9$; when $x = 0$, $y = 3$ and when $y = 0$, $x = 3$, as can be seen in Figure 3.12.

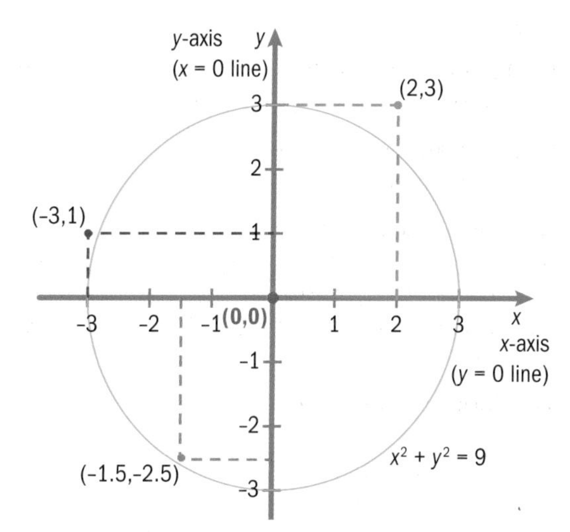

Figure 3.12: Cartesian coordinate system

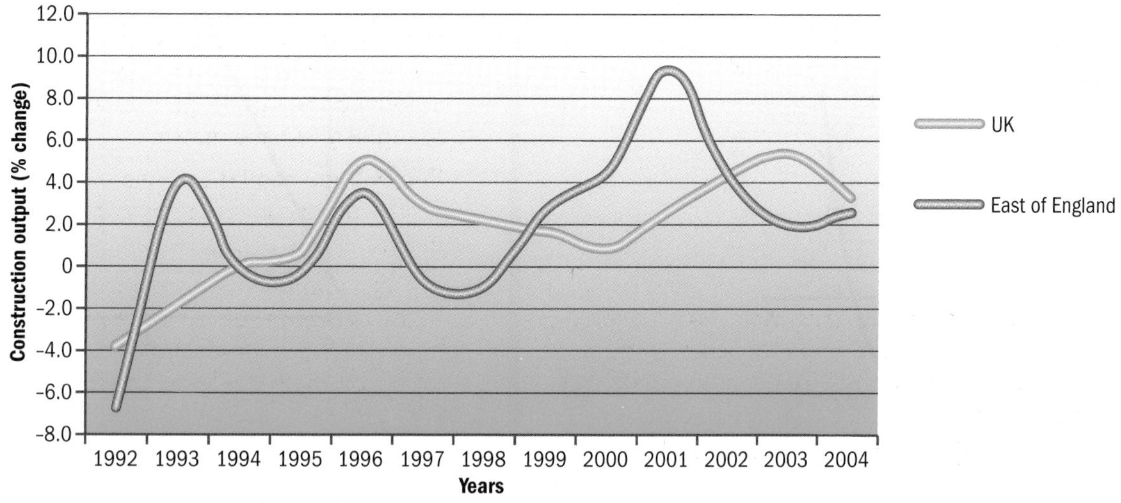

Figure 3.11: Example of a graph

Theory into practice

Figure 3.13 shows some examples of typical graphs. What do you think each graph is used for?

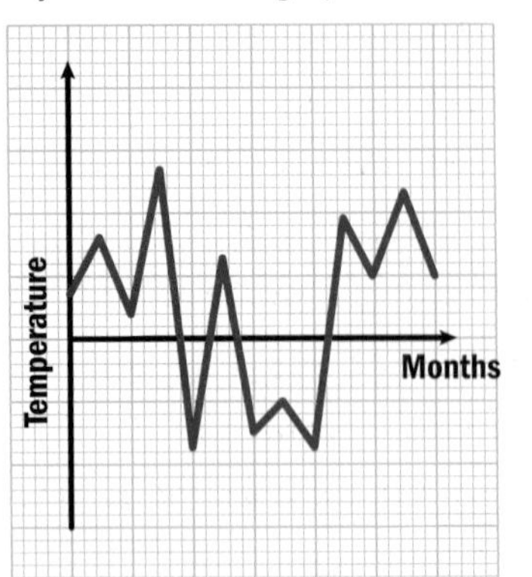

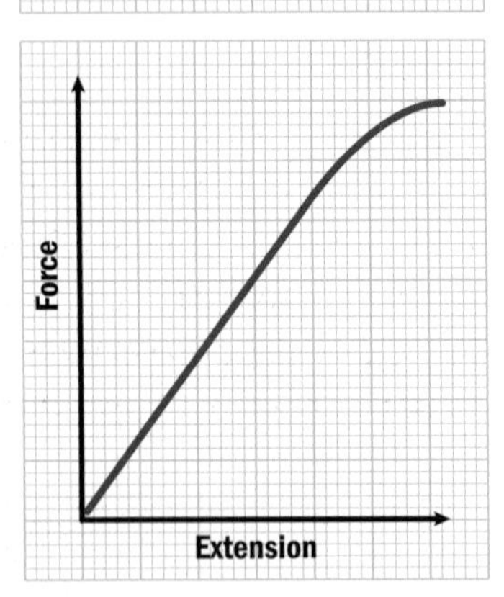

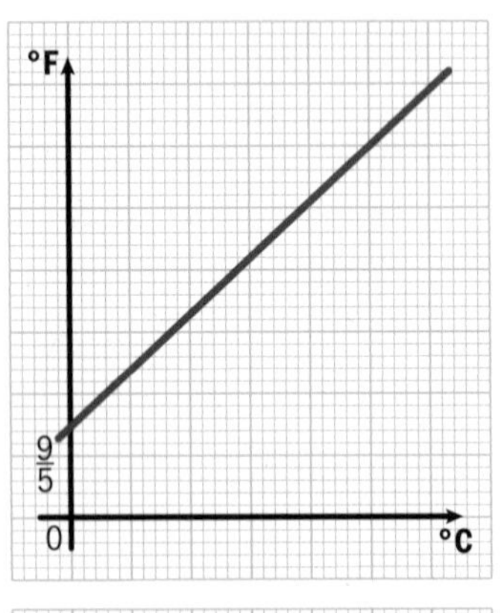

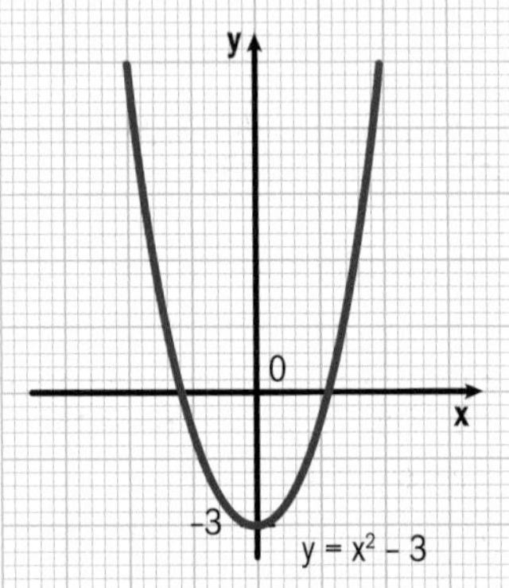

Figure 3.13: Examples of graphs

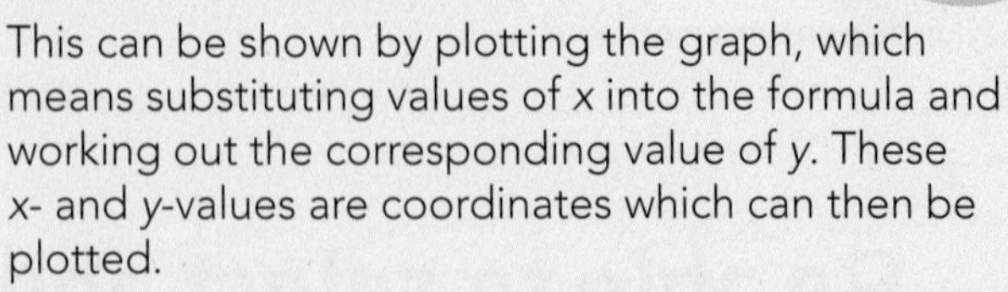

Worked example

Let's look at the equation $y = 2x + 1$.

This can be shown by plotting the graph, which means substituting values of *x* into the formula and working out the corresponding value of *y*. These *x*- and *y*-values are coordinates which can then be plotted.

A good way to work out coordinates is to use a table.

For example:

x	0	1	2	3	4	5	
+2x	0	+2	+4	+6	+8	+10	ADD
+1	+1	+1	+1	+1	+1	+1	
y	1	3	5	7	9	11	

The graph can now be plotted:

We can clearly see that the line cuts the *y*-axis at $y = 1$, but the gradient is harder to see; but for every +1 unit we move to the right, the graph goes up +2 units. This is the gradient of the line and it equals +2. Thus:

$$\text{Gradient} = \frac{\text{Vertical distance}}{\text{Horizontal distance}} = \frac{+10}{+5} = 2$$

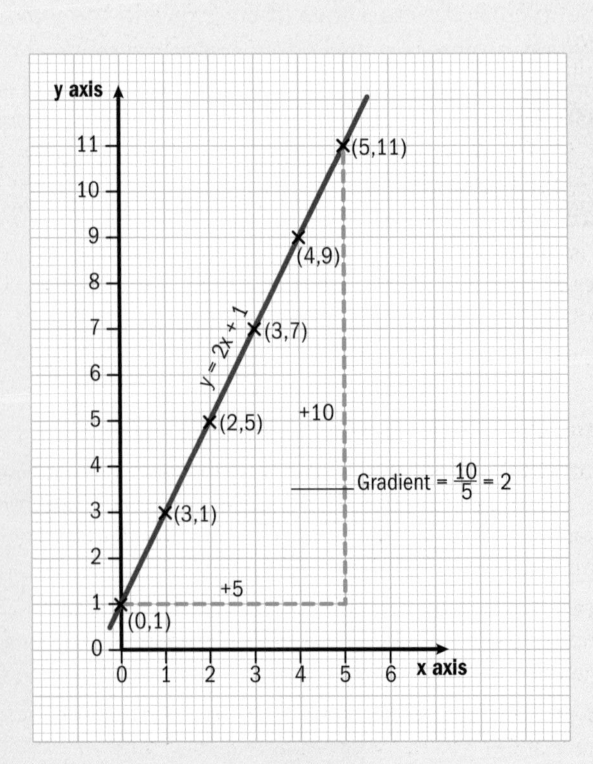

Straight line graphs

The formula for a straight line is:

y = mx + c

where m is the slope or gradient of the line and c is the distance where the line cuts the *y*-axis, called the intercept.

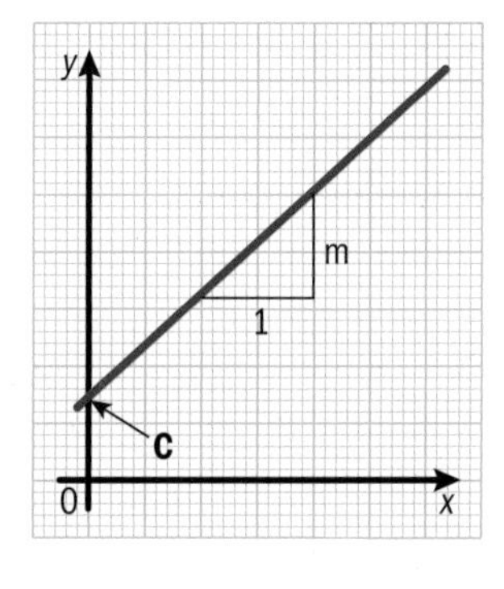

Figure 3.14: Straight line graph

A gradient can be negative or positive depending on the shape of the graph. See Figure 3.15 below.

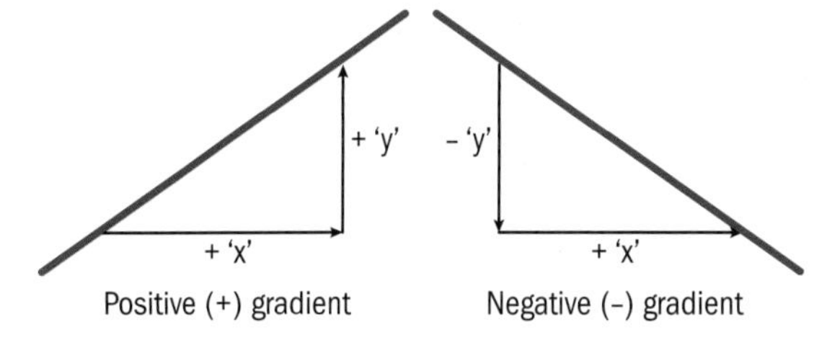

Figure 3.15: Gradients

Remember!

A line sloping *up* from left to right has positive gradient, while a line sloping *down* from left to right has negative gradient.

When choosing scales for graphs:

- Decide the range of *x* values to be plotted.
- Show the scale of the axes clearly.
- It is not always necessary to show the origin (0,0).
- Fit all points on the graph.
- Show data as accurately as possible.
- The scales on the *x*-axis and *y*-axis can be different.

Practical use of graphs

In practical work involving construction materials and their properties, tests are often undertaken to understand and prove basic concepts. If there is a linear relationship, this can be shown by plotting a 'best-fit' straight line between the experimental coordinates and then working out the equation of the straight line that is produced.

Worked example

An experiment to find the coefficient of friction between two types of metal cladding material produced the following results:

Weight W (N)	10	20	30	40	50	60
Friction Force F (N)	1.4	4.2	7.5	10.3	13.4	15.5

It is thought that both these are linked by a straight line equation F = m W + c. Show that this is the case and work out the equation of the experimental straight line.

Step 1: Plot the graph with F as the vertical axis and W as the horizontal axis.

Step 2: Draw the best fit line through the points.

Step 3: Work out the gradient by choosing two points on the line as far apart as possible, A and B. Construct a triangle as indicated and divide the vertical distance by the horizontal distance. Gradient is 0.28.

Step 4: Continue the line until it cuts the vertical 'y' axis and read off the value. This is the intercept c and in this example it scales off as –1.2.

Step 5: Collate all the data together and substitute into the standard form for an equation for a straight line.

F = mW + c

F = 0.28 W – 1.2

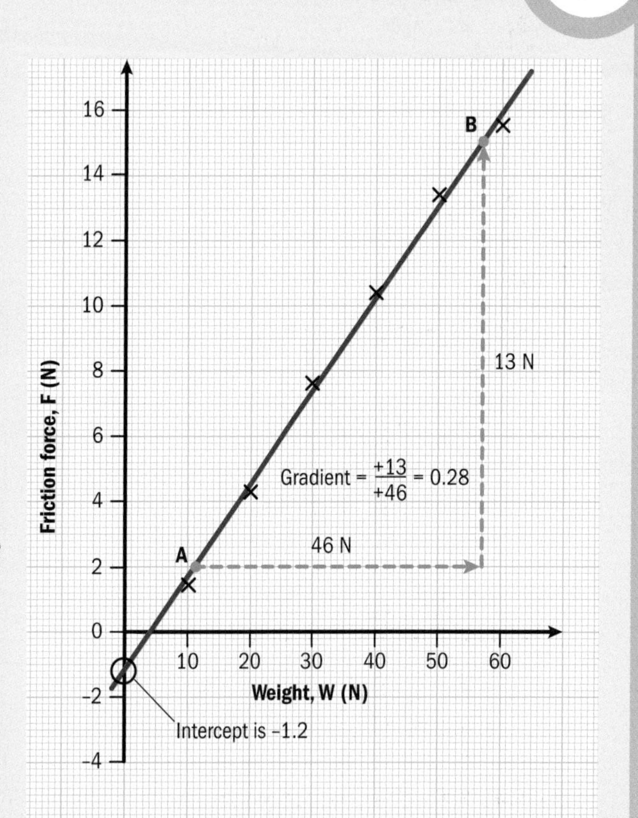

Solving simultaneous equations using straight line graphs

This method of using a straight line graph to solve simultaneous equations is often preferred by students as it is visual and uses graphical skills rather than numerical transposition.

Worked example

Find the values of x and y in the following simultaneous equations:

$x + y = 10$

$y - 2x = 1$

Step 1: Rearrange the equations into the form

$y = mx + c.$

$y = 10 - x$

$y = 1 + 2x$

Step 2: Plot the graphs by choosing suitable values for x. This may take some time by 'trial and error' to get both lines to cross (see figure).

Step 3: Find where the two lines cross, then read off the x- and y-coordinates of this point. This gives the solution of $x = 3$ and $y = 7$.

Step 4: As a check, substitute the figures back into the original equation to see if both equations balance, for example:

$x + y = 10$

$3 + 7 = 10$

Correct! And,

$y - 2x = 1$

$7 - 2(3) = 1$

Also, correct!

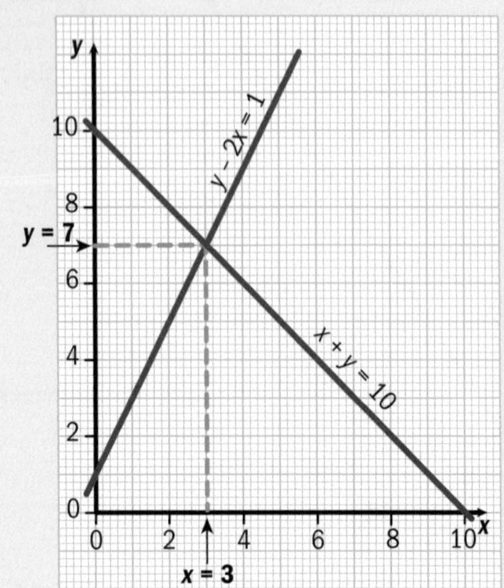

Solving quadratic and cubic equations using graphs

In a similar way, more complex quadratic and cubic equations can be solved graphically.

Worked example

Find the solution to the equation $x^2 - x - 2 = 0$.

Step 1: Set the equation to y as the subject: $y = x^2 - x - 2$

Step 2: Work out the y-coordinates for a range of selected x-coordinates by substituting them into the equation.

x	–3	–2	–1	0	1	2	3	4	
$+x^2$	9	4	1	0	1	4	9	16	ADD
$-x$	3	2	1	0	–1	–2	–3	–4	ADD
–2	–2	–2	–2	–2	–2	–2	–2	–2	ADD
y	10	4	0	–2	–2	0	4	10	

Step 3: Plot the graph and see where the line $y = 0$ cuts the curve $y = x^2 - x - 2$ to find the solutions to the equation.

The solutions are $x = -1$ and 2 when $y = 0$.

Step 4: Check your answers by substituting back into the original equation.

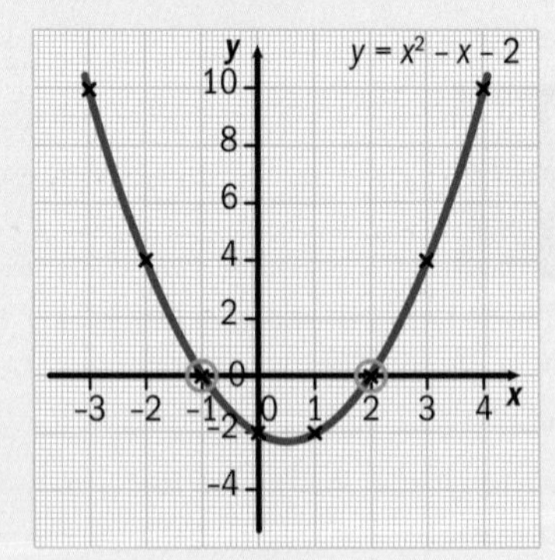

When $x = -1$

$x^2 - x - 2$

$= (-1)^2 - (1) - 2$

$= 1 + 1 - 2 = 0$

When $x = 2$

$x^2 - x - 2$

$= (2)^2 - (2) - 2$

$= 4 - 2 - 2 = 0$

Therefore, both values of x are correct.

Solving simultaneous linear and quadratic equations

This graphical method can be used for more complicated problems, such as trying to find what values of x and y satisfy simultaneous linear and quadratic equations, like the one solved previously by algebraic methods on page 76.

The graphical method involves plotting the two lines on the same graph and identifying where they cross, which will give you the solution.

Worked example

Find, by plotting a graph, the solution to the two following simultaneous equations:

$y = 3x + 1$

$y = x^2 + 2x - 1$

Step 1: Plot a graph of the quadratic $y = x^2 + 2x - 1$

x	–4	–3	–2	–1	0	1	2	3
x^2	16	9	4	1	0	1	4	9
+2x	–8	–6	–4	–1	0	2	4	6
–1	–1	–1	–1	–1	–1	–1	–1	–1
y	7	2	–1	–2	1	2	7	14

Also on the same graph plot the graph $y = 3x + 1$ to the same scale.

x	–4	–3	–2	–1	0	1	2	3
+3x	–12	–9	–6	–3	0	3	6	9
+1	+1	+1	+1	+1	+1	+1	+1	+1
y	–11	–8	–5	–2	1	4	7	10

Step 2: Identify where the two lines cross and read off the solutions for x and y at each of these points.

The graphical solution of this is shown opposite.

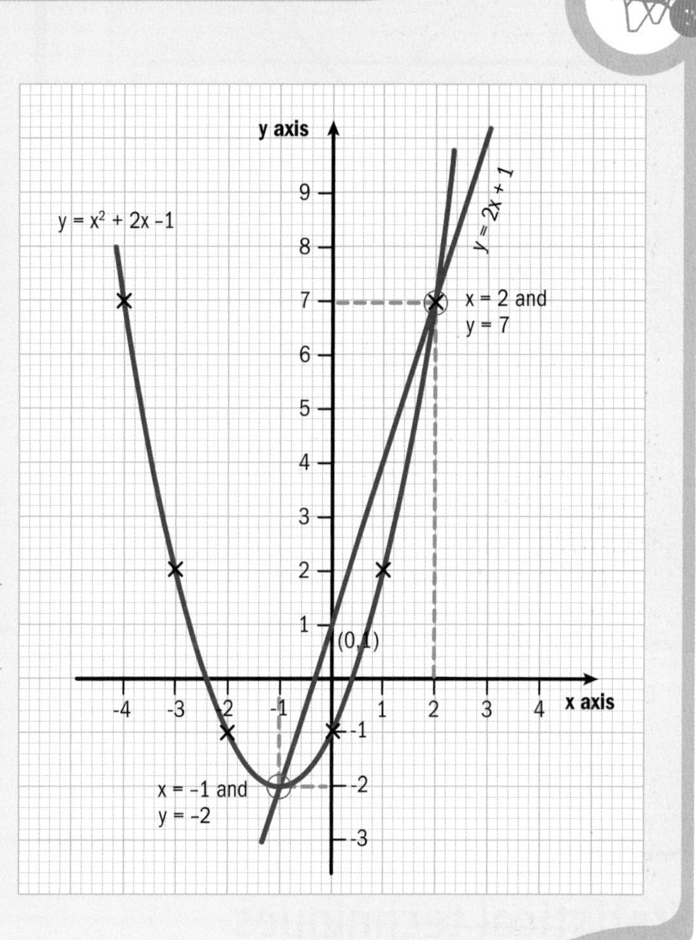

Activity: graphical techniques

1. Study the following graphs and determine the straight line equation of each line.

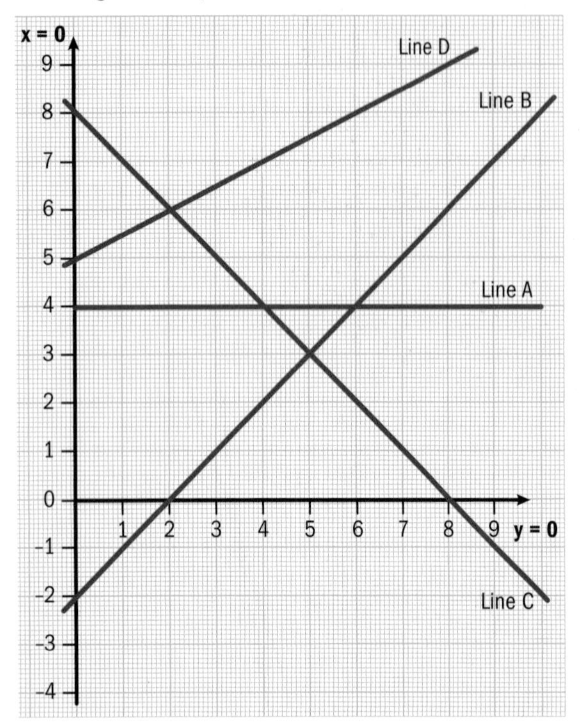

2. Solve the following equations using the graphical method:
   - Simultaneous equations $4x - 3y = 1$ and $x + 3y = 19$

     Plot values of x from 0 to 6 in increments of 1
   - Quadratic equation $2x^2 + x - 3 = 0$

     Plot values of x from –2 to + 2 in increments of 0.5

Statistical techniques

Statistics involve the collection, preparation, analysis, presentation and interpretation of data. The construction industry gathers and analyses data to enable it to make decisions on future plans, strategies and investments. The data may be collected from primary sources such as questionnaires or face-to-face interviews, or they may be taken from secondary sources such as published books or reports.

The National House-Building Council (NHBC), for example, collects information on new house prices, the size and structure of the house-building industry, house building by house type, and building times for new homes. Government departments with an interest in the built environment such as the Department for Communities and Local Government (DCLG) produce a wealth of housing statistics, for example monthly house prices, and annual land-use change and housing densities.

Presentation of statistical information

Statistical information may be displayed in tabular or visual form. Diagrams are used to show the general pattern of the data including the maximum or minimum values and the spread of data across the different categories. The most commonly used diagrams are pictograms, pie charts, bar charts and line graphs.

Pictograms (See Figure 3.16) use simple pictures to show the information. With this type of diagram the results are more visually engaging compared with other diagrams. Generally, pictograms are for straightforward 'whole or half number' statistics.

House buiding company	No. of house sales 2004–05
Brookway Homes	
Bestend Properties Ltd	
Countryfield Developers	
EFT Design & Build	
Surefire Property Investments	
Key 1000 New Start Homes	500 New Start Homes

Figure 3.16: Example of a pictogram

A pie chart is shaped like a circle and subdivided into a number of sectors each representing one category of the data. The size of the sectors are in accordance with the values they represent; thus the angles at the centre of a pie chart are determined as a fraction of a whole circle. There are several variations of a pie chart including the simple pie chart, exploded pie chart and the multi-rim pie chart (See Figure 3.17).

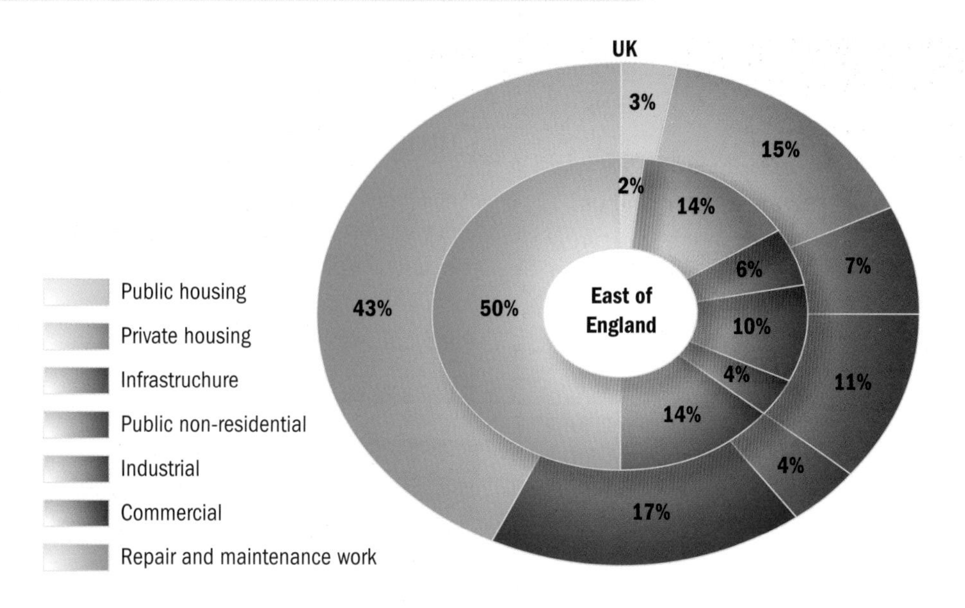

Figure 3.17: Example of a multi-rim pie chart

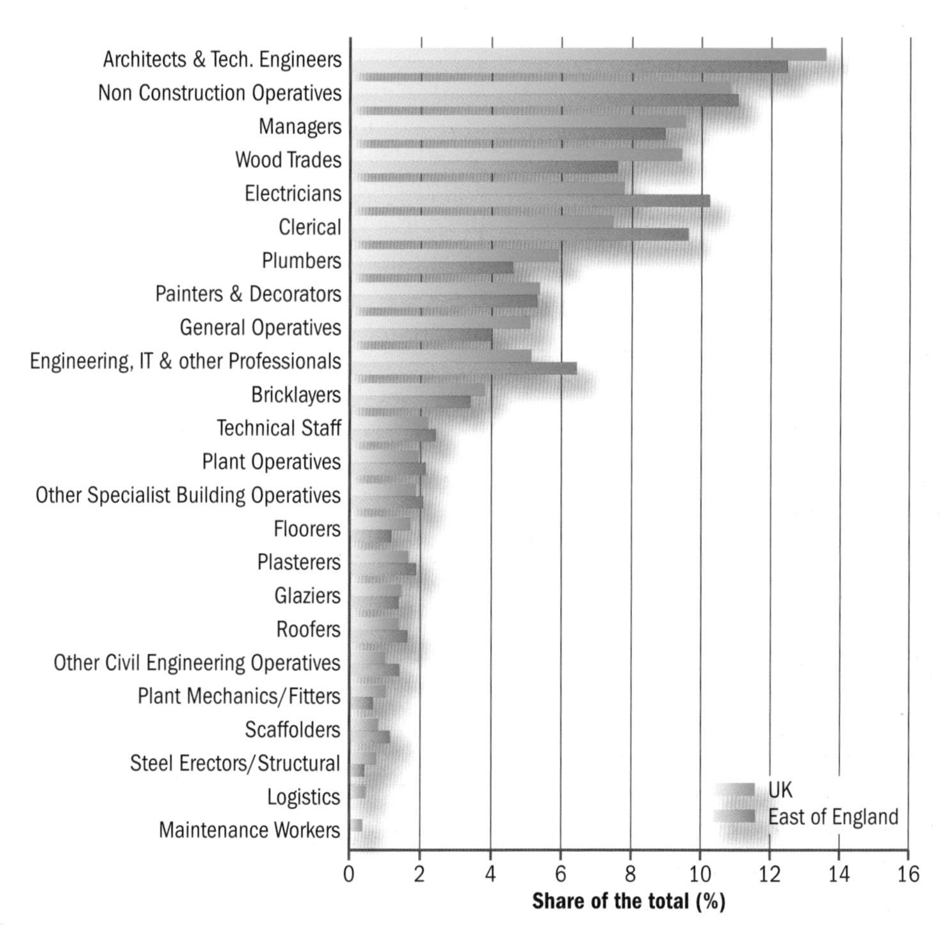

Figure 3.18: Example of a bar chart to show construction employment by occupation

Bar charts consist of data in the form of vertical or horizontal bars of equal width. These are very commonly used in construction. Their heights or lengths vary depending on the quantity they represent. Bar charts may be:

- single category bar charts
- multiple category bar charts
- sequential bar charts, such as Gantt charts that show progress against time.

Statistical definitions

Discrete data

With discrete data only whole number values are used for the data items; intermediate values do not exist. This is seen in examples such as the point scores in a test or the number of people who do a particular job.

Continuous data

With continuous data all values are possible for the data items. Continuous data is used in such instances as work study analysis and looking at the time taken to perform a given construction task.

Frequency

Frequency is the total number of times a data value occurs. In Table 3.7 the 'No. of choices made' is the frequency.

Mean

Mean is the total of the numerical values divided by the frequency. For example, if a company is currently running five building contracts for a total sum of £3.5 million, then the mean value of contract is £3.5 million divided by 5 giving £0.7 million.

Mode

Mode, also known as the modal average, is the most popular choice from a given set of options or values. In Table 3.7 the mode is 'American Shaker' because it is the most popular customer choice.

Table 3.7: Example of a summary table

Kitchen style	No. of choices made
Country Cottage	1
Urban Class	3
American Shaker	8
Regal Classic	2

Median

Median is the value found in the middle when all the data are arranged in ascending order. It is often quoted as the data value corresponding to 50% of the 'cumulative frequency' (see below). The spread of values about the median can be similarly expressed through the following terms:

- Lower quartile – the data value corresponding to 25% of the cumulative frequency.
- Upper quartile – the data value corresponding to 75% of the cumulative frequency.
- Inter-quartile range – the difference between the upper quartile value and the lower quartile value.

Cumulative frequency

A cumulative frequency distribution is used to estimate the median average and the spread of values about the median. Table 3.8 shows the test results of 30 applicants for a senior construction management post, with the third column showing the cumulative frequency.

Table 3.8: Data summarised in a frequency table

Test score mark	Frequency	Cumulative frequency
0–10	1	1
11–20	4	5
21–30	12	17
31–40	8	25
41–50	5	30
TOTAL	30	30

A cumulative frequency curve can be plotted from the information in the table, as seen in Figure 3.19. From the curve we see that the median score was 29; 50% or 15 applicants scored 29 or higher. We also can see that the lower quartile was 22 and the upper quartile was 38. Therefore, the spread of results about the median was 38 – 22 = 16 marks; 50% of applicants were within + or – 8 marks from the median value.

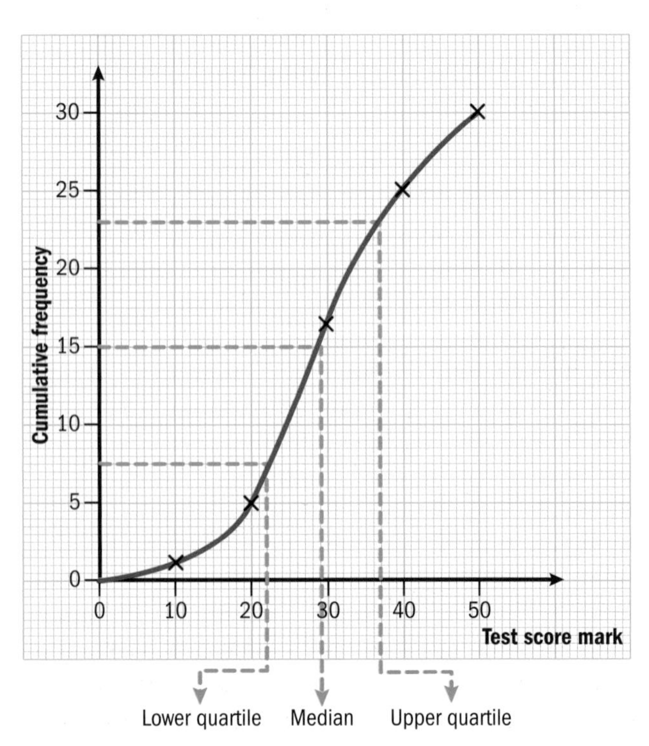

Figure 3.19: Cumulative frequency curve

Mean strength

Worked example

Fifty concrete cubes were tested for compressive strength (N/mm²) during a large concrete pour for a road bridge and abutments. The following results were obtained:

20	27	21	26	22	25	24	23	32	27
28	27	29	28	27	20	29	28	27	26
30	29	28	27	26	25	30	29	28	27
24	23	35	28	34	29	33	30	32	31
25	32	27	31	22	29	33	30	28	29

Calculate the mean strength based on the test results.

The amount of data is large, so to make it more manageable we separate it into six classes and 'tally up' the results that fall into these classes. We also need to determine for the 'class' group its mid point. The data is set out as follows:

Class interval	Class midpoint (m)	Frequency (f)	$m \times f$	
19–21	20	3	$20 \times 3 =$	60
22–24	23	6	$23 \times 6 =$	138
25–27	26	14	$26 \times 14 =$	364
28–30	29	18	$29 \times 18 =$	522
31–33	32	7	$32 \times 7 =$	224
34–36	35	2	$35 \times 2 =$	70
		$\Sigma f = 50$	$\Sigma(m \times f) = 1378$	

Mean strength =

$$= \frac{\text{total (frequency} \times \text{class midpoint)}}{\text{total frequency}} = \frac{\Sigma (m \times f)}{\Sigma (f)}$$

$$= \frac{1378 \text{ N/mm}^2}{50} = 27.6 \text{ N/mm}^2$$

$\therefore$ mean strength is 27.6 N/mm²

NB Σ means 'sum of'

Standard deviation

Standard deviation is a method used to see how all the data values distribute themselves about the mean. It is used to provide information on how tightly or loosely the values are clustered around the mean.

Let's look at an example to help illustrate standard deviation. If the debris of 100 crushed cubes was piled up in a line from the lowest to the highest value, the shape of the heap of crushed concrete might look something like the blue line in Figure 3.20. What is seen is the physical spread of results for the whole set of cubes which forms a 'bell-shaped' curve which is known as a 'normal' distribution of results.

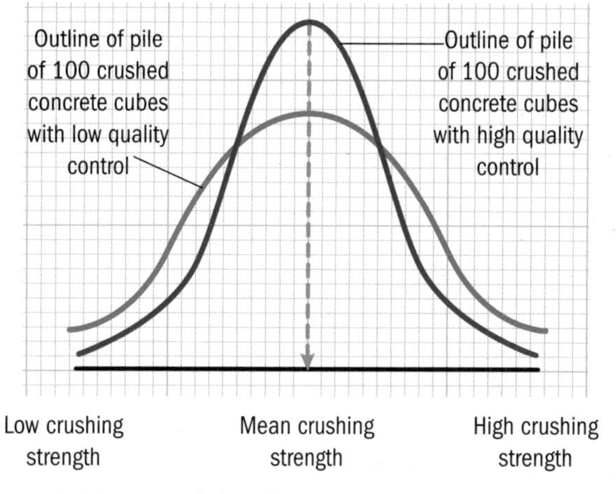

Figure 3.20: Normal distribution curve

In 'normal' distribution, one standard deviation is the measured distance from either side of the mean such that 68% of all the values tested lie between these two measured distances. This can be seen more clearly from the normal distribution diagram in Figure 3.21.

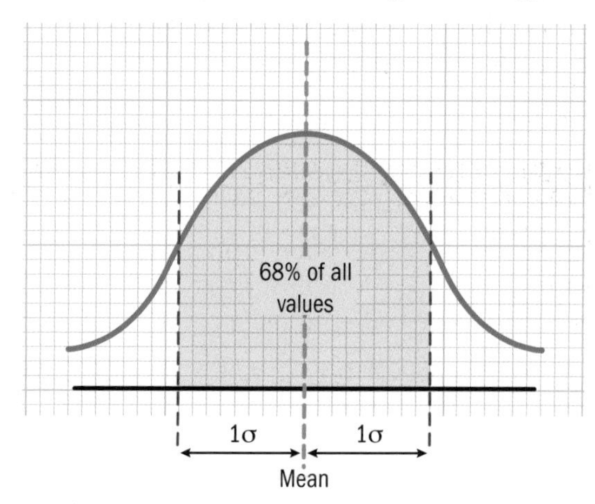

Figure 3.21: Area under normal curve

Remember!

Standard deviation is usually given the symbol, σ (sigma), and tells you how much your data values vary from the mean. It uses the same units as the original data and it is calculated by using all the data values in the set.

How to work out standard deviation

Standard deviation is worked out from the following formula, although it is best to work it out in tabular form.

$$\text{Standard deviation} = \sqrt{\frac{\Sigma f(x - x_m)^2}{\Sigma f}}$$

Worked example

Class strength interval [N/mm²]	Class strength mid-point [N/mm²] (x)	Frequency(f)	Deviation from mean $(x - x_m)$	Deviation squared $(x - x_m)^2$	Total deviation squared $f(x - x_m)^2$
19–21	20	3	–7.6	57.76	173.28
22–24	23	6	–4.6	21.16	126.96
25–27	26	14	–1.6	2.56	35.84
28–30	29	18	1.4	1.96	35.28
31–33	32	7	4.4	19.36	135.52
34–36	35	2	7.4	54.76	109.52
		Σ50			Σ616.40

Let's continue with the previous worked example, where 50 concrete blocks were tested for compressive strength (N/mm²) during a large concrete pour for a road bridge and abutments. We can now work out the standard deviation and see how the results are clustered around the mean value, which we worked out to be 27.6 N/mm².

Step 1: Tabulate the deviation from the mean $(x - x_m)$, which is the difference between the mean value and the mid-point class value ∴ 20 – 27.6 = –7.6.

Step 2: Tabulate the deviation from the mean squared value $(x - x_m)^2$. This ensures that all values become positive: $(-7.6)^2 = 57.76$.

Step 3: Multiply the frequency by the 'deviation from the mean squared' to give the total for the class: 3 × 57.76 = 173.28.

Step 4: Repeat for all other classes, then total up the final column to give a value of 616.40.

Step 5: Calculate the standard deviation where:

Standard deviation (SD)

$$= \sqrt{\frac{\Sigma f(x - x_m)^2}{\Sigma f}} = \sqrt{\frac{616.40}{50}} = 3.5 \text{ N/mm}^2$$

Therefore, we can conclude that the mean strength of the 50 concrete cubes is 27.6 N/mm² (from previous Worked example) and that 68% of the whole sample fall within + or –3.5 N/mm² of this mean value. This gives a clear indication about the range of results and allows factual comparisons between other sets of results. In this case, the quality of the concrete can be monitored throughout the whole project (see Figure 3.20).

Activity: Statistical data

1. In each of the following examples select the most appropriate method of presenting statistical data from the following studies:
   - The causes of reported accidents on construction sites in the past year in order to compare the main categories.
   - The labour, plant and materials used on six of a company's London construction sites to compare main areas of expenditure.
   - The experimental results of tests on an insulation material to compare the thickness of the material with the temperature difference across the internal and external faces.
2. The following data was recorded on tests of a new metal alloy to be used in a system of internal metal studwork.

Tensile strength N/mm^2	Frequency
101–105	5
106–110	9
111–115	19
116–120	25
121–125	18
126–130	4
	80

Using this data calculate:

- the median strength to 1 d.p.
- the inter-quartile range
- the mean strength to 2 d.p.
- the standard deviation.

Describe and compare the results in **b** and **d**.

Assessment activity 3.4

BTEC

1. Solve the following algebraic equations using graphical methods:
   - the simultaneous equations for values of x and y:

     $x + y = 5$

     $y = 2 + 3x$
   - the quadratic equation for x:

     $x^2 - 5x + 4 = 0$ P3
2. Your company fabricates and supplies composite steel cladding roof panels with rigid insulated cores. The manufacturing department has done some research and has found that the production costs of the panels vary with the thickness of the rigid insulation in the core as shown in the table:

Thickness, T (mm)	0	50	100	150	200	250
Cost per panel, C (£)	70	145	220	295	370	445

   - Plot a neat labelled straight line graph of C vertically against T horizontally to a suitable scale.
   - Using the graph estimate the thickness of insulation required if the production costs are to be no more than £250 per panel. P7
3. Go to the Health and Safety Executive website and carry out research into construction accidents in the UK by looking at their accident data:

   a Identify the most common causes of accidents on construction sites over the past year such as failing from heights, being hit by a vehicle, being hit by falling object, etc. Then produce a pie chart showing the spread of causes and from it state the modal cause.

   b Study the fatality statistics for the past ten years and plot a bar chart showing the annual rate against the time in years. Determine the mean number of fatalities over this period and state whether there is an upward or downward trend compared to this mean value. P8
4. Bestend Properties plc is carrying out a quality audit on two of its finished housing projects, 'The Lawns' and 'Meridian Waterfront'. The company produced a questionnaire for all new purchasers and asked them to rate their satisfaction with the quality of the build. The scores of the survey, are reproduced below for each of the developments; the higher the score, the more satisfied are the residents.

   By calculating the mean and standard deviation from the results to both surveys compare and contrast the quality of the build as perceived by the residents. M3

Satisfaction questionnaire responses from residents

Satisfaction score	Response from 'The 'Lawns'	Response from 'Meridian Waterfront'
1–20	6	4
21–40	10	13
41–60	22	22
61–80	8	11
81–100	2	1

Grading tips

For P3 you should use graphical methods to solve linear and quadratic equations. Make sure you clearly indicate where the lines cross on your graph and state the coordinate values of *x* and *y* for each point. For the quadratic solution clearly identify on your graph where the curve cuts the *x*-axis.

For P7 use graphical techniques to solve practical construction problems. You need to ensure that you draw the graph axes to a suitable scale to cover the whole page of graph paper, which will help you interpret the graph accurately.

For P8 use statistical techniques to solve practical construction problems. Your research data should be neatly presented with the correct titles and values noted on the charts.

For M3 set out your calculations in a clear table, and give each column heading the correct title. Clearly show how you have worked out the total 'frequency' and total 'deviation squared'. Remember that in this question there are no units as this survey records people's perceptions.

Charlie Steller

Estimator

Charlie is a construction estimator and has just finished working on a large refurbishment project where his company is tendering to convert a large Victorian warehouse into residential apartments. One of the biggest parts of this job was to calculate the area of floor, ceiling and wall finishes and then research into the labour and materials that would be needed to renovate and redecorate them.

Being able to use maths and having mathematical skills is a very important part of being an estimator. Most estimators work for a contractor or specialist subcontractor working alongside colleagues including construction managers, planners, buyers and quantity surveyors.

Charlie left school at 16 and got a job with a shop-fitting company as a trainee estimator. He went on to college and studied part-time for his BTEC National Certificate in Construction. He then went on to receive the Higher National Certificate in Construction.

Charlie not only has to be very methodical and good with figures but also has to have a good knowledge of construction methods and construction planning. He must ensure that he checks his figures when he is building up his quote; a simple arithmetic mistake could lose his company future business and affect their reputation.

He finds the cost of items from the buyer's catalogues and online guides. He also has to estimate labour and plant costs by building up rates of work calculated from past jobs that his company carried out. He sometimes also helps the construction planner by using maths to calculate the time it will take to do the work; this is very important if the company is to win the contract.

Think about it!

- What main skills does an estimator need to have?
- Would you enjoy this type of role?
- How much maths would you need for this job?

Just checking

1. Can you explain what a simultaneous equation is?
2. Explain Pythagoras' rule. When would you use it?
3. In the quadratic equation: $ax^2 + bx + c = 0$, what are 'a', 'b' and 'c' called?
4. Name one method for solving quadratic equations.
5. Explain what standard deviation is and how it can be used to monitor the quality of manufactured materials.
6. When would you use the binomial theorem?
7. What is the difference between discrete data and continuous data?
8. Explain the difference between significant figures and decimal places.
9. What are three techniques for carrying out numerical integration? Which is generally the most accurate?
10. Name three ways of presenting statistical information.

Assignment tips

- It is important that you know how to make good use of the calculator, particularly its various memories. Keep the instruction manual handy. In your assignment work state clearly how you used your calculator and its functions.
- When doing questions involving spatial problems such as volumes it is always a good idea to draw a neat sketch and label up all the dimensional information that you know. This will help you focus on the data that you know and the information you need to find.
- Set out all your solutions step by step and line by line. Make it easy for someone to follow your calculations explaining what you are doing. Remember to be consistent in your use of physical units.

Credit value: 10

4 Science and materials in construction and the built environment

Science and materials technology is becoming increasingly important as there is a growing focus on looking after the planet and its resources, and reducing the amount of pollution that we release into the environment. Developments in technology in areas such as the thermal properties of housing and commercial buildings, material recycling, sustainability and green technology help to achieve this. An understanding of science and materials can greatly educate us in the selection, use and properties of a material for incorporation into a design that is both aesthetic and efficient.

Materials are developing to keep pace with technological developments. The manufacturing processes involved in materials now have to encompass recycling, embedded energy and the effects of carbon emissions.

This unit looks at the materials used for construction and their engineering properties, which involves analysing the forces within the materials enabling a design to be safely constructed. The performance criteria for materials will be explored along with the manufacturing processes associated with some common construction materials. Material properties will be analysed and we will look into their modes of failure and study how that can be prevented.

Learning outcomes

After completing this unit you should:

1. know the basic factors that affect human comfort
2. understand how forces act on structures
3. know the performance criteria applicable to construction materials and the techniques used to produce such materials
4. understand construction materials and the techniques used to prevent their deterioration.

Assessment and grading criteria

This table shows you what you must do in order to achieve a pass, merit or distinction grade, and where you can find activities in this book to help you.

To achieve a **pass** grade the evidence must show that you are able to:	To achieve a **merit** grade the evidence must show that, in addition to the pass criteria, you are able to:	To achieve a **distinction** grade the evidence must show that, in addition to the pass and merit criteria, you are able to:
P1 describe the basic factors in simple scientific terms that influence human comfort in the internal environment **See assessment activity 4.1, page 130**	**M1** produce clearly worked, accurate answers for three different calculations relating to human comfort in the internal environment **See assessment activity 4.1, page 130**	**D1** analyse, in both qualitative and quantitative terms, the basic factors that affect human comfort **See assessment activity 4.1, page 130**
P2 describe how each factor is measured **See assessment activity 4.1, page 130**		
P3 state acceptable values for each factor **See assessment activity 4.1, page 130**		
P4 interpret underpinning concepts relating to structures under load **See assessment activity 4.2, page 138**	**M2** produce clearly worked, accurate answers for three different problems involving simple structures under load **See assessment activity 4.2, page 138, assessment activity 4.3, page 143**	
P5 predict simple structural behaviour from given data **See assessment activity 4.2, page 138**		
P6 identify the main performance criteria relating to the specification of a range of vocationally relevant construction materials **See assessment activity 4.3, page 143**	**M3** make and support valid decisions relating to the specification of materials for a tutor-provided application **See assessment activity 4.4, page 150**	**D2** evaluate preventative and remedial techniques applicable to the failure of materials **See assessment activity 4.4, page 150**
P7 describe the production and/or manufacturing processes for two vocationally relevant construction materials **See assessment activity 4.4, page 150**		
P8 describe the important features and properties of construction-related materials **See assessment activity 4.3, page 143**		
P9 explain how construction materials can deteriorate in use **See assessment activity 4.4, page 150**		
P10 explain the preventive and remedial techniques used to prevent deterioration of construction materials **See assessment activity 4.4, page 150**		

How you will be assessed

The evidence requirements for pass, merit and distinction grades are shown in the grading criteria grid. Evidence for this unit may be gathered from a variety of sources, including well-planned investigative assignments, practical work or reports of practical assignments. You will be given written assessments to complete for the assessment element of this unit. These will contain a number of assessment criteria from pass, merit and distinction.

This unit will be assessed by the use of three assignments:

- Assignment one will cover P1, P2, P3, M1 and D1
- Assignment two will cover P4, P5 and M2
- Assignment three will cover P6, P7, P8, P9, P10, M3 and D2.

Ken

This unit has made me examine the role of science in the construction of buildings. I discovered that science plays an important part of the design, from heat loss to the level of light entering a building.

The science we learned in this unit is also linked to the building regulations where heat loss, sound and daylight are all regulated and inspected in completed homes.

This unit has made me aware of many of the different materials and their properties that interact together to protect the building against the elements of the weather that deteriorate their structure.

Over to you

- What do human comforts mean to you?
- What do you already know about materials that are used for design?
- What are you looking forward to learning about in this unit?

1 Know the basic factors that affect human comfort

Build up

Science and structures

The application of science is vitally important when designing structures. Buildings need to stand up to all types of weather and be safe for the occupants that live, work and socialise within them.

- How does science achieve this?
- What factors need to be considered in designing structures'?
- How will you make sure that they are safe?

Thermal and air quality

There are several variables that are inevitably linked together that affect human comfort. Air quality is affected by how hot it is both inside and outside. Similarly, the amount of moisture that is present within the air will have an effect on humidity, which is linked to the amount of ventilation entering the environment.

Human bodies maintain an average core temperature of 37°C, depending upon the metabolic rate. If the surrounding temperature drops below this, a person feels cold; if it increases, a person feels hot and the body loses heat by the evaporation of sweat. In a climate such as the UK's, where average daily temperatures vary over the course of a year through the seasons, there is no constant temperature to design against. Houses now contain central heating to balance out these effects when the outside temperature is lower. Similarly, they are artificially cooled by air conditioning during periods of high temperatures.

Remember!

Air quality is affected by any pollution that is released into the atmosphere. Smoke and exhaust fumes cause smog over some of the world's largest cities, which has a significant effect on breathable air quality.

Nature of heat

To understand the design of buildings in relation to human comfort, we need to look at some specific elements of heat.

Practical temperature measurements are usually made in degrees Celsius (°C). The lower point 0°C is fixed at the melting point of ice at a standard atmospheric pressure of 101.32 kN/m^2. The upper point 100°C is fixed at the temperature of steam above boiling point at the standard atmospheric pressure. Normal design internal temperatures are taken on average as 21°C inside and –1°C outside, but will vary from continent to continent with the conditions required.

The thermodynamic temperature scale is the basic scale of temperature and is measured in degrees Kelvin (K). The unit of thermodynamic temperature is the fraction of the thermodynamic temperature at the **triple point of water**:

0°C = 273.16 K

100°C = 373.16 K

Key term

Triple point of water – the temperature and pressure at which the three known phases of a substance can exist, ice, liquid and vapour. The triple point of water is the equilibrium point 273.16 K at 610 N/m^2 which is one of the fixed points of international standard measurements of temperature.

The unit of Celsius, the degree Celsius, is by definition equal in magnitude to the degree Kelvin. A difference of temperature may be expressed in Kelvin or Celsius. The interval of 1°C is equal to the interval of 1K. In practice, 0°C is taken as 273 K.

Heat is one of a number of forms of energy whose units are measured in joules (J) which is a measure of the work done. The rates of expenditure of energy or doing work, or losses of heat, are measured in watts (W). A watt is equal to 1 joule per second:

1 joule/second = 1 watt (1 J/s = 1W)

Remember!

Using one bar of an electric fire is equal to 1000 watts of electrical energy used to heat the element.

Heat transfer

Heat energy is able to transfer from one mass to another in the following ways.

- Conduction – the passage of heat from molecule to molecule across a body, e.g. hot water in pipes.
- Convection – the bodily movement of a fluid, gas or liquid. Air expands on heating and rises, while the cooler air sinks resulting in the circulation of air e.g. heat emitter (radiator) heat rises up and is replaced by colder air below. *N.B. Only about 15% is radiated heat.*
- Radiation – the rays of heat travelling across a space, with or without matter being present, e.g. the heat from an infrared lamp or the sun's rays.

Thermal comfort in terms of activity

The more active you are in the environment, the more heat you will give off. The rate at which this heat is generated (metabolic rate) will depend on several factors, including:

- your surface area
- age
- gender
- level of activity.

Generally, the heat output from females is approximately 85 per cent of the male equivalent. The older you become, the less heat you give off.

Table 4.1 illustrates some heat outputs for various activities expressed in watts.

Table 4.1: Typical heat output of an adult male

Activity	Example	Heat output
Immobile	Sleeping	70 W
Seated	Watching TV	115 W
Light work	Office	140 W
Medium work	Factory work	265 W
Heavy work	Lifting	440 W

Source: Adapted from the CIBSE Guide (Chartered Institution of Building Services Engineers)

Clothing

The amount of clothing worn generally depends on the season and can affect thermal comfort. In summer, we wear fewer clothes – the heating is switched off in homes and the windows are often open for ventilation. In winter, we may wear some additional clothing indoors to feel warmer, particularly in older properties where there may be open chimneys, fireplaces and sash windows which do not seal effectively. The function of warm clothing is to trap a layer of warm air against your skin which heats up and makes you feel warmer, increasing your comfort.

The Chartered Institution of Building Services Engineers has categorised clothing in an attempt to produce acceptable parameters. This classification measures clothing in terms of a scale called a clo value. One clo represents 0.155 m^2 K/w of insulation to the body with typical values range from 1 to 4 clo. Table 4.2 illustrates how the amount of clothing worn affects the maintained temperature of a room.

Table 4.2: Clothing values

Clo value	Clothing	Typical comfort temperature when sitting
0 clo	Swimwear	29°C
0.5 clo	Light clothing	25°C
1 clo	Suit, jumper	22°C
2 clo	Coat, gloves, hat	14°C

Source: Adapted from the CIBSE Guide (Chartered Institution of Building Services Engineers)

Room temperatures

In order to understand how we maintain a steady temperature and a comfortable thermal indoor environment, we first need to look at some basic scientific principles of materials. These are the resistance of a material to the passage of heat and the thermal conductivity of the material in passing the heat along.

In order to maintain a comfortable room temperature, the building must be provided with as much heat as is lost through ventilation. The optimum temperature of a room will depend upon its use and amount of ventilation and will vary with the number of people present. The structural losses will depend upon the material used, the type of construction, the orientation of the building in relation to the sun and the degree of exposure to rain and wind.

The thermal conductivity is an important concept when looking at the temperature of a room. This is the amount of heat loss in one second through 1 m^2 of 1 metre thick material; with a one-degree temperature difference between the faces. The units are W/mK (watts per metre Kelvin).

This thermal image shows heat loss from a house. The areas shaded red are where most heat is escaping. Why do you think the roof in the middle of the picture is blue?

Table 4.3: K value of materials

Material	K value (W/mK)
Asphalt roofing (1700 kg/m^3)	0.50
Bitumen felt layers (1700 kg/m^3)	0.50
Brickwork, exposed (1700 kg/m^3)	0.84
Brickwork, internal (1700 kg/m^3)	0.84
Concrete, dense (2100 kg/m^3)	1.40
Concrete, lightweight (1200 kg/m^3)	0.38
Concrete block, medium weight (1400 kg/m^3)	0.51
Concrete block, lightweight (600 kg/m^3)	0.19
Fibre insulating board	0.050
Glass	1.022
Glass wool, mat or fibre	0.04
Mineral wool	0.039
Plaster, dense	0.50
Plaster, lightweight	0.16
Plasterboard	0.16
Polyurethane (foamed) board	0.025
Rendering, external	0.50
Screed (1200 kg/m^3)	0.41
Stone, sandstone	1.30
Timber, softwood	0.13
Timber, hardwood	0.15

Source: Adapted from the CIBSE Guide (Chartered Institution of Building Services Engineers)

Thermal resistivity (r) is sometimes more convenient to use than conductivity for thermal conductivity of materials. It is the reciprocal of thermal conductivity, that is:

$$r = \frac{1}{K}$$

The thermal resistivity can often be obtained from a manufacturer's data sheet on the material.

Air movement

Current building regulations dictate that air tests which test airtightness are undertaken on newly constructed domestic and commercial buildings. Airtightness is linked to the heat loss through ventilation of the property. Draught seals should be fitted to all openings to restrict thermal losses.

During the summer months in the UK, windows are open for ventilation. Resulting local air movement within a room can cause problems and discomfort to the occupants of the room. If the warm air entering a room is not mixed with the cooler air, the room becomes hot nearer the ceiling and colder at floor level. A draughty room may also cause discomfort; when cold air passes over the skin's surface, the skin's hair follicles are raised and shivering may occur as a reaction to the sudden cold environment.

Remember!

Many older properties have open fireplaces that act as natural ventilation and cause air movement within the building.

Relative humidity

Humidity is usually measured in terms of relative humidity rather than actual humidity. Relative humidity is the ratio of the current humidity expressed as a percentage. Air that is totally saturated has 100 per cent humidity – it can not hold any more moisture and so it may rain.

Humans are used to a relative humidity of between 40 and 60 per cent. Greater than this, and we start describing air as being 'close', 'humid', etc. The temperature also feels a degree or two warmer, because we cannot perspire readily into an atmosphere with a high relative humidity. High humidity can cause clothing to deteriorate and encourages the growth of moulds.

Heat loss due to ventilation

The natural ventilation of buildings, such as open windows, leads to the complete volume of air in a room changing a certain number of times in one hour. Typical air change values are given in Table 4.4.

Table 4.4: Air ventilation rates

Type of room	Air changes in 1 hour
Halls and passages	1.0
Bedrooms and living rooms	1.5
WCs and/or bathrooms	2.0

The fresh air entering the room will need to be heated to the internal temperature of the room/building. This is calculated with the formula:

Volume of room × Air change rate × Volumetric specific heat for air × Temperature difference

The volumetric specific heat for air is approximately 1300 j/m³K and is considered a constant in this formula which will give an answer in joules required per hour. This then has to be converted into watts in order to find the rate of heat loss, which is achieved by dividing the number of joules by the number of seconds in one hour, 3600 seconds (60 sec × 60 min):

$$\frac{\text{Volume of room/ building} \times \text{Air changes hour} \times 1300 \text{ Joules} \times \text{Temperature difference}}{3600 \text{ s}} = \text{Watts}$$

In heat loss calculations of buildings in most parts of the UK, the external air temperature during winter is assumed to be on average –1°C.

The design room temperatures in a building will depend on the purpose of the room. Bedrooms and passages can be much cooler than sitting rooms. It is convenient when carrying out heat loss calculations to assume an average internal temperature of 19°C – this will give a 20°C difference between the inside and the outside temperatures.

Condensation

Condensation is formed when hot, humid air meets a cold surface; it condenses onto this surface forming droplets of water vapour. Within the internal environment of a dwelling condensation can:

- cause timber rot
- encourage mould growth on walls and windowsills
- produce cold spots
- produce high humidity
- cause corrosion to steelwork
- dampen insulation, reducing its effectiveness.

If condensation occurs in a cavity wall within the solid construction of a brick, it is named interstitial condensation.

Acceptable parameters of heat loss

The acceptable parameters of heat loss, or **U-values**, is a complicated topic and you will need to refer to the Building Regulations Part L Conservation of Fuel and Power for guidance on the acceptable U-values that are required in today's modern sustainable designs.

Key term

U-value – a measurement of the rate of heat loss through a structure.

Ventilation is also clearly linked to the Building Regulations Part L that restricts the airtightness of modern structures. Forced ventilation has to be provided in the form of fans to bathrooms and cooking areas to reduce the amount of condensation produced.

Remember!

Standard units:

Celsius: degrees

Thermal conductivity (k): Watts/metre Kelvin (W/m K)

Thermal resistivity (r): Kelvin/Watt

U-value = W/m² K

Calculating thermal resistances, surface resistances, U-values and heat loss

Part L of the revised Building Regulations provides a detailed guide to the application of U-values to new and existing structures, which is a complicated process and is outside the scope of this unit. We shall look at how an elemental U-value is calculated using the definitions explained above.

In order to calculate the U-value of a construction element, for example a wall, we need to establish the thermal resistances of the materials it is built from and add to this the internal and outside resistances. The U-value is then calculated using the following formula:

$$\text{U-value} = \frac{1}{R_T}$$

where R_T is the total of all the resistances.

The U-value, or thermal transmittance, is the amount of heat lost, in one second, through one square metre of a structure, when there is a one degree temperature difference between the air inside and the air outside. The units are W/m² K.

We shall now look at how to calculate these resistances for the materials using their 'k' values and their 'r' value. The thermal resistance of a slab of building material can be calculated in two ways. In both instances, it is essential that the thickness of the material is known and that the thickness is stated in metres.

Worked example

Calculate the thermal resistance for 100 mm thickness of brickwork.

Assume the area of brickwork is 1 m².

Assume the temperature difference between the two sides of the wall is 1°C, thus 1 K.

Method 1: Using conductivity of material

$$R = \frac{t}{k}$$

Where R = Themal resisitance for that thickness of material

t = Thickness of the material in metres

k = Thermal conductivity of the material

$$R = \frac{0.100}{0.84} = 0.119 \text{ m}^2\text{K/W}$$

Method 2: Using resistivity of the material

$$R = t \times r$$

Where R = Themal resisitance for that thickness of material

t = Thickness of the material in metres

r = Thermal resistivity of that material

$$R = t \times r$$

$$R = 0.100 \times 1.19 = 0.119 \text{ m}^2 \text{ K/W}$$

Remember!

The units for the thickness in metres need to be specified. For example, if a wall is 100 mm wide, then this is 0.100 m in this formula.

The total resistance, R_T, to the passage of heat across a wall is built up from the resistance of the material, r, and at least two surface resistances which have to be taken into account in any U-value calculation. They are:

- the inside surface resistance r_{si}
- the outside surface resistance r_{so}

If there is a cavity present, the resistance of the cavity will also have to be included r_{cav}

Surface and cavity resistances are shown in Table 4.5.

Table 4.5: Surface and cavity resistances

Typical internal surface resistance	0.13 m²K/W
Typical external surface resistance	0.05 m²K/W
Typical cavity resistance 5 mm – 19 mm cavity	0.11 m²K/W
Typical cavity resistance 20 mm + cavity	0.13 m²K/W

The total resistance to the passage of heat is therefore calculated from:

$$R_T = r_{si} + \frac{t}{k} + r_{so}$$

Where R_T = Total resistance to the passage of heat

r_{si} = Internal resistance from the table

t = Thickness of the material in metres

k = Thermal conductivity of the material (W/mK)

r_{so} = External surface resistance from the table

Remember!

If you already have the material resistances for its material thickness from the manufacturer, then you can replace t/k with this R figure provided.

Worked example

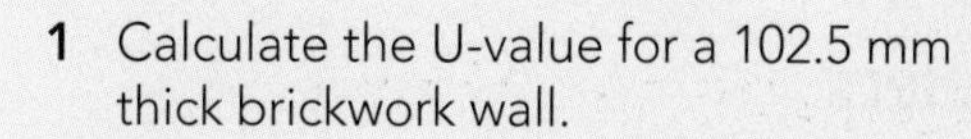

1 Calculate the U-value for a 102.5 mm thick brickwork wall.

brickwork $r_{si} + \frac{t}{k} + r_{so}$

$$0.13 + \frac{0.1025}{0.84} + 0.05 = 0.30(R_T)$$

u-value $\frac{1}{0.30} = 3.33\ \text{W/m}^2\ \text{K}$

2 Calculate the U-value for the following cavity wall construction:

15 mm dense plaster finish to inside walls

100 mm concrete blockwork medium weight

85 mm wide cavity

102 mm wide facing brickwork

$$0.13 + \frac{0.015}{0.50} + \frac{0.100}{0.51} + \frac{0.085}{0.13} + \frac{0.1025}{0.84} + 0.05$$

$= 0.13 + 0.03 + 0.196 + 0.654 + 0.122 + 0.05$

$= 1.182$

$$\text{U-value} = \frac{1}{1.182}$$

$= 0.846$

Remember!

You will need to use the table of K values – see Table 4.3 on page 122.

Theory into practice

Calculate the rate of heat loss due to ventilation for the building shown. The number of air changes in one hour is 1.35. The bungalow measures 4.5m × 3.25m in plan and has a ceiling height of 2.6 m.

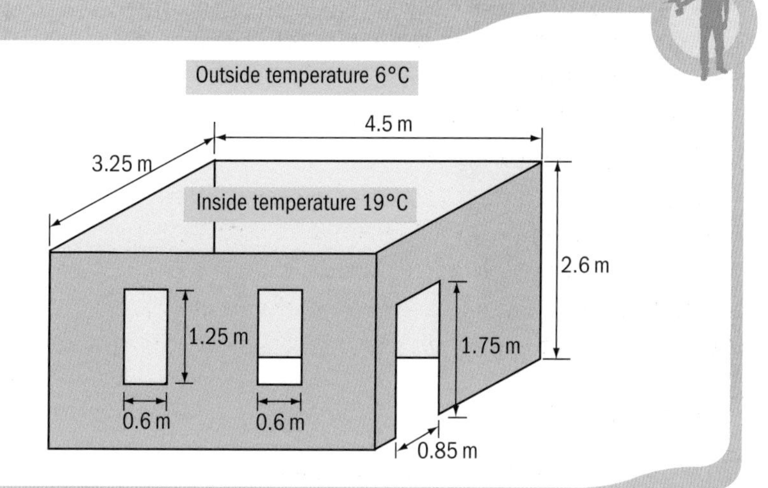

Methods used to measure thermal factors

Thermometers

A thermometer is essentially either a traditional glass type or a modern digital electronic type. The glass type consists of a tube filled with a thermally sensitive material which has a calibrated scale next to the tube. When the temperature rises, so does the level of the liquid in the glass tube which then can be read off a scale.

A digital thermometer works by using a resistor which is sensitive to electricity; the warmer it is, the less electrical current passes through it resulting in a digital temperature read out.

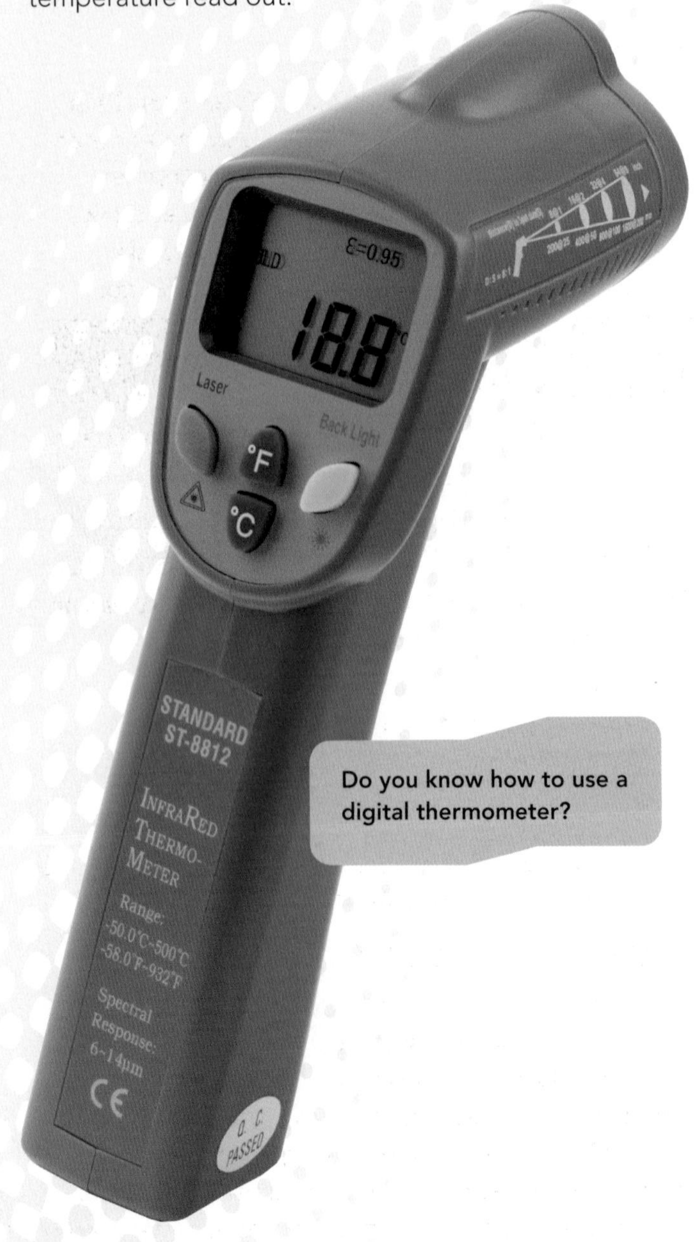

Do you know how to use a digital thermometer?

Hygrometer

Hygrometers are used to measure relative humidity. They use capacitors or resistors that react to the amount of moisture in the air which produces an electrical current which is then read off as a percentage.

Anemometer

Anemometers are used to measure air movement. A fan is attached to an electrical motor. If wind or air movement pushes the fan around, then a current is generated which produces a readout of wind speed.

Sound

Nature of sound

Sound is an aural sensation caused by pressure variations in the air, which are always produced by some form of vibration, for example the movement of air as a wind. In essence, sound is a form of energy, which is transmitted through the air as a pressure wave (see Figure 4.1); these fluctuations may take place slowly, such as with atmospheric pressure change, or very rapidly as in ultrasonic frequencies.

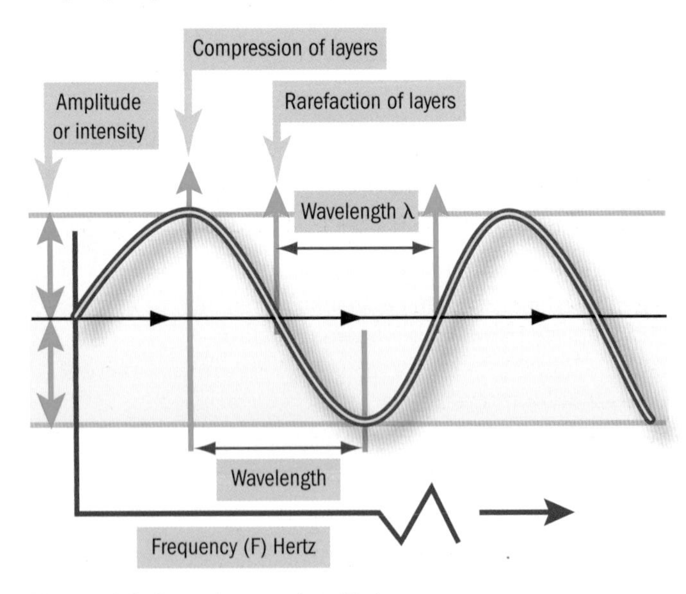

Figure 4.1: Sound waves in still air

Sound waves are like a ripple on a flat body of water. They radiate out from the source and slowly dissipate until the flat surface resumes again. This ripple produces a series of compressions (compactions) and rarefactions where the air pushes from molecule to molecule transferring the energy pulse, as shown in Figure 4.2.

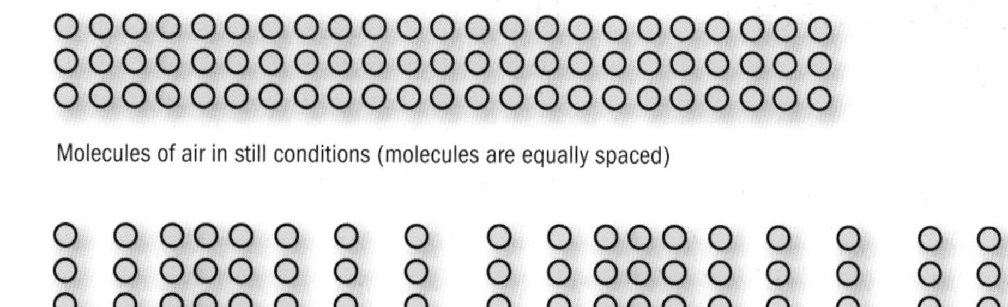

Figure 4.2: Still air and the effect of sound waves in air

Nature of hearing

The ear is divided into two parts: the outer and the inner ear. The outer ear contains the cartilage that forms the external ears and a tube that directs sound into the inner ear. The eardrum is at the bottom of this tube which vibrates when sound reaches it. Connected to the eardrum is a series of bones called the ossicles. A mechanism passes the sound as vibration via these bones into the inner ear. Here a liquid vibrates which contains tiny hairs which move with the sound vibration. When they move this is transmitted to the brain in the form of an electrical signal which is interpreted as sound. Having two ears gives humans the ability to detect from where sound is coming in stereo.

Measurement of noise

In selecting a scale to measure the intensity of sound it is necessary that the method chosen reflects the way in which the human ear responds to sound. Two important aspects need to be considered.

- The minimum sound that the ear can actually respond to, which is expressed as a value of 10 to the power of –12 watts/m^2 @ 1000 Hz. The audible range for those with good hearing is 20–20,000 Hz.
- The Weber/Fechner law suggests that the response of the ear to sound intensity is logarithmic, the bel (B) is the unit used to measure the energy of sound, and the start point of the bel scale is 10–2 W/m^2.

A 1 bel increase in energy gives an intensity of 10 × 10 and so on.

Sound levels

Sound is commonly measured in decibels (dB); 10 dB is equal to 0.1 bel (B). One bel is a measurement that is too large to use with modern instruments so we use decibels instead. The Control of Noise at Work Regulations 2005 contains a set of decibel levels where hearing protection must and should be worn.

Lower exposure action values	Upper exposure action values
Daily or weekly exposure of 80 dB	Daily or weekly exposure of 85 dB
Peak sound pressure of 135 dB	Peak sound pressure of 137 dB

Source: HSE, 'Noise at Work: Guidance for Employers on the Control of Noise at Work Regulations 2005'.

Some typical noise levels:

- a quiet office: 40–50 dB
- power drill: 90–100 dB
- road drill: 100–110 dB

Source: 'Noise at Work: Guidance for Employers on the Control of Noise at Work Regulations 2005'.

Noise control and sound insulation

For airborne sound, the best insulation is mass – thick, dense walls do not easily vibrate. In theory, if the mass of a wall is doubled, it should give a sound reduction of 6 decibels; in practice, the actual reduction is about 5 decibels. Lightweight partitions give little resistance to sound transmissions unless built in multiple layers. Glass is a very poor sound insulator – it has little mass and is thin.

Sound deadening quilts can also be used within the construction (e.g. fibreglass quilt in cavity walling) and soft finishes can be added to the interior design. Airborne sound can be prevented from travelling to other parts of the building by completeness of construction, and any air passages around windows and doors should be sealed. A soft covering can be effective when used on the top of floors; the soft covering absorbs the sound before it gets onto the structure. In order to prevent the noise of machinery travelling through a structure, we use discontinuity within the construction in the form of anti-vibration pads.

Noise transfer

Airborne sound, produced in air, for example a voice or musical instrument, starts at the source and then travels through the air to the ear of the listener. Airborne sound may travel in four ways:

- through any openings in the structure, e.g. small holes formed by removal of services
- by direct transfer through the structure
- by a structure vibrating like a drum skin and transferring the noise
- by direct transfer along the structure, e.g. air conditioning duct – this is called flanking transmission.

Structure-borne sound originates by impact on the structure or on pipes within the structure. It can even be footsteps on the floor above. The sound travels through the structure to be heard in other parts of the building.

Remember!

Standard units:

decibels (dB)

Hertz frequency (Hz)

The velocity of sound in air is more or less constant, varying between 330 m/s and 340 m/s. In more dense materials sound travels faster, not because of greater density but because of the greater elasticity.

The wavelength multiplied by the frequency is equal to the distance travelled by the sound in one second (s). Thus:

Velocity = Frequency × Wavelength

$$V = f\lambda$$

where V = velocity

f = frequency (Hertz)

λ = wavelength (metres)

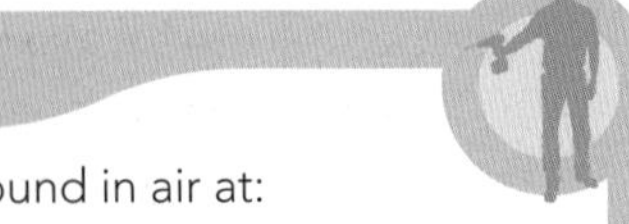

Theory into practice

Calculate the wavelength of sound in air at:

- 20 Hz
- 50 Hz.

The velocity of sound in air is 340 m/s.

Often, especially in concert halls, we need to establish how long it takes for a sound to die away as we may need to increase the amount of absorption on the walls, floors and ceiling of the structure. The following formula tries to establish the time it takes for a sound to die by 60 db:

$$t = \frac{0.16V}{A}$$

where t = reverberation time in seconds

V = volume of room

A = the existing area of absorption in m^2

Worked example

Calculate the reverberation time for a hall which has a volume of $1500m^3$ and the rear wall of $150m^2$ is covered with sound absorbing curtains

$$t = \frac{0.16 \times 1500}{150}$$

t = 1.6 seconds

Methods used to measure sound levels

The sound level meter works by using a microphone which analyses the level of the sound using electronics and converts it to a digital scale. Sound level meters differ in types from instantaneous level readers to time data logging meters which record sound levels over a set period.

Illumination

Nature of vision

Vision is a sensation caused in the brain when light reaches the eye. The eye initially treats light in an optical manner, producing a physical image which is then interpreted by the brain in a manner that is also psychological, as many optical illusion games show.

The eye contains a convex lens which produces an upside down image on the retina at the back of the eye. When relaxed the lens is focused on distant objects; to focus on closer objects, the eye muscles contract to increase the curvature of the lens. The amount of light entering the lens is controlled by the iris, a coloured ring of tissue, which expands and contracts with the amount of light present. The colour of the iris varies between individuals.

The image on the retina causes chemical changes in the receptors which then send electrical signals to the brain via the optic nerve. A large proportion of the brain is dedicated to processing this information. The initial information interpreted by the brain includes the brightness and colour of the image. The stereoscopic effect of the two eyes gives further information about the size and position of objects and enables us to judge distances. The eye is a remarkable tool that can detect a single torch beam over a mile away.

Simple colour rendering

Our eyes contain three colour receptors which have the ability to detect a rainbow of colours. The three colour receptors, red, blue and green, create the colours that the brain interprets. Visible light, or white light, from the sun can be split into the primary colours using a glass spectrum; these are the seven colours: violet, indigo, blue, green, yellow, orange and red.

Ultraviolet (UV) light has wavelengths just less than those of violet light. It is non-visible to the naked eye and is emitted by the sun and other objects at high temperature. UV radiation keeps the body healthy but overexposure can damage skin and eyes. The Earth's atmosphere protects the planet from overexposure to UV radiation. Infrared radiation has wavelengths slightly above those of red light. It can be felt as heat from the sun and other heated bodies.

Need for daylight

Our body clock needs to see daylight in order to set our patterns of sleeping and waking. People on night shifts have to reverse this trend with their work patterns. Natural daylight does influence certain chemicals within our brains that resist depression and anxiety; in other words, daylight can make a person feel good. When there is lack of daylight, some people can suffer from a syndrome known as seasonal affective disorder (SAD). This occurs during the winter months when the sun is low in the sky and the levels of light are often obscured by clouds. Light boxes which mimic the sun's natural wavelengths and frequencies are used to produce artificial light to stimulate the brain's chemical balance. Of course, the obvious need for daylight is to see! We can see in the dark, but our eyes are more attuned to daylight.

Measurement of lighting

More commonly we buy our light bulbs in watts, which is a unit of electrical power and will help us determine the amount of electrical energy it will use. However, the light amount is measured using the Lux (lx), which is the unit of luminance. The Lux measures the intensity of light and is used in modern light measurement meters. Table 4.6 illustrates some lighting levels in Lux.

Table 4.6: Lighting levels measured in Lux

Light source	Lux
Moonlight	Less than 1
Brightly lit drawing office	500
Sunlight	32,000–100,000

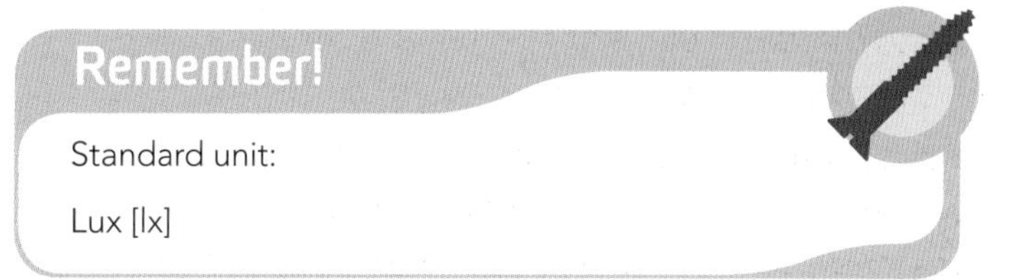

Acceptable parameters for natural and artificial light

The British day measured from summer to winter provides an average of 5000 Lux. To make sure enough daylight is entering our homes year round, we need to check that a proportion of this 5000 Lux is entering our homes through windows and openings. A figure of 1–2 per cent is considered a reasonable amount.

The CIE (*Commission Internationale d'Éclairage*) has produced the following formula in order to take into account the internal surfaces of the room that reflect light:

$$\text{Daylight factor at any point in room} = \frac{\text{Internal illuminance at that point}}{\text{External illuminance from unobstructed sky}} \times 100\%$$

Worked example

Calculate the illumination in Lux at a desk in an office where there is a requirement for a 6 per cent daylight factor. The external bright sky has been measured at 7500 Lux.

$$DF = \frac{\text{Int. illuminance}}{\text{Ext. illuminance}} \times 100\%$$

$$\text{Int. Illuminance} = 7500 \text{ lux} \times 6\%$$

$$= 450 \text{ Lux}$$

The eye can detect a wide range of light levels, but vision is affected by the range of brightness visible at any one time. Highly intense light can glare. Glare is the discomfort or impairment of vision caused by an excessive range of brightness in the visual field. Glare can be caused by lamps, windows and painted surfaces being too bright compared with the general background.

- Disability glare lessens the ability to see detail. It does not necessarily cause visual discomfort.
- Discomfort glare causes visual discomfort without necessarily lessening the ability to see detail.

Methods used to measure light

The light meter uses an electrical current generated by photosensitive electronics that reacts to the amount of light hitting its surface; this measurement is converted into a digital reading.

Assessment activity 4.1

BTEC

You are a trainee building services engineer The architect of a new house range has asked you to assist with some design work associated with simple scientific factors to provide an acceptable internal human comfort level.

Describe the basic factors that influence human comfort inside the homes in simple scientific terms. **P1**

Describe how each of these factors can be measured. **P2**

State what would be the acceptable values for each of these factors. **P3**

The architect is not convinced of your findings and has asked for proof. Produce clear and accurate answers to the following three calculations relating to human comfort in the internal environment.

1. Calculate the U-value for the following construction:
   15 mm plasterboard and skim
   100 mm lightweight block
   100 mm cavity filled with mineral wool
   100 mm brickwork
2. Calculate the illumination in Lux at a desk in an office where there is a requirement for a 4 per cent daylight factor and the external bright sky has been measured at 5000 Lux.
3. Calculate the reverberation time for a village hall which has a volume of 750 m^3 and a rear wall of 80 m^2 which is covered with sound-absorbing tiles. **M1**

Look closely at the factors that affect your human comfort (not just the three which you have explained before). Undertake an analysis of these and discuss how they interact both in quality and quantity. **D1**

Grading tips

For **P1** you will need to describe a range of basic factors that would influence your comfort if you were sitting in a room at home; make sure you use some simple scientific terms.

For **P2** you will describe how each factor is measured.

For **P3** you will state the acceptable values of each factor; make sure you use the correct unit of measurements.

For **M1** you are required to produce accurate calculations to the thermal, lighting and sound factors that have been illustrated.

For **D1** you will look at the human comfort factors and do an analysis of the quantity and quality of each. For example, if you increase the amount or quantity of heat in a room what effect does this have on the quality and what other factors are concerned, e.g. humidity?

PLTS

Planning and carrying out thermal comfort research, you will help develop your skill as an **independent enquirer**.

Functional skills

By reading and summarising information and ideas from different sources, you can improve your **English** skills in reading.

2 Understanding how forces act on structures

Structural members

Struts and ties

All structures have forces acting upon them. Newton's laws state that there must be an equal and opposite reaction to every action and so the strut is the compressive force that pushes against the weight of the structure. Figure 4.3 illustrates the strut working in position on a canopy roof. The weight of the roof is pushing down on the strut which has to resist this and pass it on to the wall. Ties carry tensile forces and are stretched by a pulling action. Figure 4.4 illustrates a TV mast which has wire ties to anchor it against the pulling force of the wind.

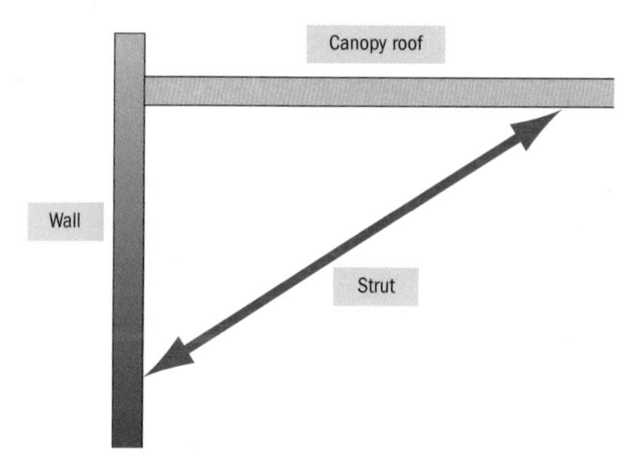

Figure 4.3: A strut

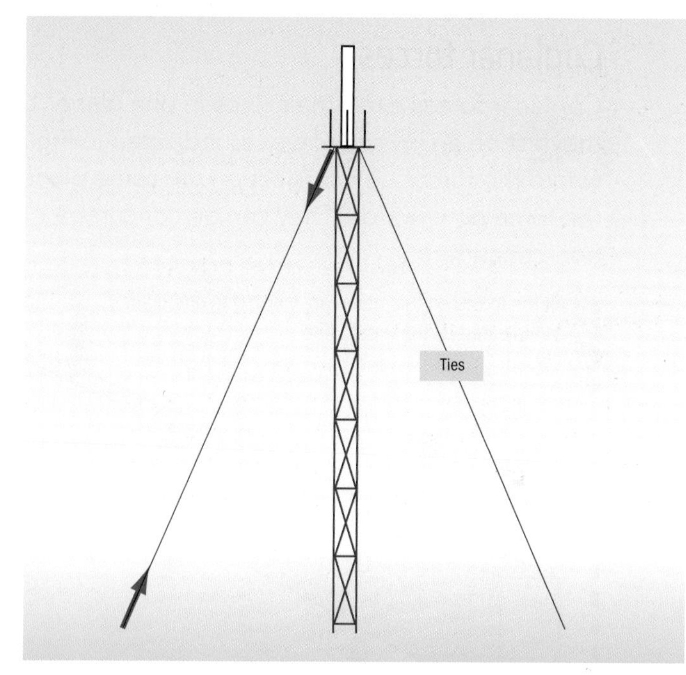

Figure 4.4: Ties

Beams and columns

Beams span between supporting walls. In spanning this distance, they have two forces exerting upon them – tension and compression – which are dealt with later in this unit. When supported at its ends, a beam tends to sag in the middle (see Figure 4.5). Beams are normally deeper in cross-section than columns, and can be lightened by turning into girders, which increases the depth and takes out some voids within the webs. These are often known as vierendeel beams.

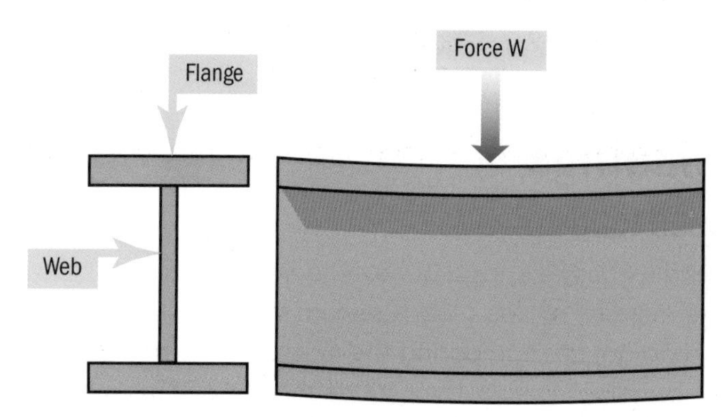

Figure 4.5: A beam

Columns are, in essence, vertical beams as they have the same 'I' section. However, they tend to be squarer in cross-section than beams. Columns mainly carry a vertical load downwards to the supporting foundation and are connected to and support beams either by bolted or welded connections. The beams hold the column into position at each storey. Columns size is very important structurally; if they are too long and thin, they may distort and bend under loading conditions. Columns can also form the vertical part of the portal frame as we shall explore under frames.

Walls and frames

Walls can be classified as either load bearing or non-load bearing infill panels.

Load bearing walls are deemed structural elements as they carry a structural force safely down to the foundations. These types of wall tend to be solid in construction and are made up of denser materials; they are often manufactured from solid brickwork or reinforced concrete. Non-load bearing walls, as the name suggests, do not have any structural importance. They are normally decorative lightweight infill panels, often constructed out of timber or lightweight blockwork.

A frame is a structure formed from many of the elements that we have discussed above, columns, beams, struts and ties. These elements can be connected together in different arrangements to form various framed structures as follows:

- an arrangement of columns and beams to form grid skeleton structures
- an arrangement of struts, ties and beams to form truss frame structures
- an arrangement of columns and beams to form portal frame structures.

Frames can be complex structures to design and analyse. Software programs make the process easier but require interpretation of the results. Thus, it is often best to leave this to a structural engineer.

Loadings

Dead loads

Dead loads are ones that have no live application. The majority of dead loads remain static or stationary and do not change during the normal operation of the building. That is, they are neither added to nor removed during the life of a building. The dead weight of the structure is a dead load; for example the self-weight of a beam.

Imposed loads

Imposed loads, also known as dynamic loads, are live loadings which are added to or removed during the life of the building. People are a good example of this; for example, if you fill a theatre with people, you increase the live loading on the floors. Water storage tanks also vary with the weight of water within them; similarly, full filing cabinets exert a point load on the floor structure. Live loads are difficult to determine in structural design as factors of safety of their force levels have to be considered.

Wind loads

Wind loads can be classified as a live load as can snc and rain as they too exert a live load on a structure. Wind loads within the UK vary with the location and exposure of the structure. The leeside of a hill will have less wind loading than the prevailing side. The western UK tends to receive prevailing wind loads. With structural design, buildings have to be anchore against the uplift created by the force of the wind and made stiff enough to resist distortion from wind pressure against one face.

Forces

Sir Isaac Newton stated force to be: 'The product of the mass of the body and the change of rate of velocity caused by the application of the force.' As Newton discovered, all objects on the Earth are helc down with the force of gravity. Therefore, force can be defined as: 'That measurable and determinable influence tending to cause motion of a body' (*OED*)

At sea level, gravitational acceleration can be measured at 9.81 metres per second (m/s^2). So all objects are weighted down with this force, which ha a measurable size of 9.81 times its weight. Therefore the units of force can be expressed as:

Weight of an object in kilograms × 9.81 m/s^2
= 1 Newton (N)

Coplanar forces

Coplanar forces exert their force in one plane. That they act on an invisible layer as indicated in Figure 4 where the forces upwards act on the same plane as the force downwards. They can be concurrent or non-concurrent.

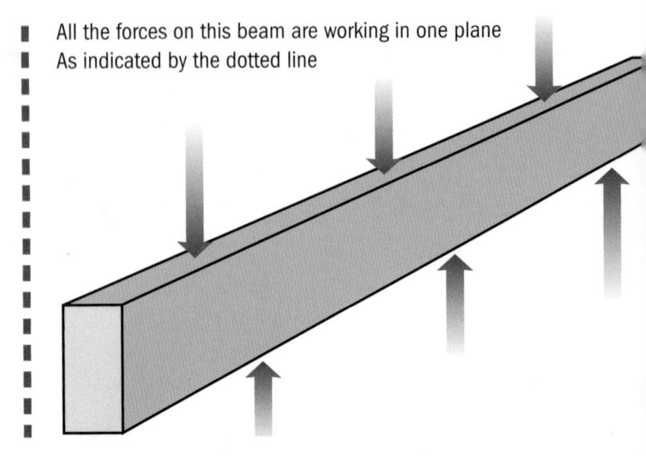

Figure 4.6: Coplanar forces

Concurrent and non-concurrent

Concurrent is a collection of two or more forces that join at a common point of intersection. They can either pull away from a common point or point towards it, as Figure 4.7 illustrates. Each arrow represents a force.

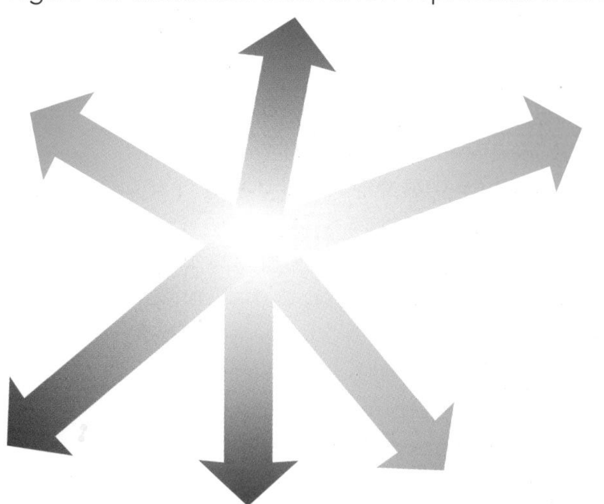

a) All the forces pull away from a common point

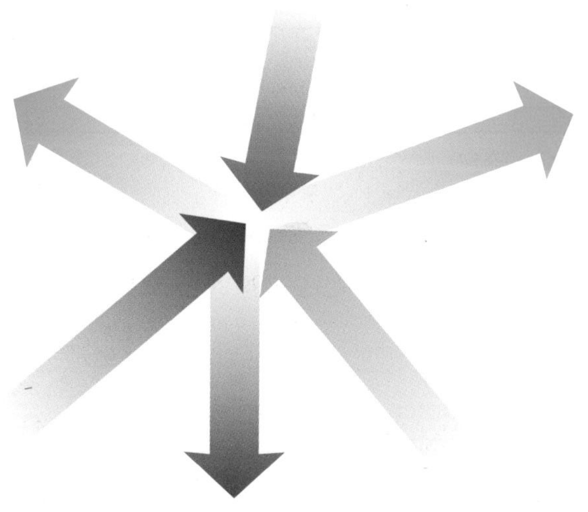

b) Forces here pull and push at a common point

Figure 4.7: Concurrent forces

Non-concurrent is simply a mixture of forces that do not intersect at a common point, as shown in Figure 4.8. They consist of a number of forces or vectors whose magnitude would involve great effort to try to calculate in structural design.

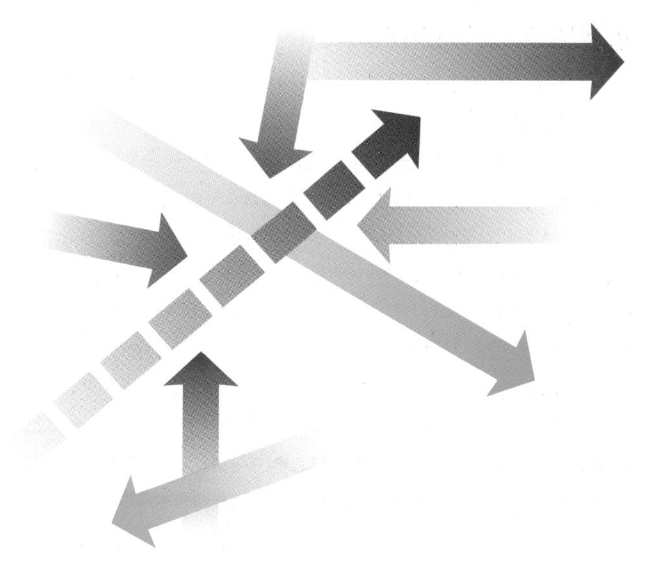

Figure 4.8: Non-concurrent forces

Load configurations

Point loads

Point load acts at a particular point. Figure 4.9 illustrates a simple beam upon which point loads act on the top of the beam. These could be heavy machinery, water tanks or aerial masts on a roof. Equal to these point loads are forces of the beam that must push upwards with equal strength to place the beam into **equilibrium**.

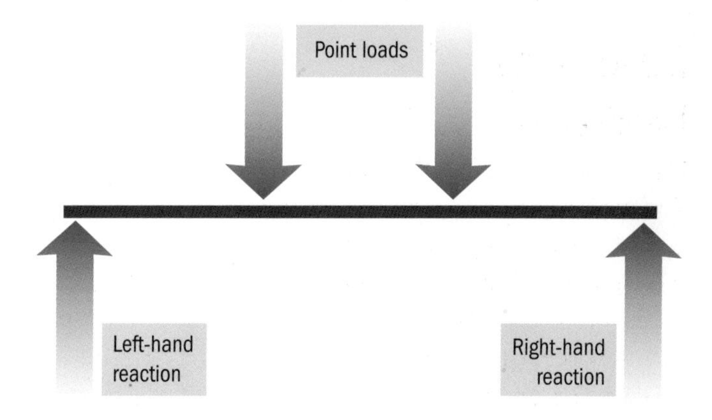

Figure 4.9: A simple beam with two point loads upon it

Key term

Equilibrium – when all the forces are balanced. For example, upward forces are equal to downward forces and clockwise forces clockwise equal to anticlockwise forces.

Uniformly distributed

Uniformly distributed forces are considered, in design terms, to be evenly placed along a structure such as a beam. These can be part of the live loads of a building; for example, people, furniture and the self-weight of the structure. As people and furniture can move around a building, it is far easier to design a building on the basis of uniformly distributed forces. Figure 4.10 illustrates how this force is represented.

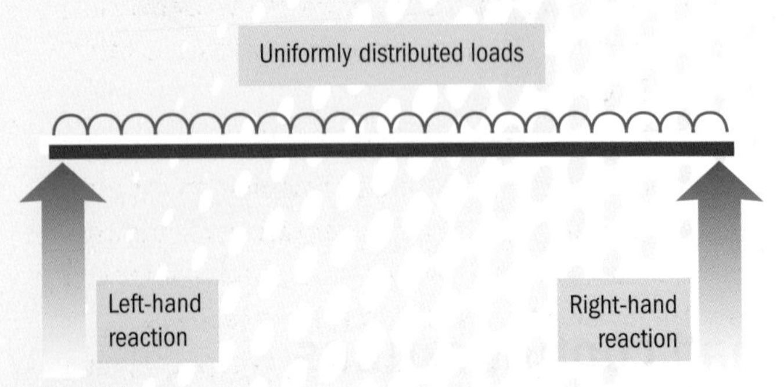

Figure 4.10: A simple beam with a uniformly distributed load upon it

Stresses

Compression, tension, bending and shear

Compression is formed when a vertical force pushes down on a structure; for example, a cube. The fibres within the cube have to push back with an equal force to resist being crushed and/or deforming under the load – see Figure 4.11.

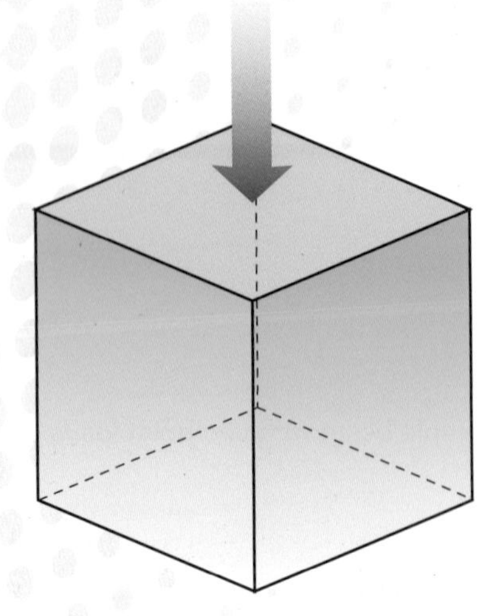

Figure 4.11: Compression

Compression and tension are often found within the same structure, such as a beam. The fibres within the material in Figure 4.12 are being pulled apart under the tension force shown by the arrows. The fibres have to resist the tension to avoid being torn apart.

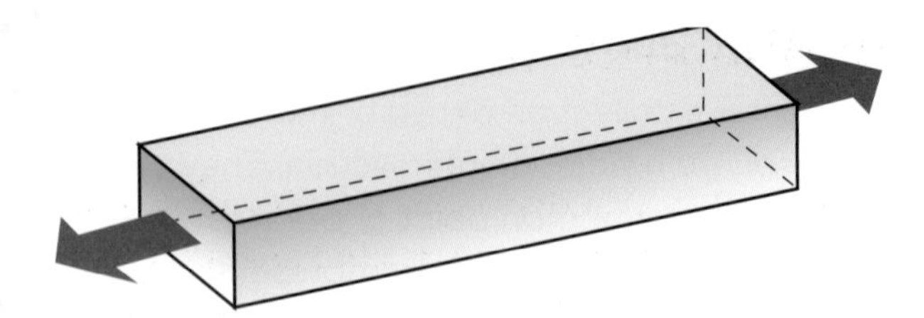

Figure 4.12: Tension force

In order for a structure to be in equilibrium, anticlockwise moments must equal clockwise moments. A force that causes bending is known as a bending moment. Figure 4.13 illustrates this with a piece of string. The piece of string is tied onto a beam. The length of the string (L) has an applied force at point A and would rotate clockwise about point A. To counter this, the wall has to restrain the end of the beam with a downward reaction force at A equal to force F in an anticlockwise direction.

Bending moment = Force × Distance acting

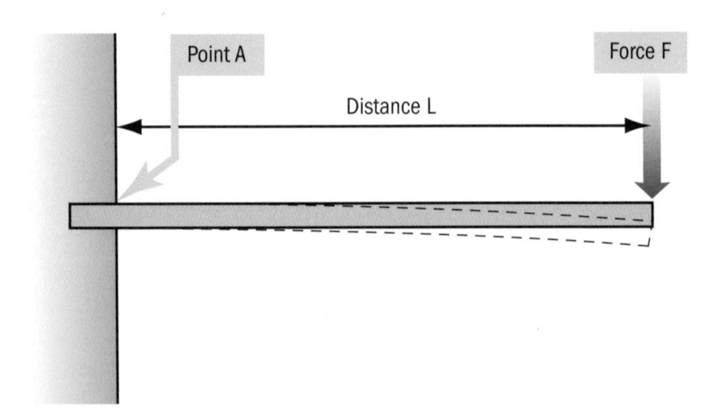

Figure 4.13: Bending moment

With shear force, forces are pulling apart the two pieces which are held together by welding or bolted connection. The bolts or welds which are being put under these forces will break across their width, shearing off at that point between the two pieces of structure – see Figure 4.14.

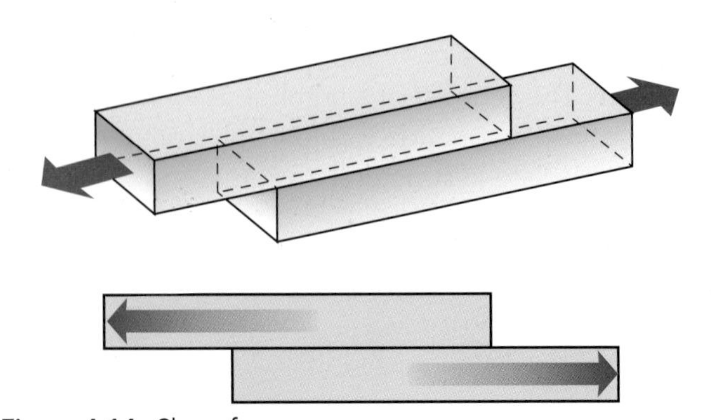

Figure 4.14: Shear force

Calculations

Stress

When a structure is subjected to a force of any type, the fibres of the material transmit the load from section to section throughout the length of the structural member. Such a system of internal transmission of these forces is termed stress. The intensity of stress is defined by the following formula:

$$\text{Stress} = \frac{\text{Force}}{\text{Area}}$$

The units used to describe stress are: N/m^2; kN/m^2; N/mm^2; kN/mm^2

where N = Newtons; kN = kiloNewtons.

Worked example

1 A steel tie with a cross-section measuring 25 mm by 25 mm is subjected to an axial pull of 100 kN. Calculate the normal tensile stress.

$$\text{Stress} = \frac{\text{Force}}{\text{Area}}$$

$$\frac{100\ \text{kN}}{25\ \text{mm} \times 25\ \text{mm}} = 0.16\ \text{kN/mm}^2$$

2 A concrete cube, of sides 200 mm is tested in compression; the failing load was 650 kN. Calculate the normal compressive stress at failure.

$$\text{Stress} = \frac{\text{Force}}{\text{Area}}$$

$$\frac{650\ \text{kN}}{200\ \text{mm} \times 200\ \text{mm}} = 0.02\ \text{kN/mm}^2$$

Strain

When a piece of a structure is loaded by a force and is placed under some measure of stress, some changes in its properties are bound to take place. This change may be dimensional in its length, section or shape. An object placed under such changes is said to be in a state of 'strain'. The effect of load (force provided by loading the member) can therefore develop in the fibres of the member as both stress and strain simultaneously.

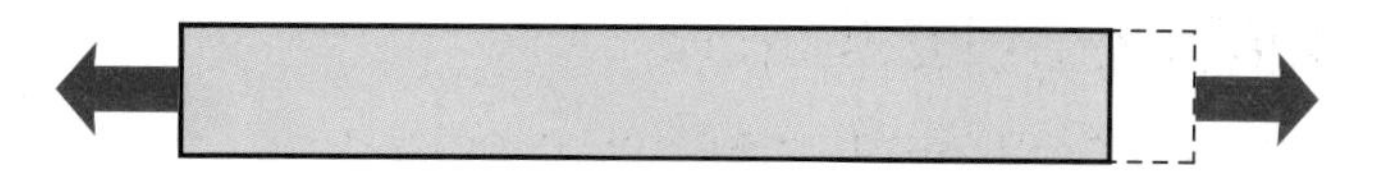

As the bar is pulled an increase in its length occurs as the dotted area indicates

Figure 4.15: Tensile strain

There are three types of different strain:

- Tensile strain – this is where a force pulls the material from both ends. As Figure 4.15 illustrates, a force is pulling from both ends of the bar. This is stretching the bar by an amount indicated by the dotted lines.
- Compressive strain – this is where the object has a force which is compressing or crushing the material which leads to a reduction in length. This is shown in Figure 4.16, where the force applied from the top has compressed the cube to the dotted line.
- Shear strain – as you can see from Figure 4.17, the force from the left has pushed the anchored cube to the right causing shearing strain as it distorts out of shape.

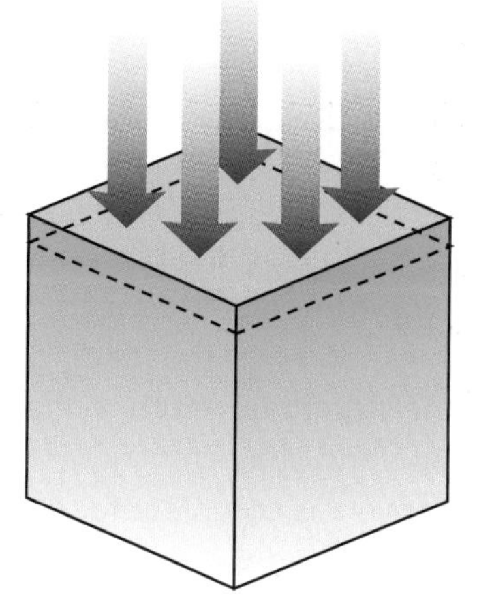

Figure 4.16: Compressive strain

Strain is calculated using the following formula:

$$\text{Strain} = \frac{\text{Extension or reduction in length}}{\text{original length}}$$

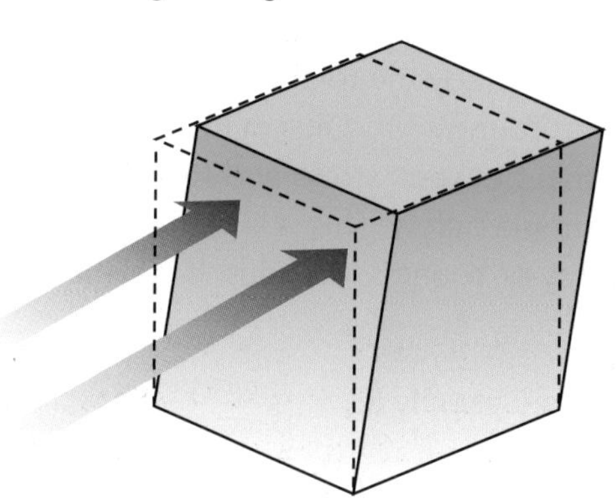

Figure 4.17: Shear strain

Worked example

1 A steel bar 1.25 m long was subjected to a tensile force of 150 kN and extended by 1.5 mm. Calculate the strain which took place.

$$\text{Strain} = \frac{1.5\text{ mm}}{1250\text{ mm}} = 0.0012$$

2 A 100 mm cube of timber was test loaded so that the compressive strain was 0.0014. Calculate the reduction in length that took place.

$$0.0014 = \frac{\text{Reduction in length in mm}}{100\text{ mm}}$$

Therefore, reduction in length = 0.0014 × 100
= 0.14 mm

Remember!

Always use the same units above and below the formula and don't forget strain has no units!

Modulus of elasticity

If you take a rubber band and stretch it under a load and then release the load, the rubber band returns to the original properties it started with. Similarly, in the diagrams on p. 135 when the loading condition is released, all the structures return to their original dimensional shape and characteristics. This is known as elasticity. The recovery is complete for most materials, as long as the load does not exceed the material's elastic limit, which varies between materials. However, if their elastic limit is exceeded, then the material permanently deforms when the stress is removed. The material's elastic limit is measured using Young's Modulus of Elasticity.

Many materials used in construction are elastic and it is useful to study the relationship between stress and strain. Young's Modulus of Elasticity measures the relationship between stress and strain and is expressed as the formula below. This is known as the Modulus of Elasticity or Young's Modulus. It is denoted by the symbol 'E'.

$$\text{Modulus of Elasticity (E)} = \frac{\text{Direct sress}}{\text{Direct strain}}$$

Remember!

Strain has no units to quantity it thus the units used to express Young's Modulus are the units of stress, e.g. N/mm², kN/mm².

Worked examples

1 In a tensile test on a 9 mm diameter steel bar, a load of 20 kN caused an extension of 0.072 mm on a 50 mm length of bar. Assuming that the limit of elasticity had not been reached, calculate the value of Young's Modulus.

$$\text{Stress} = \frac{\text{Force}}{\text{Area}} = \frac{20\text{ kN}}{\pi r^2} = 0.314\text{ kN/mm}^2$$

$$\text{Strain} = \frac{0.072\text{ mm}}{50\text{ mm}} = 0.00144$$

$$\text{Young's Modulus (E)} = \frac{\text{Direct stress}}{\text{Direct strain}}$$

$$= \frac{0.314}{0.00144}$$

$$= 218.06\text{ kN/mm}^2$$

2 Calculate Young's Modulus in compression for a timber beam 150 mm × 150 mm in cross-section, which reduced in length by 0.015 mm on a timber test for a 1250 mm test sample, when a load of 35.8 kN was applied.

$$\text{Stress} = \frac{\text{Force}}{\text{Area}}$$

$$= \frac{35.8\text{ kN}}{150\text{ mm} \times 150\text{ mm}}$$

$$= 0.00159\text{ kN/mm}^2$$

$$\text{Strain} = \frac{0.015\text{ mm}}{1250\text{ mm}} = 0.000012$$

$$\text{Young's Modulus (E)} = \frac{\text{Direct stress}}{\text{Direct strain}}$$

$$= \frac{0.00159}{0.000012}$$

$$= 132.5\text{ kN/mm}^2$$

Factors of safety

The factor of safety (FoS) is built into a structural design as an element of margin for over engineering. The FoS is a margin that can be used in the future. This over engineering is used to extend life expectancies, for example, in aircraft design, and to allow future adaptation and change of use of a structure. FoS also compensates for inconsistencies in different batches of materials, including deterioration of the material over the lifespan of the building.

The FoS is a number that can vary depending on the material and the circumstances in which it is used. To find it we divide the ultimate stress of the material by the permissible working stress.

Worked examples

1 The tensile strength of a bar is 395 N/mm². If the maximum permissible stress is 145 N/mm², calculate the factor of safety

$$\text{Factor of safety} = \frac{395\ \text{N/mm}^2}{145\text{N/mm}^2} = 2.72$$

2 A concrete beam is required to have a permissible stress of 14 N/mm², with a factor of safety of 3. Calculate the ultimate compressive stress of the concrete.

$$3 = \frac{\text{Ultimate compressive stress}}{14\ \text{N/mm}^2}$$

$$\text{Ultimate compressive stress} = 3 \times 14 = 42\text{N/mm}^2$$

Simple beam reactions for point loads

Figure 4.18 illustrates a simply supported beam with reactions at each end left and right and two point loadings of 50 and 100 kN. We shall calculate the value of the end reactions. Before undertaking this calculation, we need to recap some of the basic laws associated with this type of structure:

- The sum of the forces down equals the sum of the forces up.
- Anticlockwise moments equal clockwise moments

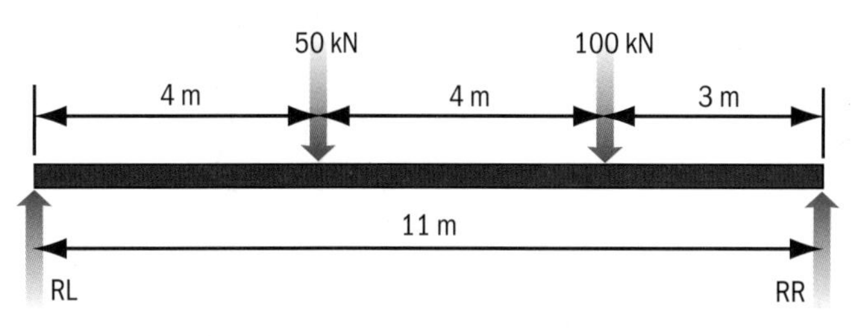

Figure 4.18: A supported beam

Worked examples

Calculate the left and right reactions.

If we start at the left-hand reaction:

Anticlockwise moments = Clockwise moments

$$RR \times 11\ \text{m} = 100\ \text{kN} \times 8\ \text{m} + 50\ \text{kN} \times 4\ \text{m}$$

$$RR \times 11\ \text{m} = 1000\ \text{kN/m}$$

$$RR = \frac{100\ \text{kN/m}}{11\ \text{m}}$$

$$RR = 90.91\ \text{kN}$$

Similarly, if we start at the right-hand reaction:

Anticlockwise moments = Clockwise moments

$$100\ \text{kN} \times 3\ \text{m} + 50\ \text{kN} \times 7\ \text{m} = RL \times 11\ \text{m}$$

$$650\ \text{kN/m} = RL \times 11\ \text{m}$$

$$RL = \frac{650\ \text{kN/m}}{11\ \text{m}}$$

$$RR = 59.09\ \text{kN}$$

Check RR + RL = sum of loads

59.09 + 90.91 = 150 (50 + 100)

correct

Simple beam reactions for uniformly distributed loads

Figure 4.19 is of a simple uniformly distributed load; the total load is calculated on the beam and is deemed to be acting at the centre of the beam.

If we start at the left-hand reaction

Anticlockwise moments = Clockwise moments

$$RR \times 8\ \text{m} = (10\ \text{kN/m} \times 8\ \text{m}) \times 4\ \text{m}$$

$$RR \times 8\ \text{m} = 80\ \text{kN} \times 4\ \text{m}$$

$$RR = \frac{320}{8} = 40\ \text{kN}$$

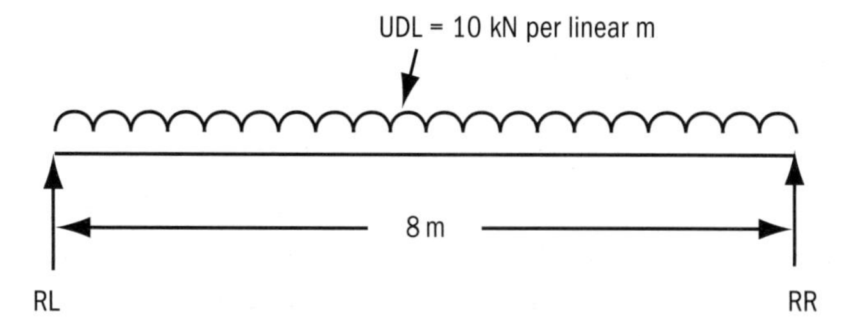

Figure 4.19: A uniformly distributed load

If we divide the total weight by two, then this should equal the end reactions, which it does.
Therefore RR = RL.

Graphical methods

Triangle of forces

The triangle of forces is a useful rule that can be used when a force can be represented by three sides of a triangle. For example, in Figure 4.20 the weight is being pulled down by gravity but is held in position by the two other forces. As long as the three forces are balanced or in equilibrium, they can be represented by a triangle. Any unknowns can be identified by scaling the sides of the diagram.

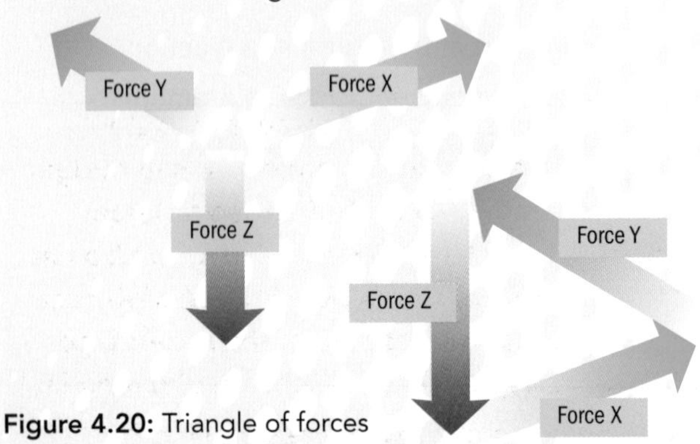

Figure 4.20: Triangle of forces

Parallelogram of forces for simple frames

When two forces act from a known point they can be represented graphically as a parallelogram by drawing in the resultant diagonal. This diagonal represents the combined force of the original two forces. Figure 4.21 illustrates this.

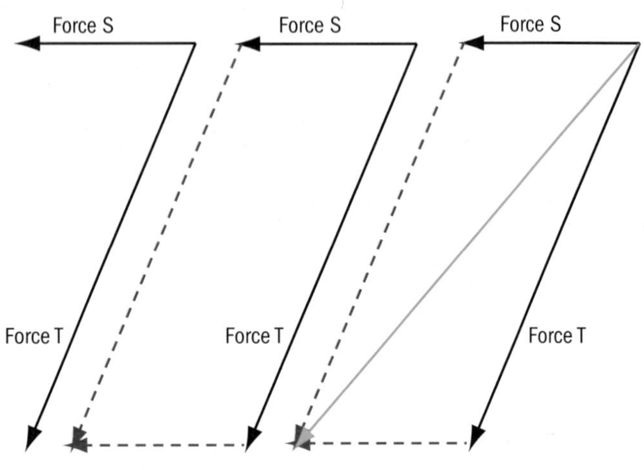

Figure 4.21: Parallelogram of forces

Assessment activity 4.2

1 The company you work for has asked you to work as the structural engineer's assistant. They have decided to test your knowledge of structures and have asked you to give descriptions with a definition of the following as well as where it would appear in a structure under load.
   - **a** Shear
   - **b** Tension
   - **c** Compression
   - **d** Stress
   - **e** Strain **P4**

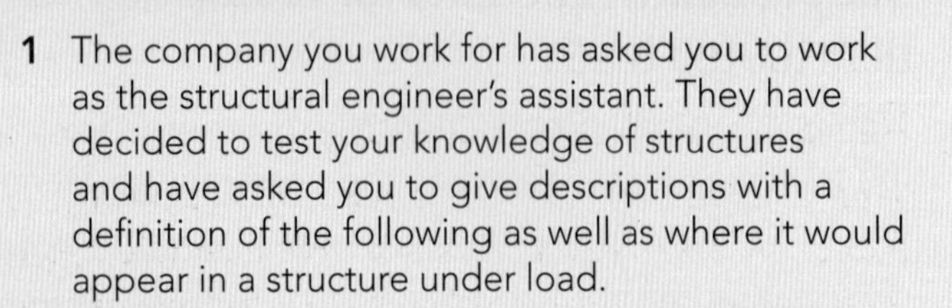

2 Predict the simple structural behaviour of the following:
   - **a** A steel portal frame overloaded at the bolted joint with the column.
   - **b** A concrete lintel only has an end bearing of 50 mm above a window at each end.
   - **c** The amount of stress that a concrete foundation can take is 30 N/mm². You know that a future loading will be 50 N/mm². What will happen to the concrete? **P5**

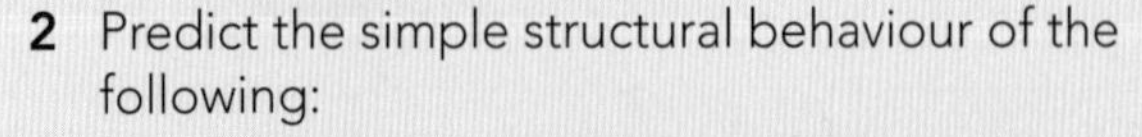

3 Produce clear and accurate answers to the following problems involving simple structures under load:
   - **a** A tie rod 25 mm diameter is subjected to a pulling force of 200 kN. What is its tensile stress?
   - **b** A concrete foundation 600 mm × 600 mm × 600 mm supports a force of 50 kN spread evenly over its surface. What is the stress in the concrete?
   - **c** A steel bar 1.50 m long was subjected to a tensile force of 125 kN and extended by 1.25 mm. Calculate the strain that took place.
   - **d** Predict the value of the end reactions for a UDL beam 10 m long with a UDL of 50 kN per metre. **M2**

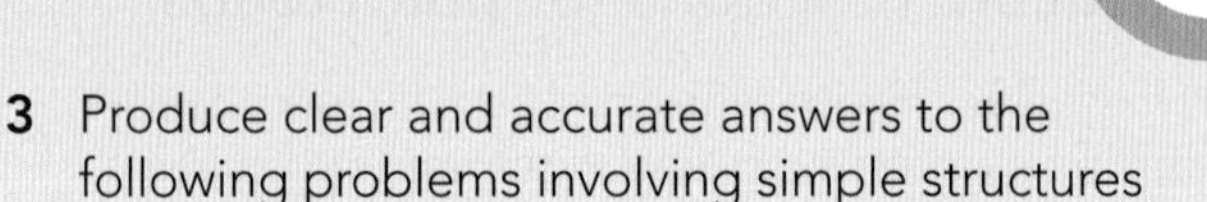

Grading tips

- For **P4** give a definition for each term and describe how it relates to structures under load.
- For **P5** predict simple structural behaviour from given data.
- For **M2** use formulas to calculate answers of stress, strain and loadings.

PLTS

Reviewing progress, and acting on the outcomes will help develop your skills as a **reflective learner**.

Functional skills

By identifying and using mathematical methods to solve problems, you can improve your **Mathematics** skills.

3 Know the performance criteria applicable to construction materials and the techniques used to produce such materials

Criteria for specification

Fitness for purpose

Materials are normally specified by a quality standard, either a British Standard, a European Standard or a standard from any other recognised body. The material has a set '**fit for purpose**' which is recognised worldwide and these quality standards make it so that the same product can be purchased globally at the same specifications.

Key term

Fit for purpose – a material must be able to be used in the context of the design. For example, glass in windows is required to let light through; any other material would not be appropriate.

There are occasions when a substandard material is fitted to a structure; this may be a temporary repair or an inclusion without the owner's knowledge. Substandard materials obviously have no guarantees.

Visual appearance

In design, the visual appearance of materials means a lot to the designer or architect as well as to the client. Visual appearance can also be linked to the texture of the materials used. Brickwork is a good example, where appearance can be altered by the use of coloured mortar in the joints which makes walls much more attractive to the eye. Light also plays an important part as reflectance and shadows can alter a building's appearance during the day or time of year. The market is moving away from the heavy construction materials that we have used for thousands of years and is rapidly developing greener, less dense materials, which also are thought to be more attractive.

Costs

Project costs are always a consideration in material specification and selection. If a project is over budget, then material specifications are changed to lesser quality ones in order to reduce costs. Quality also has a relationship with cost. Higher-quality materials cost more, but if they can be afforded at the initial stage, this will save money in the long-term life of the project.

There are two questions to ask when looking at material costs:

- Do you spend efficiently now and incur higher maintenance costs in the future?
- Do you spend more on quality now and incur lower maintenance costs in the future?

Resistance to degradation

Wear and tear on materials is linked to the quality and often density of the material. For example, high traffic areas, such as school corridors, require denser materials that resist wear and tear better. Degradation can occur as a result of:

- vandalism
- wind
- rain
- frost
- sunlight's harmful UV rays
- air pollution
- age of the material.

Design plays an important part in resistance to degradation; for example, ensuring that weathering drips are placed in the right place to avoid water stains on external brickwork. Material selection must take into account the location of the material and the usage of the environment surrounding it.

Ease of installation or use

A material that requires skilled trades people to install it will obviously prove expensive in its use. For example, glass curtain walling requires specialist design and installation. The key to the extensive use of a material is its ease of installation, the reduction in wastage as a result of installation and the ability to use semi-skilled labour to install it. Coupled with the ease of installation must be the ease of maintenance in the future. How easy will it be to replace a material that requires remedial works because of damage needs to be considered at the selection stage of the design. A material may be cheaper but incur massive financial costs when it has to be replaced at the end of its life or during accidental damage replacement.

Environmental implications

Green issues are now very important in the selection of construction materials, especially in housing construction. The amount of **embedded energy** that is contained within the material must be carefully considered along with the amount of carbon that has been released during manufacture.

Key term

Embedded energy – the amount of energy that has been used to produce the material. It is often expressed in terms of how much carbon has been released into the atmosphere during its manufacture and transport.

Environmentally friendly products that are renewable or contain a percentage of recycled materials should be considered over other materials in order to reduce the effect of global warming.

Sustainability and recycling potential

A material must now encompass sustainable elements and should be produced with regard to the environment so that it does not use up valuable resources. Timber products are great examples of sustainable materials that have little impact on the environment; they can be grown relatively quickly and their waste products can be recycled into other timber-engineered products. Listed below are some sustainable materials and their uses:

- Cedar boarding used as a cladding material to the external faces of houses is an attractive, sustainable and cost-effective method of providing an environmentally friendly product.
- Green planted roofs can be used as an alternative instead of using finite resources. Green roofs live and breathe and in urban areas provide a green environment to relax in.
- Materials recycled from demolition such as bricks and concrete can now be crushed and graded to provide hardcore fill to make up levels. Steel can be recycled back into the production process by melting down and reforming.

COSHH considerations

The Control of Substances Hazardous to Health (COSHH) Regulations 2002 must be considered with regard to:

- chemicals used in the manufacture of materials
- the use of chemicals to treat materials
- the chemical additive part of a material.

Asbestos is now banned due to the harm the fibres cause when inhaled. There are a number of products that require careful control for inclusion within buildings. Some foam insulation products require trained operatives to install them.

Chemicals such as solvents are steadily being replaced by water-based products, such as water-based gloss paints. Solvents harm the environment both in production and during use in modern products. Care should be taken to ensure that the manufacturer's data sheet is read and, if required, a chemical replaced with one that produces less harm to the environment.

Compatibility

Compatibility is important when choosing materials in terms of both costs and time. Some materials are incompatible with each other. For example, when certain metals are combined with water they react to give galvanic corrosion. The only way to prevent this reaction is to apply a surface finish to one of the metals. Removal of water is another control method that can be used to prevent the reaction, so the metals might be used internally rather than externally.

Remember!

To find out if two metals are likely to react with each other, check the anodic index, which is a chart of metals. The closer together the metals are on the index, the less likely there is to be a reaction. Selecting metals far apart can cause reactions.

Production and manufacture

Limes

Lime is an additive that is added to mortars and plasters to make them more workable and easier to spread. Lime mortars were used for many years before the modern plasticisers were introduced while lime washes were used to paint the outside of houses to give the walls a white appearance.

The raw material of suitable limestone rocks is excavated from a quarry and transported to the processing factory, where it is crushed down into a finer material. The limestone is then fired in a kiln where, at 900°C, a chemical reaction turns it into quicklime. The resulting material is then graded and processed into various types of lime and finally stored in silos for bulk delivery or bagged as a product.

Cements

The constituent raw materials in cement are limestone or chalk mixed with clays and fine sands. These materials are excavated in mass from quarries using large excavators and hauled to a manufacturing plant. The location of the cement works will depend on whether the cement uses chalk or limestone as its raw material. Limestone is broken up by crushing while chalk is processed with a water wash mill. Both these materials are separately combined with clay slurry and passed into a ball mill. The slurry is fed from large storage tanks into a kiln where the cement materials are heated; a reaction takes place and cement clinker is formed. The product is cooled and stored where it is ground into a fine powder – Portland cement – and stored in silos, where it is either bagged or bulk delivered.

Aggregates

There are two types of aggregates: crushed or uncrushed. Both types are excavated from the earth. Crushed aggregates are excavated from rock which is crushed, graded and stockpiled before use. Uncrushed aggregates are excavated from water-deposited beds, and may require washing to remove any contaminants.

Aggregates are used as:

- the basis for concrete manufacture
- the basis for road tarmacadam
- a decorative landscape material
- a drainage trench filter material
- pipe bedding
- fine aggregate sand for mortars
- hardcore fill to make up levels.

Concrete

Concrete is made up of several constituents:

- Water – this must be free from any contamination
- Fine aggregates – sands and fine gravels that fill the small voids within concrete.
- Coarse aggregates – larger 20 mm diameter.
- Cement – this provides the strength and binding agent. Various additives can be combined with the concrete to make it do different things, for example self-level, delay setting and increase waterproofing.

The design of concrete is very complex, and you will need to refer to the 'Design of Normal Concrete Mixes' for a detailed explanation of this process. Briefly, the raw materials are mixed within batching plants from specialist suppliers. Various strengths of concrete are available and depend on the variations in the mix and the use of the concrete, for example roadways or foundations.

The process of the hardening of the concrete is through the hydration of the cement, where the cement reacts with the required amount of water that has been added and hardens over a period of 28 days to reach its design strength. The categories for specifying concrete vary with use and you should refer to a supplier's website for information on this aspect.

Remember!

Concrete slump is a measure of the workability of concrete so it can easily be poured and moved. This is normally specified by the engineer.

Gypsum plasters

Gypsum plasters are manufactured from gypsum rock which contains calcium sulphate. Quantities of the gypsum rock are excavated and taken to crushers that reduce it to a fine powder. This is then heated and water driven off to produce various grades of plaster. After a final grinding, the product is packaged into bags. When this material is mixed with water it sets hard.

Gypsum products can be used in several ways:

- as a plaster to cover walls
- combined into a plasterboard material
- used within gypsum floor screeds
- as plaster mouldings.

Remember!

Workability is the key to using plasters as they will start to set from the moment water is added and mixed. A great deal of skill is required in their application.

Timber

Timber has been used in buildings for thousands of years. The production of timber stems from managed forests where selected mature trees are cut and processed within a saw mill. At this stage, the timber contains a high moisture content. Air or kiln drying of the timber reduces its moisture content and greatly adds to its strength. Various timber products can then be produced from the dried timber when its moisture content drops below 20 per cent.

Timber can be classified into two categories: softwoods and hardwoods. Hardwoods differ from softwoods in that they contain two types of cells: those that transfer sap and those that provide strength to the tree. Hardwoods such as oak, beech, ash and walnut take longer to grow. Softwoods include Douglas firs, many of which are grown in Scandinavia and Russia.

Timber gains its strength from the direction of the grain – splitting across the grain is much stronger than down the grain. Because of the inconsistency of timber it has to be stress graded for different uses, such as:

- floor joists
- ceiling joists
- roof rafters
- hip rafters
- roof trusses.

TRADA literature illustrates the various timber grades and their use and the typical spans that timber can be used for.

Remember!

TRADA – the Timber Research and Development Association.

Metals

The main metal used in construction is mild steel, and we will look at its manufacture in detail. The UK steel industry uses two processes for the manufacture of steel: the arc furnace which is powered by electricity and the basic oxygen converter. The raw materials used to produce steel are molten iron, coke and fluxes.

The arc furnace process uses cold materials to start with. A vessel is charged with scrap steel which contains a lot of recycled materials. The vessel lid is closed and electric probes are dropped down inside. When the power is switched on an electric arc forms and produces heat. This melts the mixture; other metals are added as required to produce the correct quality steel. Oxygen is blown into the vessel to purify the steel. The secondary process forms the final product.

This basic oxygen converter process uses molten iron which has been produced in a blast furnace. Molten iron is poured into a vessel along with some scrap steel and a lance blows pure oxygen through the mixture. This drives off any impurities, which float on the surface as a slag that is then removed by scraping it from the top. The pure steel is then taken to be processed into its solid form of ingots, billets or continuously poured into its final product and rolled into shape during a secondary process. This whole process takes about 40 minutes.

There are many uses of metal in construction, including mild steel reinforcing bars, stainless steel wall ties, lead roof coverings and flashings, mild steel lintels, structural steel frames and galvanised roof straps.

Paints

The production of paints is a complicated chemical process. Paints consist of many ingredients, such as:

- solvents – to thin the paint but are now being replaced by water
- driers – to increase the speed at which paint dries
- pigments – to produce a variety of colours binders – to bind and hold together the paint to produce the quality finish.

These ingredients are mixed in varying quantities to produce a variety of paints, including gloss paints, emulsions, two-part epoxy paints, lacquers, varnishes and stains.

Paints are normally stored in cans or drums until use and can be sprayed or painted by brush or roller onto surfaces.

Bricks

For thousands of years, people have been manufacturing bricks. In the UK, the principal ingredient of bricks are clays, which vary in colour. Clay is excavated using a shave cutter and then transported by conveyor belt to the process plant. Here the clay is processed so it can be moulded or extruded. Extruded bricks are cut by wires from a continuous length of extruded clay.

After processing, the bricks are air-dried and stacked ready for firing in a kiln. This firing process hardens the brick and gives it its unique colour. Various applications of sand can be added to the surface faces of the brick to produce a variety of colours. Wire-cut bricks and moulded bricks produce different textured faces. When the firing process is complete the bricks are stacked to produce brick packs that are evenly distributed in colour, and finally shrink wrapped for transportation.

Plastics

Plastics are complex organic compounds produced by polymerisation. They can be moulded, extruded, cast into various shapes and films, or drawn into filaments and used as textile fibres.

The manufacture of plastics is a chemical-based process. It starts with the conversion of crude oil into a variety of chemical products, such as petrol, oils and gases. Ethane and propane are derived from this process using a cracking tower and the resultant is ethylene and propylene. A catalyst is added and the resulting powder is a polymer-based substance. Various additives are then added and the mixture is melted and formed into pellets. These pellets are manufactured into plastic products, such as uPVC window frames.

Plastics do use a finite resource of oil, and their production is less than environmentally friendly, but they are an economical solution and can be recycled into other fibre products. Many of the commercial items we consume today contain a high level of plastic.

Liquids (especially water)

The production of water for use in construction is the same as for drinking water and the quality of the water must be ensured before it is mixed with materials. Water is normally obtained from boreholes, rivers and reservoirs and requires treatment to remove particles from it and has chlorine added to prevent microbiological decay. It is then supplied around the UK by gravity or pressure pumped by a pipeline distribution system.

Assessment activity 4.3

BTEC

1 Describe the main performance criteria that the following construction materials must meet to provide a minimum standard specification:

   a a facing brick in an outer wall

   b cement mortar

   c concrete foundations. P6

   Describe the important features and properties of these three construction materials. P8

2 A designer of new homes in your town has visited the college and given a talk on materials specification especially on damp areas of construction near to the ground. Undertake a specification for materials to be used for a wall below dpc level, including the foundations. Validate your material choices with supporting arguments. M2

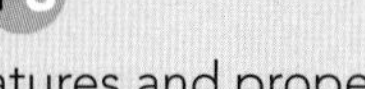

Grading tips

- For P6 describe the performance that each material will need to meet to withstand normal use in place.
- For P8 describe the important features, e.g. colour, and properties, e.g. strength, of three materials identified in P6.
- For M2 look at the location of the materials and explain what you would use and why.

PLTS

Organising your time and resources while prioritising actions to achieve assessment completion will help develop your skills as a **self-manager**.

Functional skills

Writing extended documents on the specification of materials will help develop your writing skills in **English**.

4 Understand construction materials and the techniques used to prevent their deterioration

Properties

Properties of materials vary between materials. Below we describe the common ones that can be applied to many of the current construction materials that we use on today's modern buildings. With regard to dimensions of materials, we now use the metric system for the **modular coordination** of buildings such that materials fit into the dimensions specified by an architect or a designer.

Strength

Strength is the amount of stress, tension or weight that a material can endure before it starts to deform. Strength varies between materials.

- Concrete has great compression strength but very weak tension strength and will require reinforcing with steel bars to prevent its failure by cracking.
- Timber is strong parallel to the grain but is weak perpendicular to its grain and can be split easily.
- Steelwork copes very well with tension, compression and shear, but loses strength in a fire and will eventually buckle and fail if it is not fire protected.
- Brickwork and blockwork are relatively strong on compression but can be pushed over laterally and rely on a strong bond with the mortar that holds the bricks and blocks together.
- Glass requires much support to help it resist the force from the wind. Secondary support is required when large panes are used; these also have to be toughened by a heating process to prevent them shattering.

Elasticity

As we have seen, this is the amount of stress that a material can take before its elastic limit is reached and the material permanently distorts causing dimensional change. Plastics are very elastic by their nature; concrete does flex, but it is not an inherent elastic material; glass can bend in large panes and return to its original shape; while steelwork is very good at retaining its shape under load and has a high elastic limit.

Porosity and water absorption

Porosity refers to the number of pores, or air pockets, that are present within a material. It is linked closely to the density of a material, which means the heavier a material per unit volume, the fewer pores there are within the material. Lightweight blocks have an **air entraining agent** within them that captures many tiny bubbles of air which act as an insulator within the block. Therefore, the lightweight block is a very porous material.

Key terms

Modular coordination – the grid lines on which the designer lays out a building. For example, a window opening must be worked into brick courses to avoid excessive cutting of the material.

Air entraining agent – this forms bubbles within the concrete which capture air that is used to insulate the concrete from extreme cold.

Porosity of a material is normally tested by weighing its mass then soaking it within a container over a set time period. Reweighing its mass after soaking will provide a measurement of how much water has been absorbed into the material. This can then be expressed as a percentage.

High-density materials may not be porous, for example steel work does not absorb any moisture and so can be said to be waterproof. It can be used to hold water or prevent its passage, an example being steel roof-sheeting panels.

Thermal and moisture movement

Thermal movement of materials from one season to another must be taken into account when designing buildings. All materials expand with heat and contract with cold; some have high rates of expansion such as copper and plastics. Expansion and contraction joints must be provided for in brickwork and concrete where large areas of these materials are used. These joints allow for the expansion of the material during summer and its contraction during winter months. This is especially true in climates that experience a wide variety of temperatures.

Moisture movement is especially obvious in timber-based products, which are **hydroscopic**. This can cause expansion and deformation when dried timber is placed into unheated houses and then the heating is switched on. The result is shrinkage and distortion or cracking of the finished product; for example, plastering shrinks as it dries out and cracks along junctions with other surfaces.

Thermal and electrical conductivity/ resistivity

Thermal and electrical resistivity, or the ability to restrict the passage of heat and electricity, are closely linked. For example, copper is a very good conductor of electricity and is used in electrical wiring because it can be bent; similarly, it is low in thermal resistance and can easily transmit heat so is used for hot water pipework distribution. Plastics are a non-conductor of electricity but have low resistivity thermally and again can be used for hot water distribution. As such, they are used effectively to insulate the copper conductors in electrical cabling.

Any metal has the ability to conduct electricity; the only other element that can conduct this energy is water. Great care has to be taken with electrical wiring to ensure that metal materials are not made live to the touch where wires pass through them.

Thermal transmittance (U-values)

As we saw earlier on in this unit, U-values are a measure of how thermally efficient a construction is to the passage of heat. The lower the U-value, the more resistant the material is to let heat out through the fabric. The more air that is trapped within the construction, for example a wall cavity filled with insulation quilt, the more thermally efficient a structure becomes. Since a U-value is based on the resistance of a material and the temperature difference across the structure, it is a good measure of thermal efficiency.

Durability

Durability is the material's resistance to surface scratching or abrasion, or how hard wearing it is. Durability also refers to the life cycle of a material in use. Denser materials will have much more hardwearing properties than lighter materials. Glass is very durable against the weather and is scratch resistant, but is prone to breakages if forced suddenly. Brickwork is very durable as long as the joints are maintained and a good quality facing brick is used. Concrete has a long life span but weathers poorly, becoming dirty from the atmosphere.

Workability

Workability is best expressed using the concrete as an example. Concrete has to be able to flow so it can be poured from the cement mixer. It has to be workable so it can be **compacted**. Many factors affect the workability of concrete, including:

- the amount of water within its mix
- the amount of cement within the mix
- the inclusion of any plasticisers
- the shape of the aggregate.

Key terms

Hydroscopic – the ability of timber to easily absorb moisture from the air.

Compacted – concrete is vibrated either by mechanical means or by hand to remove all the air bubbles.

Density

Density is how heavy something is for the volume it occupies, and can be expressed as:

$$\text{Density} = \frac{\text{Mass}}{\text{Volume}}$$

The standard against which other materials are judged is: 1 m × 1 m × 1 m of water, that is, 1 m^3 weighs 1000 kg, or 1 metric ton. So the density of water is represented as '1000'.

Remember!

The unit of density is kg/m^3.

Specific heat capacity

The specific heat capacity of a substance is the change in the number of joules of heat energy when the substance is heated or cooled by 1 Kelvin without a change of state, such as the melting of ice. This is related to thermal capacity which is a measure of how much of a structure can store heat energy. It is expressed as:

Thermal capacity = Mass × Specific capacity

The quantity of heat required to produce a change in temperature in a body of a specific mass and specific heat capacity is given by:

Quantity of heat = MC TD

where M = mass (kg)

C = specific heat capacity

TD = temperature difference

Viscosity

Viscosity, also known as thickness, is the resistance of a liquid or substance to flow. Viscosity is related to workability; the thicker a product becomes during its application, then the harder the product is to use and apply. Many construction products have a limited workability before final setting occurs. Moreover, because of gravity, viscosity of construction materials is important in the vertical and horizontal plan. Movement joint sealants then require different viscosities in their use and application.

Water has a very low viscosity and easily flows under the influence of gravity. Liquid plastics and silicones have a high viscosity and are able to be moulded before setting occurs.

Deterioration and failure

The following are just some of the common types of failure of construction materials. Many materials have a certain life span and simply wear out and have to be replaced or maintained.

Corrosion

Corrosion occurs mainly with ferrous metals that contain iron and results when iron, water and air come into contact. When these three are present, rust forms on untreated metals turning it an orange colour which stains and runs down surfaces. Corrosion on bolts has to be avoided as this is how structural steelwork is mainly held together. For this reason, bolts are generally manufactured from stainless steel.

Electrolytic action

As we have seen earlier, certain metals in the presence of water and other dissimilar metals have an electrolytic action in which one metal corrodes. Care has to be taken when choosing metal roof coverings, such as lead, copper and zinc, and the type of fixings used to hold the materials down onto the roof structure, otherwise you might inadvertently cause long-term corrosion. Many steel-cable suspension bridges contain thousands of steel wires that must be regularly inspected for corrosion to avoid the possibility of a catastrophic failure.

Fungal attack

Fungi attack timber-based materials in the form of wet rot or dry rot. Dry rot is caused by fungal attack; wet rot is the natural decay of the timber through excessive high moisture content which could be caused by a roof leak. Both lead to structural failure of load bearing timber. Dry rot can travel across concrete and brick walls and is very destructive. The only remedy in both cases is to remove the infected timber and the fungus.

Damp, dark, unventilated conditions will encourage the growth of mould on surfaces that are cold spots and surfaces frequently wetted, for example the base of old single-glazed window frames.

Insect attack

Insect attacks mainly come from wood-boring beetles, such as death watch beetle or house longhorn beetle. They bore into timber and eat away at the substructure causing severe structural damage, the timber loses its strength internally and eventually crumbles to powder.

Woodworm – holes left by wood-boring beetles

In 2004, termites were accidentally released into Devon, a warmer area. These are tenacious timber eaters and can easily wreck a timber building.

Frost attack

Frost occurs when temperatures fall below freezing which causes the freezing of water vapour onto a surface leading to ice crystals to build up. Frost can attack external brickwork at low levels where the brickwork often stays wet, especially below the damp-proof course. Older, more porous brickwork absorbs water into the surface which then freezes when the temperature drops. The freezing of water causes expansion as ice crystals form. This expansion leads to the face of the brickwork shelling off and spalling the brick.

Chemical and sulphate attacks

Chemical attacks occur on stones in the form of acid rain, which gradually wears away the surface of the stone and removing any carved features. Chemical attacks can also occur as traffic fumes contain high emissions from the burning of fuel. In city centres they gradually build up a layer on buildings which can cause discoloration and damage to certain building materials.

Sulphate attack is a type of chemical attack, occurring primarily on areas where sulphates can react with the cement in concrete. Sulphate attack can occur on concrete foundations where water collects sulphates from the soil or hardcore fill and attacks and weakens the foundations of the building. Sulphates can also be found in coal-fire chimneys where the combustion of coal leach through the mortar joints of the exposed chimney and weaken the brickwork structure.

Efflorescence

Efflorescence is caused by the effect of water moving through a material, for example brick and concrete. As the water migrates through the material, it dissolves soluble salts within the material. As the water evaporates, its salt crystals are left on the surface of the material. Over time, these gradually build up causing a white area of salt on the outside of the material which is obvious to the eye. This type of cosmetic damage generally appears on new buildings and will eventually clear once all the salt is washed out and the building face treated.

Ultraviolet (UV) attack

Ultraviolet (UV) radiation contained in the sun's rays affects the colours of materials by eventually bleaching out and fading the original colour with time. Timber can especially be affected by UV attacks and will turn into a grey, grainless, aged state if not treated.

Stress fatigue

Continual stress on structures will eventually lead to the permanent deformation of a structural member. This is very rare in building structures but with ever-taller skyscrapers planned, the stress effects from metal fatigue must now be considered from the force of the wind. Maintenance of the building envelope is essential; if one fixing breaks and is not repaired, then the wind can take hold and complete roofs and wall panels may peel away. Therefore, roof sheeting fixings are a weak point that must be designed by manufacturers to resist upload forces from negative air pressures on leeward sides of buildings.

Role of water in failure mechanisms

As we have seen, water has a role in failure from its expansion or freezing which causes damage to brickwork. Water, in the form of excessive rain can overflow guttering systems and enter a building causing internal damage. Water also acts under capillary action and can seep through the smallest of cracks. Design of buildings must take into account the force of gravity on water. As water washes down the faces of buildings, it causes staining from the dirt contained within our inner-city environments.

Efflorescence on brickwork. Have you noticed this on any buildings in your area?

Preventative techniques

Preventative techniques to control the types of deterioration are listed in Table 4.7. You should undertake some further research on these in order to satisfy the requirements of the distinction criteria.

Table 4.7: Techniques to prevent deterioration of construction materials

Deterioration	Damage	Prevention
Corrosion	Rust	The use of stainless steel Sacrificial anodes Full painting programme
Electrolytic action	Metal breakdown and loss	Use metals which are similar and will not react with each other Isolation
Fungal attack	Wet and dry rot	Treated timber Ventilation Removal and replacement of infected timber Prevent moisture entering timber
Insect attack	Structure of timber eaten away	Woodworm chemical treatment Timber treatment using pressure impregnated chemicals
Frost attack	Shelling of outer surface of brickwork	Use class B or A engineering bricks below damp-proof course Good design High specification facing brickwork Mortar joint maintenance
Chemical attack	Breakdown of limestone	Replacement of stone with harder material Treatment of surface of stone
Sulphate attack	Breakdown of chemical bonds of cement	Use of sulphate-resistant cement Repointing using sulphate-resistant cement mortar
Efflorescence	Formation of salt crystals	Wash down and remove salt Surface treatment with chemicals Quality specification on bricks Known source of sands
Ultraviolet (UV) attack	Colour fading	UV fixed colours resistant to UV light
Stress fatigue	Metal fracture and failure	Over design on fixings Factors of safety High strength stainless steel fixings
Role of water in failure mechanisms	Water staining Frost attack Efflorescence Dirt build-up	Good design Use of weather drips Overhangs to direct water Regular maintenance of guttering overflows and downpipes

Assessment activity 4.4

BTEC

1 You have been asked by a client to produce some information on the research associated with materials that are going to be used on their new commercial factory facilities. Describe the production and/or manufacturing processes for two construction materials that would be used on such a development. P7

Explain how these two materials can deteriorate with use. P9

Explain the preventative and remedial techniques that could be used to prevent deterioration of these two construction materials. P10

2 Consider the following three materials that are proposed for the construction of an inspection chamber:

a Facing brick weight dry 0.75 kg, weight wet 1.15 kg

b Engineering brick weight dry 1.25 kg, weight wet 1.26 kg

c Concrete block weight dry 2.5 kg, weight wet 4.5 kg

Make valid decisions using calculations as to which you would select for the material to construct the inspection chamber. M3

3 The following preventative techniques are proposed to extend the life of the materials. In each case, evaluate how the technique accomplishes this – would it be successful or not? You may have to undertake some research on each treatment.

a Chemical preservative pressure treatment to softwood timber

b Class B engineering brickwork below damp-proof course

c Colour-fast upvc cladding 

Grading tips

For P7 you will need to describe the manufacturing and production process for two materials.

For P9 you will need to find out and describe how the two materials can deteriorate.

For P10 explain how the two materials can be prevented from deterioration and what remedial work may be required.

For M3 you need to look at water absorption when deciding what materials to use for the construction of the inspection chamber.

D2 requires that you look at the technique and analyse if it will work to extend the life of the material.

PLTS

Providing constructive support and feedback to others within your group will help develop your skills as a **team worker**.

Functional skills

Identifying the main manufacturing process from a range of documents will help develop your reading and comprehension skills in **English**.

Rebecca Stevens

Structural Design

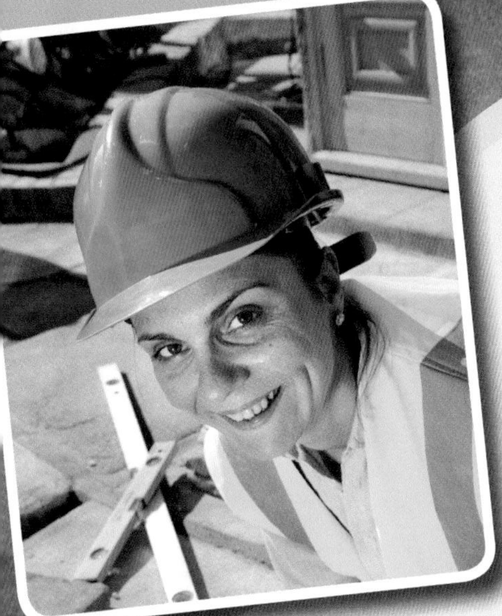

Rebecca went on from her National Diploma to take a Degree in structural design. This is a very mathematical subject, but is an area which she enjoys as computers are now used a lot to design steelwork for buildings. After receiving her degree, she was offered a job within the structural design office of an engineering consultancy that specialises in structural steel building frameworks.

Rebecca has had to become a member of the Institute of Structural Engineers so she can obtain a professional status and be recognised as a qualified structural engineer. This also enables her to obtain professional indemnity insurance against the design work she undertakes.

The work Rebecca undertakes is very taxing and it has taken her a long time to qualify as it is very mathematical in the understanding of loads and how structures react, but it has been worthwhile in achieving this as she is recognised by other professionals and her comments are often used to change architects' designs.

Rebecca has to take the architect's drawing, which gives an idea of size and the design concept, and then design steelwork to hold up the structure safely. This involves using the skills of measurement, calculations, prediction and analysis of data so a solution can be found that has a degree of safety built into it. The design also needs to take into account the fact that clients often change loads placed on buildings; for example, loading a factory floor with a new machine.

Rebecca uses structural software programs to undertake the analysis of the structure. Firstly she has to enter the loadings for the building and reference to standard loads. Secondly she has to enter the spans that the architect requires. This produces a print out that then has to be checked before working drawings for a steel manufacturer can be produced.

The structural engineer plays an important part in the realisation of an architect's vision for a building in producing a structural design that is unobtrusive and does not detract away from the aesthetics of the building.

Think about it!

- Would you enjoy this type of work?
- What skills would you need in designing steelwork?
- Would you need to have a mathematical brain?

Just checking

1 What are the three methods by which heat can be transferred?
2 What does a U-value measure?
3 Why are air changes in a building essential to understand?
4 What effect does condensation have on a building?
5 Name one way of reducing airborne sound.
6 Why do humans require daylight?
7 What is the unit of measurement for light?
8 Where do tension and compression exist in a concrete lintel?
9 What is a shear force?
10 What happens when a material's elastic capacity is exceeded?
11 Why are factors of safety important?
12 Name four ways in which a building material can be degraded.
13 Why is sustainability important?
14 How can we accommodate the thermal movement of brickwork panels?

Assignment tips

- Most of the material manufacturers will have some information on the manufacturing processes of their material or would be willing to supply it for educational purposes. Contact one such supplier.
- When thinking about the factors that affect your human comfort, imagine you are at home and consider all of the things that would make you feel uncomfortable in your home. These are the factors that you need to write about in the assessments.

Credit value: 10

5 Construction technology and design in construction and civil engineering

Construction technology and design involves planning and working with materials to construct buildings. In today's world, buildings are becoming more complex, so an understanding of construction details along with an ability to communicate your ideas effectively is essential if a project is to be satisfactorily designed, planned and translated into reality.

This is a core unit. It is intended to encourage you to develop your understanding of designing construction projects and other factors that affect this process through structured plans and by recognising the contribution of other members of your design team. There are many opportunities for you to develop your understanding of the work while working with others and within a team.

Learning to use technical language and the correct terminology of the items and structures that you will design is essential if you are to make good progress in this unit. You will use this ability to specify requirements and express ideas effectively. Building projects pass through various stages during design and construction including while works are in progress. You will learn to accommodate changes and integrate them into revised plans so that the overall project can continue and the other members of your team are kept informed.

Learning outcomes

After completing this unit you should:

1. know the factors that influence the design process
2. be able to communicate ideas between various members of the design and production teams
3. know about construction methods
4. be able to translate construction details into written and graphical instructions.

Assessment and grading criteria

This table shows you what you must do in order to achieve a pass, merit or distinction grade, and where you can find activities in this book to help you.

To achieve a **pass** grade the evidence must show that you are able to:	To achieve a **merit** grade the evidence must show that, in addition to the pass criteria, you are able to:	To achieve a **distinction** grade the evidence must show that, in addition to the pass and merit criteria, you are able to:
P1 describe the factors that influence the design process **See Assessment activity 5.1, page 165**	**M1** explain for a complex project the operation and effectiveness of the RIBA Architect's Plan of Work **See Assessment activity 5.1, page 165**	**D1** evaluate the effectiveness of the RIBA Architect's Plan of Work in terms of teamwork and the introduction of design changes after construction has started **See Assessment activity 5.1, page 165**
P2 explain the roles and responsibilities of the design team **See Assessment activity 5.1, page 165**	**M2** compare the methods recommended for communicating design changes to other members of the design team **See Assessment activity 5.2, page 170**	
P3 explain the roles and responsibilities of the production team **See Assessment activity 5.2, page 170**		
P4 describe the legal implications that could arise from miscommunication **See Assessment activity 5.2, page 170**		
P5 produce written communications between members of the design and production teams **See Assessment activity 5.3, page 173**		
P6 describe construction methods using relevant terminology **See Assessment activity 5.4, page 177**		
P7 create specifications for construction details, providing suitable instructions for the construction team **See Assessment activity 5.4, page 177**	**M3** interpret tutor-provided construction details, using recognised technical and architectural terminology **See Assessment activity 5.5, page 180**	**D2** appraise a set of instructions that represent design modifications to the original contract **See Assessment activity 5.5, page 180**
P8 produce sketch designs, plans, elevations, sections and details using standard conventions and symbols **See Assessment activity 5.5, page 180**		

How you will be assessed

The evidence requirements for pass, merit and distinction grades are shown in the grading criteria grid. Evidence for this unit may be gathered from a variety of sources, including well-planned investigative assignments, practical work or reports of practical assignments. You will be given written assessments to complete for the performing element of this unit. These will contain a number of assessment criteria from pass, merit and distinction.

This unit will be assessed by the use of two assignments:

- Assignment one will cover P1, P2, P3, M1, M2 and D1
- Assignment two will cover P4, P5, P6, P7, P8, M3 and D2.

Marie

This unit has made me examine the role of the designer and the design process that has to be followed so a structured framework exists and every member of the design team knows what their roles and responsibilities are. I did not realise how important the RIBA Plan of Work was in structuring the design process. I've learned that it contains a set of stages that have to be chronologically followed.

This unit has made me aware of the factors that have a major influence on the design process, from financial to social and legal constraints.

Over to you

- What does the design process mean to you?
- What do you already know about the RIBA Plan of Work?
- What are you looking forward to learning about in this unit?

1 Know the factors that influence the design process

Build up

The design team

There are more people involved in a building's design than just the client and the architect. In groups, discuss how the following people contribute to a building project

- Planning consultant
- Structural engineer
- Quantity surveyor
- CDM coordinator
- Health and safety consultant
- Building services engineer.

Stages of design process

Most buildings are complex structures. They represent many things and can be a type of statement by architects and designers who want to leave a legacy of their work for the communities that use them. Buildings allow us to function in many ways; as schools, hospitals, airports and railway stations, among others. Buildings provide shelter and, through their use of materials, style of build and appearance, represent our cultural values. Even a simple dwelling or house has many components and is designed to meet everyday activities and provide safe shelter. Buildings are incredibly expensive to construct and over their lifetime need to be modified to enable them to be adapted to change of use. They also need considerable amounts of maintenance if they are to be preserved and their upkeep allows them to be used effectively. There are many stages of the design process that make up the final building.

A lot of work goes into deciding whether a building project should be built. When a client asks an architect to design a building, an appraisal of the criteria for success is completed to determine whether the building can go ahead. This is known as assessing the **feasibility**. When the feasibility is completed, much more work goes into designing the building before any construction work takes place; this is known as the pre-construction phase. When all the pre-construction work is complete, the drawings and specifications are passed by the local authority building control so that the actual construction and site work can begin.

Key term

Feasibility – deciding whether the building is either practicable or will proceed.

Need for and benefits of a structured framework for design

Planning for a coordinated and structured approach is necessary so that all the key players – architects, architectural technologists, landscape architects, structural engineers, services engineers and facilities managers – are aware of the developments of the design as it proceeds. Imagine a big building project, such as the building of a motorway bridge across a river. The initial design may begin so that the architects can start to draw up the structural form that the bridge will take. As work proceeds on the exploration of the ground conditions, the data and information will need to be updated which, in turn, will allow the architect to amend the plans. Naturally, if these changes and amendments are to be accommodated, all the key players will have to understand the framework and how these changes affect all the other players in the process of building a sophisticated structure.

For most complex works in construction the RIBA Plan of Work is chosen as the one to follow (see Figure 5.1). This framework has the advantage of being easily understandable and the key personnel will all understand their roles within it.

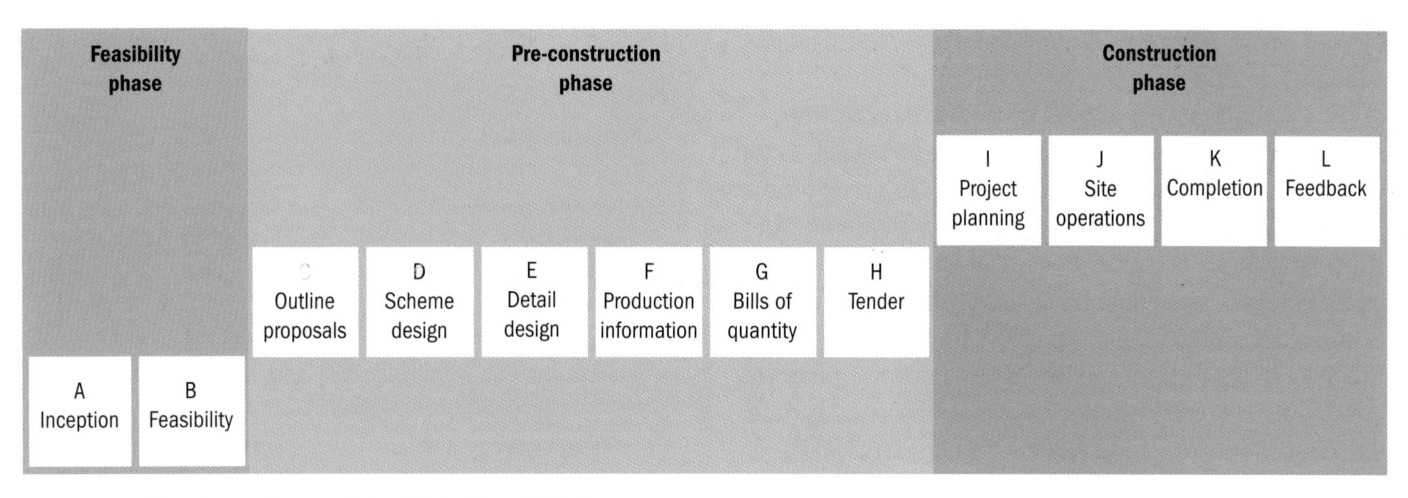

Figure 5.1: The three phases of the RIBA Plan of Work.

Take it further!

The RIBA plan is usually accepted as the most suitable plan, although many other types of plan are acceptable for smaller building projects or for instances where the RIBA plan may be too inflexible when conditions change frequently. For this unit, and for the work that we are going to do here, we will consider the RIBA plan to be the most suitable method of working, but you should be aware that in other circumstances other plans may be chosen. Do some research on the Internet and find other plans of work. When would you use these?

Characteristics of individual stages and factors affecting them

At the **inception** and feasibility stage of the plan, the architect will work closely with the client to determine and prepare the requirements of the building project. Architects will provide an appraisal and recommendation so that the client can determine the form in which the project is to proceed, ensuring that it is feasible both in technical and financial terms.

After the inception, the outline proposal and scheme design stage begins. At this stage, the architect will usually have **sketch plans** of the layout, design and construction in order to obtain approval from the client on the outline proposals and accompanying report. To complete the brief and decide on particular proposals, including planning arrangements and appearance, constructional method, outline specification and cost, the architect will then draft plans and drawings for submission to the local authority **building control** to obtain all approvals including planning consent.

Key terms

Inception – an event that is a beginning, a first part or stage of subsequent events.

Sketch plans – an architect or designer develops a sketch plan according to the needs and wants discussed with a client at the consultation stage. This is the time when the designer is creating, refining and rejecting ideas as appropriate.

Building control – the local authority department responsible for checking that the building project proposal complies with various legislative controls and regulations.

Once the outline is approved, the architect will have to obtain final decisions on all matters related to design, specification, construction and cost. During the detail design stage, working drawings are created and the full design of every part and component of the building is finalised; the architect is able to draw together the construction team. Team meetings will take place to discuss and decide materials, finishings, services, contributions by specialist firms and a range of other matters relating to the finalising of the building. Changes can be accommodated at this stage, although they may result in increased costs due to the scheme being planned on an early proposal. It is also important that any changes or modifications to the scheme are noted and accommodated so that everyone can be updated and work from the latest drawings.

During the production information stage, the preparation of product information used in the building, the drafting of **bills of quantities**, **tender documents** and project planning materials such as the programme of works showing duration of activities and the time taken to create the building itself are all put together. This is a very important stage in the process and particular care must be taken to ensure accuracy of the work so that the contractors undertaking the building work have all the necessary information and specification to complete the work to the appropriate standard. Drawings required at this stage include a **location plan** of where the work is located, a layout drawing of the construction site itself and a general arrangement drawing that shows the layout of the work to be done. From these drawings, a series of schedules and specifications will be drafted to provide any necessary additional information.

Specifications are dealt with in more detail later on, but their link with the bills of quantities is important. Bills of quantities are prepared by a quantity surveyor who reads the drawings and determines the quantities and amounts of materials needed to complete the construction work. Any part of the building that is not yet finalised or has missing information can be allocated a **provisional cost** or a **prime cost** to cover an amount for the work to take place. If the architect has not yet appointed a construction contractor or team to complete the works on site, discussions at this stage will begin to determine a company or contractor to appoint who is capable of undertaking the work.

The **tendering process** allows the client an opportunity to present a batch of work or a construction contract to contractors. Sufficient time must be allowed in the planning of the work for this stage to take place – typically around four weeks. During this stage, the contractors learn about the complexity of the project, the stages of construction, the limits and the constraints anticipated within the work so that they can calculate a realistic and accurate tender submission. The contractors then put in a tender, or quote, for the works and the client chooses the contractor best suited and able to do the work.

Key terms

Bills of quantities – a list of quantities produced to a standard that is used to price construction tenders.

Tender documents – a document or series of documents that offer to do the work at a price – contractors tender for work by providing an estimate of how much they will charge for the works to be completed.

Location plan – a document that is used to guide and identify where the work takes place; a map so that the contractor knows where the site is located.

Provisional cost – an allowance or a provisional sum that is allowed when the actual cost of something is not yet known.

Prime cost – when a cost is already known and an allowance has been made to meet this cost.

Tendering process – when an architect may ask several contractors who are interested and are capable of completing this work to provide an estimate of how much it will cost.

Sometimes a pre-tender meeting may be held with contractors and the complexity and the details of the work can be communicated so that the contractors, the architect and the client can satisfy themselves that they are all capable of completing the work within the appointed timescale and budget. A letter of invitation to tender can then be issued together with all relevant drawings, specifications and bills; and, the contractors can then visit the site of the works and determine the cost of the project. Tenders are returned to the architect at an agreed date and time. Initial comparison of the tenders received from different contractors takes place by the architect and sometimes this includes the client or their representative. This analysis of the tenders usually results in the lowest priced tender winning the contract and being appointed as the contractor.

For some work selective tendering may take place, where an architect and quantity surveyor invite contractors that are either known to them or have an established reputation for completing similar work to tender. Again, the cheapest quote is most commonly selected in this process.

After the tender is agreed, the project planning stage begins. This is where the work to produce the building starts on the building itself. Contract documents are prepared and signed. At a project planning meeting, the architect will usually clarify any undecided points at this stage and agree any further contractual points.

Contractors draft a **programme of works** that illustrates the milestones and the total duration of the work. Key milestones include taking possession of the site, dates of project progress meetings and other key dates that are achieved during the lifetime of the contract.

The contractor is expected to sign the contract documents at this stage, which include:

- a copy of the contract
- a full set of construction drawings
- bills of quantities
- specifications
- a register of drawings
- **site diary** and associated report forms.

Once the site is officially handed over to the contractor, construction-related operations on site can begin. The site is now the responsibility of the contractor who has to comply with all legislative requirements and legal constraints. The contractor should be informed of any rights of way, preservation orders, protection requirements and any other environmental issues that need to be considered in terms of the work and the activities on site in the months ahead.

Interim certificates are issued at regular intervals during the contract period based on a quantity surveyor's assessment of the work completed so far. These certificates form the basis of the calculation necessary to prepare and authorise stage payments from the client to the contractor.

The contractor has a duty to ensure the appropriate site supervision. A clerk of works will check on behalf of the client and the architect that the contractor is complying and building the project to the appropriate standards of materials and workmanship. Samples are taken of various materials, sometimes by a specific request from the architect and other times by established practice. For example, concrete is regularly tested by way of a **slump test** to ensure workability when it arrives on site, and its strength is assessed at regular intervals by testing to destruction samples after the concrete has been placed. Brickwork panels are sometimes erected to enable the architect and the client to see what the brickwork will eventually look like. Tiles, blocks, panels, etc. may also be subject to a request that the contractor builds a **mock-up** panel for viewing by the architect and the client so that the quality and the desired finish of the work can be ascertained and ensured.

Key terms

Programme of works – a bar chart which shows the duration and the resources needed to complete the job. It shows how long the work will take and how long the contractor intends to spend on each task.

Site diary – a record of relevant information about the progress of the construction works completed by the contractor.

Interim certificates – Issued by the architect to say that some of the work has been completed. Contractors can then be paid for the work that they have done so far.

Slump test – a test of the concrete's workability that shows that the concrete is of the right quality and can be poured with the correct results.

Mock-up – a sample panel or erection created by contractors and subcontractors as a sample of what the work will look like when it is finished.

The contractor's duties at the on-site stage are to work diligently on the construction works adhering to all relevant health, safety and welfare legislation. The contractor is also expected to maintain a site diary to record relevant information about the progress of the construction works. Typically, a site diary will include information on:

- weather conditions
- visitors on site for the period
- any deliveries of materials to site
- progress of work to date
- on-site personnel including subcontractors
- any comments and notes taken about the work undertaken
- discrepancies and any inconsistencies in contract documents.

As the building starts to take shape, the contractor may be required to hand over part of the building to the client. The actual date of handover is planned and any outstanding issues relating to the construction works can be determined and a solution found. The period of notice required varies from site to site, but adequate notice is required in order to prepare the area and any other supporting documentation. The architect will usually insist on inspecting the area and determining that the materials and workmanship is to the appropriate standards, that all services and equipment are functioning appropriately and effectively and that the 'as-built' record drawings are a true record of the actual building. The building manual can be simple for a domestic dwelling or could extend to a complex and detailed manual of several volumes for a more intricate commercial building.

A certificate of practical completion is issued by the architect to the contractor which then enables the contractor to claim monies due for the construction work. It also addresses any defects and snags that have arisen so far. At this completion stage, the contractor has effectively completed the construction stage and no longer has responsibilities for the insurance of the building or its works. When all defects or amendments or outstanding issues are addressed, the architect will issue a final certificate and the account will be adjusted for **variations**, subsequent instructions and fluctuations in labour and materials prices and/or costs.

Key term

Variations –items that were not in the client's original budget and therefore are additional to the contract; for example obstructions encountered within the ground during excavation work.

The final part of the RIBA plan is to analyse and evaluate the progress of what was expected with what was actually delivered. Architects will need to determine what has happened in order to achieve better performance in the future. Typically, an analysis and evaluation could include the following:

- what the client thinks of the completed building
- whether the building functions effectively
- which parts of the design were particularly successful or problematic
- if the design process could have been undertaken differently and have provided a better service to the contractor or the client
- what relationship and communication existed between the design team and the construction team and how they could be improved in the future
- if the contractor met all performance targets in a timely and effective manner
- if the design process ran smoothly
- if the job was profitable and the contract fulfilled.

Stages of design

Using the RIBA plan requires two processes to complete the construction project: an integrated design approach and an integrated team.

Integrated design has all the members of the team, such as the technical planning, design and construction teams, look at the project objectives, building materials and systems. This is done because the design relies on the different perspectives and expertise of the various area specialists involved. Building design in practice also requires an integrated team process in which the design team and all affected stakeholders work together throughout the project phases to evaluate the design for cost, future flexibility and efficiency of the project. They also have to take in the overall environmental impact and look at how the occupants' quality of life will be affected. The process draws from the knowledge pool of all the stakeholders across the life cycle of the project, from defining the need for a building, through planning, design, construction, building occupancy and operations. The end result is a high-performance building.

Often a range of considerations will need to be included in a building design in order to make sure it suits the many requirements of society. Frequently, there are legislative controls that insist buildings conform to regulations such as the Disabled and Disabilities Act, which ensures that buildings are accessible by wheelchair users or people with specific disabilities. Designing a suitable building means making sure that the finished product is:

- accessible – all heights, clearances, entrances, exits and services are integrated to address the specific needs of disabled people
- aesthetically pleasing – the physical appearance and image of building elements and spaces are pleasing to the eye and act in harmony with the environment in which the building is located
- cost-effective – building elements are selected and utilised on the basis of life cycle costs as well as basic cost estimating and budget control
- functional/operational – the building delivers the functional dimensions that it was designed to do, including spatial needs and requirements, system performance as well as durability and efficient maintenance of building elements

- productive – relates to the occupants' physical and psychological well-being and includes building elements such as air distribution, lighting, workspaces, systems and technology
- secure and safe – a safe and secure environment for occupants and assets from hazards and inclement weather patterns
- sustainable – utilises and maximises environmental performance of building elements and strategies over the life cycle of the building.

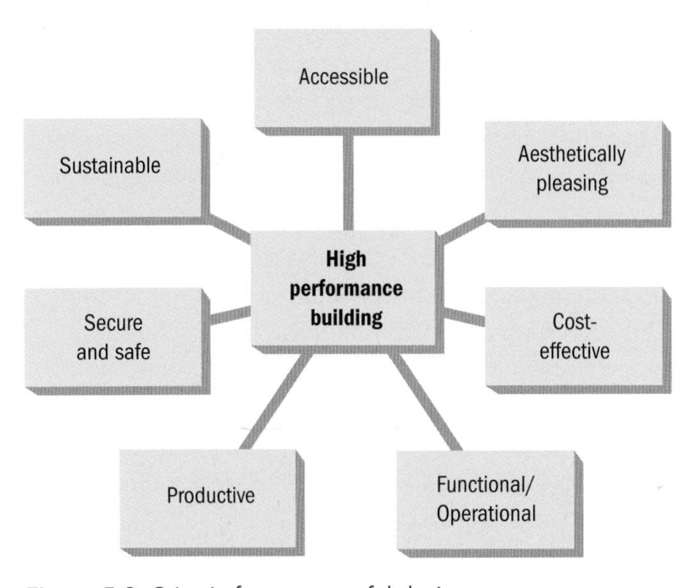

Figure 5.2: Criteria for a successful design

Implications of financial, legal and environmental constraints for design team

Designers must consider a range of factors that will have an impact on all stages of the design process, from the initial idea of a project through the various pitfalls and constraints until the building is finished and handed over to the client. Many of the constraints will revolve around the availability of funding and finance; ideas will be conceived, but the cost implications may mean that the client has to accept a different proposal.

Besides the limits of available finance, any design must satisfy the requirements of planning legislation. Building regulations are principally about workmanship and quality. Building projects will need to comply with these regulations if they are to consume a maximum amount of energy, provide adequate light, acoustics and a suitable place to live and work.

The design should also fit in harmoniously with the environment as much as possible. More than ever before, sustainability and ecological factors need to be considered, especially by the designer who should advise and guide clients to the integration of a project and the natural environment.

Construction (Design and Management) Regulations

The Construction (Design and Management) (CDM) Regulations 2007 are aimed at reducing the large numbers of serious and fatal accidents and cases of ill health which occur every year in the construction industry by improving the overall management and coordination of health, safety and welfare throughout construction projects. CDM has placed responsibilities on key team members, especially those involved with the design of construction works. Not all construction projects are affected by these regulations.

According to the CDM, the client's key duties, as far as is reasonably practicable, are to:

- select and appoint a competent planning supervisor and principal contractor
- be satisfied that the planning supervisor and principal contractor will allocate adequate resources for health and safety
- be satisfied that designers and contractors are competent and will allocate adequate resources when arranging the work on the project
- provide the planning supervisor with information relevant to health and safety on the project
- ensure construction work does not start until the principal contractor has prepared a satisfactory health and safety plan
- ensure the health and safety file is available for inspection, after the project is completed.

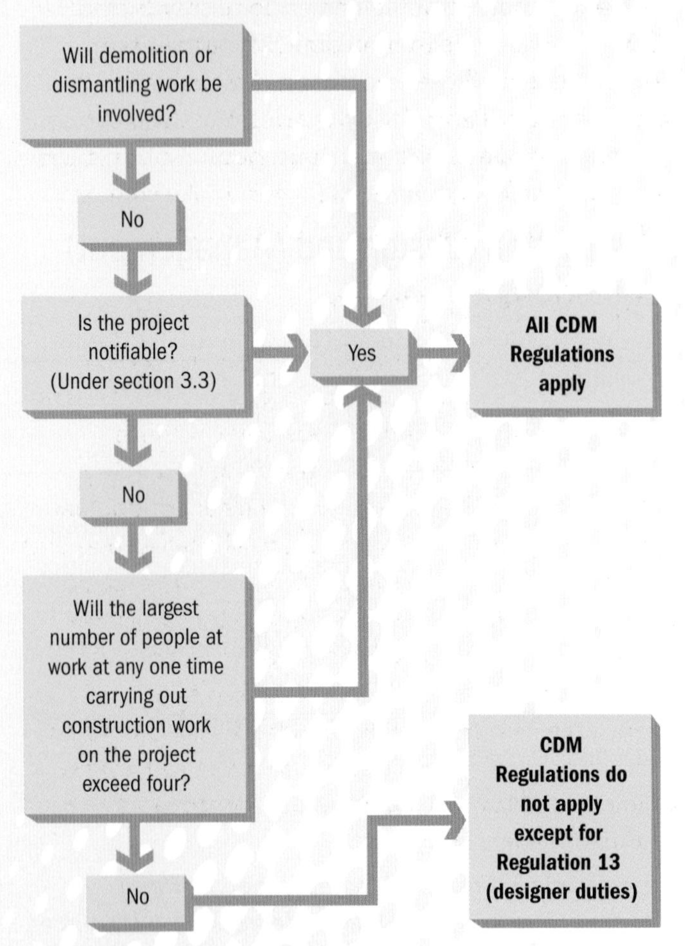

Figure 5.3: Do the CDM Regulations apply to the construction project?

The planning supervisor has to coordinate the health and safety aspects of the project design and the planning to ensure that designers comply with their duties – in particular, the avoidance and reduction of risks involved in the construction of the project. Designers must cooperate with each other for purposes of health and safety and help prepare a health and safety plan before arrangements are made for any principal contractor to be appointed.

Planning supervisors are usually qualified professionals with several years' experience in developing construction projects and hence can advise the client on the health and safety plan before the construction phase starts. If the CDM Regulations apply, the planning supervisor will also notify the Health and Safety Executive and ensure that the health and safety file is delivered to the client at the end of the project. Planning supervisors also give advice, if requested, to the client on the competence and allocation of resources by designers and all contractors. This advice can also apply to contractors appointing designers in other forms of contracting arrangements and for projects that do not use the RIBA Plan of Work. The designer needs to be as proactive as possible in discussing their duties with the clients. Within the context of design, the building project should be designed to avoid risks to health and safety so far as is reasonably practicable and at least to reduce risks at source if avoidance is not absolutely possible. During the design development, the designer must consider the hazards and risks to those constructing and maintaining the structure.

Designers must ensure that the design includes adequate information on health and safety so that they can pass this information on to the planning supervisor in order that it can be included in the health and safety plan. Above all else, it is the designers' responsibility to cooperate with the planning supervisor, and any other designers involved in the project to achieve the safe and efficient completion of the project.

Contractors also have obligations under the CDM Regulations. The principal contractor's key duties are to develop and implement the health and safety plan and ensure that only competent and adequately resourced contractors carry out any work where it is subcontracted. It is also the contractor's responsibility to ensure the coordination and cooperation of all other subcontractors; findings of risk assessments and details of high-risk operations that may be carried out during the construction process need to be shared. The main contractor also has a duty to pass this information on to the planning supervisor for inclusion in the health and safety file.

Activity: CDM Regulations

You work for a small architectural designer's practice that has been approached by a speculative developer to handle and manage a house building project. The small housing project consists of the demolition of a single-storey warehouse that used to house chemicals and household detergents followed by the erection of six new semi-detached houses. There is a preservation order on two mature oak trees adjacent to the site. Work is expected to last only 90 days. The houses that you are designing are standard layout, three-bedroom, semi-detached that will include an en-suite to the master bedroom, a family bathroom, a kitchen, separate dining room and a living room. An integral garage will be included which will need separation and require fire resistant materials where the garage makes contact with the house itself.

As the designer, decide how you will begin to plan and organise the work ahead. Consider these questions:

- Do the CDM Regulations apply?
- Are you familiar with relevant health and safety legislation, including CDM Regulations?
- What do you need to do regarding construction practice relevant to the CDM Regulations?
- Are there any special or specific measures that you need to apply regarding the approved codes of practice for asbestos, lead, substances hazardous to health, etc?
- Are you conversant with British Standards relevant to construction safety, e.g. scaffolding, steel erection, demolition, etc?
- Are you satisfied that you have sufficient health and safety training and practical experience?
- Are you familiar with the concept of risk assessment for the construction industry?

Contractors perform many duties on site and thus they are in the best position to ensure that site activities are carried out safely. Contractors must ensure that subcontractors have information about risks on site and that all workers and operatives working on the project have adequate training and a suitable induction to site procedures.

In essence, the role of the contractor under the CDM Regulations is to monitor health and safety performance by those persons on the construction site. Contractors and site operatives must comply with any site rules which may have been set out in the health and safety plan and they must make sure only authorised people are allowed onto the site.

The design team

The building's architects and designers need to be clear about what a client needs and then work closely with the construction team to ensure that these needs are realised into a fully functioning building. A typical design team establishes a clear contractual relationship between all parties and allocates roles and responsibilities for specialist activities in the process of completing the project. Sometimes construction projects are so complex that the whole team is very large, comprising people with specialist skills and experience who all contribute to the finished product. The exact requirements of each construction project vary with the type of building and availability of labour in the region.

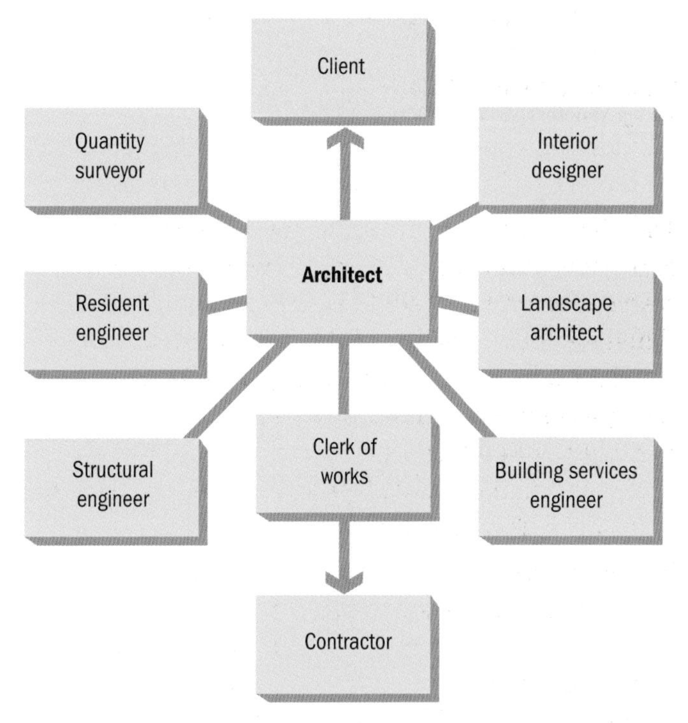

Figure 5.4: The architect sits at the centre of the design team

Architects

Architects usually need to have the following skills:

- an ability to identify and articulate a client's brief to meet both user and longer-term needs as well as society's concerns for sustainable development
- an understanding of the client's perspective and an ability to communicate effectively with each member of the construction team
- an understanding of relevant legislation and its potential effect on programme cost and quality of design
- an understanding of health, safety and welfare legislation and its implications on design and construction
- an ability to construct the team and to coordinate and integrate the work of the team members
- an ability to communicate effectively with the client along with an understanding of reporting methods.

Architectural technologist

Architectural technologists usually work in partnership with architects to establish contract procedures and administration of work. Architectural technologists have a deep understanding of construction technology and can translate ideas into sound, accurate drawings that a contractor can follow. They draft out drawings and develop construction-related details so that the project can be completed. The architectural technologist needs to be able to produce precise and accurate details of construction technology since quantity surveyors will need to read these drawings so that the bills of quantities and the specifications can be drawn up.

Interior designer

Interior designers are sometimes used on contracts where the internal finish and décor is important or of a prestigious nature. Some high-profile house building projects or city-centre apartment conversions have been developed in conjunction with interior designers who can add considerable value to a project.

Landscape architect

The landscape architect, is sometimes contracted to design the external environment of the project. As with interior design, the landscape can be enhanced by a specialist to improve the completed project.

Frequently, speculative housing projects will benefit from a landscape architect who will design open spaces, leisure areas and land drainage and so improve the appearance of the development.

Clerk of works

The clerk of works is employed directly by the client, who will want assurance that a contractor is producing a building that meets the specification in terms of both materials and workmanship. The clerk of works also reports to the architect on progress of the construction works. Clerks of works do not issue instructions and do not have authority to impose variations or changes to the design but will need to inspect the works as it proceeds and hence need to visit the site frequently. On some larger sites, a clerk of works will have a resident office and spend their entire working time there.

Structural engineer

Structural engineers determine the design of load-bearing elements of the building and ensure that each component is designed to safely withstand the loads that are imposed on the building. Structural engineers typically work for the client but are frequently engaged by the architect to inform and supervise the design and installation of structural elements as work proceeds. They work very closely with the architectural technologist and the principal contractor.

Resident engineer

Resident engineers are based close to the construction works, on the construction site itself. They report back to the structural engineer and the architect on matters relating to the structure and the load-bearing components that have been designed by the structural engineer.

Quantity surveyor

Quantity surveyors accurately determine the amount of materials needed to build the project. They prepare a bill of quantities establishing a record of all the materials needed and identify all the information necessary to draft out a specification of the works. The quantity surveyor can then advise and guide the architect or the client on the cost of the job, check tenders and evaluate any costs as work proceeds.

Building services engineer

Building services engineers design and implement a range of items into the project that improve the resources and the quality of the building. Building services and the effective use of heat, light, acoustics and other electrical appliances have seen significant improvements in recent years and many of these items, such as lifts, escalators, air conditioning and heating and ventilation systems, need to be integrated into the contract drawings at an early stage to avoid conflicts of space or to ensure that the design can accommodate machinery, plant and equipment.

Facilities manager

Facilities managers maintain the building once it is built. Their duties include the maintenance of the physical structure by undertaking works such as painting, decorating, easing doors, replacing carpet and floor finishes and so on. Their role also includes the essential duties to ensure that the building functions; examples of this are the cleaning of the building, the replacing of lamps and fittings, checking and repairing heating controls and other such essential items that are required in an occupied building.

Interactions between team members

There are many interactions between the design and construction team members that occur during a project's life cycle, including:

- client/architect – they discuss the design brief and different solutions to produce an agreed design and specification
- architect/structural engineer – they interact about the structural feasibility of the agreed design and come up with any amendments as necessary
- architect/quantity surveyor/client – they discuss the client's budget and the costs of various schemes
- principal contractor/quantity surveyor/architect – they discuss, amend or negotiate tenders and quotations for work
- principal contractor/architect/quantity surveyor – they agree on the final account on completion of the work.

Assessment activity 5.1

BTEC

1. Outline the factors you think will influence how buildings are designed. What factors need to be considered by the design team? P1
2. Explain the roles and responsibilities of the architect, quantity surveyor and structural engineer. P2
3. Describe how the RIBA Plan of Work operates. Then, explain the merits of using this plan. M1
4. Analyse and evaluate the effectiveness of the RIBA Plan of Work in terms of teamwork and how it works with the introduction of changes to design once construction has started. D1

Grading tips

For P1 you will need to select at least three factors that need to be considered by the design team, for example, the size of the land for the project.

For P2 examine the roles of some members of the design team.

For M1 you will need to obtain a copy of the RIBA current Plan of Work, and explain in your own words how it works, and what its merits are.

For D1 you will need to analyse and discuss in detail the Plan of Work in terms of handling a design change after work has started.

PLTS

By planning and carrying out research and then appreciating the consequences of changing design decisions, you should develop your skills as an **independent enquirer**.

Functional skills

Presenting information on the RIBA Plan of Work in a logical sequence clearly will help to improve your **English** skills in writing.

2 Be able to communicate ideas between various members of the design and production teams

The brief

Sketch plans and designs

Architects draft out a sketch plan that initially shows the client what can be expected from the design. Sketch plans have to take account of a range of constraints and influences on the proposed building such as town planning zoning restrictions, green belt or conservation areas, protection of natural reserves or local wildlife and other environmental issues such as sustainability.

Figure 5.5: From sketch plans to the brief

After considering the client's requirements

When clients approach architects for advice and guidance they are frequently not aware of the multitude of influences and constraints that make a building either possible or improbable. A well-defined brief is rare and members of the design team have to liaise with each other to frame up and develop this brief within the parameters of the client's requirements, which usually takes a number of meetings or consultations to develop.

To aid design

Architects test out their designs by creating cost models and working out the feasibility of the design within the client's budget. Visualisations, frequently made from computer-generated models, are created to provide an opportunity for the client to understand what the building will look like if it were to proceed. The brief moves from an initial idea to a cost model which is then modified to take account of financial constraints. During this period, the adopted brief is subject to **cost analysis** and is confirmed as either acceptable or rejected as the revised brief and moves on to a point where the team can work out the specification of the building project.

Key term

Cost analysis – the process of providing estimates of what a building project's costs and benefits are likely to be; then comparing these estimates and making a judgement as to whether it is acceptable to proceed to the construction stage.

The decision-making process

Factors contributing to design decisions

The factors that contribute to making design decisions include costs, sustainability, environment and build-ability. These are considered by the architect when taking the initial brief from the first exploratory sketches to a worked-up model that can be presented to the client for approval. Clients will naturally want to get value for money and will have ideas on how they wish the completed building to look and to perform. The architect and the design team will be able to consider how each of these requirements influences the final outcome of the project.

Some architectural practices take a methodical view of such criteria by scoring each factor and totalling up these scores to select the winning proposal. Others may wish to consider all options in descriptive terms including seeking out the opinions of neighbours to develop the notion of how the building will be accepted into the local area by the people who will work or live around the project.

Influence of decisions on the final project

Alterations, modifications and variations to the original brief can be accommodated despite the time and extra money that they may cost. An architect can issue a variation order to change the specification for a number of reasons, including:

- statutory requirements or lawful changes brought about by legislation
- omissions or additions that were overlooked in the initial briefing documents and consultation
- an error while compiling the bills of quantity that had not been picked up until construction work started.

The architect must issue the variation order in accordance with the contract; it can only be a modification, not a change that represents something different from what is in the main contract. It cannot be something that radically changes the nature of the work, requires a different contract or asks that the contractor undertake something completely unreasonable in the course of the contractor's main duties.

Legal aspects

Legal position of each member of a design team

The legal position puts the client at the head of any hierarchical chain. The client is effectively the employer and, so long as they comply with the legislative constraints, they ultimately decide on the acceptance of the design and whether to go ahead with the building project. Clients, however, sometimes place responsibilities on the architect, the design team and the contractors to promote sustainable construction methods, utilise local labour availability and promote social agendas for change.

Clients can be individuals who wish to build a house, a speculative housing project to sell on for investment purposes or a commercial or industrial building to carry on their business. Local authorities, municipal governments or government agencies can also be clients; they typically require much bigger structures such as roads, highways, airports, schools, hospitals and so on.

In order to practise, architects are required by law to hold qualifications that show that they understand design principles and can build an adequately strong structure and comply with all aspects of legislation such as planning, building regulations, Construction Design and Management Regulations and the Disabled and Disabilities Act. They are also expected to promote sustainable construction methods by 'designing in' relevant innovatory techniques. Moreover, architects have to have adequate public liability insurance to protect any third parties from any defects or problems associated with their work.

It is very important for an architectural technologist to have an understanding of technical guidelines, structural design and scientific principles as well as an awareness of legal frameworks. Architectural technologists have to carry compulsory insurance and have adequate cover should there be any liability to third parties in the event of a problem with the design of a building project.

Landscape architects have assumed greater status in the past 30 years or so. During this period, the work of a landscape architect has become more relevant to the overall design of a project. As these have become more complex, so too have the landscaping and the external environment become more complex and demanding. As a consequence, landscape architects have to comply with all aspects of legislative constraints and carry public liability insurance in case of defects or implications to third parties.

Structural engineers occupy a specialist area in construction and have very specific legal requirements. They have to be adequately qualified and belong to a relevant professional body to practise. They also have to carry adequate public liability insurance in case issues are raised with the standard of workmanship or adequacy of design with the building. Service engineers that are responsible for building services are specialists in specific areas such as heating, ventilation, acoustics, telecommunications and lift or escalator installation. Each of these areas may have its own engineers or designers who typically will be employed by the architect who will check to see that relevant services are integrated into the final building. Services engineers are expected to carry public liability insurance in case of issues being raised with the design and installation of their services.

Facilities managers will require information about the building when it is complete, although it may well be desirable to have this information available as the design and construction work takes place. They are responsible for the maintenance and servicing of the building once the project is complete. Facilities managers are usually employed by the building owner or quite possibly the building tenant and as such their legal position is different from the other members of the team.

In the preambles to a contract, a mechanism for calculating the amount of monies due for variations and other changes to the contract conditions will have been established. These include pricing within the bills of quantity or a day work rate for completion of work. Variation orders are issued by architects as an instruction for additional or modified work and as such are considered under the terms of the contract an architect's instruction.

RIBA and the Chartered Institute of Architectural Technologists (CIAT) have standard documentation to guide and direct members to use appropriate documents for communicating with other members of the team.

Theory into practice

Find out more about CIAT by visiting its website. (www.ciat.org.uk), There, you can find sample letters and documents used by architects, including:

- letters of appointment
- town planning and application forms
- start of detail design stage of RIBA Plan of Works
- invitation to tender letter
- return letters to successful and unsuccessful tenderers
- requests for information forms (RFI)
- architect's instruction form (AI)
- interim certificate (IC)
- practical completion.

Rights of client

The client has a right to expect that the architect, as their agent, will comply with relevant legislation and act ethically and in their best interest at all times. The client can reasonably expect that their rights should enable the building project to be completed through:

- provision of an adequate brief to enable the architect to develop this idea from outline sketch to a series of completed technical drawings that meet the needs of the building
- accommodation of reasonable changes to the brief during design stages or during construction
- advanced establishment and identification of fees and charges to ensure that there are no disputes and concerns of over-charging by the architect
- progress of the project at an appropriate and adequate pace that keeps delays to a minimum
- adequate communication by the architect with all interested parties and anyone who needs information to affect the work safely and speedily
- the architect's disclosure of any conflict of interest which could affect the project, its outcomes or the personnel involved
- the disclosure of any defect, once discovered, so that adequate and appropriate changes or modifications can be made.

As members of the Royal Institute of British Architects (RIBA), architects also subscribe to a code of practice. RIBA will deal with matters of unacceptable professional practice or aspects of professional incompetence and can fine, reprimand or strike off any architect who it feels has fallen short of the RIBA expectations and has not provided the client with an appropriate service. In some cases, RIBA will also provide a mediator service or appoint an adjudicator to resolve disputes. They will also provide information about the availability of expert witnesses or arbitrators who could provide technical information or reports that could be used to agree a settlement or advocate compensation.

Theory into practice

Find out more about RIBA by visiting its website (www.architecture.com).

Alongside the RIBA sanctions, a negotiated settlement for any disputes is preferable to any imposed penalty. Most architects will offer full and frank explanations to the client of any relevant information and enter into agreements on the legally binding nature of arbitration or adjudication.

Damages

'Damages' is the legal term applied to any breach of a contract between parties. In construction there are many contracts between clients, contractors and subcontractors which contain clauses relevant to workmanship, responsibilities, costs, and the duration or time allocated to enable the delivery of the services or product. Contracts are used to determine the legal position and obligations of each party, although in some cases disputes arise and may need the services of the legal system to arrive at a decision as to who is responsible and what action can be taken.

Once a court arbitrator or an adjudicator has arrived at a decision as to who is responsible for a breach of contract, there will have to be a calculation of what damages can be awarded, which is most commonly in the form of monetary terms. Construction contracts frequently contain a liquidated damages clause, and it is important to note that any award should reinstate the damaged party to a position where they would have been had the contract been fulfilled.

Liquidated damages

Liquidated damages are a predetermined method of establishing what will be paid in the event of a party not fulfilling their part of a contract. Some clients build in a liquidated damages clause to encourage the contractor to deliver the building project on time. These clauses state a fixed amount paid for each day or week that the building is not completed over the agreed date. The important thing to note about any liquidated damages clause is that it must not penalise either party and that the damages must be reasonable. Several legal cases have established that this must be a genuine estimate of any loss and any unreasonable penalty would be unenforceable.

Negligence

Negligence in construction means the failure of a person to act in accordance with the responsibilities that are reasonably expected of them. This definition can apply equally to contractors, subcontractors and designers.

Architects can be proven to be negligent if they do not comply with the respective legislative constraints and do not fulfil their obligations to the initial brief; hence it is very important that they work closely with the client to establish their exact needs and discuss outline plans and sketches.

Health, safety and welfare

The health, safety and welfare duties of all the parties to the contract are set out in the Health and Safety at Work Act 1974; everyone has a duty of care with regard to themselves and others. There is also a separate section within this Act applicable to manufacturers and designers whose products must be intrinsically safe in their design and use. This also applies to material suppliers who import goods into the UK. The Health and Safety at Work Act also lays down several rules that an employer must follow with regard to their employees' health and safety. (For more information on the Health and Safety at Work Act, see Unit 1 Health, safety and welfare in construction and the built environment, page 8.)

Environment

Environmental and green issues are more and more prevalent in today's society due to the impact of global warming and this is now steadily filtering through various regulations such as the Building Regulations. There are many pieces of environmental legislation that can be applied to construction projects – from the careful tipping of waste materials to burning rubbish on site, from pumping waste water off-site to the safe disposal of contaminated ground (see also Unit 2 Sustainable construction).

CDM

The CDM Regulations state the roles and legal responsibilities of the client, planning supervisor, designer and principal contractor in the running of the pre- and post-contract situations. The key to the CDM Regulations is to keep everyone informed about the risk of the design and construction. Architects and designers now have to assess their final designs for risks to the occupants and for maintaining the building; for example, windows can be made to tilt and turn so they can be cleaned from the inside of a building and not from height externally.

Assessment activity 5.2

1. You have just started work as a junior in a site office. The senior contracts manager would like to test your knowledge and has asked you to explain the roles and responsibilities of the on-site production team. P3
2. An architect's instruction has gone missing; what could be the legal implication of this? P4
3. Compare the recommended methods of communicating design changes to the members of a design team. M2

Grading tips

For P3 you will need to identify then describe the members of the production team, this is the team that will produce the building from the design.

For P4 describe the legal implications that could arise from miscommunication.

For M2 you will need to compare the methods available for communication; for example advantages and disadvantages of each.

PLTS

By working towards goals, showing initiative, commitment and perseverance in achieving this P4 criteria, you will learn to be a **self-manager**.

Functional skills

Taking part in discussions on the roles of the people in the design team will help your skills in speaking **English**.

3 Know about construction methods

Characteristics

Each construction method has distinct characteristics which are mainly focused on the type of material that is used. The following materials each have different construction method characteristics:

- brickwork and blockwork
- structural steel such as portal frame, skeleton frame and multi-storey
- pre-cast concrete
- in situ concrete
- timber framed
- composite which is a mixture of many materials
- various cladding materials such as stone, clay, steel and glass.

The function of the proposed building has a great effect on the characteristics of the materials used to build it. Each of the above materials can be applied to different types of buildings including:

- commercial high specification offices
- factory business units
- industrial production
- government offices
- armed services facilities
- domestic housing.

Applications and limitations of traditional and modern methods

Traditional

In the UK, brickwork has been traditionally used for buildings. Traditional building methods have a unique and historical relationship with much of our UK heritage; many old buildings are constructed from handmade bricks and are hundreds of years old. Modern technology has to be adapted when refurbishments are undertaken on such buildings. Now, a combination of brickwork and blockwork to form the external walls of the structure is used.

The space between them forms a cavity which is filled with insulation to increase the thermal efficiency of the overall construction. The traditional wall can only be applied to three- to four-storey buildings before the base thickness needs to be increased in order to take the greater loadings from above. External walls constructed in the traditional method, therefore, have a limitation with height and need to be restrained at some point for stability. This is achieved at the intermediate floor level and at the roof level through a physical connection with floor joists and with the roof ceiling joists. Gable ends are secured to the roof using straps.

Other types of traditional building incorporated a green oak frame which was used to carry the loads of the structure onto which local materials were placed to form walls and roof. Thatched roofs and lime-washed rendering are also traditional methods. Elements of these methods have been taken into modern construction with timber and rendered panels.

Modern

Modern buildings incorporate columns and beams made of concrete and steel which have revolutionised the development of high-rise structures. Concrete on its own provides high compressive strength but low tensile strength; steel provides high strength reinforcement. Developments in concrete technology have enabled components to be pre-cast off site and then installed on site. Roofs are now made of steel sheeting which rests on cladding rails and enables much larger spans to be achieved without the use of additional columns.

The traditional timber oak frames have developed into the modern timber-framed construction which replaces the inner skin of brickwork or blockwork. Modern timber-framed buildings can be erected much faster, have greater insulation properties, can be used with any design shape and have excellent environmental and green characteristics. The internal surface of the timber frame has to be lined with plasterboard to increase fire protection while a damp-proof membrane has to be installed to act as a moisture barrier to prevent internal condensation forming.

The development of the portal frame shape, which can be manufactured from concrete, timber or steel, has brought about a radical development of small to large single-storey industrial units. With the portal frame shape, the units can be erected speedily, and can incorporate a variety of external envelope types. The limitations of steel-framed buildings is that they require some form of fire protection to the steel as it loses 50 per cent of its strength when the temperature of a fire reaches over 500°C.

Influence on design

Both traditional and modern construction methods have influences on design as do the heritage and history associated with the building approaches. This is further reinforced by the **listing** of buildings along with the formation of conservation areas in places of natural beauty.

Key term

Listed building – a building of special architectural or historic interest in the UK. Alterations to these buildings cannot be made without consent or careful consideration.

Structural steelwork has enabled larger spans and greater heights to be reached. Inner cities now often have skyscrapers along their skyline. With the advent of post-tensioned concrete, reinforced with steel, larger spans that can bear more weight have been established.

Timber glulam beams, where a number of strips of timber are glued together in a mould and finished into beams, can now mimic the oak-framed buildings of the past. The winter garden building at Sheffield is a classic example.

Glass and glazing systems have a tremendous impact when coupled with the modern framed structures and can be used to great effect and be hung off a steel frame as a complete wall of glass.

Multiple construction options

Multiple construction options is a modern technique that enables a series of multiples of one module to be constructed. The modules are constructed off site and then are simply bolted together and finished on site. This is sometimes referred to as volumetric construction and has many advantages over traditional techniques, including:

- speed of construction
- savings on on-site labour
- savings on-site preliminaries
- no limitations due to weather conditions

- high-quality factory finishes
- services integrated into the design.

Many modern hotels are built on this principle where the bathrooms are constructed as modules and simply craned into position.

Variety of construction options

Primary and secondary requirements of a design

The primary requirement of a design is the structural frame of the proposed structure. As we have seen, there are several methods of producing a primary element. These also can be combined to form composite structures. For example, a design based on a steel-framed structure can be produced. The intermediate floors can be built using floor steel profiled decking, which is then used as a mould for a concrete floor; this is also an example of composite construction.

The secondary requirement of a design is the building envelope that hangs off the frame. This can take many forms such as:

- profiled steel cladding sheets
- profiled steel composite panels with insulation sandwich
- curtain walling in glass
- glazing spider systems
- traditional brick cladding
- cedar boarding.

The list can be extensive and technological advances mean that new materials are continually being developed.

Terminology

Design and construction use a large number of technical words. Table 5.1 lists some of the main ones.

Table 5.1: Construction technology

Terminology	Definition
Construction and design	
Portal frame	the shape formed by the use of two columns and two beams that meet at an apex and form a portal shape
In situ	in place supported by formwork, i.e. concrete works which are cast in situ
Superstructure	the visible works above ground
Substructure	the foundation and works below ground level that cannot be seen
Building envelope	the coverings of the building
Architectural	
Grid centres	the dimensional setting of a building using a grid format of either squares or rectangles
Working drawings	the final set of detailed drawings that enable the principal contractor to construct the building
Plan of work	see the detailed section explanation on page 157
Column	a vertical support
Lintel	horizontal support member over openings
Legislation	
Enforcement	the process of enforcing the legislation, e.g. the HSE
Prosecution	the court system of punishment
Acts or regulations	the law and subsequent regulations produced by the EU and the UK
Health, safety and welfare and environmental factors	
EPA	Environmental Protection Act
Wastage	the product left after processing a material
Risk	the level of harm associated with a hazard
Environmentally friendly	a chemical or process that does not harm the environment

Assessment activity 5.3

BTEC

Information flows from a design office to the construction site and vice versa.

Produce two examples of written communications, one from an architect to site and the other from the contracts manager to the architect. P5

Grading tip

For P5 produce a written communication; this could be a letter, a drawing register or an architect's instruction. You could pick one topic and then discuss it in writing from both parties.

PLTS

By generating ideas and exploring possibilities for a form of communication, you should develop your skills as a **creative thinker**.

Functional skills

Using different writing styles in producing an architect instruction will help develop your writing in **English**.

4 Be able to translate construction details into written and graphical instructions

Sketch designs

Sketch designs are a series of sketches, often three-dimensional, that illustrate the designer's interpretation of the client's brief. Several sketches may be drawn from which the client can then decide which parts or design they like and these can be combined into a final sketch design which is then approved and moved on to the next process. Often the local authority and other agencies can be involved in outline planning permission.

Thinking point

The information produced and illustrated by the designer must be accurate, clear and easy to understand by the contractor or the builder undertaking the work on site.

Divide yourselves into teams and discuss what could happen if the sketch design was not accurate.

Drawings

Construction drawings

Drawings are the main medium by which designers convey their ideas to builders and need to show all the relevant information clearly. The ability to work from drawings is a well-developed skill practised over time; and each team member needs to seek out the information that they require from the drawing. For instance, a building control officer will be interested in the materials and want to know that the completed job will comply with the existing regulations, whereas the builder will be most concerned with how they will be able to complete the job and the logistical problems of arranging all the components in a reasonable and workable sequence.

ISO drawing sizes are standardised – see Unit 8 Graphical detailing, page 299 for the dimensions of the most commonly used sizes.

Before creating drawings and beginning work on a project, it is customary practice to establish what is required from the work such as:

- title of each drawing and date that it will be needed
- contents of each drawing
- what scales are to be used for each drawing
- what size sheets are to be used
- referencing system to be adopted
- tolerances and style of the content.

Construction drawings are important documents and a register of their issue will need to be kept. Everyone concerned can be assured that they are using the correct version of the drawing and any amendments have been correctly assimilated by the team.

Location drawings

Location drawings show where the work will take place (see Figure 5.6) and are the first drawing that most people see of the project. Location drawings include references to site plans, floor plans, sections and elevations; they give an overall impression of the whole building or project.

Figure 5.6: A location drawing

Site plan

Site plans show the overall layout of the site and locate the buildings, roads, landscaping and datum used for the development (see Figure 5.7). Frequently, they are drawn at a scale of 1:200 or 1:500.

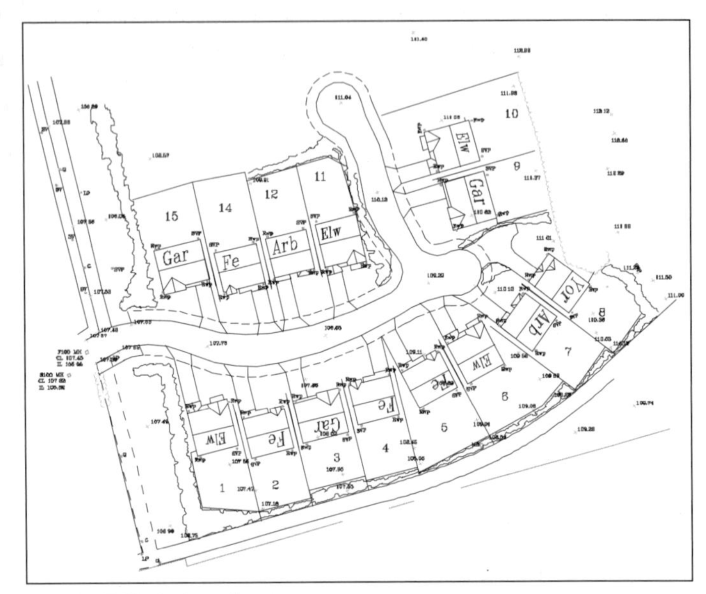

Figure 5.7: A site plan

Floor plans

Floor plans usually show the layout of rooms, doors and windows. They are illustrated by various scales – 1:100 is the most common although several instances may be found where 1:50 or 1:200 is used (see Figure 5.8).

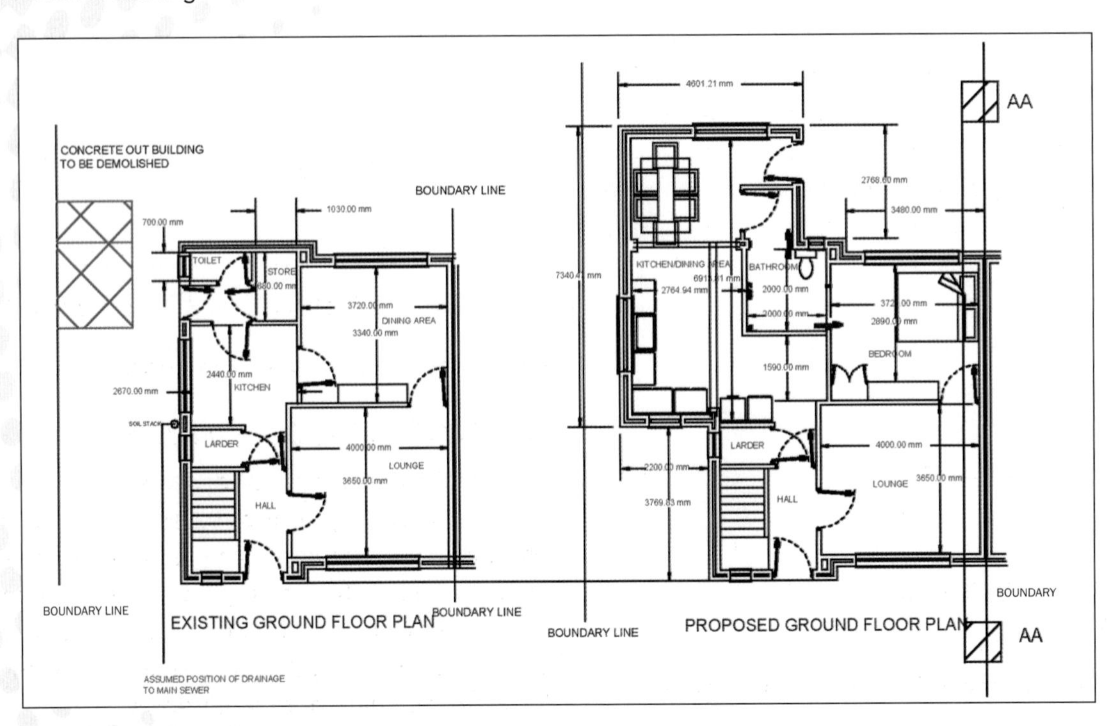

Figure 5.8: A floor plan

Section drawing

A section drawing gives a vertical view of the building and shows technical details that could not be seen on a plan or floor layout (see Figure 5.9). The scale used for vertical sections varies, but frequently 1:50 is used; while sections or drawing of individual components, a scale of 1:10 or full size could be adopted.

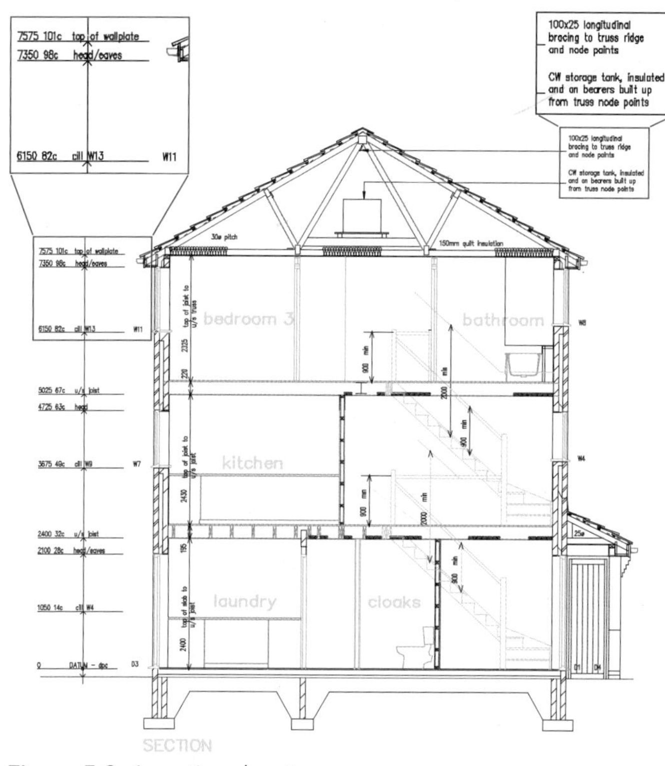

Figure 5.9: A section drawing

Construction detailing

Construction detailing means providing clear details about each component. On a scaled drawing at 1:500, many of the intricate features of the components and their interface will not be seen; hence a more detailed section will need to be provided. In Figure 5.10, the foundations can be seen on the main cross-

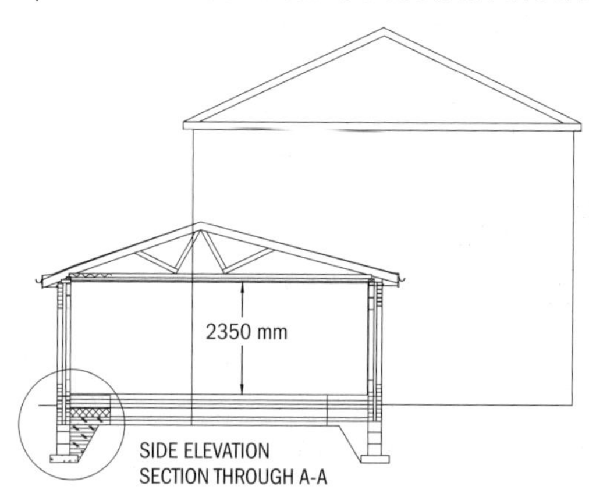

Figure 5.10: Foundations are clearly visible on the main cross-section

section, but the details and the actual components cannot be easily extracted. A section to show this part of the works is necessary; the construction detailing can be seen in Figures 5.11 and 5.12 and has a more appropriate scale with dimensions shown to help the contractor to carry out the works more accurately.

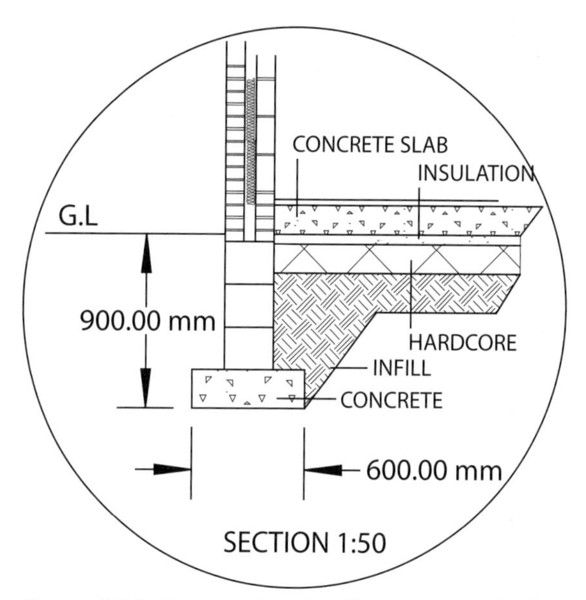

Figure 5.11: Construction detailing showing the foundations

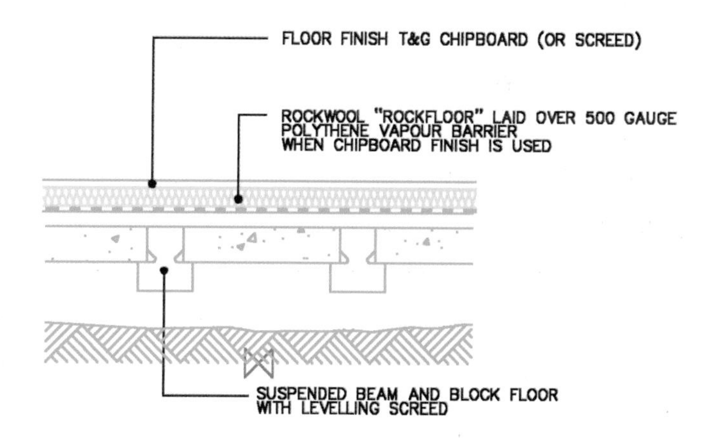

Suspended Ground Floor
Concrete Beam and Infill Block

Figure 5.12: Beam and block floor construction detailing

Door and window schedules

Door and window schedules consist of a list of all the door or window types needed for the project. It makes ordering the materials much easier (see Figure 5.13). On the schedule each piece is given the same numbered reference as it has on the drawing. Ironmongery can be added onto the schedule so any one can know what has to be fixed to the doors or windows.

WINDOW SCHEDULE				
Window Number	**Boulton + Paul Reference**	**Size W × H mm**	**Birtley Lintel Ref × length mm**	**Catnic Lintel Ref × length mm**
W1	HN10C	488 × 1050	see D1	
W2	N10C	488 × 1050	see D3	
*W3	HN12C	488 × 1200	CB70 × 1500	
*W4	HN12C	488 × 1200		
*W5	C312CC	1440 × 1050	CB70 × 2100	
W6	212C	1200 × 1200	CB70 × 1500	
W7	210C	1200 × 1050	CB70 × 1500	
*W8	HN10C	488 × 1050	CBEV50 × 750	
*W9	C213CC	1200 × 1350	CB70 × 1500	
*W10	212CC	1200 × 1200	CBEV50 × 1500	
*W11	212CC	1200 × 1200	CBEV50 × 1500	

Figure 5.13: A window schedule

Processes and procedures required to obtain planning consent

There are numerous Acts of Parliament and legislative controls that need to be considered for larger developments; for most residential and small developments, the main ones to consider are:

- Town and Country Planning Act 1990
- Town and Country Planning (General Development Procedure) Order 1995
- Town and Country Planning General Development Order 1988
- Town and Country Planning (Fees for Applications Amendments) Regulations 1997
- The Planning (Listed Buildings and Conservation Areas) Act 1990.

Local authorities have a wide range of powers to control construction development in their area and will develop plans for the area in a number of ways. The local authorities have to protect the area from developments that would have an adverse effect on the natural and built environment while considering the need to allocate land and areas for development by entrepreneurs or to satisfy local needs.

Theory into practice

Visit the website of your local authority. Check out the planning department and the services that it offers.

As well as the power to control the class of buildings, local authorities 'zone' areas of land for industrial, residential, conservation and natural beauty. Hence, in these areas building designs may also be controlled so that new buildings will be in harmony with the existing surroundings. Most of these conditions have to be considered for new developments or buildings. Several other conditions exist for change of use and adaptation of buildings. There are some projects that do not require planning consent, which include:

- small developments such as extensions to semi-detached and detached houses of not more than 70 cubic metres or 15 per cent, whichever is the greater, up to 115 cubic metres in industrial buildings or up to 25 per cent of the total volume or 1000 square metres of floor space
- agricultural buildings or buildings on agricultural land
- temporary construction site accommodation
- erection of garden walls, fences and gates
- certain developments undertaken by the local authority.

To obtain **planning permission**, an application must be made to the planning department at the local authority offices. Early discussion between the architect and the department is nearly always beneficial in order to rule out or take into account any objections or obvious conditions that will apply to any application. These discussions often advise and guide the architect as to whether a full application is likely to pass through the planning committee and approval to proceed be granted.

Key term

Planning permission – this is the formal consent from the local authority allowing you to build. You must not build without planning permission (except in the case of the exemptions listed above).

In order to gain outline planning permission, the application will need to be supported by the following items:

- four copies of the application forms available from the local authority
- one copy of a certificate confirming the ownership of the land on which the development will take place
- one copy of the site plan or a block plan indicating the position of the development.

After the application has been submitted, the planning committee may grant outline planning permission for the development to proceed to full planning permission. In this case, in addition to the information stated above four copies of every floor plan and four copies of every elevation will be required. These drawings should show any relevant features of the site, the location of any existing and proposed building and the external finishes such as the colours and textures of brickwork, type of curtain walling and so on.

The planning authority must provide a decision on all applications within eight weeks of the application being made unless there is an agreement to extend this for some reason. The planning committee sits every month throughout the year and grants or refuses planning permission by letter. Sometimes the committee will state that permission will be granted subject to a condition or improvement to the application. By all accounts, this condition will need to be reasonable, fair and relevant to the development concerned; an example of this would be the committee insisting that the development include sufficient car parking space for all occupants of the building. Outline planning permission will remain current for three years and if full permission is granted, construction work must begin on the site within five years.

Appeals can be brought should an applicant be dissatisfied with a decision. These decisions can be heard in the first instance by the local authority and then by the Secretary of State and, finally, if the applicant is still dissatisfied, by a court of law.

Building Regulations apply to most new buildings and many alterations of existing buildings, whether domestic, commercial or industrial in England and Wales. The Building Regulations exist to promote quality workmanship and to ensure the health and safety of those people that use the building. Building Regulations have differing levels of compliance so that buildings are built fit for purpose – for instance, applying Building Regulations for houses means ensuring one level of heat retention that would be inappropriate for shops and retail outlets which use large glass areas for shop displays and have a much different maximum amount of heat loss.

Building Regulations promote:

- standards for most aspects of a building's construction, including its structure, fire safety, sound insulation, drainage, ventilation and electrical safety (added in January 2005)
- energy efficiency in buildings
- the needs of all people, including those with disabilities, in accessing and moving around buildings.

Assessment activity 5.4

The architect and the project manager have been discussing the different construction methods that could be used for producing modern housing.

1. Describe two modern construction methods that could be used for building this housing estate. P6
2. The project manager eventually convinces the architect to use one of the modern construction methods. Produce a written specification for a method using timber that could be issued to the construction team. P7

Grading tips

For P6 you will need to describe a modern method of construction; look for a sustainable method employing timber or a technological product.

For P7 produce a specification that should contain as much information as possible including, British Standards, any other standard, manufacturer's installation details etc. so the un-informed person knows all about the product or method of construction you are using.

PLTS

Identifying questions to answer and problems to resolve on construction methods will help develop your skills as an **independent evaluator**.

Functional skills

Producing a specification by reading and understanding technical documents will help improve your reading skills in **English**.

Specifications

In addition to the contract drawings, which convey an image of what the finished product should look like, a **specification** provides essential information to the contractor about the materials, the finish and the workmanship required of each element of the building. The specification is therefore complementary to the drawings and it is a very important part of the documentation necessary to complete the works.

Key term

Specification – a written document that expresses information from the drawing in technical terms. It helps the reader to understand what the drawings represent.

Quantity surveyors typically write the specification, so it is essential that they have a command of both construction technology and linguistic skills to express themselves in writing. Building legislation and current standards, codes of practice and requirements for the quality of the finished product are all criteria that need to be considered if designers and specification writers are to work closely together to ensure that the harmony between drawings and specifications is maintained.

The specification is used by the following people.

The design team – as their proposals are being drafted and developed, the specification is used as a design tool. Notes, comments and initial thoughts on the suitability and durability of materials in the construction can be maintained so as to determine whether the building project can proceed in the present form. This then has an impact on the standards adopted by the design team and the methods of quality control expected of the work completed on site.

The quantity surveyor – the specification is used to translate the bills of quantity into detailed descriptions of all the components in the building project. It will be read by contractors who wish to tender for the project, which will enable an estimation of the time and resources needed to complete the works. Measured items are accounted for methodically and systematically on the specifications; thus there should be no doubt as to what is required and what the completed product should look like.

The construction team on-site – the specification is used as a reference to what the designer instructed them to build. The contractor will use the specification as a guide to the tests and quality control mechanisms that ensure the building is fit for purpose.

Well-crafted and explicit specifications make clear the extent of the works, show information that cannot easily or readily be shown on a drawing and include references to current standards and practice. Specifications also provide a useful explanation of the materials to be used and the standards of the quality expected within them. Included in the description is often any caveat or restriction on the use or installation or conditions during which the work can be carried out, such as time factors, responsibilities, curing times, application periods, which could not be effectively described or shown on a drawing.

The specification forms a very important part of the contract documents and includes a considerable amount of information. Correctly written specifications should be:

- in agreement with the drawings and the other contract documents. Cross-referencing is essential if they are to be read in conjunction with existing drawings
- logically developed and straightforward to use. They should follow a reasonable and easy-to-understand format that shows each component in turn and in an appropriate sequence
- fully descriptive allowing the reader to clearly ascertain the details of the components as described
- specific, precise and free from ambiguity; thus, not leading to any chance of misinterpretation
- concise and to the point avoiding any superfluous language or unnecessary descriptions.

The specification writing is about describing the end results rather than the means of achieving this end result. In other words, it relays information about the construction details, but it does not lay down how the construction work should take place. This is usually a matter for the contractor, except where there are health, safety and welfare issues. In those cases the specification should remind the contractor of the requirements for the safe and successful installation of a particular component in line with current legislation.

Specifications should be written to work from general principles and then move into specific elements of each component. A typical sequence is to consider the subject, then move to a description of the circumstances and finally to consider the size of the element. Descriptions taking this format usually relate specific instructions better than wordy, rambling explanations.

There will inevitably be a degree of jargon and/or technical expressions that specification writers assume the reader is capable of either understanding without explanation or is able to seek their meaning from suitable sources. However, specifications should use only words in common usage and that are universally accepted so that contractors are readily able to understand them. They should also avoid writing specifications in legal terminology.

Standard specification clauses exist in databanks and databases which can be very useful. The advent of computers, word processors and spreadsheets has aided this development and precious time and energy can be saved by using readily available software that allows the use of standard specifications.

Performance specification

Specifications can also be used to provide information to a manufacturer or supplier of components, for example subcontractors who supply doors, window frames, staircases, air conditioning, telecommunication equipment and so on. For these purposes, the specification will state what is needed in terms of performance. Examples of performance specifications state such things as the fire resistance, the acoustic performance, the dimensional tolerance and the durability of the component rather than a full description of the way the component is used in the building. Bricks, for example, can be specified as type, size, colour and texture and then several manufacturers can be approached to see if they have a product that fits the performance of what the architect is looking for so that the architect can choose the one that provides best value on behalf of the client.

Sample specification

Specifications help and guide the builder. They provide information that cannot readily be seen on the drawing. We're going to look at a sample specification for the house from Figure 5.10. Let's start with the foundations. The specification lists the ground conditions, the local authority building control officers and the dimensions of the trenchblock that the designer has specified for use in the footing.

Sample specification: Foundations

Assumed ground class IV (firm sandy/firm clay) to be checked on site and confirmed by Building Control. Depth of foundations to Local Authority approval. Concrete strip footings minimum 200 mm thick projecting 150 mm to each side of load-bearing walls 275 mm thick. 'Trenchblock' to be used below ground level.

Now, let's consider the walls. Here the components are explained in terms of the dimensions, thickness and the nature of how they go together. Knowledge of construction technology is important as references about the insulation filling necessary to comply with the Building Regulations and how the steel ties that bind both skins of the external walls together will be built into the structure need to be included. Important items such as the damp proof course and the filling made to cavities at low level along with the closings at junctions such as windows and doors will need to be explained so the contractor can complete the work to the appropriate standard.

Sample specification: External walls

275 mm cavity walls, 100 mm facing bricks, 75 mm cavity with blown fibre insulation. 100 mm medium density concrete block inner leaf, dry lined with 9.5 mm plasterboard on dabs, seated around all junctions with floors, walls and ceilings.

200 mm long stainless steel butterfly wall ties at maximum 450 mm vertical centres and 750 mm horizontal centres. Cavities filled with weak mix concrete up to ground floor level. Horizontal damp proof course 150 mm above ground level and to vertical cavity closures. Cavities closed at eaves, verge and all openings with insulating blockwork/dpc to reduce the risk of cold bridging. Flexible closers to be used vertically at all junctions with party walls.

In Figure 5.10, the floor is not immediately obvious and there is vital information required to allow the contractor to construct the components missing from the illustration in conventional drawings. Hence the specification would then include a description of the main things that the designer has included.

Sample specification: Floor

Powerfloat finish to 100 mm thick concrete slab on a 1200 gauge polythene sheet. Damp proof membrane on 50 mm sand blinding on 150 mm clean and dry well-compacted hardcore base. Damp proof membrane to lap damp proof course at edges and external walls.

So, from the specification we can gather that the ground floor is made of concrete and has damp proofing resistance built into the construction. Let's now consider the upper floors which are of a different type of construction and will need their own specification. The underside of the upper floor which also forms the ceiling of the room below is also specified here with the ceiling finished with 12.5 mm plasterboard and finally skimmed with gypsum plaster.

Sample specification: Upper floors

19 mm tongue and groove flooring grade moisture resistant type 2/3 chipboard on softwood joist. Double joists under partitions with 38 mm thick minimum solid strutting at mid span for joists spanning 2500–4500 mm. 2 number rows equally spaced for spans over 4500 mm. 30 × 5 mm galvanised mild steel straps at 2 m centres with 100 × 50 mm sw noggins fixed to external walls. Galvanised mild steel joist hangers at party walls. 12.5 mm plasterboard and skim to soffits.

The roof specification is very important and includes information on the pitch of the roof, the tiles and the roof covering. Details that the builder will need to be able to construct the roof according to the standard of workmanship that the designer wishes to incorporate are also included. Reference to bracing, mild steel straps and additional timbers included for strength is identified along with the finish to the underside of the ceiling in the upper rooms.

Sample specification: Roof

Concrete interlocking roof tiles on 38 × 25 mm preservative treated softwood battens on sarking felt. Gangnail trussed rafters to BS5268 Part 3 1985 laid at 600 mm centres with 100 × 25 mm sw diagonal bracing to BS5268 Part 3 30 × 5 mm galvanised mild steel straps at maximum 2 m centres with 75 × 50 mm sw noggins fixed to 3 no trusses and external wall at ceiling and roof level.

150 mm fibreglass quilt insulation laid between ceiling members with an additional layer laid at 90°. Ceiling finish 12.7 mm plasterboard and plaster skim. Eaves ventilation equivalent to a continuous 10 mm slot with proprietary insulation restrain. 25 mm eaves vent to sloping soffits with ridgevent.

Assessment activity 5.5

BTEC

1. The architect still is not convinced over the proposed specification for the modern method of construction and has asked for clarification of some further details.

   Produce a small sketch which includes a plan and elevation view and a cross-section with annotations of your chosen modern construction method. **P8**
2. You have been handed a set of timber-framed section details. Evaluate these using technical and architectural terminology. **M3**
3. The architect has issued several instructions that require design modifications to the modern method of construction that has been chosen. These are:
   - an increase in the storey height of the buildings
   - a change in orientation of the layout of the scheme.

Appraise what the consequences will be of these instructions. **D2**

Grading tips

For **P8** you will produce a sketch design or illustration for your chosen method of modern construction. Reference to manufacturers' websites would provide some design illustrations on modern construction methods.

For **M3** evaluate a design in terms of using the correct technical language that would be industry standard.

For **D2** you will need to appraise a set of instructions that represent design modifications to the original contract.

PLTS

Organising time and resources and prioritising actions to answer the task will help develop your skills as a **self-manager**.

Functional skills

Producing a design drawing by using a CAD package will help improve your skills in **ICT**.

Stephen Johnstone

Architectural Design

Stephen works within the design office of the local authority architecture division that produces work for the authority's capital budget works; he started as a modern apprentice and has risen up to the position of architectural technologist. Stephen's main workload has been in detailing the drawings of the structural elements that the chief architect has designed and laid out.

Stephen now has a full appreciation of what a client's brief entails and is only just starting to get involved in the pre-drawing phase of architecture. A client has an idea that has to be turned into a working drawing so it can be built on site by the skilled operatives. Through every stage there are constraints and factors that can affect a design, from the land itself to the demands of the planning authorities.

Stephen has wanted to get involved directly with clients and asked if he could run his own small design contract. The chief architect has agreed and is letting Stephen design, run and complete a £50,000 refurbishment on a school's reception area.

Stephen warms to this challenge and starts by undertaking a measured survey of the reception area so he has a detailed dimensioned drawing to discuss with the client. Stephen produces a final design that is presented to the governing body of the school for approval. Stephen also makes a small scaled model so they get a three-dimensional view of what it will look like.

Using the approved tendering process, Stephen then appoints a contractor who runs the contract through to handover to the client. When the project is finished, the chair of the governing body writes a letter of thanks to the chief architect; he was extremely pleased with Stephen's running and controlling of the project which was of excellent quality and brought in on budget.

Think about it!

- What is a client's brief?
- How would you turn the brief into a viable design?
- What additional skills would you need to run a contract?

Just checking

1 What are the stages in the RIBA Plan of Work?
2 Explain the advantages and disadvantages of using the RIBA plan.
3 List four members of the design team and identify their main functions.
4 How does an architect differ from an architectural technologist?
5 Define the term 'specification'.
6 What are the characteristics of a well-written specification?
7 What are the duties of a quantity surveyor?
8 To whom does the clerk of works report and what is the function of the role on construction sites?
9 How does an invitation to tender differ from a selective tender?
10 What documentation is necessary to support a planning application?
11 What is the difference between building regulations and planning consent?
12 What scale is most suited to a location drawing?
13 List six components shown on a floor plan layout drawing.

edexcel

Assignment tips

- Many architectural websites and the NBS provide model specifications to help you formulate your specification.
- Local authority planning websites contain recent planning applications which will contain written and drawn specifications.
- Many of the professional associations will provide details of roles and responsibilities within the design and construction teams.

Credit value: 10

6 Building technology in construction

The modern construction industry requires technicians who understand current construction methods and have an appreciation of traditional materials as well as a knowledge of plant and equipment that will be required to put these components together.

Buildings have evolved over time from simple shelters to the complex and sophisticated structures that we see around us today. Combinations of traditional materials such as brick, mortar and timber form the basis of most buildings, but the increased use of modern materials such as structural glass, plastics and prefabricated components is found in abundance as buildings become more efficient, integrate greater building services systems and other features to make them a more appropriate place to live and work. The challenge for the builders is to combine sustainable building technology with the ability to deliver high-quality buildings for society. The need to plan and organise this work, including the use of skilled labour, is necessary if this challenge is to be met.

Learning outcomes

After completing this unit you should:

1. understand common forms of low-rise construction currently used for domestic and commercial buildings
2. understand foundation design and construction
3. understand the techniques used in the construction of superstructures for low-rise domestic and commercial buildings
4. understand the implications of issues and constraints on building construction.

Assessment and grading criteria

This table shows you what you must do in order to achieve a pass, merit or distinction grade, and where you can find activities in this book to help you.

To achieve a **pass** grade the evidence must show that you are able to:	To achieve a **merit** grade the evidence must show that, in addition to the pass criteria, you are able to:	To achieve a **distinction** grade the evidence must show that, in addition to the pass and merit criteria, you are able to:
P1 explain the different forms of low-rise construction currently used for domestic and commercial buildings **See Assessment activity 6.1, page 191**	**M1** justify the selection of suitable materials and techniques for use in the construction of substructures for low-rise domestic and commercial buildings, for two different tutor-specified scenarios **See Assessment activity 6.2, page 204**	**D1** evaluate the environmental performance of modern materials and techniques used in the construction of substructures for low-rise domestic and commercial buildings, for two different tutor-specified scenarios **See Assessment activity 6.2, page 204**
P2 explain how the procedures used in subsoil investigation provide information for the design of substructures **See Assessment activity 6.2, page 204**		
P3 describe the principles of foundation design **See Assessment activity 6.2, page 204**		
P4 explain the methods used to construct different types of foundation **See Assessment activity 6.2, page 204**		
P5 explain the principles of superstructure design **See Assessment activity 6.3, page 215**	**M2** justify the selection of suitable materials and techniques for use in the construction of superstructures for low-rise domestic and commercial buildings, for two different tutor-specified scenarios **See Assessment activity 6.3, page 215**	**D2** evaluate the environmental performance of modern materials and techniques used in the construction of superstructures for low-rise domestic and commercial buildings, for two different tutor-specified scenarios **See Assessment activity 6.3, page 215**
P6 describe the techniques used to construct and finish the component elements of a superstructure **See Assessment activity 6.3, page 215**		
P7 explain the implications of environmental issues and legislative constraints on building construction **See Assessment activity 6.4, page 222**	**M3** evaluate three pieces of legislation applicable to the construction process in terms of the relevance of the legislation and the stage at which each applies **See Assessment activity 6.4, page 222**	
P8 explain the purpose of the various parts of the infrastructure required to support the construction process **See Assessment activity 6.4, page 222**		

How you will be assessed

The evidence requirements for pass, merit and distinction grades are shown in the grading criteria grid. Evidence for this unit may be gathered from a variety of sources, including well-planned investigative assignments, practical work or reports of practical assignments. You will be given written assessments to complete for the assessment of the grading criteria for this unit.

This unit will be assessed by the use of three assignments:

- Assignment one will cover P1, P2, P3, P4, M1 and D1
- Assignment two will cover P5, P6, M2 and D2
- Assignment three P7, P8 and M3.

Brady

The UK contains many different types of buildings all of which at some time may require work undertaking on them. I did not realise that you often have to match traditional materials with modern sustainable methods. I also now realise that building work has to follow legislation to ensure that it is carried out safely and with protection for our environment.

This unit has made me aware of the methods used for the foundation design that hold up a building safely as well as the techniques used over ground to provide a superstructure.

Over to you

- What do you already know about low-rise construction?
- What does the term 'superstructure' mean to you?
- What are you looking forward to learning about in this unit?

1 Understand common forms of low-rise construction currently used for domestic and commercial buildings

Build up

Prefabrication

Prefabrication of a building is becoming more popular. In groups, think about the following questions:

- How would you explain 'prefabrication'?
- Would you use this method for the build of high-quality homes?
- Would you consider this a sustainable method of construction and why?

Forms of low-rise construction

Prefabricated construction including timber frame

In the UK, timber-framed houses have been built for over 300 years, but they represent a modern method of working. Now builders would use more modern construction techniques which enable timber panels to be crafted in workshops and then taken to site for assembly. This development, along with the use of modern materials and the increased use of insulation to comply with legislation on heat loss, has made homes more efficient. The reduction in on-site time has also made timber-framed housing more attractive to builders who can offer better homes at lower prices.

The use of timber-framed panels has also changed the function of some of the elements of the structure. Some items that were at one time load bearing are now no longer used to support other parts of the structure.

Steel frame

Steel frames use a series of beams and columns and may be composite in design, that is, a mixture of two technologies. Steel frames structures might contain concrete floors and steel supporting beams.

Portal frames

A portal frame is a shape that looks like a **portal**. It has a low pitched roof member that is attached to a column at either end. Portal frames can be constructed from several different materials: timber, steel or reinforced concrete. Generally, though, modern portal frames are constructed from steelwork that enables a high strength-to-weight ratio, which gives a large economical span. Concrete portal frames are generally pre-cast within a factory environment that ensures consistent quality; then they are assembled on site using a mobile crane. Timber portals are constructed using **glulam** construction and stainless steel fixings. With all three methods, cladding rails hold the portals together along with eaves beams.

Key terms

Portal – a large gateway or doorway.

Glulam – a process of gluing timber strips together to form a solid beam within a mould.

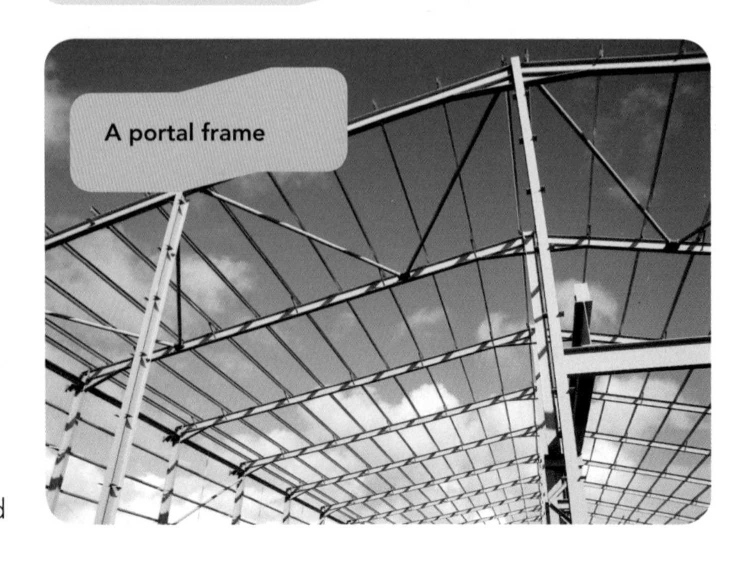

A portal frame

Portal frames have the following advantages:

- large spans can be achieved
- speedily erected
- factory-produced quality
- easily extended and adapted
- good height-to-strength ratio
- enable lighter foundations than traditional construction of brick and block
- aesthetically pleasing when timber portal frame is used
- are able to be recycled when the building comes to the end of its life.

Their disadvantages and limitations include the following:

- require some form of coating to prevent rust
- require fire protection coating
- expensive to bend
- external coatings have to be maintained.

Concrete frames

Concrete frames are generally divided into two forms: in situ and pre-cast. In situ concrete frames, which consist of beams and columns, are constructed by pouring mixed concrete into a formwork. Once the concrete has set and attained enough strength, the formwork shuttering is removed. Pre-cast concrete may require some of the beam-to-column connections to be cast on site or bolted to form a rigid structure.

The advantages of concrete frames include:

- in-built fire protection
- mouldable into any shape
- high compression strength
- no secondary finish required
- faster construction period using pre-cast
- variety of surface finishes.

Their disadvantages or limitations include the following:

- require initial support
- require use of crane
- poor strength on tension
- assembly requires highly skilled workforce.

A concrete frame

Load bearing and non-load bearing

Timber-framed buildings use panels typically made from a 100 mm × 50 mm softwood carcass with plywood or other sheet materials nailed to it to give it strength. The timber panels are load bearing and support the roof, while any **cladding**, such as brickwork, acts only as a decorative facing. Internally, the panels act as partitions, simply dividing the interior into useable space; they do not support the structure and are therefore non-load bearing. Moreover, the internal partitions may need special fixings and fittings if they are to be used to support other items such as cabinets or other wall-mounted items.

Key term

Cladding – the lightweight material that forms the enclosure of the building. Cladding is usually lightweight because it does not have to carry the structure.

Single-storey buildings

In single-storey commercial premises, the main structure is frequently made from a structural steel frame. This frame provides the anchor points for rails or cross rails – known as purlins – that support aluminium sheeting which has been formed and shaped into a corrugated profile. Aluminium cladding uses significant amounts of insulation material to enable the building to comply with relevant heat loss legislation and it provides some stiffening of the frame to wind forces, however, it is mainly considered to be non-load bearing. In most portal frame buildings, the first two metres of the walls are made from traditional masonry (brickwork/blockwork) walls, which provides additional security at low level.

Low-rise buildings

As buildings increase in height, they need additional bracing and strengthening. Low-rise commercial buildings of up to three storeys are typically made from structural steel in standard sections that provide an open-plan area. These areas can be further split into smaller spaces by the occupier and tailored to their needs using demountable partitions and non-load bearing elements.

Detached housing

Many designs for detached houses in the UK have been simplified and standardised to National House-Building Council (NHBC) standards. The rising standards have led to an increased number of features being incorporated into most detached houses, including en-suite bathrooms, home offices, fire and intruder alarms.

Take it further!

The NHBC (National House-Building Council) is the industry body responsible for setting and maintaining construction standards for new homes in the UK. Carry out some research on the Internet and find out about some standardised detached housing designs.

Terraced housing

In Victorian times, terraced housing provided volume housing at cheap prices as well as the ability to accommodate large numbers of people in a given area. Modern designs use a similar format of structural walls coupled with roofing systems that stretch across multiple properties with appropriate fire and privacy breaks to create individual dwellings.

Roofs

Roofs can take many forms ranging from flat, gabled, monopitch and hipped; they can also accommodate multiple pitches and features. Traditionally, roofs have been built on site by carpenters using large sections of timber and creating the features as work proceeds. More recently, the use of trussed rafters has meant that the roof is crafted using a series of manufactured triangular frames that are then assembled in situ. The use of trussed rafters has resulted in improvements in the standard of workmanship and thermal performance. The use of smaller, more economic sections of timber has also made roofing more affordable.

Flat roofs are cheaper to construct than pitched roofs. Their affordability has led to them being used for smaller contracts, such as small extensions to existing domestic dwellings. Flat roofs need to be constructed correctly or they may suffer from a great deal of condensation. For bigger projects, the use of large sheet materials or rolled aluminium profiled sheeting with correctly positioned **vapour checks** has been successful.

Short-span roofs can be made from timber, whereas longer-span requirements will need stronger members made from either steel or concrete. Short-span and medium-span roofs are cheaper to construct.

Key term

Vapour check – impermeable barrier or membrane used to prevent moisture passing through the structure.

Activity: Roof components

Draw a traditional pitched roof with a hip section and a dormer window. Identify the components of the roof, using the correct terminology for each part.

Theory into practice

The span of a roof is dependent on the roof loadings and the use of the building, for example if clear, uninterrupted areas are required with no columns. This may be required within a warehouse where large forklift trucks operate. Generally, the wider the span, the deeper the beams required to carry the load.

Implications of different forms of construction

Traditionally built buildings are labour intensive with most of the construction being done on site. The process is expensive and can be difficult to monitor quality control. Various trades people are required such as a ground working crew, a bricklaying gang, several joiners or carpenters, roof tilers/slaters, plasterers and then the services engineers, such as plumbers and electricians, to fit out the interior services. It also requires that materials are delivered to site which then have to be stored ready for use.

Prefabrication

Prefabrication involves manufacturing and constructing components such as kitchen units, staircases and window frames in a factory. The components are then taken to the site where they are assembled. Using prefabricated components reduces the amount of time spent on site compared with the traditional method of constructing everything on site from scratch.

There has recently been a shift towards prefabrication of single or multiple components. This has not only increased the speed of on-site erection for the completed units but has also resulted in improved performance. This is due to the fact that a high level of quality control can be maintained in workshop conditions before the completed components are taken to site for assembly.

Take it further!

In 2003, Forest Bank Prison was built near Manchester. The use of prefabricated cells allowed the prison to be built in reduced contract time since all the cells were designed to be constructed off site, prepared with all the necessary fittings and taken to site for assembly on a prepared base. Later, when all the cells were bolted together, services were connected and secure doors were fitted. Research on the Internet; see if you can find information about other prefabricated buildings.

Using prefabricated components improves the quality of the finished product and also speeds up work on site. However, the success of prefabrication within the construction process depends on the following:

- A well-designed construction site layout –This typically relies on the construction site manager having a detailed knowledge of the operations and the layout of the construction site to ensure that each component is integrated at the appropriate time.
- Correct sequencing of the construction works – This relies on the designer and the construction site manager working together to ensure that the correct plant and components are delivered to the site in the right sequence.
- Use of prefabricated components wherever possible – careful consideration of the structure and the design brief will enable the designer to identify all components of the project that are suitable for prefabrication. The construction site manager should be consulted as they are also likely to have detailed information on the practicalities of delivering large items to site and whether it will be feasible to assemble and install large items on site.
- Use of plant at all stages of construction – wherever possible, skilled craftspersons should use machinery to help them assemble and install components correctly. The reduction or elimination of as much physical labour as possible is a key feature in improving the economic viability of prefabricated housing and building projects.

A prefabricated house

Non-load bearing elements such as kitchen cabinets, door frames, bedroom furniture and many other components are now frequently made off site and taken to the building project for assembly and installation rather than being built in situ. This prefabrication of elements can also be used for some load bearing components; standard forms of structural steel sections can be cut, bent and shaped to specifications obtained from the design drawings and specifications for offices or other steel frame buildings. The structures are then taken to site and placed on prepared slabs and linkages and assembled in sequence from the ground upwards.

While there has been a significant improvement in the standard of workmanship and the speed at which buildings using prefabricated components can be built, there are some limitations to the use of prefabricated elements. Usually, not all components are built off site and there will always be a need for some minor alterations as work proceeds. This is especially true of instances where the design has been agreed in advance. If the design needs to be changed, then it may be easier to make any alterations required on site while work is taking place.

Where the whole building is completely prefabricated, variations are possible. Care must be taken to ensure that the designers are consulted to check that the removal of or amendment to any component does not adversely affect the structural integrity of the building.

Buildings

Houses and flats

Houses are typically small structures of one or two storeys with simple components such as bricks, mortar and timber used to make floors, roofs, windows and so on. Plastic resins and other synthetic materials have become available in recent years and have made the **thermal performances** and noise exclusions of houses significantly more efficient. Improvements in the comfort features of houses, such as integrated kitchens, built-in bedroom furniture and building services, have meant that our houses have become increasingly sophisticated structures. These still have to comply with legislation and regulations that place high standards of performance on both the quality of the finished product and also the workmanship of the assembly of these components.

Flats and apartments are individual dwellings that are grouped together in a common structure, making the construction and assembly of these units more economic than if using the land for only one unit.

Theory into practice

Working with a partner, list all the different types of buildings you can think of. Explain how they differ from each other.

Warehouses and light Industrial units

Warehouses and light industrial units are commercial buildings. Most likely, they are **open-span buildings** with steel frames that are clad with aluminium or functional materials which allow space to be enclosed both economically and safely.

Key terms

Thermal performance – the ability of a material to retain heat in the structure.

Open-span buildings – these are typically built using a skeleton frame, which allows an open-plan floor area that can be divided into smaller spaces. This clear floor space is particularly well-suited to offices where the final floor plan can be changed to suit any client who wishes to move into the building.

Retail

Retail outlets can be purpose built or adapted from existing buildings. Newly built premises often take the form of large open-span, portal frame buildings which provide a large floor space; for example, a supermarket or a low-rise department store. Some smaller shops and retail outlets are modified from houses. These need special works to adapt them from a residential property into a commercial retail building. The requirements for fire resistance, noise reduction and ventilation are stricter for commercial buildings than they are for domestic dwellings, and any modification will need to take account of these if the adaptation is to be successful.

Theory into practice

Working with a partner, list all the things that you would need to change in order to convert a house into a small retail shop. Will the types of commodity the shop sells make any difference?

Offices

Offices use large open-span buildings and many employers choose to create smaller working spaces based on their functional requirements; for example, secure areas for exchange of money or small private units for individual discussions and meetings. Small office buildings can be built in much the same way as housing. Once there is a requirement for more than four storeys or for larger spans, a more robust method of construction will be necessary, perhaps including the use of structural steel rather than bricks, mortar and timber.

Assessment activity 6.1 P1

Explain the different forms of low rise construction currently used for domestic and commercial buildings. P1

Grading tip

To achieve P1 you will need to identify and describe the different forms of construction used for each of the building types.

PLTS

Asking questions and extending your thinking on methods of construction will help develop your skills as a **creative thinker**.

2 Understand foundation design and construction

The choice and selection of the building type will rely largely on the functional requirements of the building and the use to which it is put. In order to determine the suitability of the ground and its capacity to support the building, a thorough and detailed site investigation is necessary so that the foundations and the fabric of the building can be selected.

Subsoil investigation

Site survey and subsoil Investigation

The main objective of a site investigation is to examine the ground conditions so that the most appropriate type of foundation can be selected. A typical site investigation begins by establishing the geology of the site with a **desk study** and a **walk-over survey**; it continues with an examination of the **geotechnical properties** of the ground. There is a variety of techniques for a direct investigation and the selection of appropriate methods for given locations.

Key terms

Desk study – an investigation of information about a piece of ground undertaken by reviewing existing records.

Walk-over survey – visiting the site to enable the surveyor to match the information from the desk study to what they see in the field. Experienced surveyors can get a feel for ground conditions by undertaking a walk-over survey.

Geotechnical properties – how soil is likely to perform when imposing loads on it or what will happen when water is removed to allow work to take place.

Site investigation involves gathering all the information on ground conditions which might be relevant to the design and construction of a building on a particular site. On a site intended for low-rise development, the desk study is the first stage of investigation. This involves checking existing records such as geological maps of the site, utilities records, local historical archives, etc. The desk study, along with the walk-over survey, will (hopefully) give sufficient information about the ground and groundwater conditions, and the problems that they may pose for the construction and the finished building. The walk-over survey:

- checks the accuracy of desk study information
- obtains any additional information required to ensure that the building can be constructed safely and within budget.

Take it further!

Where can you find information about soil or site conditions? Use the Internet to see what information is available through utilities companies and the local authority.

There are many cases where a site investigation is specific and tailored to a particular site. Generally, all site investigations should provide the following information:

- classification of soils
- soil profile
- soil parameters.

Classification of soils

During the ground investigations the soils with similar engineering behaviour are classified into broad groups. The simplest classification used by geologists is:

- rock
- granular soils such as sands and gravels
- cohesive soils such as clay
- organic soils such as peat
- fill or made ground.

These broad soil classifications coupled with simple tests to determine **soil parameters** or to detect the presence of chemicals harmful to construction materials are normally sufficient for low-rise buildings.

Key term

Soil parameters – how the soil will react to building work and imposing loads. The soil can be expected to carry a certain amount of weight depending on its parameters and its characteristics.

Soil profile

Ground investigations identify the levels of the various soil or rock types on the site by recognising the boundaries between them. This is known as the soil profile and builds up a picture of what the ground looks like under the surface of the Earth. Boundaries between the various soil types are not always distinct, which means it is difficult to tell what conditions will actually be found when digging into the earth.

Soil parameters

Engineering design and calculation may require finding out the soil parameters especially ground that has difficult soil such as clays since these soils perform variably; hence the need to quantify the load imposed by a building and the ability of the ground to support this load. In many situations, soil parameters may not be required, for example where conditions are particularly good (such as rock, which in most cases will support buildings very well), or particularly bad (such as peat, which should be excavated away to deeper, better load bearing ground).

Planning and carrying out the site investigation

Desk studies and the walk-over survey are both indirect investigations and are normally cheaper than direct investigations which involve drilling into the ground, extracting samples and then laboratory testing of samples to determine their properties. These operations require planning if they are to be fully effective. The stages of a full investigation are as follows:

Stage 1: Carry out a detailed desk study and walk-over survey. Identify the probable ground and groundwater conditions. Locate areas on the site that are likely to cause problems such as areas of fill, old hedgerows and trees, mineshafts, low-lying ground, etc.

Stage 2: Make initial designs for the structure and the site. Work out the positioning of the proposed structure so as to avoid as many problems as possible. Design structural forms with special regard to anticipated ground hazards such as peat or excessive water conditions. Make preliminary estimates of the type of foundations required; determine the position of critical slopes and retaining walls.

Stage 3: Plan the direct methods required for the site investigation. Identify the depths of investigations required at different locations around the site. **Boreholes** should always penetrate completely through **made ground** or in-filling. Identify suitable in situ and laboratory testing methods for the expected soil conditions. Decide on the number of exploratory holes and the sampling and testing frequency, making allowances for the presence of unforeseen ground or groundwater conditions.

Stage 4: Keep records of the investigation. Record the basis of the planned site investigation and the expected ground conditions. The specialist contractor who carries out the work will then know if the ground conditions they encounter are unforeseen, in which case they may have to alter the scope of the field or testing work.

Ground investigation techniques

Many different methods of ground investigation are available for direct site investigations. It is important to select methods which:

- will work in the particular ground conditions expected at the site
- give the information required to resolve any expected construction problems
- give information that will help resolve the structural design calculations
- keep within the financial constraints of the construction process.

There are a number of techniques involving the excavation of or drilling into the earth. Let's look at some of the methods available for site investigation

Trial pits

Trial pits are extremely valuable if the depth of the investigation is limited to approximately 5 metres or less, which is a depth that can conveniently be excavated using a back-actor or 360 degree slew hydraulic excavator. Trial pits are particularly useful in the investigation of sites intended for low-rise construction which have foundations that are generally 0.4– 0.5 metres wide and 1–3 metres deep; so the depth of the investigation is 3–5 metres.

Routine trial pit records are normally kept as a 'log' with the soil profile recorded as the hole is dug. This type of excavation is also relatively cheap and offers quick results, but care needs to be taken with exposed excavations, as some pits may need shoring with timber supports or hydraulic earth props. Furthermore, open excavations are dangerous and should be clearly marked or cordoned off to prevent anyone falling into them.

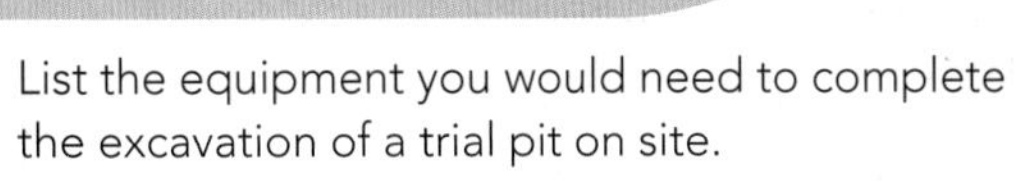

List the equipment you would need to complete the excavation of a trial pit on site.

Auger holes

Auger holes, unlike trial pits, do not allow the soil to be examined in situ. Auger holes are normally made by hand turning an auger drill into the ground although some light power tools may also be used. Typical auger holes are 75–150 mm in diameter. Samples of the soil are collected at the surface when the earth is displaced by the **helical auger flight** and the information is recorded and the descriptions collated to allow an engineer to produce a cross-section drawing which represents the conditions below ground.

Key terms

Boreholes – holes sunk into the ground to extract soil samples at differing levels. The information is recorded as the holes are drilled so that the design engineer discovers at what depth each soil is found.

Made ground – any ground that has been artificially made from material placed from previous works, for example layers of stone compacted and laid to form a level surface ready for construction work.

Helical auger flight – the technical term for the large spiral drill of the auger drill.

Ground investigation techniques

Window sampler

A window sampler is a steel tube usually about 1 metre long with a series of 'windows' cut in the wall of the tube which are used to view or take specimens of the soil. It is driven into the ground by a lightweight percussion hammer, then extracted using hydraulic jacks. Samplers come in a range of diameters with the largest diameter driven in first, followed by a smaller diameter; this allows a sample from the bottom of the hole to be taken. This process continues with the use of smaller diameter samplers until an adequate depth is investigated. The results are recorded as an ongoing process up to a depth of approximately 8 metres.

Boreholes

In the UK, 150 mm or 200 mm boreholes are normally made using light percussion equipment, which is portable and is often towed behind a four-wheeled drive vehicle. The main advantage of light percussion drilling is its ability to make deeper holes in a wide range of ground conditions; this is necessary for civil engineering projects, medium-rise construction and low-rise construction built on poor bearing ground. This method, however, is considerably more expensive than shallow trial pits and auger holes.

In clay soils, the borehole is made by dropping a hollow tube – a claycutter – into a hole so that the clay becomes lodged in the tube's base. The claycutter and its contents are then lifted carefully to the surface.

In granular soils, a hollow tube with a flap valve is surged or pressed in the ground using water pressure. Soil drops out of this water/soil mixture and is collected. Material taken from the drilling tools is usually also retained as small 'disturbed' samples which can affect the results of the soil analysis. Ideally, undisturbed samples are better since they show what the soil is actually like. However, disturbed samples can be analysed in a laboratory, which is more costly but more accurate in determining the soil's load bearing capability.

In cohesive soils, samples are taken at intervals of 1.0–1.5 metres by driving a sample tube into the bottom of the hole. These undisturbed samples are often taken to a laboratory for analysis. On site, the supervisor records data which is relevant to the sampling such as weather conditions, ground conditions, etc.

Boreholes should penetrate all deposits unsuitable for foundation purposes such as unconsolidated fill or material that needs further compacting such as peat, organic silt and very soft, **compressible clay**. Borehole depth depends on the **stress distribution** under the foundation and the depth requirements should be reconsidered when the results of the first borings are available. It is often possible to reduce the depth of subsequent borings or to confine detailed and special exploration to particular strata.

Key terms

Compressible clay – clay that can be compressed or compacted to increase its strength or load bearing capacity.

Stress distribution – this is how the foundation distributes the load of the building. A very wide, flat foundation will support more load than a narrow strip foundation.

The maximum number of boreholes will depend on the complexity of the local geology and the planned construction project, and it may change as more information becomes available from the early investigations. The location of boreholes depends on the nature of the site with additional boreholes being drilled at problem areas and at locations near the site of the proposed structure. Where practicable, boreholes should be located along **grid lines** at regular intervals to enable **section drawings** to be produced.

Thinking point

The cost of the site investigation is around 0.5–1 per cent of the capital cost of the construction contract. It is considered good practice to invest adequately at this stage of the construction. What is 'capital cost'? Do some Internet research and find out!

Radon gas

Radon is a radioactive gas that occurs naturally and is found in many locations in the UK, with areas such as Cornwall and South Wales having higher concentrations of the gas. It has no taste, smell or colour and is produced from the decay of uranium that is found in small quantities in all soil and rocks. Concentration levels of radon gas can build up in enclosed spaces and be found in excavations during construction work. Environmental health officers are appointed by the local authority to determine safe levels of radon gas in existing properties; the Health and Safety Executive has the task of monitoring how contractors deal with radon gas during construction work. Contractors have a duty of care under the Health and Safety at Work Act to provide a safe and healthy place of work; Building Regulations require that buildings and buildings extensions constructed after 2000 in radon-affected areas have protective measures installed during construction. To prevent the build up of radon gas, barriers are built into the structure. This barrier must cover the total area of the foundation of the building.

Thinking point

Building Regulations are appropriate for England and Wales. Separate documents exist for Scotland and Northern Ireland. Do some research and find out what these documents are called.

Groundwater conditions

Groundwater is found below the ground in the spaces and cracks between soil, sand and rock. The general level of water in the ground is known as the **water table** – the level of the water table is of interest to builders since their operations will be affected by the presence of water. Digging into the ground can cause groundwater to fill up any trenches and excavations.

Construction companies have to plan how to remove or deal with water during construction; groundwater conditions are significant because they can affect construction in a number of ways:

- A high water table can lead to extra costs – from de-watering techniques, increased support for trenches and excavations, ground stabilisation requirements, etc. – and make construction more difficult.
- The presence of chemicals in groundwater, such as acids and sulphates, can lead to damage of foundation concrete and other materials used in the substructure .
- A high groundwater table implies that **pore-water pressure** in the soil is high, which usually means that the soil is weaker. As well as influencing foundations, high pore-water pressures will adversely affect the stability of slopes and the pressures on **retaining structures**.

Key terms

Grid lines – an imaginary series of lines running north–south and east–west which allows designers and engineers to plot key positions on site.

Section drawings – a profile of the ground using the information from the boreholes next to one another – that way engineers can predict what happens to the ground between each borehole.

Water table – the level of water found in the ground during excavations.

Pore-water pressure – the pressure that water exerts as it moves through the fissures and cracks in the ground.

Retaining structures – walls and buildings that hold back or support the earth.

Theory into practice

What are the implications for health, safety and welfare of personnel working on site in excavations where water is present? What can be done to minimise risks?

Foundation design

Foundations carry the weight of the building, transferring the load safely to the ground below. They spread the weight of the building, and hence the load, to an acceptable level of force exerted on the ground. The **safe loading** of the building will need to be determined by calculating the capability of the soil to carry the load. Reference tables in the 'Building Regulations Document A 1/2 Minimum Width of Strip Foundations' allow designers to select foundations of known dimensions and performance matched to the soil characteristics.

Key term

Safe loading – the solution calculated from the soil's ability to carry a load plus a factor of safety.

Impact on foundation design

Foundation design is complex and the following factors need to be considered fully in order to select the correct foundation.

- Loading – the imposed load from the building including the deadweight of the structure and any live loads exerted from wind forces.
- Water table – the usual choice is to remove the water found in excavations either as a temporary solution or to install land drains and remove the water on a permanent basis.
- Contamination – water and nitrates in the ground combine and produce acid that erodes concrete over a period of time. Sulphate-resistant concrete is specified for these purposes but care is required both in its handling and its placement to ensure that it reaches its compressive strength in good time.
- Bearing capacity of the soil – following a thorough site investigation, it should be accurately determined how much force the soil will support. The bearing capacity of the soil can be enhanced or improved with ground stabilisation techniques or further excavated to lower levels of earth that will support the load with better bearing capacity.
- Cost – the respective costs of any options should be explored and taken into account by the design engineer.

Principles of design and factors affecting choice of foundations

The selection of a foundation for a particular project depends upon:

- the type of building structure
- soil conditions
- external or site constraints, e.g. the size and shape of the site layout
- type of foundations available (see Figure 6.1).

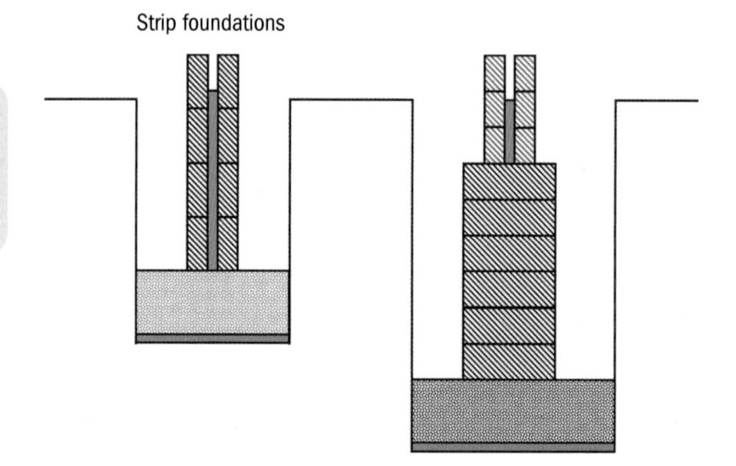

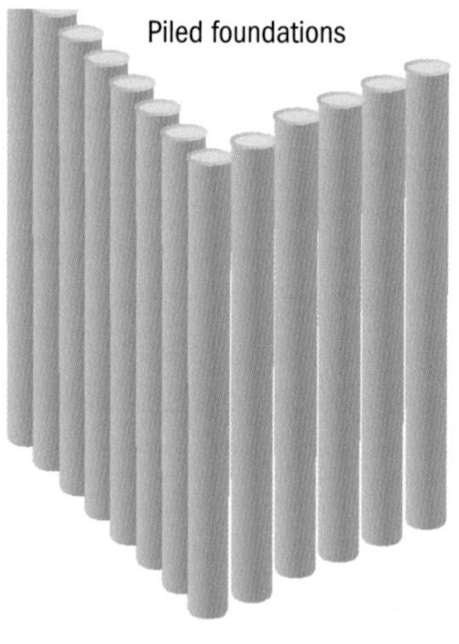

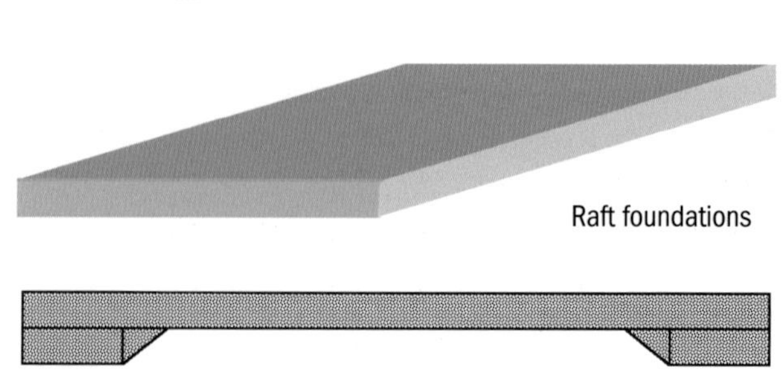

Figure 6.1: Strip, short-bored piled and raft foundations

Low-rise buildings, such as most houses or domestic dwellings up to about four storeys with load bearing walls, generally allow line loads from walls to be distributed to the earth below via strip foundations.

Theory into practice

Which types of foundation are most suited to a portal frame, low-rise flats and a simple detached house? Which of these foundations are suitable for more than one application?

Heavyweight buildings and those with a requirement for open plan spaces, such as offices, result in heavier loads that are usually delivered to the foundations in isolated points via columns from a structural framework.

The two main characteristics of soil conditions that affect the choice of foundation are its strength, or the maximum loads that can be carried without failure, and excessive **settlement**. Normally, the settlement of a soil relates to spread foundations such as strip, raft and pad foundations whereas the strength of the soil relates to piles.

Soils such as peat, silts, soft alluvial soil and filled ground have very poor load bearing and settlement characteristics; while poor cohesive soils such as clays are very sensitive to changes in moisture content. In most cases, these types of soil prove unsuitable for bearing loads and the excavation of these unsuitable soils and the siting of foundations on better load bearing ground will prove to be the most satisfactory solution. There is also a range of techniques that can be employed to 'stabilise' the ground and improve the load bearing capacity.

Cohesive soils are weak compared to sands and gravel but have reasonable bearing capacities. They are also prone to long-term settlement and, with this in mind, foundations should be of an appropriate size. As with poor soils, cohesive soils are sensitive to changes in the moisture content and are prone to swelling and shrinkage when the water content of the soil is altered.

Non-cohesive soils such as sand and gravel are generally stronger and have a higher bearing capacity and low **compressibility** compared with cohesive soils. Settlement usually occurs instantaneously during the construction of the building. This type of soil may prove acceptable to any of the foundation types selected such as strip foundations, piled foundations or raft foundations.

Key terms

Settlement – the way that soil reacts to having a load placed upon it – usually the soil 'sinks' a little due to the extra weight placed upon it, although this settlement should not affect the building a great deal.

Compressibility – the ability of soil to compress and withstand a load imposed upon it.

Rock is generally the strongest material on which to put a foundation and its safe load bearing capacity is not usually a major consideration. Some rocks, however, such as chalk will behave more like weak soil especially when the moisture content is high.

Contaminated soils can be problematic for the foundation and the other substructure works. A more detailed investigation is normally made in order to determine the extent and concentration of the contamination. Only after these are known should the overall strategy for the site be decided: heavily contaminated sites may be costly and impractical to clean up but may prove suitable for light industrial use. Typical examples of contaminated land are:

- landfill sites
- gasworks sites
- sewage farms and works
- scrap yards
- industrial areas.

The two most common approaches to the problems associated with contaminated ground are either removal or capping. Removal of the soil is a permanent way of approaching the problem, but it is costly and disposal to licensed tips can be awkward because of the nature of the material. Capping involves the sealing of the material by a layer of clean material of approximately 1 metre deep. This effectively places the contaminated soil in a zone that is not readily accessible. Although consideration must be given to the effects of placing material in such proportions, capping is usually the most economical and common form of treatment.

Theory into practice

Use the Internet to find out how you could dispose of contaminated ground or materials. What legislative controls are in place to ensure safety when disposing of contaminated materials?

When foundations need to be built on poor ground two solutions can be considered:

- to excavate the soil away until ground of good bearing capacity is found
- to improve the density of the soil so that it can accept the load of the building.

These techniques normally employ large pieces of machinery that can have an effect on the site and its constraints. The nature of making adjustments to the ground conditions makes these techniques suitable for isolated or relatively exposed sites where there is little risk of damage to adjacent buildings.

Vibro-compaction of poor soils has been employed for more than 30 years. It consists of a large vibrating poker that is vibrated into the ground on a grid pattern of approximately 2–3-metre centres. Sand is then pumped in through the poker as it is extracted to fill the void. The vibration rearranges the particles of the soil to make them denser. The treated soil is then suitable for light-use spread foundations such as strips and rafts. The foundation must be designed to accept local soft spots that have been missed by the treatment.

A similar concept to vibro-compaction, vibro replacement is more suitable for poor cohesive soils such as peat and silt. The technique requires that stone columns are installed through the poor soil on a grid pattern of 1–3-metres over the site. Holes are drilled in the earth using a poker similar to the vibro-compaction treatment or a suitable piling rig and, with the installation of the stone columns, the ground is compressed; the density increasing and some of the load is transferred to a lower level of the ground that can support it more effectively. Over the past 20 years, dynamic compaction has been used on poor soils on remote sites. The technique uses a large crane to drop weights of approximately 20 tonnes from a height of approximately 25 metres on a grid pattern across the site. Most sites need about three to five passes before a satisfactory density is achieved. There are problems with the use of this method, most notably the restriction on the site conditions: it can only be considered where the dropping of a 20-tonne weight will have little or no effect on neighbours.

Piling

Piling is used in situations where the ground is unsuitable for heavy loads, in particular where the loading of the building is concentrated on a few points. Piles transmit the load of the building to strata beyond the depth of the practical reach of spread foundations. There may be a high water-table level where spread foundations could be employed but where piling may provide a cheaper answer than lowering the water table. Piled foundations may also provide a satisfactory solution where a site is restricted and spread foundations or their excavation may cause problems to adjacent buildings.

Piles can add tension to resist uplift, which might be caused by wind loads, or buoyancy conditions from a fluctuating water-table level, and are designed to take vertical loads. They can also accommodate some horizontal movement; if this horizontal movement is expected to be of a sizeable nature, raking piles are usually used.

Piles are classified as either replacement or displacement. Replacement piles remove the earth and place another material – typically concrete – in the hole created by the boring tool. Displacement piles are driven into the earth by a piling hammer, by doing so, the soil is displaced. The following are the main types of pile:

- bearing piles which transmit the building load directly to solid strata
- friction piles which rely on the frictional resistance to the ground. They are used where safe bearing strata cannot be reached
- consolidated piles which are used in a situation where the ground is weak or waterlogged strengthening the overall nature of the ground
- sheet piles which are used to contain earth and prevent movement that would result in a weakening of the natural foundation. They are more frequently used for temporary works but can be permanent.

The method of driving and sinking the piles depends on various things, including site conditions, space and height available, proximity of buildings, avoidance of vibration and number of piles required (see Figure 6.2).

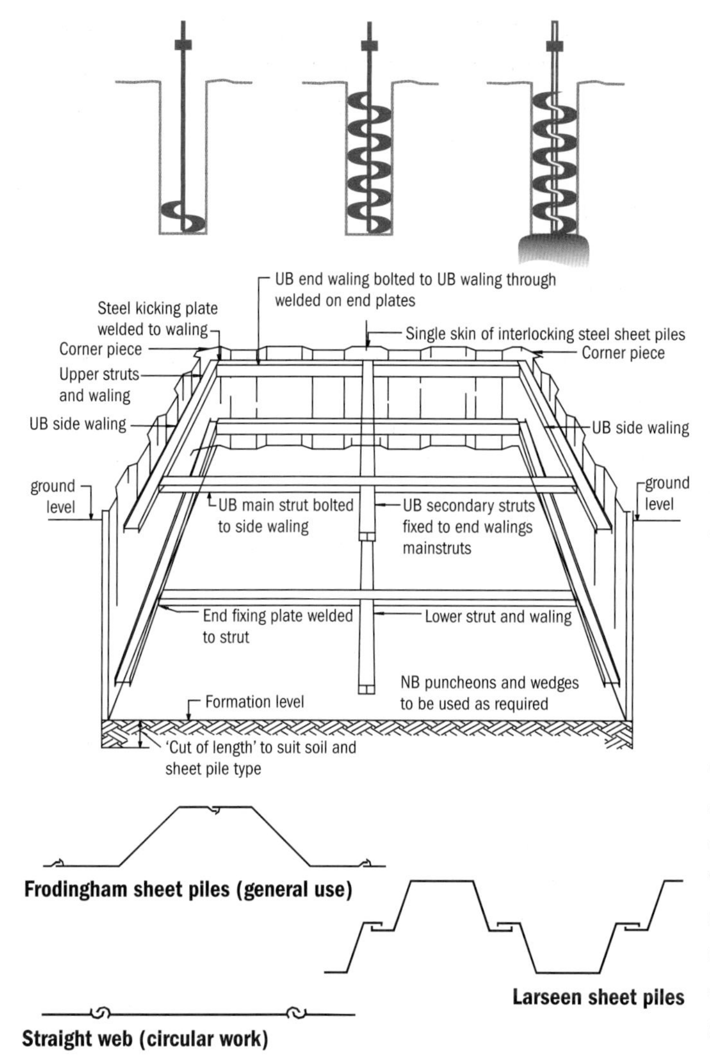

Figure 6.2: Types of piles

Timber piling

Timber is a suitable material for piling. However, the length and sectional area are limited in them. Timber piles are quite rare nowadays; frequently used for small projects adjacent to water such as piers, jetties and landing points; they are prone, however, to rotting if alternately wet and dry conditions exist.

Steel is a material that is flexible and versatile; it can be cut and welded fairly easily on site. It is generally used in two types of piles: bearing and sheet. Sheet piles will be discussed in more detail later. Load bearing piles are usually universal beams or columns, which provide a good profile for the required friction in the ground. They can be either single or multiple; the use of multiple sections helps creates stiffness and rigidity in the section used.

Concrete

Piles made out of concrete are suitable for unrestricted sites where long piles do not pose handling problems. Concrete piles require minimum site work; once driven into place, the pile only needs stripping of the top to expose and integrate the reinforcement to the main structure.

Pre-cast concrete piling requires an extensive site investigation in order to determine the subsoil conditions to avoid problems such as those mentioned above. These piles are generally cheaper than those cast in situ and control on quality is very good. Protection to the head of the pile may be necessary to allow the pile to be driven, but this is more for practical purposes than appearance since the top of the pile is cut away. Some disadvantages do exist – the pile could be deflected by obstructions or be too long to be driven.

However, concrete piles can also be driven using short lengths which are easier to manipulate and transport on site. These piles are made of concrete shells, which are driven by a steam hammer onto a special concrete or steel shoe; this becomes a watertight concrete tube in which the main reinforcement is inserted. This method can be used in fairly unstable soils such as running sand, made up ground and gravels. In situ concrete piles are created using a sealed tube and mandrel; the tube is sunk to the correct depth by a piling rig before the concrete and reinforcement is introduced. Care must be taken to ensure that the level of concrete is above the bottom of the sleeve since the earth pressure can cause 'necking' or 'waisting' at low levels. An alternative is to drive the pile with a steel cylinder weight that has a semi-dry concrete mix at the bottom. When the pile reaches the required set, the plug is forced out and the concrete and reinforcement is introduced to form the completed pile.

In situ excavated concrete piles use a single flight auger drill on the end of a Kelly bar to bore a hole into the earth. Extending the Kelly bar can increase the depth. The spoil is removed by retracting the drill to the surface. This method is generally slow and is restricted to depths of up to about 13 metres. Using a continuous flight auger, the earth is removed, which gives a belated indication of the subsoil conditions. Some piles may have an added advantage of being hollow through which the concrete is pumped under pressure as the drill is retracted. This creates two positive features: first, the need for a lining in cohesionless soils is removed;

second, the concrete is forced into the surrounding subsoil which provides a good frictional interface.

A piling rig with auger drill

Piles can also be under-reamed to provide increased base for end bearing piles. However, under-reaming is a slow process requiring a stop in the drilling to change the tool for the under-reaming operation. In some clay soils it may prove more profitable to use a deeper, straight-sided shaft. Consideration should be given to arranging the piles into a pattern that might include clusters of piles grouped together with a pile cap.

Structural requirements, effects of and precautions against subsoil shrinkage, ground heave and differential settlement

Foundations, whether they are piled, raft, strip or pads, should be sufficiently deep and robust to withstand the effects of being in the ground. Acid and sulphate resistant concrete should be used in all foundations to prevent acidic attack and corrosion through nitrates and other impurities found in the ground.

The effect of water in the ground should not be underestimated. If the level of water rises, any building of lightweight construction has the potential to 'float' or be lifted by this rising level of water. This movement, which is likely to represent differential forces on the building at various points, will lead to cracking and possible failure or rupture of the building components.

Subsoil shrinkage

The stability of the ground should be considered, bearing in mind that any temporary excess amount of water in the ground may well subside in times of dryness. A prolonged lack of water on the ground could lead to subsoil shrinkage and cracking. Foundations that are subject to ground conditions such as this may also crack, move and deform if they are not designed with sufficient density to resist these forces in the ground.

Ground heave

In very cold conditions, ground heave due to the water level freezing needs to be considered. The effect of water freezing in the ground will result in expansion which could lead to differential movement in the building.

The effect of water freezing in the ground can also 'crush' elements of the building that are not sufficiently protected. As a consequence, compressible materials are placed in the excavations and foundations to enable the ground to take up this differential movement without affecting the structure.

Differential settlement

Buildings will exert differing levels of point loads on the earth at their foundations that sometimes leads to differential settlement. These loads will result in a twisting and buckling of some of the structural elements of a building as it settles. Naturally, the foundations will have to be robust enough to deal with these movements and the inclusion of flexible joints which allow the building to accept these movements will need to be incorporated into the design.

Excavations

Excavations up to 5 metres depth

There are significant dangers in excavating earth and working in the ground which will need to be considered in the foundation design and the method of constructing the foundations. Most buildings up to three storeys high will have foundations of less than 5 metres deep while excavations in excess of 5 metres deep are specialist operations with specific health and safety requirements for the contractor. In order to place the foundations at a suitable level, earth is excavated to load bearing strata. Building control officers from the local authority are usually involved in determining the suitability of the depth of foundations and can advise builders on local ground conditions through site investigation and soil analysis.

Excavations in the ground can be very dangerous. Earthwork support is essential if personnel are to work safely in excavations. In 2005 alone, there were several casualties and deaths in construction resulting from earthwork collapse during excavations. Safety of those operatives working in trenches must be paramount when designing suitable foundations and therefore, wherever possible, the depth of a trench should be kept to a minimum. To protect other people in the area, trenches should be cordoned off and clearly marked with suitable warning signs.

Water elimination

As we have seen, fluctuations in the ground water table can have a detrimental effect on the foundation and other building components; rising, falling or continuous variation in the water level can have a disastrous effect on the ongoing construction works. Therefore, eliminating water using temporary pumping and removal to enable the work to proceed safely may be necessary.

Temporary supports and health and safety issues

As earth is excavated, the ground becomes more unstable. Therefore, it is necessary to support all surrounding ground to prevent any sudden or unplanned collapse. Until very recently, timber shores or planks were used to support the ground while the excavated material was carted away and the foundation concrete poured. Modern methods of supporting the ground include sheet piling or purpose-made 'boxes' of steel braced with hydraulic or mechanical jacks that enable operatives to work safely while there is any danger of the earth crumbling or collapsing into the trench.

Excavation and earth moving plant

Hand excavation in all but very small and shallow excavations has now been replaced by the use of mechanical excavators. Before an excavation takes

An earthwork support prevents the trench from collapsing

place, the area should be examined both on site and from any available drawings to see if there are any signs of underground services. Cable avoidance tools (CAT) scanners that send ultra-sonic rays through the ground can detect underground services prior to the excavation. In any event, mechanical diggers and plant should not operate within 500 mm of any services, which means that most excavations near to services should be finished off by hand excavation.

Foundation construction and design

Domestic foundations are commonly formed from either a 600 mm wide concrete strip or a 450 mm wide concrete trench excavated to 1 metre below final external ground level. However, there are several alternative foundation methods such as in situ rafts or a number of types of piled foundations, such as pre-cast concrete, shell or continuous flight auger, all of which support the load bearing walls and ground floors of a building. The Building Regulations 'Approved Document A – Structure: 2004 edition' requires minimum depths for strip foundations; these are 750 mm in clay and 450 mm in other soils. In practice, foundations are usually 1 metre or so deep.

The main requirement regarding loading and ground movement in relation to domestic buildings of the Building Regulations for foundations is set out via Approved Document A; For foundations, the main requirement is that they are designed and constructed so that all the loads that are applied to a building are safely transmitted to the ground without impairing the building's safety.

Prior to construction, the ground substratum should be checked to ensure it has adequate bearing strength to support the foundations and is free of possible conditions that may affect the integrity of the foundation. These may include swelling, shrinkage, slippage or subsidence that could be caused by geological conditions or other factors such as mining, landfill or moisture take-up from tree roots. If some of these conditions exist, it may be possible to overcome them by the use of specialist foundation techniques of rafts or piling with suspended floors in conjunction with ground stabilisation such as grouting of old mine workings or vibration compaction of filled or made-up ground where, due to economic, speed and safety factors, deep conventional foundations would not be viable.

Construction techniques used for strip, pad, raft and pile foundations

Foundations should be constructed in sufficient depth of ground and with appropriate materials to ensure that they adequately support the building. Approved Document A of the Building Regulations contains guidance about the design and construction of plain in situ strip foundations for domestic construction up to three storeys in height. The design is based upon a number of assumptions:

- The concrete mix is ST2, GEN 1 or a ratio of 50 kg Portland cement, 200 kg fine aggregate and 400 kg coarse aggregate.
- The concrete mix is designed accordingly for soils that contain aggressive elements/chemicals such as sulphates.
- The ground and level that the foundations are formed on has no major strength variation, e.g. rock and clay that may cause uneven settlement problems.
- The total load applied to strip foundations does not exceed 70 kilo-Newtons (kN) per metre length and the width of the foundation is based upon Table 10 in Approved Document A.

The Building Regulations require that, where the proposed foundation design or conditions do not follow the above guidelines, they are designed upon proper structural principles with necessary calculations to prove the adequacy and the stability of the design.

The Approved Document guidance regarding the design and construction of strip foundations requires the following points to be adopted:

- walls are positioned centrally on foundations
- the foundation concrete has a minimum thickness equal to the foundation projection on either side of the wall or a minimum of 150 mm
- at the ends of walls or around projections such as chimney breast/piers that there is a foundation and that the projection is maintained
- the foundation bottoms are level and the changes in foundation levels are accommodated by steps not exceeding the foundation concrete thickness
- at foundation steps the concrete is continuous and overlaps the lower section by twice the height of the step.

For the first time, the 2004 Building Regulations also provide minimum foundation depths. These are 0.45 metre for all strip foundations (to prevent risk of frost) and 0.75 metre in clay soils; however, the Regulations acknowledge that these depths may need to be increased to suit local circumstances. Prior to this, Approved Document A did not give a specific minimum depth but required it to be at a depth where the ground is not subject to movement. This has been generally interpreted by most local authority building control surveyors as 1 metre below finished external ground level in average soil conditions. However, foundations may need to be deeper below the zone of influence from trees. Foundations also need to be provided with other precautionary measures such as clay board to avoid damage by ground movement. Conversely, it would be pointless to excavate a foundation trench to 1 metre depth if solid rock is encountered, as this would not be subject to heave or subsidence.

Environmental issues

The Building Regulations 'Approved Document H – Drainage and waste disposal: 2002 edition' requires that precautions are taken for drains passing under buildings or within their zone of influence This is because damage to the drains could cause leaks or blocks which might cause local settlement of the building's substructure. Approved Document H states:

- Where foundations are excavated alongside existing drains, etc. they should be excavated to a level below the influence of existing or proposed drain level.
- Conversely, if drains are excavated below a foundation zone of influence or within a metre, the drainage trench should be filled with concrete up to the foundation's level of influence.
- Where drains pass through the building's substructure and foundations, an allowance should be made for the settlement of the building or drains. This is usually achieved in two ways, either by building the pipe solidly into the wall with flexible joints and rocker pipes either side, or lintelling over the pipe leaving at least a 50 mm gap around the pipe and filling the gap with a flexible inert material such as mineral wool or sheeting covering the opening over with rigid flexible material. Also, the pipe should be bedded and surrounded with a minimum thickness of 100 mm granular material such as pea gravel.

Buildings near public sewers should have foundations that are constructed and positioned not to influence or damage the sewer. They should be at least 3 metres away from the sewer if the pipe is 225 mm or 3 or more metres deep. The purpose of this is to provide working space around the sewer. Smaller adopted sewers may be built over with the agreement of the water authority, but this should not exceed a 6-metre length or cover any access points.

Rafts and pad foundations

In the 1940s and 1950s, **raft foundations** were quite common, particularly beneath the prefabricated, pre-cast concrete or steel buildings erected during the years following the Second World War. Most of these houses were built on farmland where the soil was generally of modest to high bearing capacity. Rafts (or foundation slabs as they were sometimes called) were often used because they were relatively cheap, easy to construct and did not require extensive excavation (trenches were often dug by hand).

Key terms

Raft foundation – a slab that supports the building over a large area.

Pad foundation – a mini-raft similar in function to a raft but not connected to other pads that support structural members.

In modern construction rafts tend to be used:

- where the soil has low load bearing capacity and varying compressibility – this might include loose sand, soft clays, fill and alluvial soils (soils comprising particles suspended in water and deposited over a flood plain or river bed)
- where strip or **pad foundations** would cover more than 50 per cent of the ground area below the building
- where differential movements are expected
- where subsidence due to mining is a possibility.

Flat slab rafts, which have no perimeter or internal beams, offer a number of advantages over strip foundations:

- no trenching is required
- simple and quick to build
- less interference with subsoil water movement.

They are generally suitable for good soils of consistent bearing capacity and have also been recommended in some mining areas. However, detailing of rafts needs careful thought; for example they may be subject to frost attack around the edges, the edges themselves are exposed, and there is the risk of cold bridging around the perimeter. These rafts will flex if ground movement is considerable, so the superstructure needs to be designed accordingly.

Shallow rigid rafts for one-storey, two-storey and three-storey housing can be cheaper than piles. On poor ground, the raft must be stiff enough to prevent excessive differential settlement. This usually requires perimeter and internal ground beams to help stiffness and minimise distortion of the superstructure. Some overall settlement of the house will inevitably occur, but differential settlement should be kept within acceptable limits.

On filled sites, depending on the fill depth, rafts can be a cost-effective alternative to piling. They can also be used on sloping sites as an alternative to stepped strip foundations. A well-compacted (in shallow layers), graded granular fill can form a suitable base. Designing the fill and the raft is obviously specialist work and many speculative house builders would probably prefer 'tried and tested' stepped strip foundations. In the UK, rafts have to be designed on an individual basis by an engineer so they can satisfy the requirements of the Building Regulations. Although, engineers are advised to consider local tradition with regard to raft design.

Assessment activity 6.2

BTEC

1. The architect on your project is unsure why so much money has been spent on site investigations. Explain how the procedures used in subsoil investigation provide information for the design of substructures. P2

   The architect now requires a briefing on the design of the foundations. Describe the principles of foundation design for the architect to understand. P3

   The architect requires a method statement for the construction of the foundations. Produce a statement on how the following foundations would be constructed:
   - strip foundation
   - piled foundation
   - raft foundation. P4
2. You have been asked to plan the substructure of a single detached house which is to be built on an area known to suffer from mining subsidence. The use of two types of foundations are being considered: wide strip and raft. Justify the selection and use of each of these foundation types including the materials used. M1

   Try to think of all the factors involved in the use of foundation and substructure materials. Evaluate the environmental performance of modern materials and techniques used for the substructure construction foundation types from M1. D1

Grading tips

For P2 you will need to link the various subsoil investigations to why they provide information for substructure design.

For P3 explain the principles of foundation design and how it supports the factors found in any subsoil investigation.

For P4 you will need to explain the methods of constructing the three foundations. Include some illustrations to support your explanation.

For M1 justify the use of each type of foundation design, that is list the advantages or disadvantages of each.

For D1 evaluate the environmental performance of the two foundation types listed in M1.

PLTS

Planning and carrying out research while appreciating the environmental consequences of various foundation forms will help develop your skills as an **independent thinker**.

Functional skills

Taking part in discussions on the principles of foundation design will help improve your speaking skills in **English**.

3 Understand the techniques used in the construction of superstructures for low-rise domestic and commercial buildings

Superstructure design, factors affecting choice and construction techniques

The superstructure is the building's shell that sits on top of the substructure and forms the outer envelope of the building. The superstructure enables the building to become weather-tight and includes all the walls, the roof, the floors, the doors and the windows.

Walls

Walls are traditionally made from bricks, although nowadays house builders tend to use timber-framed buildings.

The UK, like most European Union (EU) member states, now has its own **National Annex**, published as part of BS EN 771-1:2003 to provide informative guidance on all clay bricks currently produced and traded within the UK that have an HD classification and meet the standards. This annex provides specification guidance on aspects of HD type clay bricks such as:

- dimensions and tolerance
- configuration and format
- density
- compressive strength
- freeze/thaw resistance
- active soluble salts content
- durability designations
- water absorption
- reaction to fire
- bond strength
- clay engineering and DPC bricks.

Key term

National Annex – a document supplementing the British Standards specifications that ensures materials meet quality requirements. These documents apply to all members of the EU but may vary slightly from one member state to another.

Concrete blocks have been in common use since the 1930s. Blocks can be used in either leaf (or both leaves) of a cavity wall. They are also used for internal load bearing walls and partitions. The nature of the blocks will depend, to some extent, on the nature of the insulation. Insulation is typically in the form of cavity boards and a thermal dry lining.

Theory into practice

Using the Internet to source bricks and blocks from a range of manufacturers; investigate how many colours, shapes, sizes and finishes are available.

Early blocks were often made from local aggregates, which were often industrial waste products such as breeze and clinker. Now, blocks used to form internal leaves of external walls are frequently made from aerated concrete and finished by plasterboard fixed with plaster dabs as most blocks cannot be plastered. Aerated blocks can be used for lightweight partitions and load bearing internal walls. They have been available for about 40 years and have replaced the earlier lightweight aggregate blocks.

For commercial buildings, where larger open span areas are required, portal frames are common (see Figure 6.3). This allows the use of brick and profiled aluminium sheeting for the cladding and enclosure of the structural elements.

For domestic dwellings, brickwork walls are usually constructed with a cavity to ensure that they comply with Building Regulations and offer an adequate resistance to the passage of heat energy; since building in a cavity means that the inner leaf of the wall can be built using materials that are usually high in energy efficiency. Successive legislation has adopted more stringent measures to make new buildings as efficient as possible with additional insulation and a range of different materials used to promote sustainable building projects. In some cases,

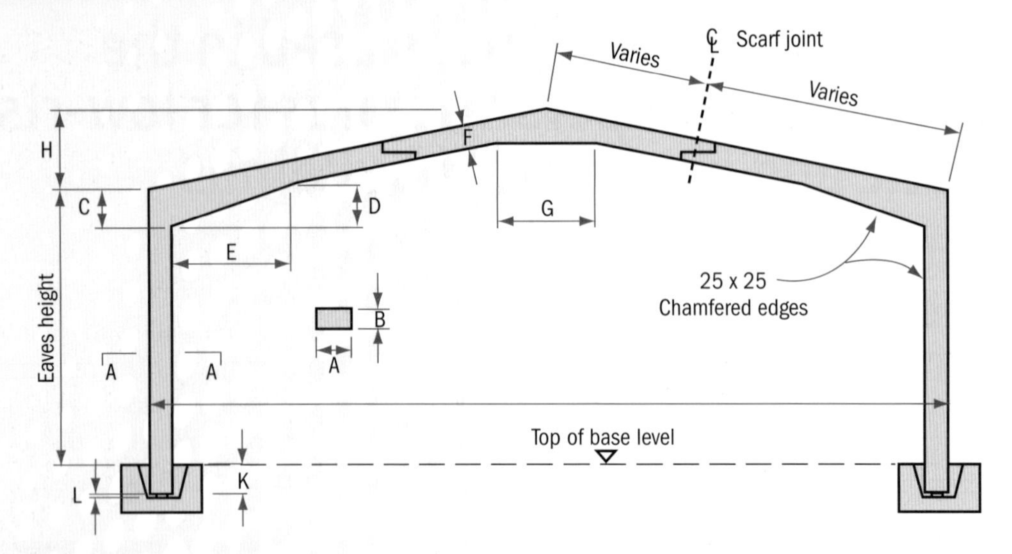

Figure 6.3: A portal frame

the inner leaf acts as an anchor for plasterboards or other materials that finish off the wall and provide a smooth finish ready for decoration.

Cavities also reduce or eliminate any excessive amount of moisture or damp from entering the property from the outside. Within the construction, a barrier to the entry of moisture usually takes the form of a damp-proof course, a thin layer of pitch polymer plastic that prevents any moisture from getting to an area. It is very important that these barriers are built in at the correct position and in no way bridged or broken so that moisture cannot pass to the superstructure from the ground. In normal conditions, the damp-proof course is set no more than 150 mm above finished ground level (see Figure 6.4).

Timber-framed housing construction

Timber frame is a method of constructing houses and low-rise buildings using structural timber, typically prefabricated in a factory and assembled on site in sequence according to manufacturers' instructions. The prefabrication of panels in workshop conditions improves quality control and timber-framed houses are renowned for their energy efficiency. Most timber-frame systems are made on large jigs, usually by hand, although in some plants automation is replacing the need for operatives. A waterproof breather paper covers the outer face of the panels partly to protect them during transport and site erection, and partly to prevent water crossing the cavity and wetting the panel once the building is complete.

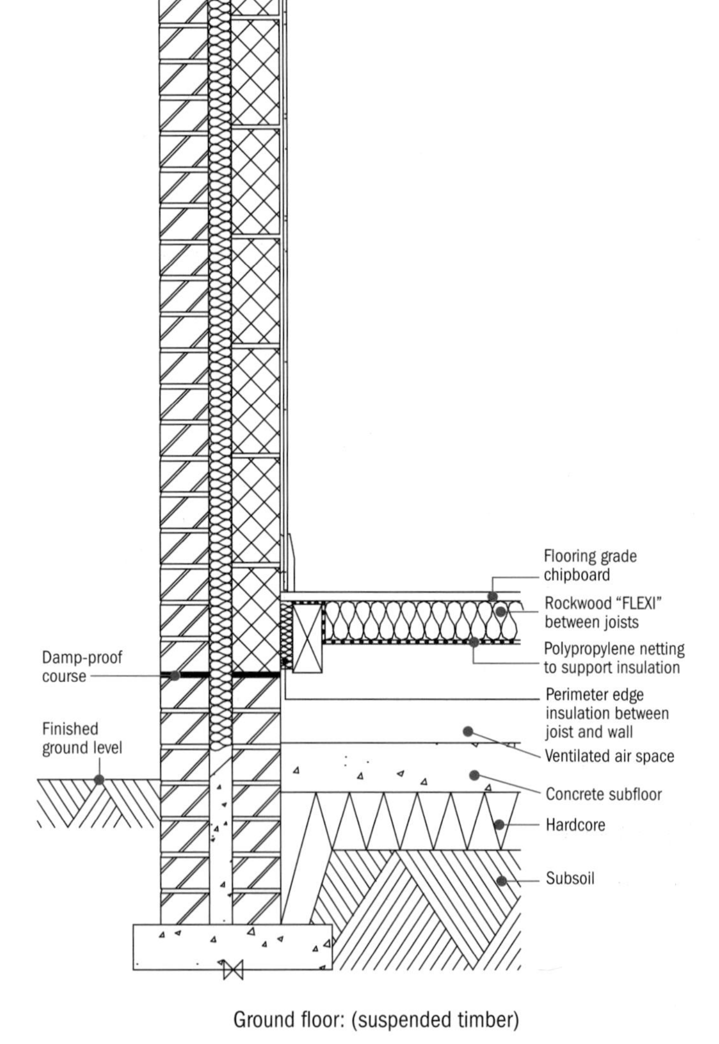

Figure 6.4: Damp-proof course

Theory into practice

Compare the advantages of building in traditional materials with materials used in timber-framed housing.

Most timber-framed houses in the UK use platform framing methods whereby each storey is assembled, and each subsequent floor forms the platform for the next storey (see Figure 6.5). External walls are usually constructed from 100 mm × 50 mm softwood studs nailed together to form a frame. Wind bracing using 18 mm plywood or other sheet materials is fixed to one side; the void between the inside plasterboard finish is filled with glass fibre insulation. A suitable cladding is then applied; the most common form in the UK is brickwork, though in some cases a lightweight cladding system to provide an external finish and keep out the weather is used. It is the timber-framed structure which carries all the building loads; the brick facing is merely a cladding.

Due to the likelihood of condensation forming inside the walls, a moisture barrier or vapour check is essential. This barrier usually takes the form of polythene sheet or foil backing to the plasterboard.

Modern timber-framed houses can provide very high levels of thermal insulation. The wall insulation is contained within the panels. If the panels are 90 mm thick, typical **U-value** is 0.35W/m2K; 140 mm panels will provide a level of about 0.25W/m2K. The internal lining usually comprises a layer of plasterboard with some form of vapour check behind. The purpose of the vapour check is to prevent moist air migrating through the panel and condensing on the cold side of the insulation.

Key term

U-value – a measure of how much heat is lost through a structure

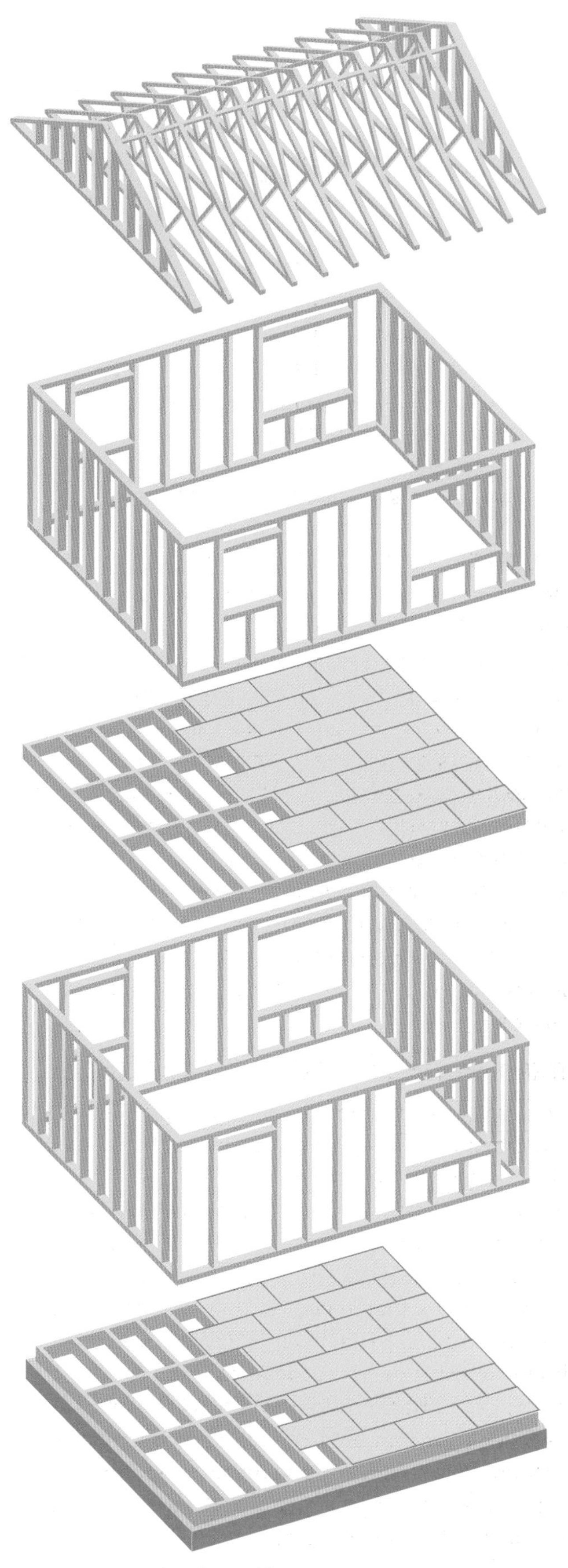

Figure 6.5: A timber-framed house

Floors

The upper floors of timber-framed houses are supported by the wall panels. No load is carried by the brick outer leaf. The floor in a timber-framed house is no different from a floor in a brick/block house. It comprises a series of joists supporting some form of boarding. Boarding is normally in the form of chipboard, strand board or ply. In some cases, the timber floor is prefabricated (assembled into panels) in the factory; in other cases, the timbers are pre-cut to length but assembled on-site.

Interestingly, the most common form of timber-frame system – the platform frame – is named after its floor system in which the joists sit on top of the head binder. The construction varies depending on whether the joists are parallel or at right angles to the panels. In the former, two header joists sit on the lower-storey panels; the inner header joist is slightly offset to provide a fixing for the plasterboard ceiling. Where the joists are at right angles to the panels, a header joist is required to provide additional support for the upper panels and to provide a fire barrier to prevent smoke and fire entering the floor void.

Concrete slabs now form the ground floor of most typical housing and domestic properties in the UK. Ground-floor slab construction is regulated to ensure that the maximum heat loss permissible through this structure does not exceed the amounts shown in the Building Regulations. To ensure that the floor complies with this requirement, insulation is built into the slab.

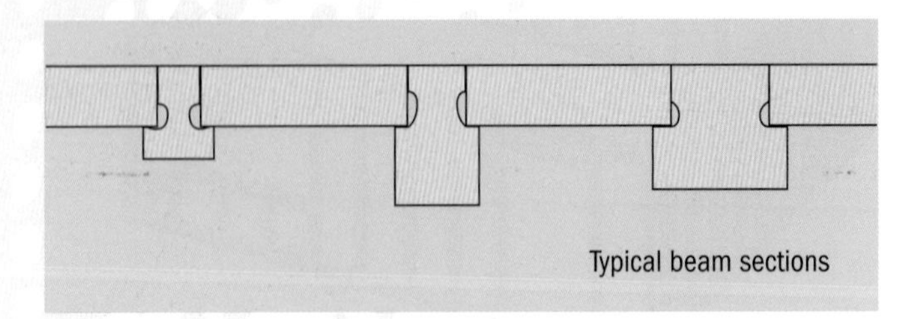

Figure 6.6: Beam and block floor systems

Suspended floors consisting of beams with blocks to infill the separating space have become very popular. These beam and block floor systems combine pre-cast concrete beams and infill blocks to produce high performance yet economic ground and intermediate floors in housing and other building types. Insulation is required to ensure compliance with building regulations. Typical beam sections are determined by the span and the spacing, so that they can carry safe working loads. Both lightweight and dense aggregate concrete blocks complying with British Standards can be used for beam and block floor construction with the following advantages:

- simplicity – the same type of blocks may be used for both walls and floors
- cost saving – long spans are achieved without intermediate support
- performance – requirements for thermal, acoustic and fire resistance are easily achieved
- reliability – effects of ground heave or shrinkage are eliminated
- versatility – beam and block systems may be used for ground and intermediate floors
- working platforms – once installed, the floor may be used as a working platform.

Roofs

Roofs need to be weatherproof and provide shelter from the elements. They also need to have strength, durability, fire resistance, heat retention and a pleasing appearance. For industrial or commercial buildings that require a large uninterrupted span, the roofing system also forms a major part of the structure; for example the roof of a portal frame building. Most roofs on domestic dwellings are pitched, although a significant number of flat roofs exist for smaller properties and home extensions. Pitched roofs may consist of rafters and purlins alone, with the purlins supported on posts, masonry walls or primary trusses. They may also consist of an arrangement in which the lower ends of each pair of rafters are tied or trussed together.

The roof construction has to limit the loss of heat from a building and in domestic construction this is usually achieved by incorporating a suitable thickness of insulation into the roof construction. However, if the roof is a cold construction, where the insulation is either between or under the rafters/ceiling joist, the roof must be ventilated to prevent condensation and the possibility of moisture damage or rot. Alternatively, a vapour permeable felt fixed in accordance with both the felt and tile manufacturers' recommendations is an acceptable alternative to ventilation in a cold roof construction. If, however, the roof has a warm construction, where a continuous layer of insulation is provided above the rafters with a suitable felt and cladding/covering, ventilation does not have to be provided within the roof as condensation and associated damage should only occur on the outside face of the felt and not damage the roof itself.

Whichever type of roof construction method is used, the Building Regulations specify a maximum U-value that the construction needs to reduce the heat loss from the roof, which is where most heat is lost in a building. There are various ways to meet Building Regulations U-values including the elemental method which adds together the overall heat loss for all elements of the building. The permitted U-values using the elemental method are as follows:

- pitched roof with insulation between rafters: 0.2 W/m^{2}K
- pitched roof with integral insulation (e.g. insulation over rafters as warm roof): 0.25 W/m^{2}K
- pitched roof with insulation between joists: 0.16 W/m^{2}K
- flat roof: 0.25 W/m^{2}K
- insulated sloping ceilings for loft conversions: 0.30 W/m^{2}K
- metal framed roof lights: 2.2 W/m^{2}K
- other framed roof lights: 2.0 W/m^{2}K.

It is usually easier to increase the level of insulation in roofs using these methods to allow lower levels of insulation to be used elsewhere in the building.

If the target U-value is used to show compliance with the Building Regulations, a higher general U-value of 0.35 W/m^{2}K is permitted for roofs. The Building Regulations, through 'Approved Document C – Site preparation and resistance to contaminants and moisture: 2004 edition' and BS 5250:2002 'Code of practice for control of condensation in buildings', require that cold roofs are ventilated to limit condensation build-up and associated long-term problems of rot, dampness and mould growth which can both damage the fabric of the building and the occupants' health. The amount of ventilation required depends upon the design/form of the roof. The amount of air passing through a gap needs to be assessed as adequate for an enclosed space, therefore ventilation is usually in the form of equally-spaced continuous gaps. These gaps might take the form of spaces between eaves or soffit casings in the structure. However, ventilation can be provided by other means such as ventilation tiles or air bricks, provided they are equally and adequately positioned within the roof construction.

The most common form of roof construction, particularly in today's housing, is trussed rafters which have a tie between the lower ends of each pair of rafters and some form of cross bracing between (see Figure 6.7). They are normally designed and made using computer-controlled design and fabrication by a specialist manufacturer. The members are usually 38 mm or 35 mm wide and are all in line with each other because the joints are made with punched metal plates. Proprietary trussed rafters are used extensively in domestic construction because they are quick and easy to erect.

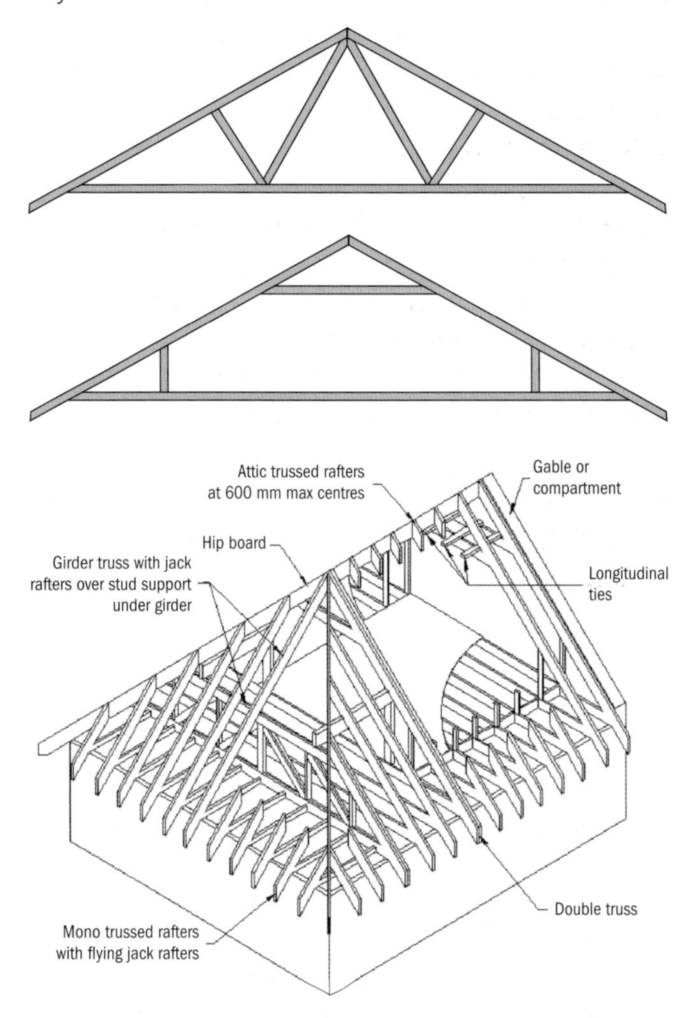

Figure 6.7: Trussed rafters

Trussed rafter roofs are designed in accordance with BS 5268-3:2006 'Structural use of timber: Code of practice for trussed rafter roofs'.

Many manufacturers have their own software to calculate appropriate truss designs in accordance with this Standard. In order to calculate the design, the manufacturers will need to know:

- height and location of building (to determine wind conditions)
- profile, span, spacing and pitch of trussed rafters
- overhang at eaves and verges
- method and position of support
- nature of roof coverings and ceiling materials
- size and position of water cisterns and loft hatches.

Under Building Regulations, pre-treatment of roofing timbers is required in certain parts of the country (in the south east, for example, to protect against House Longhorn Beetle). Some developers treat the timbers as a matter of course, although there are increasing environmental concerns regarding some of the material used. Treating the timbers usually requires some form of pressure impregnation with chemicals that are designed to guard against rot as well as insect attack.

Prefabricated trussed rafters, called attic trusses, can also be designed to accommodate rooms in the roof. Prefabricated roof panels are increasingly being used as an alternative to roof trusses, especially in timber-framed buildings as they can provide unobstructed roof spaces for occupation. The panels are similar to timber-framed wall panels and enable a weatherproof structure to be achieved rapidly on site.

Trussed rafters are sometimes purpose-designed for an individual project and constructed on site or in a workshop. These rafters are usually lapped at joints and fixed with nails, bolts or timber-connectors.

More traditional forms of roof include the use of primary trusses and site-constructed, or 'cut' roofs, using an arrangement of rafters and purlins. Primary trusses are designed to support purlins which in turn support rafters. They are of heavier construction than trussed rafters and are spaced at intervals which are normally a multiple of the rafter spacing, for example 1800 mm or 2400 mm. Purlin-supported rafters need no tie members; the rafters are notched or 'birdsmouthed' over the purlins or otherwise fixed to avoid displacement and to keep all loads vertical. The purlins are fixed to posts, walls or primary trusses and all or any of these supports are designed and built to remain in a state of equilibrium. The main reason to use purlin-supported rafter structures is usually to allow the roof space to be used.

In a pitched roof, there is a choice of insulation position: at rafter level or at ceiling level.

At rafter level, the insulation is either between the rafters or above and between the rafters. Insulation between the rafters can be designed in two ways:

- 'breathing' with vapour-permeable tiling underlay
- ventilated design (see Figures 6.8 and 6.9).

With breathing roof design, insulation fully fills the rafter space without airspace between the insulation and tiling underlay, which must be vapour permeable. If a thin layer of insulation is installed, it is recommended that an insulation/plasterboard laminate is used as the internal lining to prevent thermal bridging. This is not necessary where rafters are at least 140 mm deep and fully filled with insulation. A combined airtight-vapour control layer should be placed on the warm side of the insulation. This not only makes the ceiling convection tight but also restricts the amount of water vapour passing through the ceiling. Where cables and piped services are to be installed, the plasterboard lining may be battened out to provide a suitable duct. The services should be routed on the inside of the vapour control layer to avoid any puncturing.

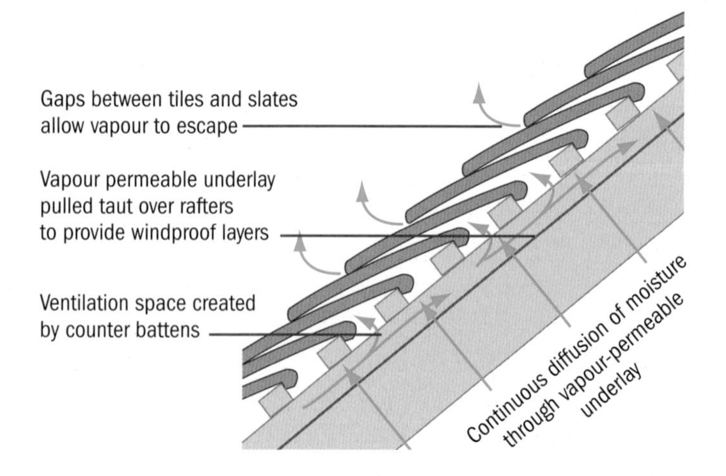

Figure 6.8: Section through roof members showing tiles and barrier

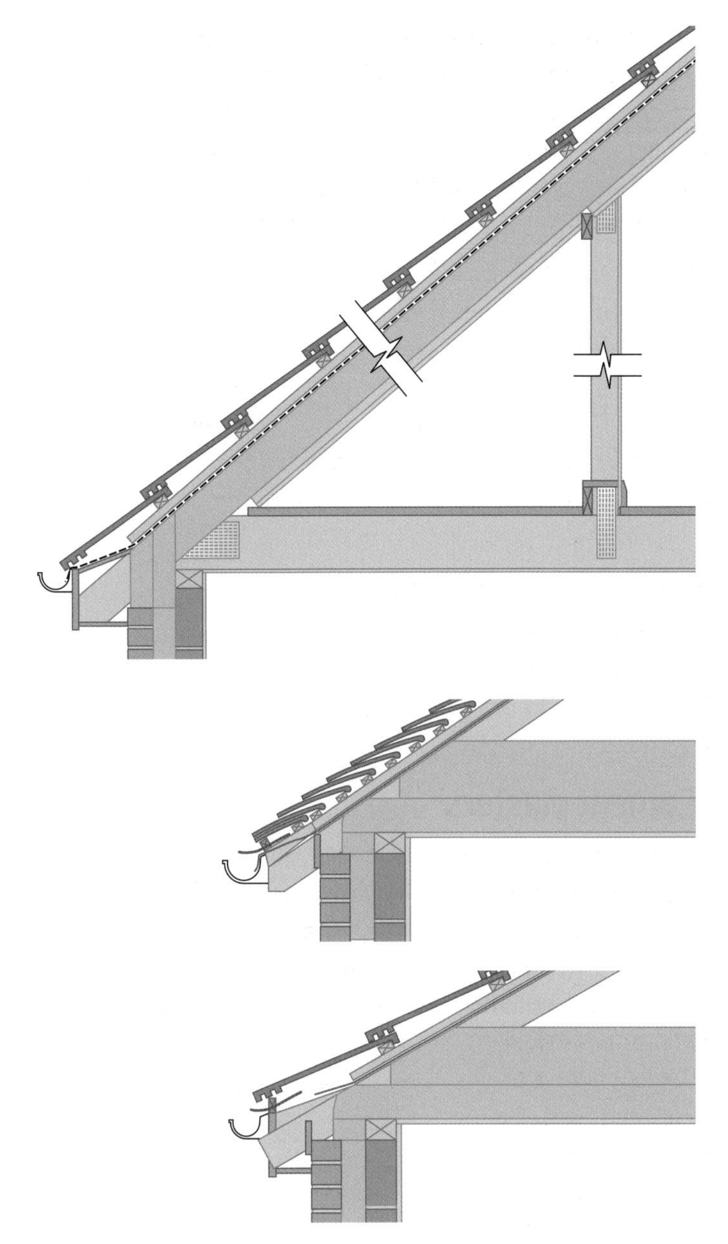

Figure 6.9: Roof eaves

With the ventilated design, there is 50 mm ventilated airspace between the top of the insulation and the tiling underlay. Should the rafter depth be insufficient to accommodate both the insulation and the ventilated airspace, an insulated dry lining is recommended which has the added benefit of minimising thermal bridging. Ventilation openings should be provided at every roof void; at the eaves, ventilation openings should be equivalent to a 25 mm continuous gap while at the ridge, the ventilation opening should be the equivalent to a 5 mm continuous gap on each side of the ridge.

Thinking point

Working in teams, discuss why ventilation is very important in roof spaces.

Stairs

Stairs are used to gain access and egress from the upper floors of a building. They are traditionally manufactured from timber and must conform to the Building Regulations. There is a standard for both the height and width of a step. Modern methods of construction have enabled prefabricated concrete stairs to be manufactured and lowered into place. These require finishing with a floor finish.

Windows

Windows were traditionally manufactured from a treated timber; the timber frames had a painting system applied to them. However, once the paint barrier was broken, the timber degraded and got swollen with moisture. With the introduction of the modern uPVC technology, warmer and draught-sealed, high-quality windows have substantially replaced the manufacture of timber windows. However, timber-engineered, environmentally friendly and high-performance windows are making in-roads into the current uPVC market.

Doors

Doors are essential for maintaining the security of a building. There are wide variety of types available with an even wider variety of uses, including:

- traditional timber-panelled, external doors
- plywood, flush internal doors
- part-glazed doors
- roller-shutter doors
- uPVC French and patio doors
- fire-rated doors.

Superstructure finishes

Factors affecting the choice of finishes

Many factors affect the choice of finishes for a domestic or commercial building design, including the following:

- cost – the budget for a project has to be maintained and often finishes of a lower quality are seen as a cost saving

- use – a school corridor will require harder-wearing finishes than a library corridor
- cleanliness – certain finishes, for example, those used in a hospital operating theatre, must be capable of being sterilised
- aesthetics – the attractiveness of a finish is often an essential item
- fashion – finishes may be chosen because they are the current fashion; for example, block paving of a driveway
- fire resistance – often finishes have to resist the spread of fire and flame
- sound resistance – the denser a material, the more effective it is at sound reduction
- health and safety – such as non-slip finishes
- function – a shower's finish has to be waterproof
- colour – you would not use red finishes in a prison cell, for example, as it promotes anger.

Early plastering

Plastering is an ancient craft. Originally, the materials and techniques used were dependent on the local geology, the materials to hand and the technology available to the local artisan.

In earliest times, buildings or structures were plastered to render them windproof and weathertight. Simple mud daubs on basic interwoven wattle structures are among the earliest means of walling. Early stone walls were rendered with basic plaster. Even timber buildings, such as log cabins, were weatherproofed with layers of plasters made of clay reinforced with straw.

As early societies developed and the first cities came into being, people recognised the decorative potential of plaster work as well as its hygienic, fire and weatherproofing qualities. A flat, hard surface was easy to clean, did not harbour dirt, was comfortable to the touch and was pleasing to the eye. It could be painted and decorated and also provide the substrate for three-dimensional decorative relief. It slowed the spread of fire, especially in timber buildings, and helped deaden sound as the needs for privacy and comfort became increasingly important.

Gypsum plasters

Gypsum is the basis of today's plaster, with few exceptions. It is a naturally occurring mineral sometimes known as alabaster when found in its soft, translucent form. When burnt at relatively low temperatures (less than 200°C), some or all the water is driven off and a highly reactive powder remains. This powder, in its hemihydrate form, is known as plaster of Paris and reacts vigorously with water to set hard and quickly.

Lime plasters

The vast majority of ancient plasters and mortars were based on lime and its various by-products. Limestone is made up of the mineral calcium carbonate. Chalk is one of the purest forms of naturally occurring limestone and is nearly 100 per cent pure calcium carbonate. Many impurities exist in most other limestone, mainly clay, and these impurities can be very important in the finished properties they give to plaster.

Cement plasters

Plasters based on cement have been common for more than 100 years. They were mostly used for external renders, often gauged with lime to produce a more workable mix. Cement-based renders need to be used with care; strong mixes with a high proportion of cement are likely to shrink and may part from the background.

Plasterboard

Plasterboard comprises a core of gypsum plaster with added aggregates and thick paper linings bonded on either side. Their lightweight and low thermal capacity means that plasterboard will warm up quickly and will help reduce the risk of surface condensation. Plasterboard linings are not suitable for areas of high humidity or areas which are permanently damp.

Plasterboard has been available since the 1920s, although its early use was mainly confined to ceilings where it proved a cost-effective and quick alternative to traditional lath and plaster. Nowadays, it is available in a wide range of grades and sizes and is used for a variety of purposes, including ceiling linings, wall linings and proprietary partition systems. It can also be used to improve thermal insulation, sound insulation, fire protection as well as provide vapour control layers. Early plasterboards always received one or two coats of plaster, but in modern construction most plasterboard is self-finished.

Wallboard and vapour control wallboard

Wallboard is used for a variety of applications, including drylining walls, lining ceilings and on stud partitions. Typical thicknesses of wallboard are 9.5 mm, 12.5 mm 15 mm and 19 mm and it is available in a range of sizes, such as 1800 mm × 900 mm, 2400 mm × 900 mm, 2400 mm × 1200 mm and 3000 mm × 1200 mm. Wallboard can have tapered or square edges with the tapered edges designed for direct decoration or skimming, and the square edges for skimming or textured finishing.

Vapour control wallboard is used where there is a risk of condensation. It is similar to wallboard except that the inner face is covered with a thin vapour control membrane.

Thermal boards and moisture resistant boards

A thermal board is a wallboard with insulation bonded on the inner face. The insulation can be of various types, including polystyrene and phenolic foam. Thermal boards often contain an integral vapour control layer to minimise the risk of condensation.

Moisture resistant boards are usually 2400 mm × 1200 mm with a thickness of 9.5 mm or 12.5 mm. They can be used for external soffits or as a base for wall tiling around showers, etc.

Dry lining

When plasterboard is used as a wall finish in place of wet plaster, it is referred to as dry lining. In modern construction there are two main approaches: bonding the boards to adhesive dabs, or securing the boards to metal channels which have been bonded to the background. British Gypsum and Lafarge, currently the two major manufacturers, offer both systems. They differ in detail but are broadly the same in terms of principle.

Theory into practice

Dry lining is quicker and cheaper than more traditional methods of plaster finishes; find out how many contractors can do this work. Use the Internet to list specialist contractors in your area.

The use of adhesive dab is the simpler of the two systems. It comprises a series of adhesive dabs applied by trowel to the wall and typically 50 mm to 75 mm wide and 250 mm long. Three 'columns' of dabs are normally required per board although 9.5 mm x 1200 mm boards usually requires four. Horizontal dabs are appliedbetween the columns at ceiling level, and a continuous band of adhesive at skirting level. When the dabs are in position the board (can be pressed and tapped into position, tight against the ceiling. It is temporarily supported at floor level by off-cuts of board. An insulated reveal board is available where there are risks of **cold bridging**. This occurs when the insulation layer within a wall or roof is interrupted by another material or is reduced in thickness. There is greater heat loss through the thinner area of insulation and that part of the wall or roof has a reduced internal surface temperature. When the warm, moist air inside the property comes into contact with the cooler surface, it is chilled and less able to carry moisture. This results in surface condensation or pattern staining of décor.

Key term

Cold bridging – an interruption in the insulation layer which leads to greater heat loss in that area. Can result in surface condensation or staining of decor.

The procedure for thermal laminate plasterboard, which is a plasterboard with an inner layer of insulation, is similar, although two plug fixings are required to ensure that the plasterboard is not distorted in the event of a fire.

In the second system, the boards are fixed to a series of metal channels bonded on dabs to the background. The channels are fixed at 600 mm centres (vertically) with top and bottom channels running horizontally. The boards themselves are fixed to the channels with special screws, typically at 300 mm centres. The boards must be well fitted to ensure there is no flow of air behind them. Failure to do this will reduce their thermal performance. Additional sealant can be provided around the edges of junctions such as window reveals and external angles. In long runs of dry lining, Building Regulations require the provision of vertical cavity barriers to limit the spread of fire. This can be formed using a continuous vertical line of dabs running down the centre of a board.

Case study: New builds

Modern requirements increasingly call for new build units of high quality and technical input, both traditionally and on a design and build basis. Ideally, you will have built up several years' experience and accomplished many successful projects, having constructed hundreds of new build units over many years to tender for appropriate contracts and seek out further work.

You will need up-to-date knowledge of Building Regulations, Scheme Development Standards, National House Building Council (NHBC) standards, etc. Imagine you have secured the contract to build a development of 44 new units. Think about the possibilities of creating a good place to live and a lasting legacy of your work that you can be proud of for many years to come.

3-bedroom, semi-detached

A large three-bedroomed semi-detached house which offers spacious family accommodation. The 'Beverley-Hills' incorporates a dining kitchen, WC and generously proportioned lounge with open-plan staircase to the ground floor. The first floor houses three bedrooms which includes an en suite to the master bedroom.

Dimensions	
Room	mm
Kitchen/Dining room	4451 × 2752
Lounge	4451 × 4833
Bedroom 1	4290 × 2423
Bedroom 2	2912 × 2183
Bedroom 3	2525 × 1812
En suite	2285 × 1675

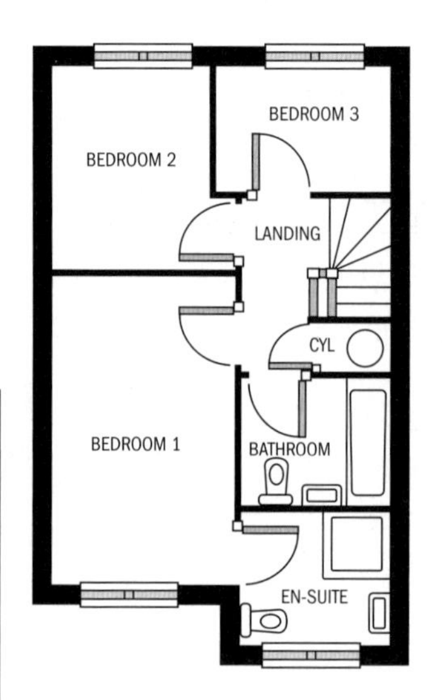

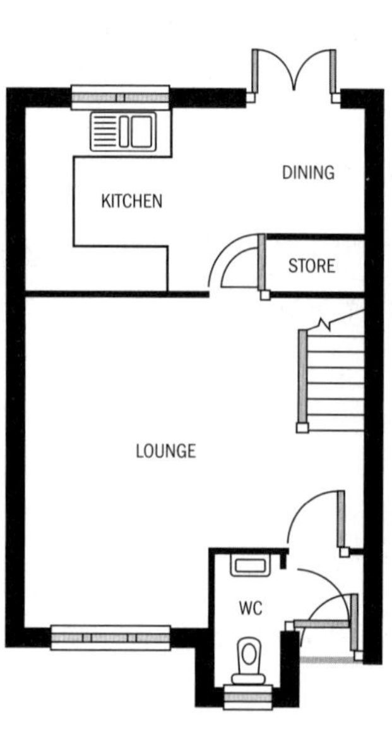

1. What kind of investigation is necessary to find out about the land that the building will sit on? How will you know that the foundation is strong enough to support the house and the loads imposed on it?
2. Having decided what type and size of foundation, how will you communicate your thoughts and decisions to the builders. Can you draw to a suitable scale all the components in the substructure so that the builder can clearly see how the building goes together?
3. What type of superstructure will you use and how will you explain to potential buyers that this type of house is safe, warm and uses good materials, particularly if they are of sustainable construction and do not cause problems for the environment?
4. What drawings and specifications are required to explain how the superstructure goes together?
5. What plant and equipment will you need to build the elements of the houses that need special equipment and tackle, e.g. the fixing of the roofs? How will you gain access to high-level work and how will you ensure that any operative working there will be safe in their work?

Assessment activity 6.3

BTEC

Your friend requires a design for a home extension and as you work as an assistant in an architectural practice they have asked you. Explain the principles of superstructure design to your friend. **P5**

During the process of producing the extension design your friend asks about the finishes and the methods of construction proposed for the roof and walls. Write a short explanation for your friend. **P6**

The following is proposed for the extension:

- three layer felt built up flat roofing
- 300 mm width blockwork cavity walls with fully filled cavity and rendered exterior.

Justify the selection of suitable materials and the technique proposed. **M2**

Evaluate the use of these proposed materials for the wall and roof and explain how their use impacts on the environment. **D2**

Grading tips

For **P5** you will need to explain in some detail the various aspects of superstructure design e.g. floors, walls, and roof.

For **P6** explain the methods available to construct the superstructure and explain what finishes are available for the inside and outside completion.

For **M2** justify the materials and methodology employed for the two superstructure items; explain the advantages or disadvantages.

For **D2** some evaluation is required on the proposed materials for the wall and roof in terms of their use and impact upon the environment.

PLTS

Setting goals with success criteria for your development and work will help develop your skills as a **reflective learner**.

Functional skills

Producing a written justification on materials and methods will help to improve your extended writing skills in **English**.

4 Understand the implications of issues and constraints on building construction

Environmental issues

Environmental impact resulting from materials and methods used in construction

Sustainability is about balancing society with the environment. There are three main threats to environmental sustainability:

- global warming
- resource depletion, such as extracting minerals from the ground, and how quickly we use up those resources
- pollution in the atmosphere due to human activities.

The most immediate of these threats is global warming and climate change leading to a rise in sea levels which could bring about flooding and damage to buildings in low-lying areas.

The construction industry uses natural products such as timber, clay, iron and other minerals and contributes to the economy and the gross domestic product – nearly 2 million people work in the industry every day. We have to consider how our activities contribute to global warming and how we replace those materials taken from our planet.

Construction activities and how homes and offices are used also impacts on sustainability issues. Designers have a responsibility to consider how we generate and deal with waste products and the use of materials that add to pollution in the environment. Higher standards and awareness of sustainability issues will bring about improved performance in energy use, lower costs and generally increase the quality of life.

The Code for Sustainable Homes has been issued by the Department for Communities and Local Government; this should promote changes in the way that designers tackle sustainability issues in new developments. The UK government believes this will become a national standard to be used by building designers and constructors to develop better homes and buildings as well as making clients and homeowners aware of sustainability issues around the home.

Developing an awareness of sustainable issues and relating these issues to construction should lead to a reduction in CO_2 emissions, better management of water disposal and an emphasis on recycling materials. Homeowners can be assured that changes made to the design and construction of a new building comply and promote sustainability and hence be aware of the improved standard and performance of the building. It is highly likely that contractors who comply with new codes of practice will use this fact in their advertising and promotional literature to demonstrate this improved performance and show the benefits of lower energy consumption and water efficiency.

Theory into practice

Use the Internet to find out whether major contractors operating in your area have a 'Sustainable Construction' policy. How do they differ from one another?

Extraction and manufacture

The extraction and use of fossil fuels is the primary source of carbon dioxide (CO_2). It also causes the majority of eco-toxic pollution, and is the prime resource depletion issue in the UK. Our economy is largely dependent on fossil fuels. Reducing CO_2 emissions is therefore by far the most significant issue when constructing buildings. Designers and constructors of buildings have a duty to produce homes, offices and buildings that are energy efficient, reduce heat loss and have built-in provisions to run appliances, such as washers, dryers, dishwashers and boilers, that display the 'Energy Saving' logo and are approved by the Energy Saving Trust. Lighting and illumination can also use low-energy light bulbs in preference to standard bulbs. During the construction process, contractors need to be aware of waste materials and understand that by operating efficiently they can contribute to reduce costs that high levels of waste bring.

Construction methods

Heavyweight or dense construction is the preferred method by many since it tends to store heat in the thick walls and materials. It may take longer and hence more energy to heat these buildings to a satisfactory level, but they will stay warmer for longer periods due to their dense constructional form. Furthermore, in summer heavyweight buildings tend to be cooler since it takes longer for any heat to penetrate or dissipate through the heavyweight construction. This effect is known as thermal mass.

Lightweight construction is used for timber-framed housing; it can be erected quickly and has an excellent thermal performance. The increased use of prefabricated designs has also improved quality control and led to higher standards. In some cases, these efficiencies have been passed through the supply chain with lightweight construction now being the preferred option for low-cost homes and social housing projects.

Recycling and waste implications

In order to conserve the Earth's finite resources, we have to be aware of the long-term implications of manufacturing materials. The manufacturing of materials needs to contain an element of recycling as does the eventual demolition of them when the life of the building is exceeded. Reinforcing bars can be recycled into steel while concrete can be crushed into hardcore fill for levelling sites. Brickwork can also be recycled; bricks can be cleaned and reused for new house construction with any damaged ones recycled as hardcore. Steelwork from buildings and any metals can be reshaped into new products during steel-making processes.

Embedded energy

Embedded energy is the amount of energy that has been used in the manufacture of a material; this should also include the transport costs of getting the material to its final point of use. It is therefore far better to specify and select materials that have a low embedded energy. Cement-based products such as pre-cast concrete and building blocks contain high levels of embedded energy from the manufacture of the cement.

CO_2 emissions

CO_2 is the greenhouse gas that is associated with the current trend in global warming. Releasing CO_2 into the atmosphere increases the average temperature of the world which has consequences not only for weather patterns but also for sea levels. Using timber-based products will help reduce CO_2 emissions; as a tree grows it absorbs CO_2 and gives off oxygen as a waste product through photosynthesis. During the processing of timber very little CO_2 is released into the atmosphere making this a very sustainable product; its waste can also be reused and turned into engineered timber.

Noise

Noise is a form of pollution that can cause environmental disturbance to residents living near construction sites that produce a high level of noise. Noise pollution from a construction site is temporary environmental disturbance which will eventually leave the environment. However, things can be done to help reduce the construction site noise. For example, where cutting and processing of materials is carried out on site, noise suppression systems should be used where available.

Case study: New homes

For a new home being built today a builder/developer will need to meet any level of the Code for Sustainable Homes. For Level 1 this means that the home will have to be 10 per cent more energy efficient than one built to the 2006 Building Regulations standards. This could be achieved by:

- improving the thermal efficiency of the walls, windows and roof reducing air permeability by improving the control of the fresh air into a home, and the stale air out of a home (a certain amount of air ventilation is needed in a home for health reasons)
- installing a high-efficiency condensing boiler
- carefully designing the fabric of the home to reduce thermal bridging and thus reduce the amount of heat that escapes between the inner walls and the outer walls of a home.

The Level 1 home will also have to be designed to use no more than about 120 litres of water per person per day. This could be achieved by fitting a number of items such as:

- 6/4 dual flush WC
- flow reducing/aerating taps throughout
- 6–9 litres per minute shower (note that an average electric shower is about 6–7 litres per minute)
- 18 litres maximum volume dishwasher
- 60 litres maximum volume washing machine.

Other minimum requirements for Level 1 include:

- surface water management – this may mean the provision of soakaways and areas of porous paving
- materials – this means a minimum number of materials meeting at least a 'D' grade in the Building Research Establishment's 'Green Guide' (the scale goes from A+ to E)
- waste management – this means having a site waste management plan in place during the home's construction, and adequate space for waste storage during its use.

After meeting the above, a builder/developer will still need a further 33.3 points to get to Level 1. So they must do other things to obtain the other points such as:

- providing accessible drying space to eliminate the need for tumble dryers
- providing more energy-efficient lighting (taking into account the needs of disabled people with visual impairments)
- providing bicycle storage
- providing a room that can be easily set up as a home office
- reducing the amount of water than runs off the site into storm drains
- using environmentally friendly materials
- providing recycling capacity either inside or outside the home.

Divide yourselves into teams and discuss the following:

- Identify which of these Level 1 elements are included in the current Building Regulations.
- Use the Internet to obtain the price of a selection of items such as a condensing boiler, a dual flush WC, aerating taps and fittings for the bathroom and kitchen.
- What other items can be built into the design to encourage recycling of waste products?

Pollution

Pollution and wastage must be controlled as these can damage the environment. Chemicals especially should be limited in use during the construction process. Care should be taken in what wastage is taken to landfill and all waste should be sorted into recycling skips. For example, some plasterboard manufacturers supply waste skips for recycling off-cuts of their products.

Legislative constraints

Building Regulations

The Building Regulations apply in England and Wales and exist principally to ensure the health and safety of people in and around buildings. They also provide for access to and around buildings and energy conservation. The regulations apply to most new buildings and many alterations of existing buildings, whether domestic, commercial or industrial.

A detailed explanation of the Building Regulations, the Building Control system and how they might affect individual building projects is provided on the Planning Portal. (go to www.planningportal.gov.uk).

Building Regulations promote the following:

- Standards for most aspects of a building's construction, including its structure, fire safety, sound insulation, drainage, ventilation and electrical safety. Electrical safety was added in January 2005 to reduce the number of deaths, injuries and fires caused by faulty electrical installations.
- Energy efficiency in buildings. The changes to the regulations on energy conservation came into effect on 6 April 2006 and will save a million tonnes of carbon per year by 2010 and help to combat climate change.
- The needs of all people, including those with disabilities, in accessing and moving around buildings. They set standards for buildings to be accessible and hazard-free wherever possible.

Health and Safety at Work Act 1974

The Health and Safety at Work Act provides the legal framework to promote, stimulate and encourage high standards of health and safety in places of work. It protects employees and the public from work activities. Everyone has a duty to comply with the Act, including employers, employees, trainees, the self-employed, manufacturers, suppliers, designers, importers of work equipment.

The Act places a general duty of employers to 'ensure so far as is reasonably practicable the health, safety and welfare at work of all their employees'. Employers must comply with the Act by:

- providing and maintaining safety equipment and safe systems of work
- ensuring materials used are properly stored, handled, used and transported
- providing information, training, instruction and supervision, and ensure staff are aware of instructions provided by manufacturers and suppliers of equipment
- providing a safe place of employment
- providing a safe working environment
- providing a written safety policy/risk assessment
- looking after the health and safety of others, e.g. members of the public or visitors to the site
- talking to safety representatives.

An employer is forbidden to charge an employee for any measures which they are required to provide in the interests of health and safety, for example personal protective equipment used during the construction process.

Employees have specific responsibilities too. They must:

- take care of their own health and safety and that of other people
- cooperate with their employer
- not interfere with anything provided in the interest of health and safety.

It is the responsibility of the Health and Safety Executive (HSE) to carry out inspections of construction sites. The powers of an inspector include:

- the right of entry at reasonable times, etc. without appointments
- the right to investigate and examine
- the right to dismantle equipment and take away substances/equipment
- the right to see documents and take copies
- the right to assistance (from colleagues or the police)

- the right to ask questions under caution
- the right to seize articles/substances in cases of imminent danger.

The HSE can visit any site and report on whether there is an infringement of the Health and Safety at Work Act. It does this by producing **legal notices** that are issued to **improve** or to **prohibit** something from continuing or to bring about a prosecution. Both employers and employees may face prosecution.

Key terms

Legal notices – these are written documents requiring a person to do/stop doing something.

Improvement – identifying what is wrong and how to put it right within a set time.

Prohibition – banning the use of equipment/unsafe practices immediately.

The Construction (Health, Safety and Welfare) Regulations 1996

These regulations cover most aspects of site safety and welfare, including the following:

- safe places of work – excavations, tunnels, work at ground level and at height, access to and from the work area
- falls – physical precautions or equipment to check a fall and prevent falls through fragile materials; scaffolding and supervision by a competent person; safe use of ladders
- falling objects – measures to prevent falling objects and/or provide covered walkways
- work on structures – prevent collapse of structures; plan demolition and dismantling and supervise such work by a competent person; take precautions with explosives
- excavations – prevent collapse and risk from underground cables and services
- drowning – rescue equipment must be immediately available
- traffic routes, vehicles, gates and doors – make these safe and provide safe access and egress onto and from construction sites
- prevention and control of emergencies – procedures for evacuating sites, fire-fighting equipment, emergency exits
- welfare facilities – provide sanitary and washing facilities; provide rest facilities; provide facilities to store and change clothing
- site-wide issues – including fresh air, protection from bad weather, lighting, cleanliness on sites, marking of the perimeter of the site and maintenance of equipment in safe and sound condition
- training, inspections and reports – all work should be carried out by people with the training, technical knowledge and experience to do so safely, or should be supervised by those with such qualifications. Before work at height, on excavations, cofferdams or caissons begins, they must be inspected by a competent person, and a written report made.

Town and country planning legislation

Towns and developments need to be planned appropriately. Planning is necessary to ensure that the buildings and environment created is fit for purpose and does not interfere with other properties around a given site. Local authorities are responsible, through their planning committees to decide whether a new building or an alteration to an existing building is suitable. These alterations could be an extension to a house or the complete adaptation of a large building into an office building.

Planning permission is not always required; internal alterations or work which does not change the appearance of the outside of a property does not require planning permission. In other instances, such as extending or altering the physical shape and size of a building, planning permission will be needed. It is not required, generally speaking, for changes to the inside of buildings, or for small alterations to the outside such as the installation of telephone connections and burglar alarm boxes. Some properties have special terms and conditions from when they were built which will require permission to make any changes to the original terms and conditions, such as the requirement for or restriction on the erection of fences around the front gardens.

Dividing a home into flats, or creating a separate home within an existing building will also need planning permission as would the alteration of a domestic building into a workshop or office building.

Infrastructure

There is a huge range of construction plant available today to assist in the construction of modern domestic and commercial buildings. Here we shall have a look at the major pieces of plant that you would expect to see when visiting a complex, active construction site.

Scaffolding

Scaffolding consists of scaffold metal tubing and timber scaffold boards (see Figure 6.10) which are all tested to achieve a British Standard under current UK health and safety legislation. For health and safety reasons, scaffolding must be erected by a qualified and competent person and cannot be adapted, moved or dismantled by any unqualified individual.

Scaffolding is a complex structural system of many components such as:

- standards – the vertical tubes
- boards – the element walked on
- ledgers – a horizontal tube
- putlogs and transoms – a tube that is built into the construction as work proceeds or acts independently
- guard rails – two rails that prevent falling
- toe boards – a vertical board to prevent objects falling off the scaffolding
- bracing – diagonal bracing that secures the structure against collapse.

You should refer to the HSE's publications on scaffolding for detailed requirements. Scaffolding must be inspected by a competent person and the Construction (Health, Safety and Welfare) Regulations should be referred to for the frequency of inspections and the records to be kept.

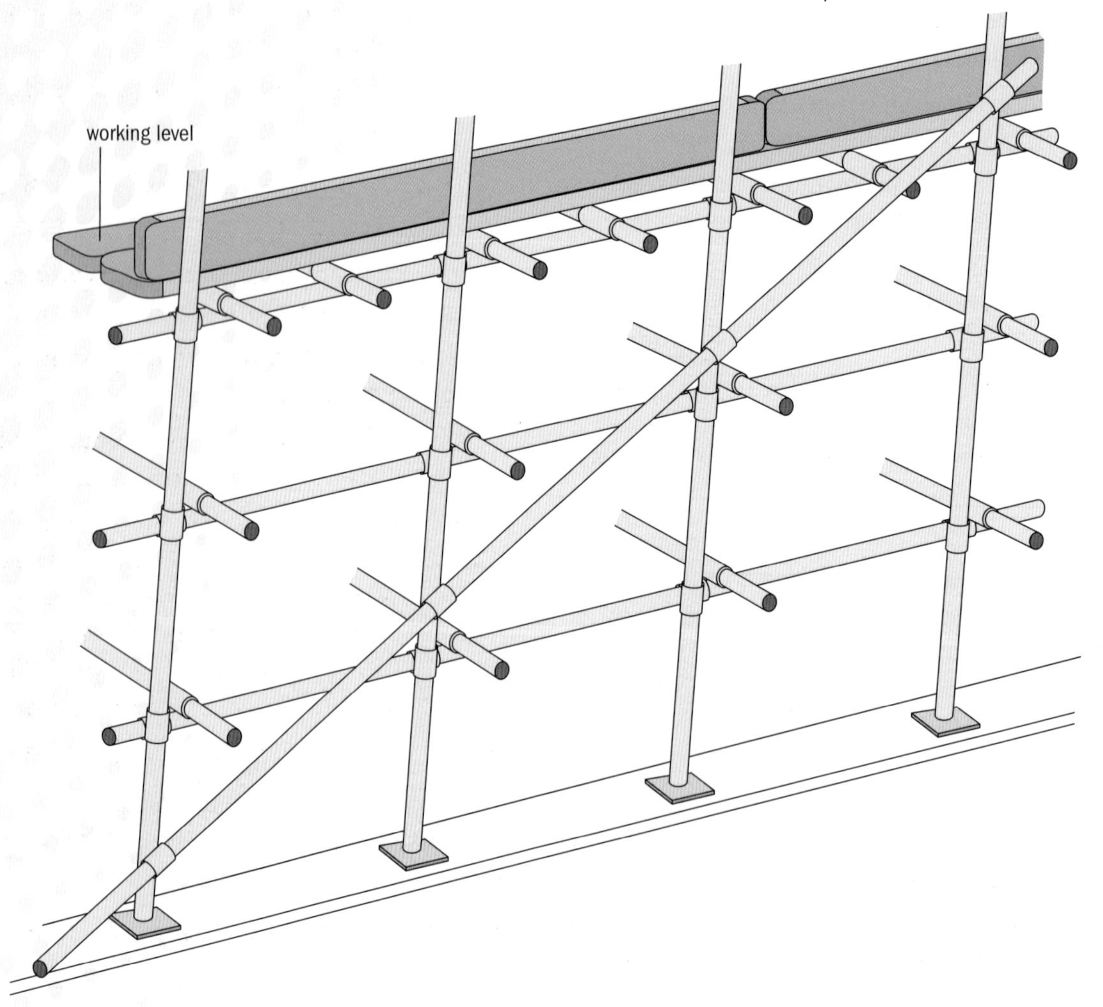

Figure 6.10: Typical scaffolding system

Trench support systems

Trench support systems are used to support the sides of excavations and can take many forms, including:

- steel sheet piling which is driven in and propped
- hydraulic props between timber shores
- steel hydraulic trench boxes.

Forklift trucks

On construction sites, forklift trucks have large off-road tyres in order to cope with the rough terrain of site conditions. This type of plant is ideal for the unloading and distribution of materials in any 360-degree location, both vertically and horizontally. Modern forklifts have telescopic booms that can be used to reach onto scaffolding platforms to deposit materials. An access platform with a roll-over guard rail system has to be provided at these points on the scaffolding.

Dumper trucks

Dumper trucks are steered hydraulically and can have tipper facilities incorporated into the front and rear hoppers. They are available in a variety of weight and volume capacities and are ideal rough-terrain vehicles which can manoeuvre over any rough ground. These machines are used for moving loose, bulky materials, such as drainage bedding.

Cement mixers and silos

Cement mixers range from a full batching plant that can produce concrete down to a small bag mixer. Cement silos and mortar producing plants should be used on larger construction sites where economies of scale can be obtained by purchasing raw materials in bulk.

Excavators

There are two types of excavator:

- 360-degree excavators
- Backhoe excavators which can only move 180 degrees.

The most common excavator is the JCB 3CX which is very versatile and can have many attachments incorporated onto its hydraulics. Excavators have two movable forms: wheeled and tracked. They are used mainly in demolition and earth excavation and removal but can also be used to lift materials if certified to do so.

Cranes

Cranage varies widely from small 15-tonne lifting cranes to large tower cranes. Their use is limited by the reach that is required and the amount that has to be lifted. A specialist hirer would provide all the necessary documentation and they must be used alongside a certified **bankssperson**.

Key term

Bankssperson – a competent person who supervises the lifting operations of the crane.

Small plant and tools

There are many small plant and tools that are regularly used on construction sites, including:

- cartridge guns – for firing nail fixings
- 110v drills – for creating holes of various diameters
- small petrol generators – for producing independent power
- petrol-driven saws – uses a rotating disc for cutting materials.

Supply of building materials for both traditional and modern projects

Many materials are supplied in construction and these are sourced either:

- from the manufacturer directly where bulk amounts can be purchased
- from a distributor such as a builder's merchant where smaller lots can be purchased.

When purchasing materials, you should always select the 'crane off load option'. This enables easy off loading of large heavy materials onto the correct location using an attachment on the delivery wagon which is, in effect, a crane. Where possible, materials should always be on pallets to prevent damage and they should be wrapped to be protected from the weather.

Prefabricated components and system building

For more on these topics, see pages 206–11.

Assessment activity 6.4

The local housing developer is considering building 30 new houses on the outskirts of your town. Explain the implications of environmental issues and legislation that will place constraints on this development. **P7**

The local residents are concerned over the noise, dust and disruption that will be caused by this development. Explain in terms of the builder what infrastructure will be required to support the construction of the houses and what the infrastructure's purpose would be. **P8**

Evaluate the following legislation in terms of its relevance and what stage of a build it applies:

- the CDM regulations 2007, Town and Country Planning Act
- HASAWA 1974. **M3**

Grading tips

- For **P7** you will need to explain what the legal and environmental constraints in constructing these homes are.
- For **P8** you will need to look at what infrastructure will be required, for example temporary roads, cranes, excavators etc., and what each does.
- For **M3** you will be required to do some evaluation on the different pieces of legislation and at what stage of a build they will act.

PLTS

Planning and carrying out research and appreciating the environmental consequences of various acts and regulations will help develop your skills as an **independent thinker**.

Functional skills

Taking part in verbal discussions on the constraints in constructing homes will help to improve your skills in **English**.

Linda Brett

Design and construction

Linda has worked for a timber framed design and construction company since leaving college with her National diploma in Construction. She gained this employment because of her interest in ICT and computer aided design. Linda's role involves producing final designs in timber framing. Linda's company provide a full service which means that they will take an architect's design and produce a timber-framed building from it right through to the finishing trades. She often has to take an architect's drawing and convert it into a panel assembly drawing that notes the loadings required for the project.

Timber-framed houses are very highly insulated homes so Linda has to purchase large quantities of insulation which fits between the timber studs internally and is covered with plasterboard. Linda also has to plan the construction and delivery of the timber framed components; the ground floor arrives before the first floor and any secondary supporting steelwork is delivered as required. She also needs to get the external finish approved by the architect or client. This takes great planning and coordination as well as communication skills which she uses to programme a work schedule that all understand.

Linda's role often extends into other aspects of the house design, such as the choice and layout of a kitchen, the selection of bathroom fittings and furniture. This means she has to have excellent design skills and be able to communicate these to a client so they know and understand what they are getting in terms of quality and layout. This often means that Linda has to work with specialist subcontractors who will undertake many aspects of the house construction that her company is unable to do, for example electrical wiring and kitchen supply and fitting.

Linda really enjoys this job as she is able to meet many different people from clients to architects and is able to see the final results of all the hard work when a client is happy and satisfied with their finished home.

Think about it!

- What is a timber-framed house? How does it differ from a traditional house?
- What is the most important skill for Linda's job?
- Have you got the skills and qualities to do this job?

Just checking

1 Identify and compare three different methods used for soil investigations.
2 List six factors of ground conditions that would affect the type and size of foundation design.
3 Sketch and annotate four different types of foundation suitable for low-rise domestic buildings.
4 Describe with the aid of appropriate well-annotated sketches and drawings, the process associated with the construction of a timber-framed house. The process should include:
   - site set up, clearance and organisation
   - material storage
   - plant positioning
   - access and safety on site
   - excavation and plant needed
   - water elimination
   - temporary support.
5 Draw a cross-section through a concrete pot and beam floor showing the structural components and the position of the insulation.
6 Compare the properties of steel and timber. Relate the properties to their performance when used as structural members.
7 Draw a vertical section through a timber-framed house showing all the key components and how the house is constructed.
8 Draw to a suitable scale a section through a portal frame building showing the structural frame and a suitable form of cladding.
9 Identify and describe three internal and three external finishes.
10 List four common forms for cladding used for timber-frame domestic housing. Identify the merits and disadvantages of each form.
11 List the plant and equipment necessary to fix a pitched timber roof on a low-rise domestic dwelling.
12 Evaluate the impact of raw materials used in the building of the substructure of a domestic dwelling. Explain how building contractors can contribute to the development of sustainable construction.
13 Analyse and evaluate the performance of lightweight construction. Explain how this differs from heavyweight construction.

Assignment tips

- The Chartered Institute of Building may have some web-based resources to explain the various roles of the production team.
- When investigating legal implications, look for penalties, delays and loss and expense claims in relation to this.
- The RIBA plan of Work may contain reference to written communications to the various members of the design and construction team.

7 Project management in construction and the built environment

The success of any construction company depends largely on how well its construction projects are planned, organised and controlled. Success on a construction project is often measured in terms of the level of profit made – did the project meet expectations, or did it fall short of the profit margin applied at the tender stage when the work was won?

Planning and organisation are essential throughout the life of the project – from the initial design and briefing stage all the way through to completion of the construction phase. The resources required to complete the project – labour, materials, plant and machinery – also need to be carefully controlled.

This unit will help you to understand the management functions and techniques of the planning process, and the importance of information technology in monitoring the project. It looks at the roles of individuals and how they interact to achieve the client's goals, which are generally to produce the project within budget, to complete it within an agreed time-scale and to an acceptable quality.

Learning outcomes

After completing this unit you should:

1. know the roles and responsibilities of, and interaction between, the parties involved at each stage of a construction process
2. understand the resources required to complete a construction project
3. understand the functions of management in the production stage of a construction project
4. be able to develop documentation for construction teams.

Assessment and grading criteria

This table shows you what you must do in order to achieve a pass, merit or distinction grade, and where you can find activities in this book to help you.

To achieve a **pass** grade the evidence must show that you are able to:	To achieve a **merit** grade the evidence must show that, in addition to the pass criteria, you are able to:	To achieve a **distinction** grade the evidence must show that, in addition to the pass and merit criteria, you are able to:
P1 identify the various stages of the construction process for a low-rise domestic or commercial building **See Assessment activity 7.1, page 240**	**M1** produce organisational charts to explain the group dynamics of team working **See Assessment activity 7.1, page 240**	
P2 describe the roles and interrelationships of the members of the building team involved in resource management, planning and production **See Assessment activity 7.1, page 240, and Assessment activity 7.3, page 247**		
P3 discuss the resources required to complete a construction project **See Assessment activity 7.2, page 242**	**M2** compare the advantages and disadvantages of resource management techniques **See Assessment activity 7.3, page 247**	**D1** compare two software systems that can facilitate planning, organisation and control processes **See Assessment activity 7.3, page 247**
P4 explain the techniques used to plan, organise and control a construction project **See Assessment activity 7.3, page 247**	**M3** discuss the factors that may have an adverse impact on planning and organisation **See Assessment activity 7.4, page 250**	**D2** evaluate a range of planning, organisational and control techniques in terms of utility and efficacy **See Assessment activity 7.5, page 258**
P5 explain the management procedures used to monitor and control resources when organising construction projects **See Assessment activity 7.5, page 258**		
P6 discuss examples of resource planning and management documentation **See Assessment activity 7.6, page 262**		
P7 create planning documentation, including bar charts, networks and schedules, for typical low-rise domestic or commercial projects **See Assessment activity 7.6, page 262**		

How you will be assessed

The evidence requirements for pass, merit and distinction grades are shown in the grading criteria grid. Evidence for this unit may be gathered from a variety of sources, including well-planned investigative assignments or practical programming work.

You will be given written assessments to complete for the assessment of this unit.

This unit will be assessed by the use of two assignments:

- Assignment one will cover P1, P2, P3, M1, M2 and D1
- Assignment two will cover P4, P5, P6, P7, M3 and D2.

Teresa

Before reading this unit I did not realise that the use of programming to control a construction project was such a vital tool in organising and coordinating project resources.

This unit has made me aware of the factors that have a major influence on the planning process, so it can be delivered on time. It has also given me an understanding of the roles and responsibilities in managing a project as well as the interactions that are required between all the parties on site.

Over to you

- What ICT skills will you need in order to produce programs?
- What other members of the design team would a planner interact with?
- What responsibilities would a planner have on a large commercial project?

1 Know the roles and responsibilities of, and interaction between, the parties involved at each stage of a construction process

Build up

The design team

The management of resources on a construction site has to be efficiently undertaken in order not to waste any time. In groups, think about the following questions:

- Why is this important?
- What resources need managing?
- What methods are available to do this?

Stages of a construction process

The stages of a construction process could also be thought of as the life cycle of the construction project: it is conceived as an idea, developed into a design, built, used and maintained and finally, when no longer of use, demolished. Figure 7.1 illustrates the life cycle of a construction project.

Design

Design is the phase that requires the most planning and control. If more resources and time are spent at this stage, the following stages will run much more smoothly.

The design stage involves looking at the overall scheme, the feasibility of the scheme, the budget for the scheme and the time frame in which to deliver the completed project. This phase often starts with a site investigation of the ground conditions of the site in general to see if it is suitable for the intended project.

The architect or designer runs this phase of the construction process. They will receive a **brief** from the client which sets out an idea of what they want constructed. The brief is then turned into a sketch design and, eventually, a final design. This process can take many months to complete. Permissions also have to be obtained from the local planning authority for planning consent to build.

Production

The production planning phase begins with the **procurement** of a contractor, the company that will physically construct the design. The next step of this phase involves combining the materials using labour resources; this is assisted by specialist subcontractors and any plant and machinery. The production planning continues until completion when the building project is formally handed over to the client. The contractor has to put right any problems within a certain period – the defects period – which is usually six months. The liability of the contractor is then completed with the issuing of a **final certificate** and the payment of all monies.

Key terms

Brief – the client's idea of what they want, which the designer has then to turn into a reality in accordance with current regulations and legislation.

Procurement – the process of finding and aquiring the expertise, labour, plant and materials needed to build a construction project.

Final certificate – this is a certificate written by the designer that releases the contractual obligations of the main contractor with the final payment of monies withheld as retention.

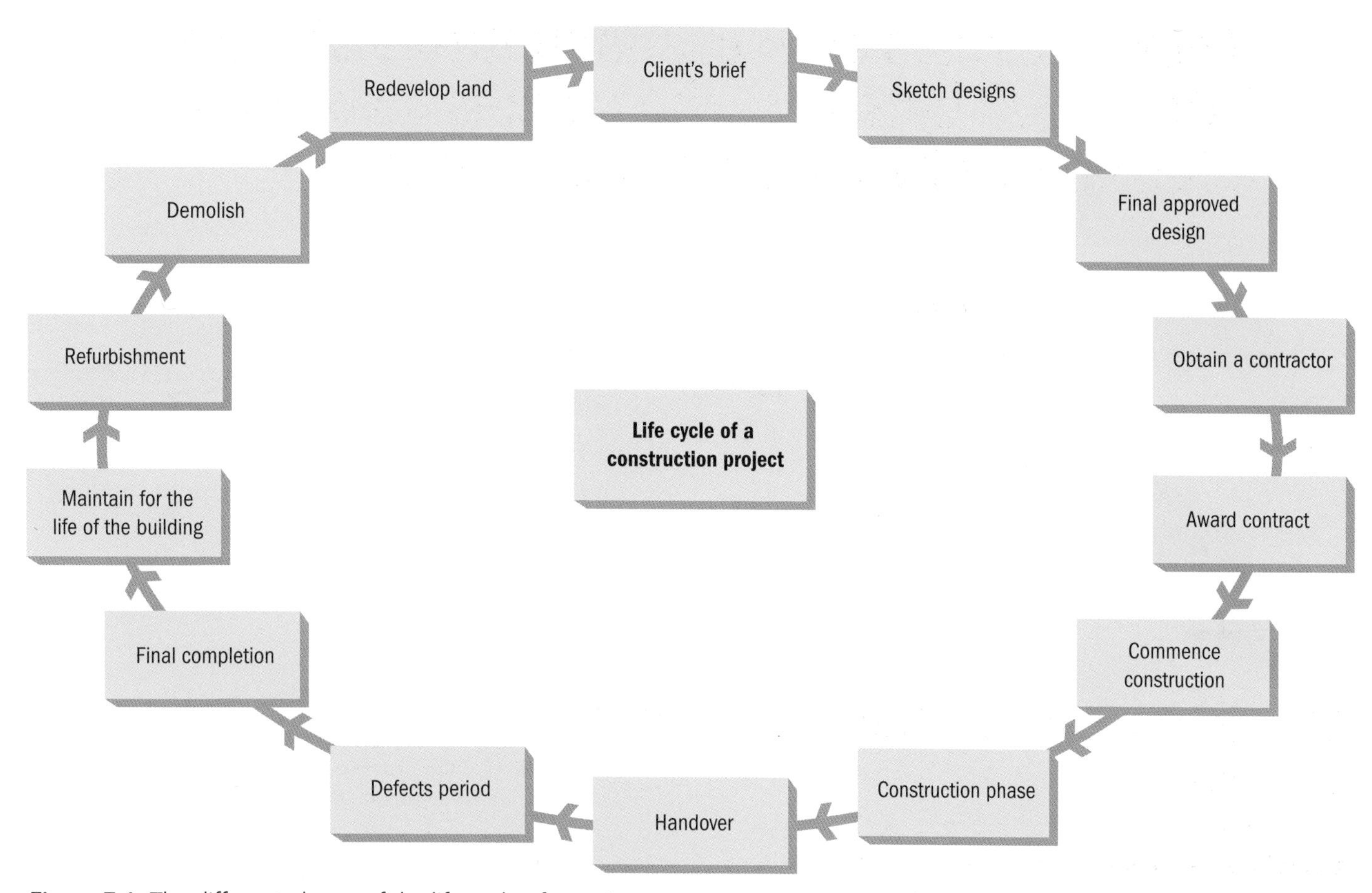

Figure 7.1: The different phases of the life cycle of a project

You will often see a rig taking soil samples in a green field

In construction, production for a typical project may involve the following sequence:

1 Site set up
2 Excavation works
3 Construction of the foundations
4 Erection of the structural frame elements
5 Construction of the walls
6 Ground floor slab
7 Construction of the roof
8 Installation of openings – windows and doors
9 Joinery installation
10 First-floor installation
11 Installation of electrical services
12 Installation of mechanical services, including plumbing
13 Internal finishes – plastering, painting and flooring
14 External work – car parks, roads and landscaping
15 Drainage and connections
16 Clean and handover

This list is not exhaustive and will vary with the type of project that is being constructed.

Maintenance and repair

The maintenance and repair phase involves the day-to-day care required to keep and maintain the operational function of the building for which it was designed. This phase will include:

- flushing and cleaning of the above and below ground drainage systems, including emptying of gullies
- cleaning of guttering to prevent blockages
- replacement of lighting elements
- renewal of roof coverings
- servicing of gas boiler
- redecorating of areas of high traffic
- oiling of door and window fittings
- servicing of air-conditioning units
- repainting of external joinery
- repairing floor finish to worn-out surfaces.

The above list is just a few of the maintenance items that have to be attended to in order to extend the life of the building. Maintenance is:

- reactive – problems are fixed as soon as the call comes in
- preventative – a regular programme of spending on repairs
- cyclical – annual or biannual items are completed, such as painting programmes.

The quality and level of maintenance depends very much on the amount of finance that is set aside to look after and service a building.

Alteration

Alterations tend to involve smaller items such as removing and repositioning doorways. Window replacements are also a good example, where an old metal window is removed and replaced with a uPVC window – this can often be accomplished within a day. There is some planning involved to relocate the occupants while the alterations are carried out.

Some alterations involve major works, such as extensions undertaken to domestic and commercial properties as a need for more space develops and major alteration or rebuilding is uneconomical. An extension can be built leaving the break-through to the main building as the last thing done so lessening the disruption.

Refurbishment

An existing building can be given a new lease of life with a major refurbishment which can potentially involve the whole building. During this stage, key elements can be updated in line with current legislation and technology; often, encapsulated asbestos can be removed, old heating systems refurbished and electrical wiring replaced with modern wiring and circuit breakers. Refurbishments can include replacing existing carpets, windows, doors and wall coverings with new ones, bringing the building up to modern standards of construction and design.

Refurbishment of **listed buildings** has to be carefully planned as like for like must be replaced. This can often mean having paint specially manufactured to match.

Key term

Listed building – building of special architectural or historic interest in the UK. Alterations to these buildings cannot be made without consent or careful consideration.

Demolition

Demolition is the process of removing the existing structure from the site in order to redevelop. Often these are structures near the end of their useful life and cannot be adapted or refurbished due to cost constraints. Recent examples of demolition are the removal of many local authority flats that lacked sufficient insulation and were damp and cold to live in.

Demolition may be instantaneous using explosives or slow and methodical using machinery to carefully cut and remove the building. There is a vast amount of planning for the destruction of a building using explosives. The building has to be structurally weakened in preparation for drilling and charging holes with explosives. The police, fire and highway authorities have to be contacted regarding road closures and notices to evacuate residents on the day of demolition.

Similarly, demolition involving the use of machinery has to be planned because of the noise, dust and volume of road traffic. The modern approach to demolition is to recycle parts of the structure. Metals can be reprocessed and brickwork and concrete can be crushed to produce a hardcore for reuse in filling materials.

Demolition using machinery

Asbestos is a hazard that has to be planned for. It must be removed by a specialist contractor before any demolition process can take place and disposed of in a licensed tip.

Feasibility studies

The Royal Institute of British Architects (RIBA) publishes the RIBA Plan of Work. This contains a detailed step-by-step planning process that can be used to control and organise the whole construction process from the client's idea to the **development of the brief** and from the final design to the construction process. (For more information on the RIBA Plan of Work, see Unit 5 Construction technology and design in construction and civil engineering, page 157.)

Key term

Development of the brief – the development of the client's idea for a design of a building or a concept; this is then extracted and evolved by the designer so it can be taken forward to the feasibility stage.

The feasibility stage is undertaken to see if the client's idea is viable. Typical questions that need answering before the project can enter the detailed design stage are:

- Is the proposed site large enough?
- Will the total cost be within the level of financial resources available?
- What constraints are there on the proposed site?
- Are there any local authority planning constraints on the site, such as the height of any proposed building?
- Is the ground-bearing capacity sufficient to support the weight of the building?
- Will the design achieve value for the client's money?
- Is the design safe under the Construction Design and Management (CDM) Regulations 1994?

These questions will then expand into more comprehensive questions upon detailed investigation of the proposed site. Money is well spent at the feasibility stage to avoid expensive mistakes during the construction phase. A detailed site investigation report will reveal any problems that a proposed site may have. Often the most expensive part of a construction project is the redesign of the foundations when the ground is excavated and problems are discovered.

The design team's planning will need to be undertaken at this stage; consultants will be needed to help with the initial feasibility studies including what roles and responsibilities and duties need to be established.

It is important to keep the client well informed at this initial stage and to seek their approvals before major financial commitments are made. Often a feasibility study may result in the project failing to enter the design stage because the project constraints are too great and the project would be too expensive to complete. In this case, another less expensive solution must be negotiated with the client or the project stops.

Figure 7.2: Design sketch

The design process

Following the completion of the client's brief and the feasibility stage, the design process begins in earnest. The initial budget for the project needs to be established once the brief has been developed. The design process may involve several different design schemes, with the client choosing the one that matches most closely their original idea, or they may pick parts of each design to arrive at a final solution. Each design is costed by the quantity surveyor and a final scheme is selected and the project budget finalised. This will then need written approval from the client. Accurate costings are important to ensure that the final construction will be within budget. A full project brief must cover all aspects of the construction proposal; all consultants must be engaged and the design team put together to start the next stage.

Procurement

Procurement is the method by which a suitable construction company is selected to undertake the work. There are several ways to undertake procurement, each requiring careful planning; the following are the most common.

- Design and build contractors – the contractor undertakes not only the construction of the project but the design phase as well.
- Negotiated contracts – there is only one contractor pricing the work and the contract sum is negotiated between the client and the contractor.
- Competitive **tender** – the client normally obtains estimates from six contractors; generally the lowest estimate wins the contract.
- Partnering –the contractor tries to reduce the client's budget and any savings are split equally between both parties.

Key term

Tender – to make a formal offer or estimate for a job.

The procurement method chosen may depend on several factors: the client's budget, the type of work, the client's historical relationship with a contractor and the location of the work. During the procurement stage, the following needs to be planned:

- the criteria on which contractors may join the competitive tender list
- the time limit contractors will be given to price the work
- the contents of the tender documentation, such as what is to be sent out in the package to price
- the length of time required to check the estimates, correct any errors and award the contract.

Once a contractor has decided to tender for a contract, they will have to undertake a considerable amount of planning in order to produce an estimate. First, they need to look at potential site constraints that might have an effect on the price by going on a site visit. The tender documents will be sent out to obtain estimates for materials and prices from subcontractors. **Method statements** may need to be prepared by the estimator and the contract documents often call for a tender programme to be submitted with the estimate; this outlines the duration of work activities on site.

The preliminaries or variable items such as site supervision, long-term plant hire and site accommodation will also need to be planned. The final planning will be the tender adjudication which is a process whereby all who would be potentially involved in the project can assess the amount of risk. The estimator, buyer and contracts manager will sit down with the managing director and assess what percentage profit they will place on the tender and any amounts for risky elements.

Pre-contract production

Production planning is essential for the smooth delivery of the completed construction project. The pre-contract production will start once the contractor receives confirmation that their tender has been successful and they have been awarded the contract; the construction company normally has to start on site within two to three weeks. There is a considerable amount of work to do, all of which will require efficient planning, organising and controlling.

The designer will supply two sets of construction phase drawings and specifications, and will set a date for the pre-contract meeting, when all the interested parties meet to establish lines of communication and provide information required, such as the start date and a health and safety plan. The purchasing department will look through the estimate to establish which materials have a long delivery period, as these will need ordering first to avoid any delays in the building process.

A **contract programme** will then be produced. This involves subdividing the tender and drawings into separate construction activities with their own schedules and setting up links between the activities so that a **critical path** is established.

At this point, the contracts manager and site manager, who will be responsible for running the project, will need to be selected. The labourers will also need to be notified of their start date on site – there may need to be some coordination with other projects they are currently working on.

Key terms

Method statements – documents which identify the methods used to price the work items, the plant and labour required for each activity.

Contract programme – often a simple bar chart, showing activities against time, which offers an effective visual representation of the construction project. The percentage of actual work completed can be plotted against work that was scheduled to be completed.

Critical path – the link between construction activities crucial for the contract to complete on time. There is no flexibility of time within these activities so any hold up in these activities will delay the final handover date to the client.

A scale drawing of the construction site set up showing temporary facilities will be produced for the site manager to work from (see page 251 for more information on temporary facilities); traffic routes, material storage areas, skip locations and the location of concrete mixing must be agreed. Lighting, power, heating and water will need to be provided for the temporary accommodation facilities such as site cabins, box containers, toilets, meeting rooms, mess rooms and drying rooms. The easiest way to do this is to establish on site the permanent services, including water, gas and electricity that will be required for the completed project and also obtain a temporary metered supply for the contractor to use.

The contract administration process will require initial planning and involves the contract documentation, instructions, materials requisition, ordering, placing subcontracts and administering the contract when it has been awarded. The construction phase health and safety plan will need to be completed before any work can begin on site. Site documentation will need organising – including setting up site diaries, **confirmation pads** and **drawing registers**. Vital to all this is communication: if the duration on site will be for a lengthy period, a site telephone will need to be ordered, along with email facilities and a fax machine.

Key terms

Confirmation pads – used to confirm verbal instructions given on site.

Drawing register – this contains all of the drawings for the project in numbered order with the latest revisions issued.

Post-contract maintenance and repair

Planning for maintenance and repair can often mean waiting for a 'shut-down' in a process so that a contractor can enter a building and work safely. This happens on many factory sites where the main activity cannot be stopped; so a great deal of pre-planning has to be done before work commences on site. For example, many oil refineries operate with specific periods where areas of production are taken off line for as short as two-week slots during which time all maintenance and repairs must be undertaken. When there are no chemical processes to shut down the maintenance and repair work can be planned on an interval basis. This means that certain items of building maintenance can be undertaken on a yearly basis, such as:

- cleaning out and flushing through guttering
- gas checks to boilers and servicing
- electrical testing of the earth circuit
- painting external joinery
- drainage inspection
- oiling window hinges
- changing of light bulbs.

The client will need to put in place resources to finance and fund this annual maintenance.

Refurbishment

Planning for refurbishment work requires some time. A company may be aware that a major refurbishment of all or part of its premises is required, but this may be delayed until a convenient time is available to undertake the task.

Refurbishment may be undertaken in phases, which will avoid disruption to the whole site, although some electrical cabling work may need to take place outside the designated refurbishment area. Handling a refurbishment in phases makes it a lot easier for the client. Smaller phases also mean shorter durations to complete them and the client will need to relocate fewer staff into temporary accommodation while the refurbishment takes place. If a client has multiple phases, they may be able to move production to another site while the work is undertaken, otherwise the disruption may be extensive.

Case study: Site management

Joel Moss is a newly qualified site manager who is working under Fred Smith, an experienced site manager. Both are responsible for running a large construction project. Whenever Fred leaves the site Joel deputises for him. Fred is frequently disappearing off site for long periods.

The site was initially set up well, with perimeter fencing and a separate compound with box containers and site accommodation. Waste skips were provided and a rough terrain forklift truck to transport materials around the site. However, wastage is increasing on site, valuable materials are having to be replaced and site labour costs are rising. The project is falling behind on programme and may incur penalties. There may be several reasons for this.

- **Divide yourselves into small groups and discuss how the situation could be improved.**
- **List what may be going wrong on this site with the control of resources.**
- **Look closely at what is not stated in the case study.**

Members of the building team

Managing director

The managing director of a construction company is responsible for overall planning and is often the owner of the company or may be answerable to shareholders. The majority of the decisions made at this level will be strategic or financial. They look at the full picture of the company's current workload, its future workload for contracts already won and possible workload for contracts being tendered for. This global view enables strategic planning on the levels of supervision required for each contract, and may involve trained personnel to supervise the company's workload.

The managing director will also be involved in the financial planning of the business, including cash flow forecasts, financing of the workload, risk assessments on complex construction projects, bad debts and the type of work the company would like to undertake. These responsibilities differ from the planning decisions that are taken at construction site level (see below).

Site manager

The site manager, or supervisor, is concerned with the day-to-day planning, organisation and control of the construction site, including the organisation of resources. (There are several methods that can be employed to accomplish these tasks which are explored later in the unit.) The site manager has to ensure that the construction project is delivered on time, to budget and to the required quality.

One of the site manager's main responsibilities is to make the best use of resources which involves maximising production and ensuring the use of the right labour skills to fulfil a task. Materials must be used efficiently and not wasted. Plant must be utilised to offset the establishment and running costs against the value of labour saved. Subcontractors must be organised and controlled to work effectively and safely within a team.

The site manager is responsible for all the operations on site, although they may delegate some of these to general forepersons who may control sections or specific trade areas on the construction site such as bricklaying, carpentry and finishes (see below). The site manager is often from a trade background with wide experience of different construction situations, knowledge and training.

Planner

The planner is a technical role usually based at the company's head office, although on complex, expensive projects, they may be employed full time in an office on site. The construction planner is generally responsible for:

- supervising contract programmes (main, monthly, weekly, daily)
- monitoring and reviewing progress
- scheduling materials delivery
- scheduling labour
- reporting procedures
- scheduling plant
- overseeing pre-contract tender programmes.

The planner has an overview of the company's whole operation and will report on the progress of the company's construction workload either to the managing director or contract team director or manager. They are expected to plan how to make the most efficient use of resources, including:

- working out how many operatives will be required on each project to enable accurate labour forecasting
- supplying the buying or purchasing department with materials schedules of what is required, when and how much to enable **economies of scale** in the purchasing of materials
- moving equipment from site to site to ensure maximum utilisation of the contractor's own equipment or that hired by the company.

Quantity surveyor

The quantity surveyor is primarily responsible for the financial planning of the construction company's operations. They deal with all financial aspects such as payments for supplies, invoicing of clients, claims for **variations** and final accounts.

Key terms

Economies of scale – the ability to buy in bulk, thus receiving greater discounts.

Variations – items that were not in the client's original budget and therefore are additional to the contract; for example obstructions encountered within the ground during excavation work.

The quantity surveyor will deal with the **cash flow** of the business and will plan the periods when the company receives payment for the work undertaken on behalf of their clients. These are called valuation dates and are normally at 30-day intervals in accordance with the contract between the company and its client.

Key term

Cash flow – the amount of money flowing into the company from clients and out of the company as payments. Money flowing in should be greater than money flowing out.

Buyer

The buyer is responsible for the purchasing and ensuring of the timely delivery of materials and plant resources. There is often insufficient space on site to store all the materials required throughout the life of the project, stored materials can become damaged, and suppliers have to be paid for them. Ideally, resources need to arrive on site just before they are needed. The buyer will need to analyse the main contract programme to obtain a set of delivery dates for the materials. Often the buyer will place a bulk order with a supplier, who will then deliver specific quantities to the site when requested by the site manager.

Estimator

The estimator undertakes tendering operations for the company which is the process by which the company obtains work. A potential client will ask a number of companies interested in carrying out their project to submit 'sealed bids' outlining the estimated cost of the work. It is the responsibility of the estimator to work out these costs, and they may also be expected to submit health and safety plans and construction programmes as part of the tender. During this process, the estimator will have a great deal of interaction with specialist subcontractors, who will be required for installations that cannot be undertaken by the company.

Site supervisor

The site supervisor, or manager, is responsible for the day-to-day running of the construction site. They deal with the site workers, subcontractors, material and plant movements and resourcing, and are expected to maintain the construction programme. On larger, complex construction sites, there may be several site supervisors for different sections of the project, who all report to a site manager. The site supervisor is also responsible for the health, safety and welfare of all the workers on their site.

Bricklayers have a responsibility to produce work that is of the correct quality

General foreperson

The general foreperson reports to the site manager or supervisor and is often trade specific, for example brickwork. There may be a general foreperson for carpentry, joinery and finishes. Their role is to assist the site supervisor with labour control, materials control and some of the trade-specific plant.

Craft operative

Craft operatives are the trade workers who have a craft background, such as joiners, bricklayers and steel fixers. They are responsible for producing work that is of the correct quality. They collectively contribute to maintaining the construction programme, and have a duty under health and safety legislation to work safely.

General operative

General operatives undertake semi-skilled works, such as the excavation of drainage trenches, working with concrete, keeping the site clean and seeing to the movement of resources.

Theory into practice

The following is a list of roles that particular members of the construction team would undertake. For each of the roles, decide which member of the team's responsibility it is. Remember, there is often an overlap between roles and responsibilities in small to medium sized companies and more than one person could undertake these activities.

- The long-term business plan of the company.
- The delivery date for a full load of bricks.
- The checking of engine-oil level on the site rough terrain forklift.
- The issuing of the monthly construction programme.
- The fixing of timber stairs.
- The pouring of concrete foundations.
- The completion of the tender document.
- The monthly valuation.

Interrelationships

Head office and site organisational charts

The head office and the construction site are set up quite differently from each other. At site level, the administration is organised specifically for the site's size and complexity. The head office, on the other hand, is organised to administer all the company's construction sites. The organisation charts shown here (Figures 7.3 and 7.4) are just one example; one company's layouts may be very different from another, for example some construction companies may have more than one office and act regionally with several offices, so the structure will be fairly complex.

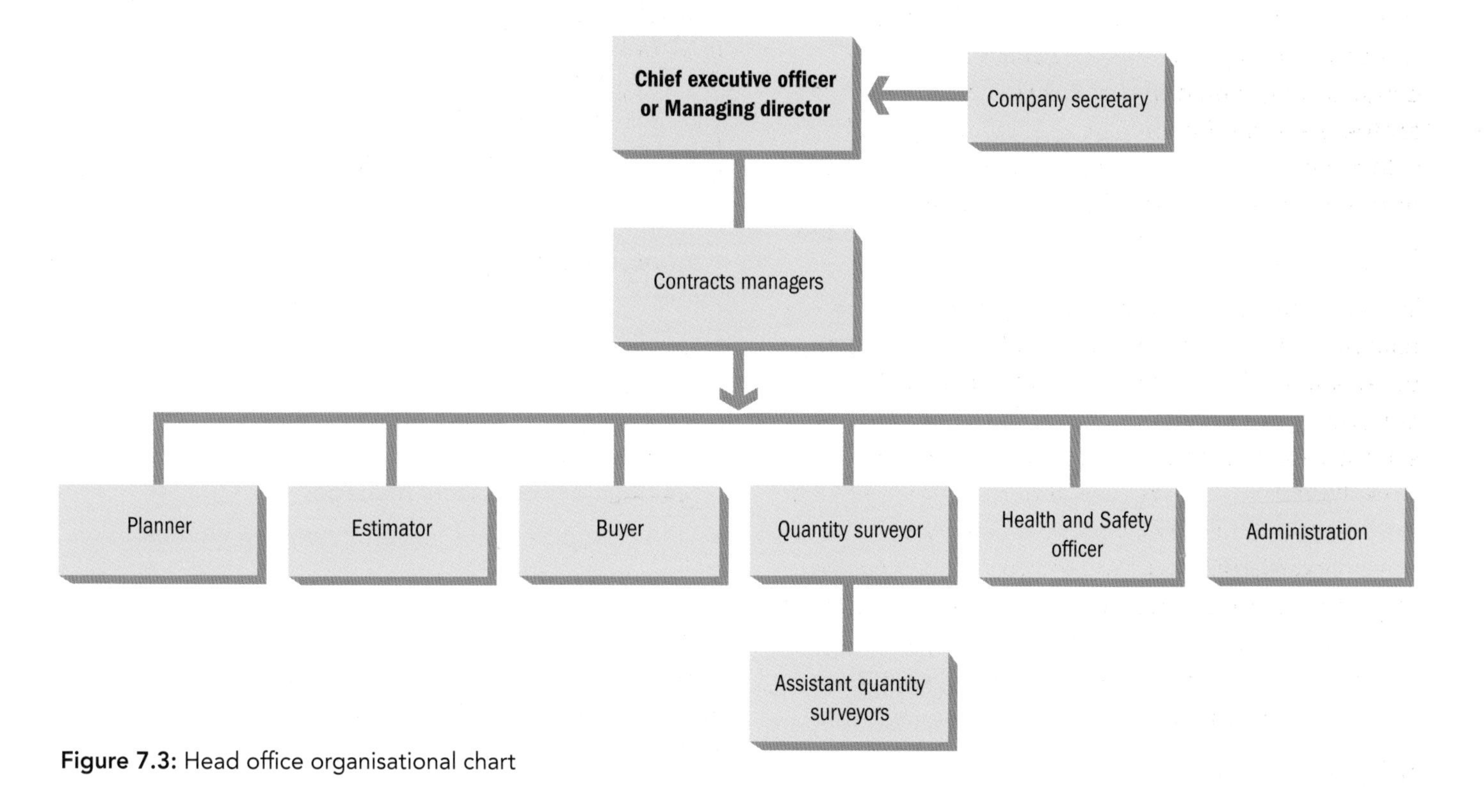

Figure 7.3: Head office organisational chart

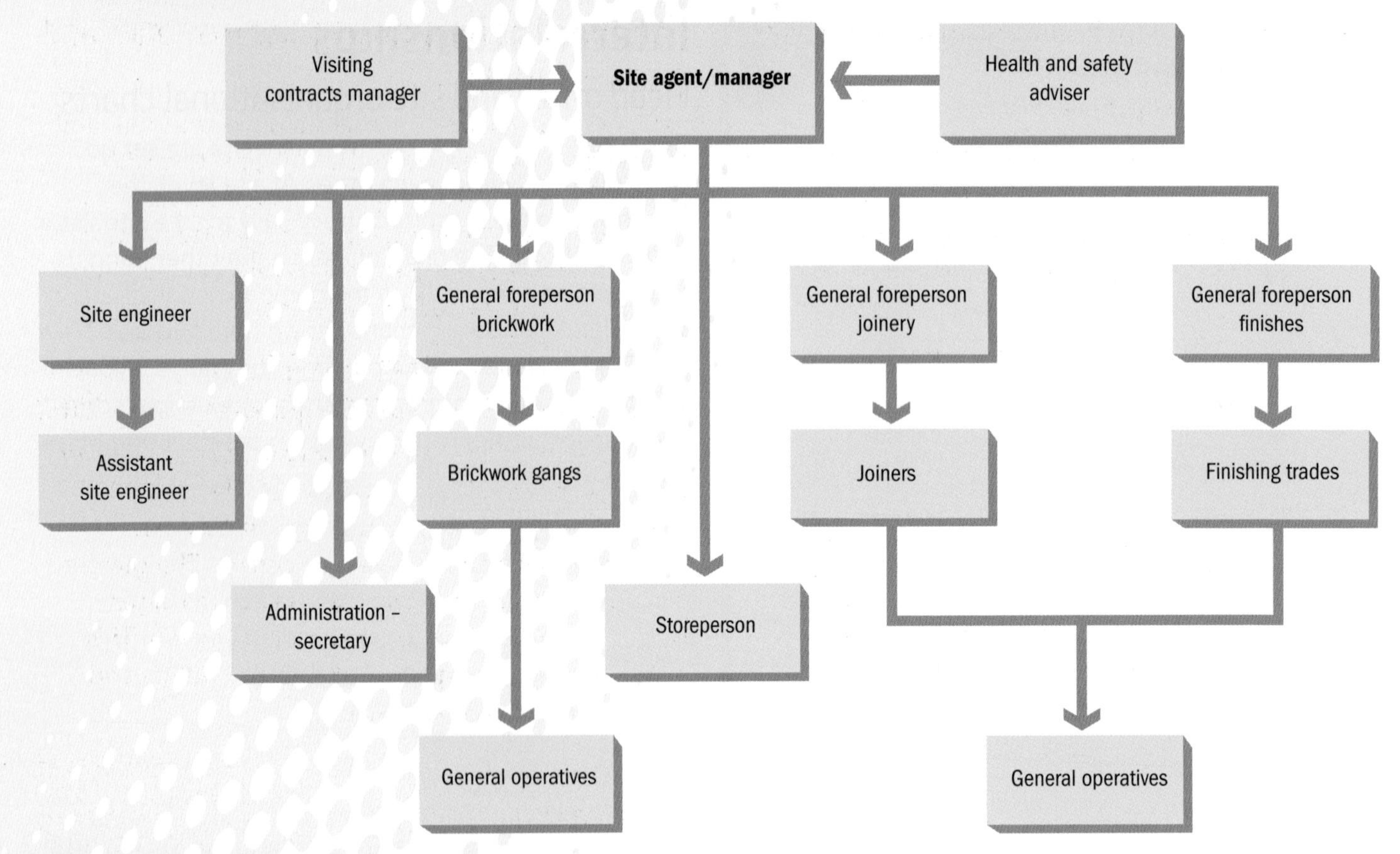

Figure 7.4: Construction site organisational chart

Team interaction and communication

Communication while working within teams is essential especially when the teams may be split between the client's team and the construction team. The client's team will initially be concerned with planning the design phase and may include the following personnel:

- architect or designer
- quantity surveyor
- planning supervisor
- structural engineer
- services engineer.

We have previously looked at the members of the construction team, namely the planner, buyer, estimator, site manager and operatives. The teams may be specialised depending on the type of work that they are involved in and may be subdivided into several teams:

- maintenance work
- interior shop fitting
- sports complexes
- shopping developments
- housing
- commercial developments.

Interaction between teams and within teams may involve several different types of communication. For example, many construction companies use 'virtual' construction sites where all design and construction information is held and each member of the team can log into the system to communicate and transfer information. Traditionally, interaction may take place between:

- the architect or designer and the principal contractor's **contracts manager**
- the contracts managers and their site managers or supervisors
- the site managers and the general forepersons
- the general forepersons and the crafts operatives and general operatives.

Key term

Contracts manager – a member of the construction team whose role is to manage several contracts. The contracts manager moves resources around as and when required, and deals with the designer and construction team on each.

The communication within the teams can take place:

- between the architect or designer through **architect's instructions**
- between the architect and designer and the **principal contractor** via verbal instruction
- between all primary parties through site meetings to discuss the construction programme and progress, including health and safety and the budget
- illustrating changes to designs through drawings
- by email with attachments.

Theory into practice

Formal communication occurs once a month either on the construction site, or in the client's or architect's offices. The purpose is to discuss the programme and progress, information required, variations, and health and safety, etc. In groups discuss the following:

- What type of communication would this be?
- Who would be present at this monthly meeting?
- How would the communication be recorded?

Communication between architect and principal contractor

The architect, often referred to as the employer's representative, is the client's representative on the construction site. The main communication between them and the principal contractor is to ensure that once the design is started, it is finished on time, to budget and to the agreed quality.

The architect will issue either verbal or written architect's instructions. Communication also takes place through drawn information from the architect issued under a drawing register which are where the latest drawing revisions are stored. Architects also lead many of the site progress meetings where all parties to the construction phase can meet face to face to discuss progress, problems and any other issues. Minutes of the meetings are recorded.

Communication between contracts manager and site manager

Communication between the contracts manager and the site manager is normally done via phone, email and fax. It will take place daily to ensure that the site manager has all the resources needed.

Communication between site manager and general forepersons

Again, communication between the site manager and the general forepersons will be daily; starting with an initial meeting first thing to establish the tasks for the day, the resources required and the deadlines to be met. Copies of any instructions received from the architect will be passed on at this point.

Communication between general forepersons and craft or general operatives

Communication between the general foreperson and craft operatives will mainly be verbal. The **ganger** in charge of a specific team, or gang, will take direct instructions from the general foreperson. The ganger acts as the control point so that the site manager or supervisor needs to talk to only one person. This will involve delivering materials to their section and utilising any lifting equipment.

Key terms

Architect's instructions – written instructions issued on behalf of the client which instruct the contractor to undertake a particular item within the contract.

Principal contractor – often referred to as the main contractor. This is a named person, under the CDM Regulations, who undertakes most of the construction work on site.

Ganger – the person in charge of a work gang or sets of gangs which may be trade specific such as a team of ground workers.

Levels of responsibility and accountability

The ultimate responsibility on the principal contractor's side rests with the company's managing director. The architect or designer appointed as the client's representative takes responsibility for the design phase and subsequent running of the construction phase. The head office's estimator is responsible for the pricing of the client's work, while the planner is accountable for the smooth running of the main contract programme for the project. The site manager is responsible for delivering the project on time, the quantity surveyor to deliver on budget, and the general forepersons to deliver quality from the trades.

Assessment activity 7.1

BTEC

1. You have been asked to produce a list of activities for a construction programme of a project. Identify the various stages of the construction process for a low-rise portal framed commercial factory unit. P1
2. The contracts manager has asked you to sort out the organisational structure for a large project. Describe the roles and responsibilities of members of the building team involved in:
   - resource management
   - planning
   - production. P2
3. The managing director of the company you work for is not sure how the team's structures are organised. Produce an organisational chart to show a team working on a typical construction site which can be used as a model for future site set ups. M1

Grading tips

For P1 you will need to identify at least 20 stages for the construction of a project. This will include pre-, post- and production phases.

For P2 you will need to examine the roles and interrelationships of members of the building team. If you look at the unit content it gives you a handy list of the building team members.

For M1 you will need to draw some organisational charts illustrating the building team set up for a project.

PLTS

Planning and carrying out research, while appreciating the consequences of team working will help develop your skills as an **independent enquirer**.

Functional skills

Producing organisational charts using a word processing package with help improve your skills in **ICT**.

2 Understand the resources required to complete a construction project

Resources

Human resources

Management resources should be identified as preliminary costs and must be included in the estimate for a construction project. Human resource supervision is essential in order to control and ensure the efficient running of the other resources of labour, plant, materials and subcontractors. A good site manager is essential to any construction project. Multi-million pound contracts may require many levels of management, at the top of which is the project manager with overall responsibility for the project.

Managers are highly qualified personnel who are usually members of the professional society for construction managers, the Chartered Institute of Building (CIOB). Holders of a CIOB qualification require training and experience before they can be accredited. The Construction Skills Certification Scheme (CSCS) also offers construction qualifications for managers through card schemes which cover health and safety issues.

Take it further!

The CSCS card schemes cover more than 220 occupations including supervisory and management roles. The card proves that the holder is competent in their work and has health and safety awareness and is widely used on construction sites – workers who do not carry a card are barred from working on site. Undertake an Internet search to find out more about:

- the CIOB
- the CSCS card scheme.

Look at both these institutions' websites to see what they undertake or what services they provide.

Direct and subcontract labour

Direct labour costs relate to employees directly employed by the company while subcontract labour refers to self-employed people or agency workers not directly employed by the company; they are not entitled to the terms and conditions that directly employed labour has, such as sickness pay and holiday entitlements.

Labour must be utilised efficiently which means other resources of materials and plant must be coordinated to avoid delays in starting work on site. Multi-skilling of labour is now a modern approach where specific trade barriers are lifted and other skills are taught to employees. This allows companies to get maximum efficiency from their workforce. To obtain the correct quality of labour, a construction company can either:

- advertise for trades people, stating the level of qualifications and experience that the employment contract requires, or
- recruit at apprenticeship level and train the employee to the required level and standard; this has the advantage that the employee will understand how the company works.

Plant and machinery

Plant and machinery play a major part in any modern-day construction process. The following is a typical list of plant that may be used on site:

- 360 degree excavators
- dumper trucks
- rollers
- vibrating plate compactors
- forklift trucks
- generators
- compressors and breakers
- drills
- cranes.

The larger items of plant from the above list will be hired, typically for a day's minimum hire; the contractor will also have to pay for travel to and from the site. It may be more cost-effective to purchase some of this plant, especially the small tools, rather than hiring on a weekly basis. This plant must be utilised efficiently; if it stands idle, it costs the contractor money.

The contracts manager must utilise plant sometimes across several sites to obtain maximum efficiency, and planning decisions will often involve considering whether the cost of hiring plant will be offset by savings in labour costs. Sometimes decisions have to be made on the grounds of health and safety under what is **reasonably practical**.

Key term

Reasonably practical – a measure put in place to prevent injuries to workers that is both sensible and sound for that particular situation as well as realistic in safety terms.

Planning decisions in the selection of plant will be based on some of the following criteria:

- the weight of what has to be lifted
- the height that a material needs to be placed at
- the distances that materials will need to be moved
- the experience of the trained operatives
- the size of the construction site
- the length of time on site.

Materials

Materials are the physical resources that are used to construct the finished project. They each have different properties and performance and will arrive on site in a variety of forms. Where possible, heavy materials should be delivered by crane offload where the delivery vehicle has a crane attached to it in order to lift and place the materials on the ground. Much larger materials such as structural steelwork and roof trusses will require a stand-alone crane. Since construction sites can be very awkward to travel around, rough terrain forklifts may be used to place materials exactly where they are required.

Safety tip

Care must be taken in the manual handling of heavy materials; any with a specific weight of over 25 kg should not be lifted by hand.

And remember: a clean and tidy site is a safe site!

It is essential to take care of materials as any damage costs will be at the contractor's expense and not the client's. Wastage equals loss of profits, thus materials need to be stored correctly on site. Cement in bags, for example, will not stay fresh for long when stored exposed to the weather. A site may require a shed to store cement and racking for scaffolding while bricks should be delivered shrink-wrapped and on pallets so they do not get damaged.

Material security is another issue. Some materials can be very expensive, such as floorboards; the higher the cost of the materials, the more thought will need to be given to their storage. Construction sites must be fenced in to prevent the theft of materials from the site and CCTV and security guards can often be deployed as a deterrent to theft. Storage compounds are an ideal solution, having separate fenced areas for materials to be stacked inside, ready for use.

On congested sites, materials may need to be delivered just before they are needed. The planner needs to schedule the materials against the contract programme to ensure that delivery dates are met by the suppliers.

Domestic subcontractors

It is impossible for a construction company to undertake all the work; domestic subcontractors will need to be hired. The reasons for this are twofold:

- It is expensive for a company to employ operatives who are specialised in a specific trade if it cannot keep them working all the time.
- The level of training required for highly skilled operatives, together with the associated costs, may be prohibitive.

Subcontractors are the ideal solution; they are highly trained and specialised in their role. They are also very effective in undertaking large areas of work activity within the contract programme. Subcontractors typically undertake the following work:

- mechanical installation
- electrical installation
- fire alarm installations
- lift installations
- CCTV systems installation
- flooring
- painting and decorating
- plastering.

Nominated subcontractors

Nominated subcontractors are those which the client wishes the main contractor to use; the client may have entered into negotiations with a nominated subcontractor to agree a specialist piece of work. The main contractor is then instructed to use this contractor and it is included as a sum within the tender documentation.

Assessment activity 7.2

The construction company you work for has been awarded a contract to build a small business development on the outskirts of the city centre. This will involve the bringing together of several different resources. You have been asked by the buying/planning department to discuss the resources required to complete the building project so it can schedule the work into the plan. P3

Grading tip

For P3 you are required to discuss the resources needed to complete a construction project.

PLTS

Proposing practical ways forward and breaking these down into manageable steps will help develop your skills as an **effective participator**.

Techniques

Production of long-and short-term programmes

The contract programme is a visual reference that can be used to control the progress of the project; the most common form being a bar chart (see page 261). Programmes are produced by taking the contract drawings and the specification and the time allowances from the estimator and dividing the contract up into specific activities. These are then represented as a 'bar' on the chart. The length of the bar represents the duration of the activity on the site. A short-term programme can be a weekly or a monthly programme and may take a particular activity and break it down into some detail for issuing to operatives on site to control progress. The programme is best displayed on site within the supervisor's accommodation.

Scheduling of material requirements

Materials require some form of organisation and usually need scheduling to enable materials to be ordered from a supplier. Often tenders consist of a set of drawings and a specification; there are no **bills of quantities**. The estimator therefore has to schedule the materials that are present in the drawings. For example, there may be a variety of window types and a different number for each, which will need adding up so they can be ordered.

Schedules are typically prepared for:

- ironmongery
- doors
- reinforcement
- windows.

Key term

Bill of quantities – a list of quantities produced to a standard that is used to price construction tenders.

Figure 7.5 is an example of a reinforcing schedule that a structural engineer has produced for a project. It shows the diameter of the reinforcing bars and their length. From this, the total weight of reinforcement can be calculated. Scheduling has to take place before materials can be requisitioned (see below) and subsequently ordered.

Simple Analytical Engineers
Contract: New College Library — Bar Schedule No: 003/147/SAT03

Member	Bar Mark	Type & Size	No of members	No in each	Total	Length of each bar – mm	Shape Code	A	B	C	D	E
Beam on centre line	1	T20	4	3	12	1571	20	385				
	2	R10	5	25	125	2168	405	196				
	3	T32	3	5	15	1916	st					
	4	R16	8	7	56	547	st					
	5	R25	6	51	306	256	405	205				
	6	T20	4	2	8	852	150	457				
	7	R10	2	4	8	567	202	600				
	8	R10	1	5	5	400	st					
	9	R25	3	4	12	150	st					
	10	R16	7	3	21	220	205	600				
	11	R16	6	32	192	1515	st					
	12	R8	4	1	4	458	st					

Figure 7.5: A reinforcing schedule

Requisitioning

Requisitioning is the process where all the finalised schedules and material requests are submitted to the buying department. Alternatively, this can be done internally within the buying department. A requisition states what material is required and when it is needed.

Ordering

Ordering is best accomplished as a head office function; ordering from site should be limited to a small amount. Requisitions can be collected together to form full loads; in this way better discounts can be obtained from suppliers by purchasing in bulk. If a site has limited space available, then parts of the order can be called off in small deliveries to site. The quoted prices must be checked against the estimator's tender figures to establish whether the price is within the contractor's budget. In this way, costs can be controlled.

Receiving and checking

When material or a piece of plant is delivered to the site, there will be documentation called the 'delivery ticket' accompanying it. A copy is normally given to the site from the supplier. The delivery ticket may state the amount and type of material delivered to the location. It is the responsibility of the contractor's representative receiving the goods to check the delivery and note any shortfalls on the delivery ticket. If any shortfalls are discovered after the delivery ticket is signed, they are at the contractor's cost.

Site handling

Site handling is a management function and involves the selection of the materials handling equipment for distribution around the construction process. Things such as height, reach and the ground conditions will need to be taken into account with the distribution equipment that is hired. The supervisor should check that the hire rate is in line with what the estimator placed in the tender. The supervisor must also ensure that the appointed driver has the required level of competency and is licensed to use that vehicle.

Storage and security issues

Materials need clean and tidy storage areas so they do not become damaged. High-value materials must be stored under lock and key on site to secure them against any loss which could cause a time delay on site. Security box containers are a solution to this problem along with a night security person. **Just-in-time deliveries** that have detailed scheduling can also help avoid storage problems of theft and damage.

Key term

Just-in-time deliveries – materials ordered to be delivered just before they are required on site.

Labour management techniques

Work and method study

Work and method study is the application of science to the movement and efficiency of site operations. In essence, it is a time and motion study used to analyse operations involving labour to seek ways of improving efficiency on site. Work and method study, if applied correctly, will save financial resources since production will be increased; hence there will be an increase in the contractor's profit margins. For example, a manager may notice that labour is being used to barrow by hand concrete into an awkward part of the site. He then undertakes a study of this involving timing each employee, and calculates that changing the method – using a concrete pump – would save both time and labour costs.

Sometimes work and method study may cause resentment among the labour force, who believe they are being unfairly watched. Thus, unobtrusive methods have to be employed to undertake work and method study effectively.

Control and organisation of labour

Labour can be controlled and organised in several ways:

- through gangers, each being in charge of a labour gang, who report to the general trade foreperson, who then reports to the construction site manager. In this way, effective supervision and control of the labour is in place
- through direct supervision in small groups. This would be undertaken on small sites with the site supervisor working directly with the labour force
- through subcontractors controlling their own labour. Often large construction sites do not directly

employ labour as all the construction work is subcontracted out

- by labour meetings at the head office between all contracts managers, where labour requirements for each site are discussed and labour movements directed between sites
- by written instructions on job cards. These contain what is required to be done and the location for each labour resource; when one is completed another is issued
- through the main contract programme planning, where the labour resource is scheduled into the master programme and labour reports can be run and issued
- through time sheets against each job. Each person submits their time sheet weekly to show where they have worked and for how long. Head office can keep control of hours by totalling the time sheets against what was expected for each work activity on the programme.

Plant needs to be utilised effectively to ensure maximum use

SIMTOP BUILDERS

Name		Trade	Week ending	/ /
Employee code			Passed for payment	

Vehicle cost	Description of work (Location)	PART 'A' To be completed by operative																		PART 'B' For office use only					
		MON			TUE			WED			THURS			FRI			SAT		SUN							
		1	1½	2	1	1½	2	1	1½	2	1	1½	2	1	1½	2	1½	2	2	JOB No.	Basic hours	Overtime @ 1½	Overtime @ double	Job bonus	Expenses taxable	Expenses non tax
Vehicle details	Basic hours																			TIME SHEET TOTAL						
	Overtime @ 1½																			BROUGHT FORWARD						
	Overtime @ doublet																			TOTAL CLAIM						
	TOTAL																				OFFICE USE ONLY			GENERAL BONUS		

Figure 7.6: A time sheet

Plant management

Hire

For most construction sites of a short duration (16–20 weeks) it is often more cost effective to hire plant for several reasons.

- It would cost more to purchase than the total hire charge over the period of use on site.
- Plant is always delivered fit for use and is tested. (Plant associations have to operate under the Provision and Use of Work Equipment Regulations (1998).)
- There are no maintenance costs involved as these are borne by the plant hirer, except for punctures on site.
- Plant can be off hired (the hire terminated) when not in use, whereas the contractor's own plant would have to stand idle.

Lease or purchase

Leasing a piece of plant is similar to hiring but under an agreement with a finance company. In this method, the contractor hires the piece of equipment for a number of years, usually three, and pays a rent per week or month. Sometimes there are restrictions, for example the number of miles the contractor can do with the equipment per year, and it is often difficult to end a lease agreement early.

Purchasing a piece of equipment is sometimes the best and most efficient method. The construction company needs to establish whether it will make full use of the equipment; if it is standing idle for long periods this will not be an effective use of investment. Before deciding to purchase equipment, the company will need to answer the following key questions:

- How will the purchase be paid for – through a bank loan or from the company's profits?
- How much maintenance and servicing will the equipment require and what are the likely costs?
- Will an operator need to be trained to use the equipment?
- How much will the equipment depreciate in value per year?
- What percentage productivity a year will the equipment attain?

Once the company has the answers to these basic questions, it will be able to establish how much a piece of plant will cost the company to buy, run and maintain, which can be compared with the cost of hiring the plant to see which is cheaper.

Take it further!

The decision to lease, hire or purchase must take into account the productivity gained, the labour savings made and the **opportunity cost** lost. Do some research on the internet and find out what the economic term 'opportunity cost' means.

Utilisation and control

Plant needs to be utilised in the most cost-effective way. Maximum utilisation can be achieved by transporting plant from site to site such that **down time** is kept to a minimum and full use of the plant is maintained. Highly trained and experienced operatives will be more productive than inexperienced ones and this factor must be taken into account.

Plant can be controlled by

- ensuring that, if it is not in use, the plant is off hired
- transferring it to another site where it can be used
- using short-wave radios to speed up communication and avoid delays
- using trained and qualified operators
- using plant sheets to list items on site
- ensuring the correct equipment is in place for the work in hand.

Key term

Down time – when equipment is not being used.

Software applications

Availability and use of software for programming and monitoring

There are several software systems available that will produce a construction programme from data from the estimator's tender.

Construction programme software can undertake many functions. Plant and labour resources and costs can be entered, and a variety of reports that track progress both in time and in financial cost can be produced. To monitor a construction programme, the progress against each activity can be entered and a report can be produced to illustrate the current status.

Assessment activity 7.3

1. The contracts manager has just returned from a surprise site visit to one of the company's construction sites. They are far from happy, muttering that it is organised chaos there. You have been asked to identify and explain the techniques commonly used to plan, organise and control resources on a construction site, so this information can be given to the managing director to improve the site's organisation. P4
2. After your report, it has come to light that the site and office are now undertaking some very effective resource management techniques. The managing director has been shown these and is quite impressed. They have asked you to continue your valuable research by comparing and evaluating the following techniques, outlining the advantages and disadvantages of each:
   - Long-term contract programmes
   - Receiving and checking material deliveries
   - Work and method study
   - Sorting waste into skips. M2
3. The chief planner of the company you work for is keen to upgrade the existing programming methods manual to a more modern computer-based approach. Undertake some research on the Internet and examine the software systems that are available.

   When you have found an example explain how this may be able to make easier the planning, organisation and control of the construction processes on the company's sites. D1

Grading tips

For P4 you will need to explain some of the techniques used to organise and control resources on a construction site.

For M2 you will need to compare at least two of the resource management techniques in terms of advantages and disadvantages.

For D1 you are required to analyse and compare two software systems, for example Microsoft Project.

PLTS

Asking questions to extend your thinking will help develop your skills as a **creative thinker**.

Functional skills

Describing by writing an explanation on the control of resources will help improve your writing skills in **English**.

Factors in the planning process

Labour availability and cost

The availability of labour for the construction process depends on many factors with geography being one of the most important. The further the construction project is from large cities and towns, the less local labour there is likely to be. Similarly, larger cities attract more work, which ultimately means higher wages to pay workers because they can pick and choose which projects to work on. Another factor is the economy. The construction business relies on a strong economy to drive new enterprises in developing new factories and commercial units. However, when the economy is booming all available labour will be working and earning high wages, so additional labour resources may not be available.

Labour skill levels

There are different skill levels within the construction industry, from labourers to skilled operatives, and each will have a specific trade, such as joinery, plastering or bricklaying. Since there are clearly defined lines between each trade, there is no overlap of roles in the working environment.

The skill level of general operatives, however, can vary as they cross over several trade areas. General operatives tend to work with the substructure items of drainage, foundations, concrete works, assisting bricklayers, cleaning and waste removal among others. Similarly, many will be able to drive different pieces of construction equipment, for example dumpers, forklifts and rollers. Often additional payments are made to general operatives who are classified as semi-skilled, for example some will be able to **power float concrete**.

The increasing shortage of qualified and skilled trades people in many sectors may often be as a result of government funding in one area and not in another. It can also be caused by technological economic booms that draw all the available apprentices into that career path instead of construction.

Motivation of labour

A clean, warm working environment is the basis for the first step in motivating the workforce. Keeping people dry, or giving them a warm room to dry their clothes is essential as is giving them a place to eat and have warm drinks.

The motivation of employees is essential in order to achieve the contract deadline completion date. Employees can be motivated in several ways with money always being a driver of productivity; bonus schemes work well as long as both company and worker receive something out of it. Incentives such as company cars and vans act as a reward for effort by employees while complicated bonus schemes that nobody understands can de-motivate employees very quickly. Many companies operate staff discount schemes and other benefits in the form of vouchers that employees can use to obtain services outside work, for example private health benefit schemes for the whole family. Team working is another motivational tool where rewards can be offered for the most productive team.

Productivity

Productivity has to be maintained on a construction site in order to achieve the construction programme's final completion date. If activities fall behind they can be brought back on schedule by the redistribution of labour resources from one activity to another. As mentioned earlier, labour resources are often organised into 'gangs'. For example, a bricklaying gang may contain three people, two of whom are bricklayers with the third person being a labourer assisting them.

Progress is monitored by evaluating the level of production achieved on site against contract programme activity; output from the gangs can then be driven to meet any shortfall if an item is behind programme. Failure to accomplish any monitoring can result in the construction programme falling so far behind schedule that the handover of the completed project is late. If the contractor fails to hand over on time, the company may have to pay **damages** to the client for every day that the programme is late.

Key terms

Power float concrete – a process whereby the surface of the concrete is machined smooth using mechanical equipment.

Damages – financial penalties for every week or part week that the programme has overrun which are usually charged for every day that the programme runs late.

Plant output rates and efficiency

A piece of construction plant needs to run at 100 per cent efficiency; plant standing idle costs the contractor revenue and is a waste of a financial resource that could have been better placed elsewhere. However, plant on site rarely runs at 100 per cent efficiency which may be for the following reasons:

- drivers have to take rest breaks
- the machine may need to have one tool removed and another type fitted
- the machine's tyre may puncture
- the machine may break down or require servicing
- a newly trained operator will be slower than an experienced one
- the machine may be old and worn and slower than a newer model
- plant size may not be correct for the type of work being undertaken.

These are just some items that will slow down the rate of efficiency on a construction site. For this reason, output rates will vary daily. The planner will therefore work out an average output rate which reflects the actual output achieved on site.

Material availability

The availability of materials has become a problem in recent years, as suppliers no longer carry sufficient stock levels due to the financial implications; materials standing in a supplier's yard tie up financial resources. Many specialised materials, such as special bricks, have to be made to order and can take many months to be delivered. This time needs to be taken into account, from the specification of the materials by the architect to the delivery periods that may have to be dealt with.

This may mean the client pre-purchasing a material and placing a provision in a tender document; when the material does arrive the contractor is paid for handling, storage and delivery.

Some materials may be in short supply which can be caused by several factors:

- strikes or disputes at the production plant
- lack of transport
- location of supplier, for example, Germany and the material has to pass through UK Customs
- breakdowns in a production line
- high demand in one country that draws supplies from another

Delivery periods

The delivery period occurs from the supplier receiving the contractor's order to the time that the material is available for delivery to site. Delivery periods can vary widely. If a material is not stored as stock in a builder's merchant's yard, then it will be delivered directly from the manufacturer to the yard when next available. If the delivery is a full load, then it may be delivered directly to site.

If a contractor is late placing orders for materials, this also may affect the delivery period. Certain manufacturers only roll so many types of certain materials, such as roof cladding, then they change the rollers for another product. The contractor may therefore have to wait for the next scheduled production run for delivery of the material.

Materials delivery

Materials can be delivered loose, for example gravels and sands, or bagged or packaged, for example plaster with the greater discounts being obtained by purchasing in bulk. The various modes of delivery include:

- by tanker
- by tipper
- by silo
- by pallets, shrink-wrapped, and **crane offload**
- on a flat bed wagon with drop sides for forklift offload
- in a van
- by chute, for example a concrete wagon.

Key term

Crane offload – a load is delivered by a crane that is bolted onto the wagon. The crane is operated by hydraulics and lifts solid material off the wagon safely onto the construction site.

An excavator

When materials arrive on site, they are often temporarily stored before being moved to the correct location. The method of delivery will depend on the type of material and how it is finally handled on site. Bricks are best delivered by crane offload on pallets or banded with forklift holes and covered in plastic wrapping. This keeps them clean and enables them to be picked up and moved. Concrete can be moved by pumping it along a hose to the point where it is required – this can be done from some distance away. Sands and gravels are best delivered in bulk and tipped onto a clean surface so they can be either shovelled up in a machine bucket or loaded into a dumper.

Take it further!

With most materials, there are packaging disposal and wastage issues, as the materials need to be unwrapped and cut to size. Do some research on the Internet and find out how some construction materials are packaged and delivered. How are these packagings disposed of?

Waste

Some of the materials utilised on site will be wasted in the course of the construction phase of the programme. For example, timber is normally delivered in lengths to the nearest 300 mm – lengths in between will require cutting on site. Wastage can occur due to several other factors, including:

- human error, such as incorrect ordering of quantity
- poor workmanship on site by semi-skilled workers
- theft which requires materials to be replaced
- incorrect lengths ordered
- wrongly specified material for the job
- not checking quantities when they are delivered
- materials getting damaged when they are moved around the site
- water damage due to poor storage.

When a tender is prepared, the estimator will include within the cost of the contract allowances for wastage. This is usually around 2.5–5 per cent, but it varies with the amount of cutting needed to make the material into the finished product. Waste can be minimised by recycling and selling what is left over; skips on site for metal, timber and cardboard will soon pay for themselves.

Assessment activity 7.4

BTEC

The contracts manager has just been to a site meeting for the new city library building that your company is currently half way through constructing. The contracts manager has been asked to find out why the contract is behind programme and you have been asked to prepare a report for submission to the client.

Discuss in your report the factors that could contribute to the poor planning and organisation on the site and, for each of the factors you have highlighted, the possible effects. **M3**

Grading tip

For **M3** you will need to discuss the factors that have contributed to the adverse delays to the programme, such as labour motivation, efficiency, the weather, etc.

PLTS

Taking part in discussions on the delays to the program will help you develop as an **effective participator**.

Functional skills

Adapting your contributions in the discussion to suit your audience will help you in speaking and listening in **English**.

Context

Finance

Finance is the most important resource on construction projects. The client has to establish the finance required to see their design through to a finished structure while the contractor has to finance the construction works on site. To assist this process, regular payments are made called 'valuations'. Each month the work is measured on site to establish how much has been completed and the architect will then issue a certificate to certify this; the certificate is similar to an invoice and states what the client must pay the contractor. From these payments, a deduction of normally 5 per cent is taken for retention which ensures that the contractor will complete in accordance with the contract conditions. The retention is not paid to the contractor until the project is handed over to the client.

Contract period and liquidated and ascertained damages

The only two dates that are on a typical contract are the start and completion dates. Should a contract overrun the completion date through the fault of the main contractor then the client may have the right to damages which are losses that the client may incur as a result of not receiving handover of the contract on time, for example loss of rental income. The damages are then charged against the main contractor's payments.

Site layout and organisation

Site layout and organisation, if done correctly, will greatly add to the efficiency and safety of the construction site. Detailed planning at the pre-commencement stage will benefit from using a site layout plan to assist the organisation and control of the construction phase works. The site layout will be governed by the size of the site and the **footprint** of the building that is going to be placed upon it, so no two site layouts will be the same.

Key term

Footprint – the shape the outline of the building leaves on the ground when viewed in plan from above.

Site layout involves organising the temporary facilities that are required to construct the site. These include:

- site accommodation
- waste removal
- temporary services
- traffic routes
- car parking
- security
- fencing
- storage of materials
- transport
- positions of cranes
- scaffolding.

Temporary facilities and works

Temporary facilities cover the site cabins required to house, feed and keep dry the labour workforce constructing the project. They are not part of the final construction project handed over to the client. Temporary supports required during construction can be steel piles to support excavations, temporary props, scaffolding, cranes and forklifts. Compound areas will need to be constructed to keep and protect materials stored inside, and the site will need temporary fencing around it to prevent harm to outsiders. Temporary supplies of power and water will be required to construct the project, along with lighting for some of the winter months.

Health safety and welfare issues

For more on these issues, see Unit 1 Health, safety and welfare in construction and the built environment.

3 Understand the functions of management in the production stage of a construction project

Management procedures

Forecasting

There are several items in a construction project that require some form of forecasting, including:

- Labour requirements – this will be required by the contracts manager from each construction site so they can work out when to bring in additional labour or transfer labour from one site to another.
- Cash flow – this will be undertaken by the construction company's quantity surveyor or the financial manager. This is necessary to predict how much revenue and expenditure is entering and leaving the business.
- Plant requirements – if the company owns several pieces of plant, then the utilisation for key items like a concrete pump or an excavator needs to be forecast so that it can be used on each site at the correct time and maximum utilisation achieved.

- The weather – the pouring and finishing of concrete outside has to be timed with good weather. Concrete can be damaged if it rains or is subject to excessive cold or excessive heat and dries out too quickly.

Planning

Contracts of some duration are normally controlled by a contract programme. The contractor is required to produce a main contract programme illustrating how the construction site activities have been planned and linked. This is used to monitor progress at each site meeting. Look at the contract programme in Figure 7.10 (page 257), which shows how the contract has been subdivided into activities with time operating left to right.

Organising

Site organisation is paramount with communication the key, both verbally and written. This can be achieved through a sequence of line managers, from the general foreperson to the site agent and, ultimately, the contracts manager or director. 'Tool box talks' involve a five-minute discussion of key issues with supervisors who then inform the rest of the workers. Discipline is therefore essential and must be monitored.

Monitoring

Progress on the project needs to be monitored through use of the contract programme; progress is plotted to establish if work is on time, in front or running behind. If work is falling behind, monitoring will catch this early so that resources can be directed to the lagging activity to bring it back on schedule.

The financial side of the project must also be closely monitored to ensure that the predicted profit for the contract is met and that a positive cash flow is maintained. This monitoring is called 'coat value reconciliation', and involves adding up all the costs against revenue at a particular time interval – to show a profit or a loss.

Controlling

Resources will need controlling to avoid wastage. High wastage cannot be maintained for long as it adds up to a loss of profit to a contractor. Controlling is undertaken by the supervisors on the site who may be aided by computer software solutions that identify what needs to be there and when. Control must be effective and directed at the site resources – anything beyond the supervisors' control is serious and should be avoided. Discussion at site meetings where items can be recorded is a very effective method of controlling the programme and progress.

Coordinating

Coordinating the subcontractors and the main contractor's labour activities is required in terms of health and safety. For example, it would be dangerous to have the roofer working above the floor layer. Coordination is needed especially with the building's services; for example, where services will be run within roof voids, or which cable or pipe work needs to go in first.

Reviewing

Reviewing provides feedback so a more viable solution can be implemented. It involves looking at all the above processes and establishing whether they have worked. For example, a risk assessment has to be reviewed to check the required resources have been allocated to the work and that the control measures are working effectively.

Organising construction projects

Site layout plan

Figure 7.7 shows a typical site layout plan within which area the contractor would construct the building. The contractor would not be allowed to work outside the boundary. It requires some logical thinking on how best to prepare the site layout plan which would be drawn to scale and, once complete, passed to the site supervisor to add all the site accommodation, box containers, traffic routes, etc. The contracts manager would need to be consulted and when detailing a site layout plan, the following need to be considered:

- traffic routes
- labour movement
- materials and plant location
- **access and egress**
- site accommodation
- storage and security.

Key term

Access and egress – entrance and exit.

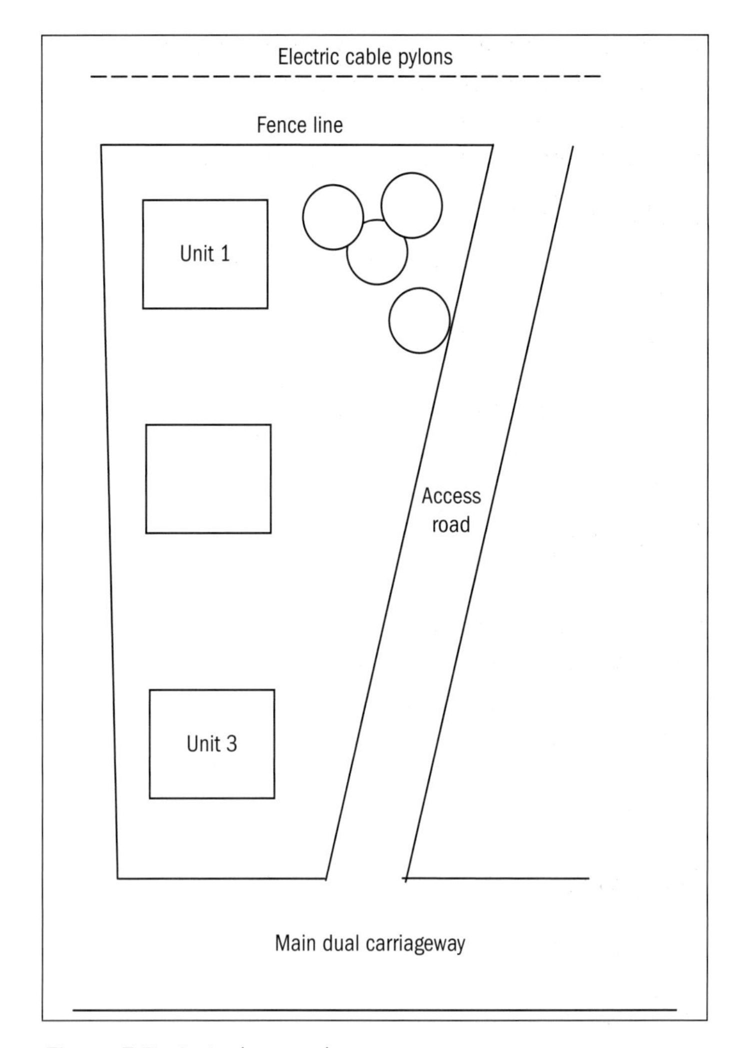

Figure 7.7: A site layout plan

Traffic routes

On a congested site it may not be possible to have two-way traffic, so a one-way system may have to be implemented, with traffic lights to control the flow of delivery and construction vehicles. Traffic routes should have a pedestrian walkway, ideally fenced, to separate operatives from moving plant and machinery.

Traffic routes should be established using well-compacted hardcore such that the surface does not become muddy and a hazard. Routes that run off site onto the highway may have to have lorry wheel washes installed to prevent mud, soil and other materials being deposited along the public highway which can be a hazard. This can often be resolved by employing a road sweeper to control dust and debris.

Labour movement

The movement of labour on site should be controlled to ensure that activities critical to the overall programme are staffed sufficiently; thus ensuring it does not fall behind programme. Moving labour on site must go with skill levels, for example a joiner could not lay bricks.

Regular labour resource meetings held at the head office enable individual construction site requirements to be planned using each site manager's labour forecasts. The contracts manager can then decide where to use and relocate the labour to make maximum use of individual skills, training and experience.

Materials and plant location

The temporary storage of materials and plant needs careful planning. Often there is insufficient space on site and arrangements may have to be made with suppliers to deliver smaller quantities as they are required. Materials need to be located close to the work area to avoid too much handling or double handling which increases the incidence of damage and wastes resources.

The location of lifting equipment, especially **cranage**, depends on the size of the structure and the point of furthest reach required. Tower cranes are often placed within the centre of a project, often within the lift shaft, as this is the central point.

Key term

Cranage – a device used for lifting, for example a static tower crane or a mobile crane.

Access and egress

Access and egress can either be to the place of work or the site entrance and exit. Access on construction sites is generally through the use of temporary works such as scaffolding, which is wrapped around the building and can be raised and lowered as required. Mobile elevated platforms and scissor lifts are another method of controlled access to certain points of construction; for example, steel erectors use them to fasten the bolted connections together safely. Access can be gained by the use of scaffold mobile towers that are climbed and incorporate a working platform.

Take it further!

You will need to refer to The Work at Height Regulations 2005 when considering access and egress. These regulations apply to all work at height where there is a risk of a person falling and being injured. You will need to consider whether an alternative method of work avoiding working at height can be employed, and if not, how work may be safely accessed by operatives working at height. Do some research, find a copy of The Work at Height Regulations 2005 and come up with some alternative methods to avoid working at height.

Site accommodation

Site accommodation includes supervisors' offices, canteen or mess room, drying room, client's offices and toilets. Larger long-duration sites may have whole banks of offices as several head office functions are moved onto site. These will all need the connection of temporary services of electricity and water and bottled gas to provide heating, lighting and cooking facilities.

The amount of space available on site may be limited so accommodation may be in a stackable cabin format with external stairs. Cabins can be delivered in two formats: those with wheels or those with jack leg units. For some, a crane may be required to place them in position.

Storage and security

On large sites security personnel should be stationed next to the entrance gates, so they can see everyone who enters and leaves the site as well as direct deliveries. On smaller sites the site manager's accommodation would be located at the gates for the same reason. This would help to reduce the incidence of theft from the site.

Storage concerns include having the right size of box containers to suit the size of materials that are being stored, deciding whether the materials need to be kept in a heated environment, and deciding on what level of security is required for the containers. Tool safes enable equipment to be securely stored overnight to prevent theft.

Health safety and welfare

For more information, see Unit 1 Health, safety and welfare in construction and the built environment.

Method statements

There are two types of method statement:

- The first gives a detailed description of how a task is carried out. It lists the resources required, including labour, plant and materials. By undertaking this process, the hazards associated with the work can be identified as well as the control measures that will need to be put in place to minimise the risks.

Cabins provide on-site accommodation

- The other type is compiled by the estimator. For a known task, the estimator will list what labour, plant and machinery will be used for work that has been included in the tender document. This system is written into a method statement which will be used by the contracts manager to establish what construction plant will be required on the project and how much it will cost.

Progress monitoring

The monitoring of work on site is essential to obtain a clear picture of progress on the master construction programme. Tracking progress will establish if the project is on schedule. If the project is ahead of schedule, then there may be too many labour or plant resources assigned to it, although there may be cost savings, for example the hire of cabins which are paid for per week on site. Being behind schedule is more of a cause for concern as this can lead to large financial penalties, both in **liquidated damages** paid to the client and the cost of increasing resources to pull back the lost time.

Key term

Liquidated damages – financial penalties paid by the contractor to the client when the project overruns the agreed completion date through no fault of the client.

Progress is plotted on the main contract programme bar chart by showing two bars for each task or activity. The first bar is the actual point in time that the project should be at and the second one is the percentage of work completed. Therefore, if the contractor is half way through a task, the progress bar should be 50 per cent – less than 50 per cent highlights an issue.

Site meetings

On large projects, site meetings are generally held once a month. They provide an opportunity for all the parties to the contract to discuss certain key issues. The meeting is normally held on site so a walk around may be arranged to look at certain problems either before or after. Typical topics for discussion include:

- information required
- clerk of works' report
- progress to date against the programme
- variations
- architect's instructions
- mechanical and electrical queries
- health and safety issues.

The site meeting has a set agenda that is sent out before the meeting. A set of site meeting minutes is taken, usually by the architect or designer. The minutes are used as a true record of what was discussed, and can be referred to if disputes arise.

Subcontractor liaison

Regular liaison needs to take place between the subcontractor's main site contact and the main contractor, who will organise weekly site meetings ensuring that all key subcontractors attend. Subcontractor liaison is an important aspect since a large proportion of work on site is subcontracted. Regular contact will help avoid any issues over services coordination between several subcontractors. Access and egress issues, the use of cranage, waste disposal and storage are just some items that need coordinating.

Site resources documentation control

There are several ways to document the control of resources on a construction which include:

- Day work sheets – these sheets record the additional resources utilised above what is listed on the contract (labour, plant and materials). The contractor's quantity surveyor will then use the sheets to work out the cost of the additional resources and produce a price for the final account to be paid by the client.
- Daily report sheet – this records the labour resources utilised on site each day as well as any instructions received and a record of what progress was made. This helps to control the level of labour on the site as numbers present have to be recorded daily.
- Time sheets – these allow workers who travel from site to site to record their hours against each contract so that the cost of their time is allocated fairly.
- Materials requisition sheets – these may be submitted by the site supervisor or the estimator to the purchasing manager in order for them to raise an order for the materials required on site. The sheets will specify the estimator's price, so a comparison can be made with a supplier's quoted price and savings made, the delivery time and the quantity required. From this information, an order can be placed.

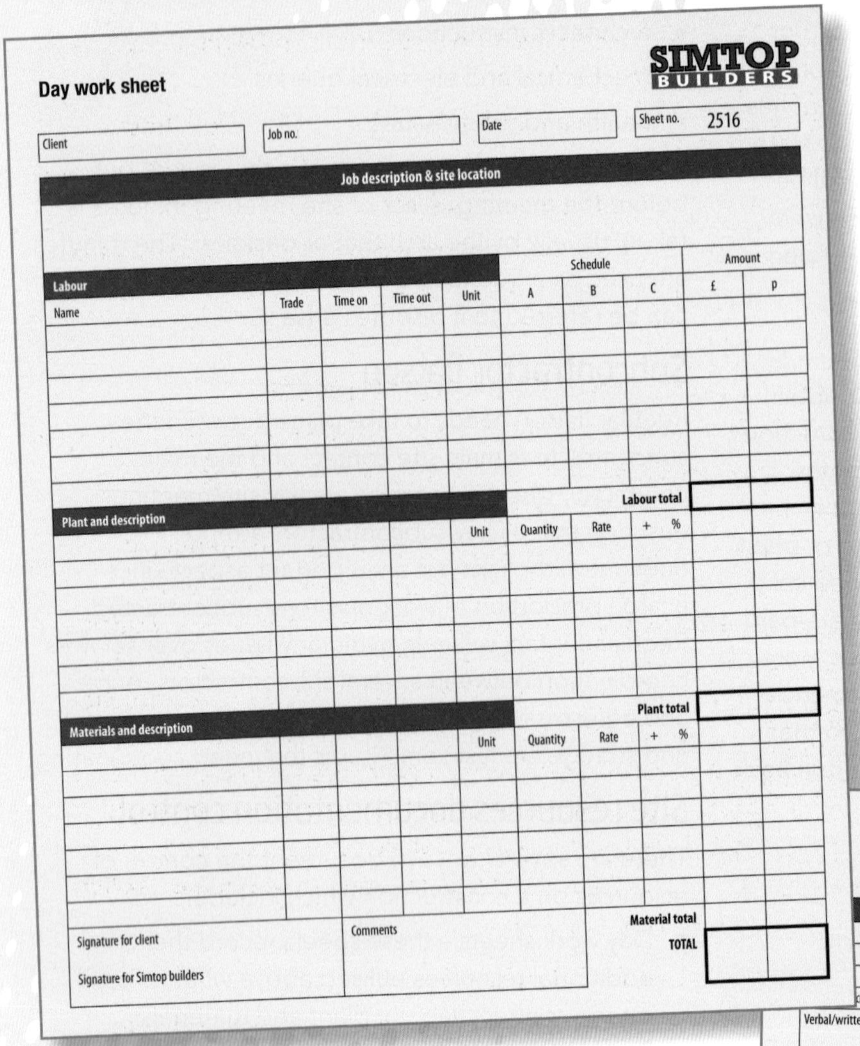

SIMTOP BUILDERS

Day work sheet

Client	Job no.	Date	Sheet no. 2516

Job description & site location

Labour					Schedule			Amount	
Name	Trade	Time on	Time out	Unit	A	B	C	£	p
							Labour total		

Plant and description				Unit	Quantity	Rate	+ %		
							Plant total		

Materials and description				Unit	Quantity	Rate	+ %		
							Material total		
							TOTAL		

Signature for client

Signature for Simtop builders

Comments

Figure 7.8: Day work sheet

For more information on resource allocation documentation, see page 259.

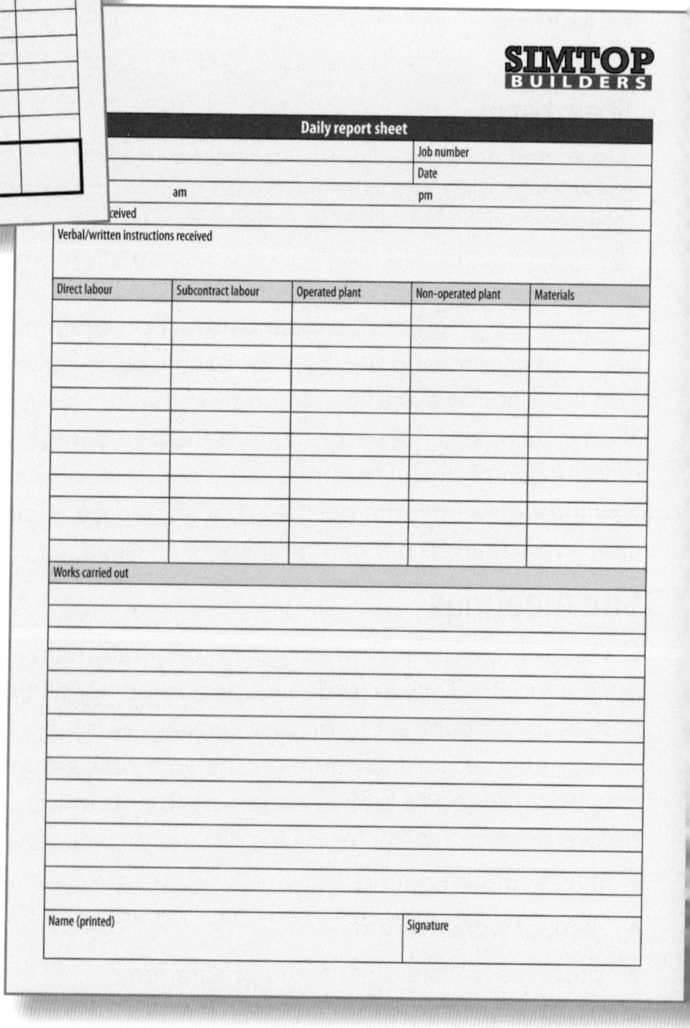

SIMTOP BUILDERS

Daily report sheet

Job number

Date

am

pm

ceived

Verbal/written instructions received

Direct labour	Subcontract labour	Operated plant	Non-operated plant	Materials

Works carried out

Name (printed)

Signature

Figure 7.9: Daily report sheet

Programmes of work

Programmes of work become vital in the management of the project during the production phase. The most commonly used programme of work is a bar chart. Figure 7.10 illustrates how the contractor can see from week to week where progress should be on site.

JEP Construction

	Mar-10	Apr-10	May-10	Jun-10	Jul-10	Aug-10	Sep-10	Oct-10	Nov-10
Set up site	■								
Excavation	■								
Foundation		■							
Substructure		■	■						
Superstructure			■	■	■				
Roofing						■			
Services			■				■	■	
Internal finishes								■	■
Landscaping									■

Figure 7.10: Contract programme

Variables

Weather

Some types of weather can hamper the progress of a project. For example, concrete working can be seriously affected by rain, wind and excessive sunshine, while brickwork can be so badly damaged by rain and frost that it has to be rebuilt.

Take it further!

The UK Met Office offers weather forecasting services for the building and construction industry, including wind speeds and wind direction for those working with tower cranes. Go to their website and find the weather forecast including wind speeds and wind direction for your town for the next week.

Construction companies use various methods to provide protection from the weather, including:

- the use of timber-framed housing – the structural part of the building can be completed and then the brickwork outer skin laid when weather permits
- covering up of brickwork with hessian or insulation during frosts
- insulation of concrete to protect it from frosts
- the complete enclosure of a building project using scaffolding and sheeting
- sheeting over concrete works using tent structures
- changing specifications, for example pre-cast concrete.

Availability of skilled labour

Availability of skilled labour often depends on the type of work involved. Recent labour shortages have been as a result of the cyclical nature of the construction industry which is linked to the growth of the UK economy. However, highly skilled operatives are rarely out of work; for example, there are few people in the UK who have the skill to thatch the roof of a listed building. To obtain skilled labour, construction companies sometimes make use of labour-only agencies that supply operatives for hire on an hourly basis. Alternatively, companies may retrain directly employed, semi-skilled labour as a stop-gap measure to allow them to make maximum use of their skilled trades people.

Labour disputes

Strikes are now largely a thing of the past due to government legislation. However, labour disputes about pay and conditions on site sometimes lead to unofficial strikes. The manager has to walk a tightrope to make sure that the workforce is happy while the company is not held to ransom and that productivity is maintained.

Confined access

Confined access may occur unexpectedly when temporary structures need to be erected around the permanent structure. Often, these will restrict access and the site manager may have to organise an alternative, safer access point. Stairway towers should be used to gain access to upper floors and the roof of a construction project.

Safety tip

Ladders, as a means of access, should not be used if there is a safer alternative.

Late design changes

Traditionally most of the design is completed before the award of the contract to the main contractor. However, on design and build projects this may not be the case and late design changes may have to be accommodated into the main contract programme. To avoid this, the programme needs to be designed with enough flexibility so it can accept design changes without too much disruption to the other activities.

Late construction information

Late construction information such as drawings or specifications can have a knock-on effect on the programme of works. It is the manager's responsibility to explain the effect that late construction information will have on the programme and progress on site. Therefore, they must make sure that all requests for information are recorded and confirmed at site meetings within the site minutes. The manager will need to regularly chase the designer or architect for outstanding information. Communication is the key to avoiding lengthy delays to this process.

Material shortages

Material shortages often occur when a certain material is in such high demand that the manufacturer is unable to supply sufficient quantities. If this happens, the buyer may have to advise the architect or designer and request that an alternative material be sourced and accepted as a specification change. Alternatively, in times of shortages, the contractor may have to collect the material from anywhere it can be supplied in order to reduce the amount of time delay on site.

Case study: Roles and responsibilites in a small firm

Stewart recently started employment at a local construction company in his home town. It is a traditional family-run construction business that is privately owned and has no external shareholders. The managing director, Colin Sherlock, makes all the company's financial and commercial decisions. Unfortunately, he fell ill very suddenly and had to take extensive leave in order to fully recover. In his absence, the other less senior partners agree to appoint a temporary managing director and elected another family member, who happens to be the firm's accountant.

After a settling-in period, the organisation and communication over important decisions started to break down with little or no direction being given by the new managing director. The contracts manager was of little help, the estimator did not know what percentage profit to add to tenders and the quantity surveyor had no understanding of the company's cash flow or its finances.

In desperation, Stewart was called in as an external consultant to try to organise everyone's roles and responsibilities, then put these into a company manual that would also be used for obtaining a quality award under the ISO scheme.

Divide yourselves into teams and discuss and identify the roles of the people mentioned above and establish what their responsibilities should be in a normal structured environment.

Assessment activity 7.5

BTEC

1. The contracts manager has asked you to review the company's site management procedures used to monitor and control resources when organising construction projects. Produce a short review. P5
2. The manager has asked you to evaluate how useful and efficient the following are:
   - contract programs
   - network analysis
   - materials scheduling. 

Grading tips

For P5 you will need to explain some of the site management procedures that can be used to control and monitor resources; see the list in the unit specification content.

For D2 undertake an evaluation on the range of the specified techniques discussing what works well and which is more efficient.

PLTS

Proposing practical ways forward, and breaking these down into manageable steps will help develop your skills as an **effective participator.**

Functional skills

Undertaking an evaluation on the management techniques will help you in developing complex sentences in **English**.

4 Be able to develop documentation for construction teams

Head office and site documentation

Schedules

Schedules are used to group a number of similar items together to make ordering them much easier. For example, a set of drawings with several doors on them can be grouped together onto one drawing which would contain the type, specification and quantities for each door.

Requisitions

A requisition is a request to the purchasing department for a specified item to be ordered. It is formally written upon an order form so the amount and supplier can be identified. Internal requisitions can be used to pull out items from internal stores.

Method statements

A method statement is a written statement of how an activity is going to be carried out. It lists all the labour and plant that will be used to perform the operation along with the methods that will be employed.

Budgets and cost plans

Budgets are essential in order to maintain costs for the client and to keep on budget. All aspects of a contract are given a cost against them; thus elements can be designed with the budget in mind. Cost plans occur at the construction phase; they are a cost control measure to ensure that the work is completed within budget.

Goods received sheets

Goods received sheets are used to record each material that is delivered on site (see Figure 7.11). Any short deliveries or discrepancies are recorded on the sheet, which is then sent to head office for processing. When a supplier's invoice arrives it is checked against the site record to ensure that the quantity is correct. If there is a problem with the delivery, then a credit notice can be requested against the invoice. In this way, costs from suppliers can be controlled and a check kept that goods received are of the right quantity and of an acceptable standard.

SIMTOP BUILDERS

Goods received sheet

Contract no:			Contract:				Week ending 19			
							For office use only			
								Reserves		
Date	Ticket no.	Order no.	Quality & quantity checked (initial)	Supplier	Qty	Description	Invoice no.	Price	£	p

Weekly plant sheet

SIMTOP BUILDERS

Contract

Contract no | Sheet no | W/C

Supplier	Item of plant	M	T	W	T	F	S	S	On hire		Off hire		Invoice ref no	£	p
									Date	Ticket no	Date	Ticket no			

Figure 7.11: Goods received sheet

Figure 7.12: Plant sheet

Plant sheets

Plant sheets work in a similar way to goods received sheets (See Figure 7.12). The on and off hire dates of plant are recorded, so these dates can be checked against the invoice to establish if it is correct. Any discrepancies can be justified with the supplier.

Job cards

Job cards state what task has to be undertaken on site and are issued to the operatives. The operative records the hours they have worked on the job card, which can then be used by the estimator to price future works as well as to monitor actual hours against those planned to see whether the workforce needs motivating to recover lost hours.

Vehicle allocation sheets

Vehicle allocation sheets are used to track the hours of each company vehicle and apportion it against the contract on which the vehicle was employed.

Bar charts

The contract programme bar chart was developed by Henry Gantt in the early twentieth century and is still one of the most popular methods used to produce contract programmes on site. The length of each horizontal bar on the chart represents an on-site time period; while the relationship between bars represents the logical sequence of work on site (See Figure 7.13).

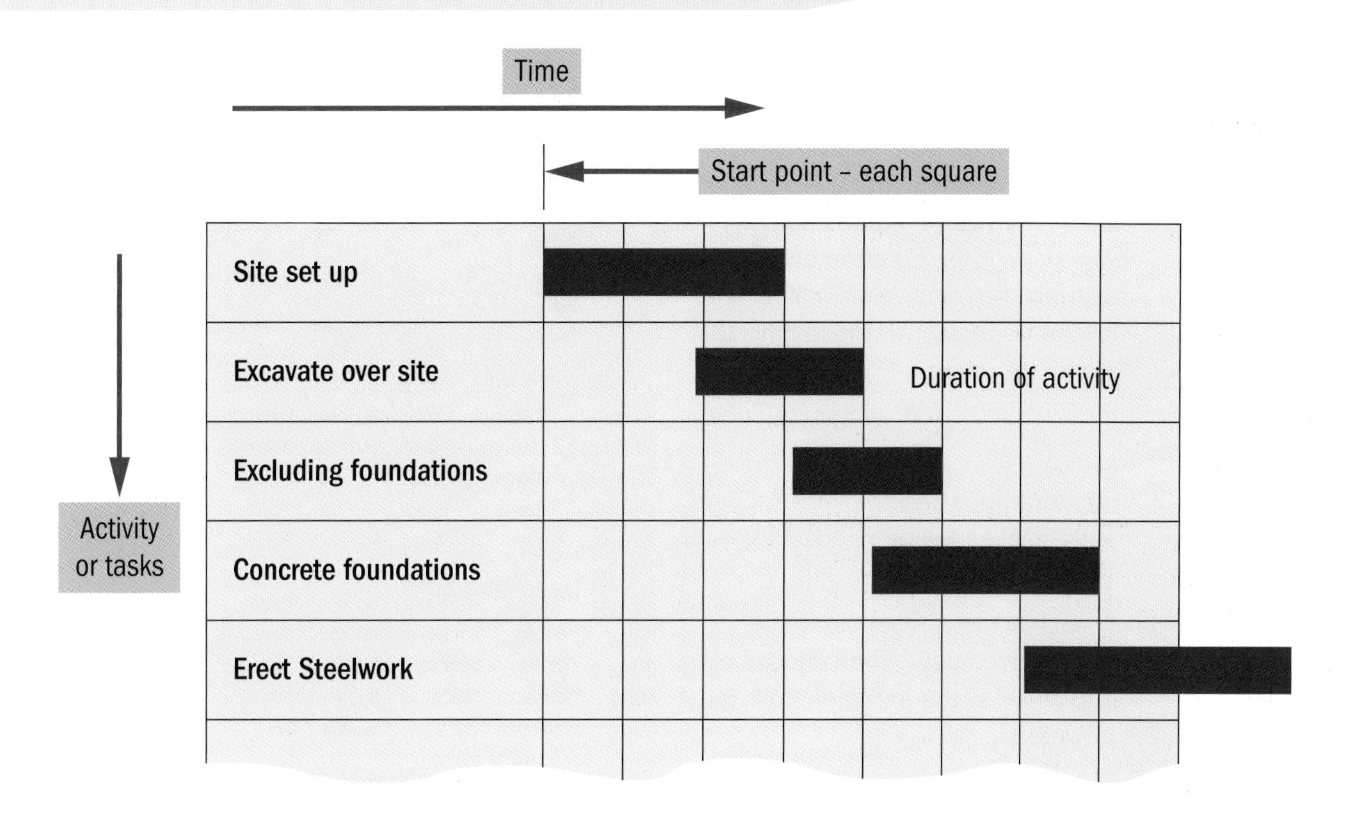

Figure 7.13: Contract programme bar chart

To produce a simple bar chart (See Figure 7.13), this sequence needs to be followed.

1. Analyse the contract drawings and specification and establish how many activities will be needed for the master programme, for example, site clearance, site strip, excavations, concrete works. On average, a medium-sized construction site will have 20 activities.
2. Find the length of each activity as set out by the estimator in the tender document. Then choose a suitable time unit to cover all the activities, such as number of days or weeks.
3. Record the logical sequence between activities so there is a working link between each.
4. Establish which activities are critical to the overall programme.
5. Plot each activity on a rough outline bar chart, begin each at their earliest start point.
6. Establish the critical path through the bar chart. Some activities can float within the critical activities, which means that their start or finish times can be delayed and will have no overall effect on the completion date; other activities are said to be critical to the overall programme and any slippage will affect the end date.
7. Adjust the non-critical activities to suit the labour resources on site.
8. Produce the final bar chart for issue to site.

On the bar chart, time runs from left to right with the commencement of the programme being the first activity and the completion and subsequent handover of the project, the last. To monitor progress, place a string line across the programme at the current date, which will show the percentage of each bar that should be completed. This will highlight which activities are ahead or behind schedule.

Networks

Networks, which are referred to as an arrow chart, are more complicated than simple bar charts. Each arrow on the network represents an activity, just like the bars on the bar chart. The description of the activity is placed above the arrow, with the duration of the activity below it. As with the bar chart, time still works from left to right, with the first activity representing the commencement of the programme.

To produce a simple network, the following sequence should be followed.

1 Analyse the contract drawings and specification and establish how many activities will be needed for the programme.
2 Find the length of each activity as set out by the estimator in the tender document. Then choose a suitable time unit to cover all the activities, such as number of days or weeks.
3 Record the logical sequence between activities so there is a working link between each of the activities.
4 Draft out the network diagram using nodes and arrows.
5 Write on the description and durations of each activity.
6 Calculate the left-hand side of the circles' earliest start times right through the network; where two arrows finish at one node, take the highest value calculated starting at zero.
7 Work backwards from the completion node, taking the lowest value where two arrows enter a node until arriving back at the commencement, which is zero.
8 Identify the critical path and mark this in red – this is the path where the left-hand and the right-hand figures in the circles are the same value.
9 Produce the final network for issue to site.

Assessment activity 7.6

BTEC

1. As an assistant site manager you need to know what documentation is required to be used on site and at head office. Discuss some examples of such documentation for management and resource planning. P6
2. Your tutor will issue a list containing activities, durations and sequences for a low rise domestic building project. Produce a bar chart for this project. P7

Grading tips

For P6 you will need to describe some examples of site and head office documentation. Look in the specification content for some examples.

For P7 you are required to draw out the bar chart accurately and in the correct sequence.

PLTS

Generating ideas and exploring possibilities in sequencing activities will help develop your skills as a **creative thinker**.

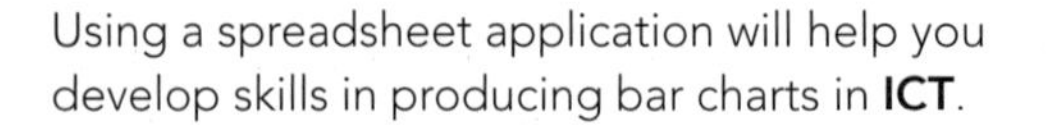

Functional skills

Using a spreadsheet application will help you develop skills in producing bar charts in **ICT**.

Christopher Phillips

Project manager

Christopher is an experienced project manager for a land development company that specialises in leisure and recreation facilities. He had worked as a site manager for a number of years before stepping up to his current post where he is responsible for five of the company's development sites. Christopher is involved in all aspects of the project and even has to appoint an architect to commission the design and production drawings. His role is to organise the smooth running and operation of the developments, from the site investigations to the tendering and appointment of a contractor to undertake the work.

Christopher loves the job. There are so many different people and roles involved within a construction project that no two days are the same. Having to run five projects at the same time means Christopher has to manage time efficiently to stay on top of all developments. Christopher also has excellent IT skills in project management and uses contract programs to coordinate information, contractors and monitor progress on all of his sites.

Christopher has targets to be met; he must deliver the projects in on time, stay on budget and produce the right quality for the client. This means he often has to make decisions that could have a financial affect on a project. Christopher has to have excellent personal skills in communication and chairs several meetings a week with different contractors and clients. These meetings have to be recorded and minutes taken so a true record is kept. Luckily, Christopher has an outgoing personality and gets on with people that he interacts with and is approachable and open to ideas.

Christopher really likes the work that he does and enjoys the feeling of completing a project that had some difficulties which have been overcome to produce a quality project for the client.

Think about it!

- **Would you enjoy such a job?**
- **What qualities would you need?**
- **How would you cope with the pressure?**

Just checking

1. Name four principal members of the construction phase team.
2. What interaction occurs between the contracts manager and the site supervisor?
3. List the phases of a construction project.
4. Name three ways in which maintenance can be undertaken.
5. What does procurement involve?
6. Identify the four principal resources needed to fulfil a construction contract.
7. What factors will affect the plant requirements on a typical project?
8. What factors will need to be taken into account when producing a construction programme?
9. What affects the output of a piece of construction plant?
10. What would influence the layout of the temporary organisational structure on site?
11. What decisions need to be made when a piece of plant is to be hired or purchased?
12. What functions do management undertake during the project?
13. In what way does a goods received sheet control materials?
14. How can the labour workforce on site be controlled effectively?
15. How can wastage on a construction site be minimised?

Assignment tips

- Ask a local construction company if they are able to provide organisational charts of site teams, which could help you tackle this assessment work.
- Take a look at a local construction project to help identify examples of the resources that are required to complete a contract.
- Internet-based progress photographs could provide examples of the stages of the construction process for a low-rise or commercial building.

Credit value: 10

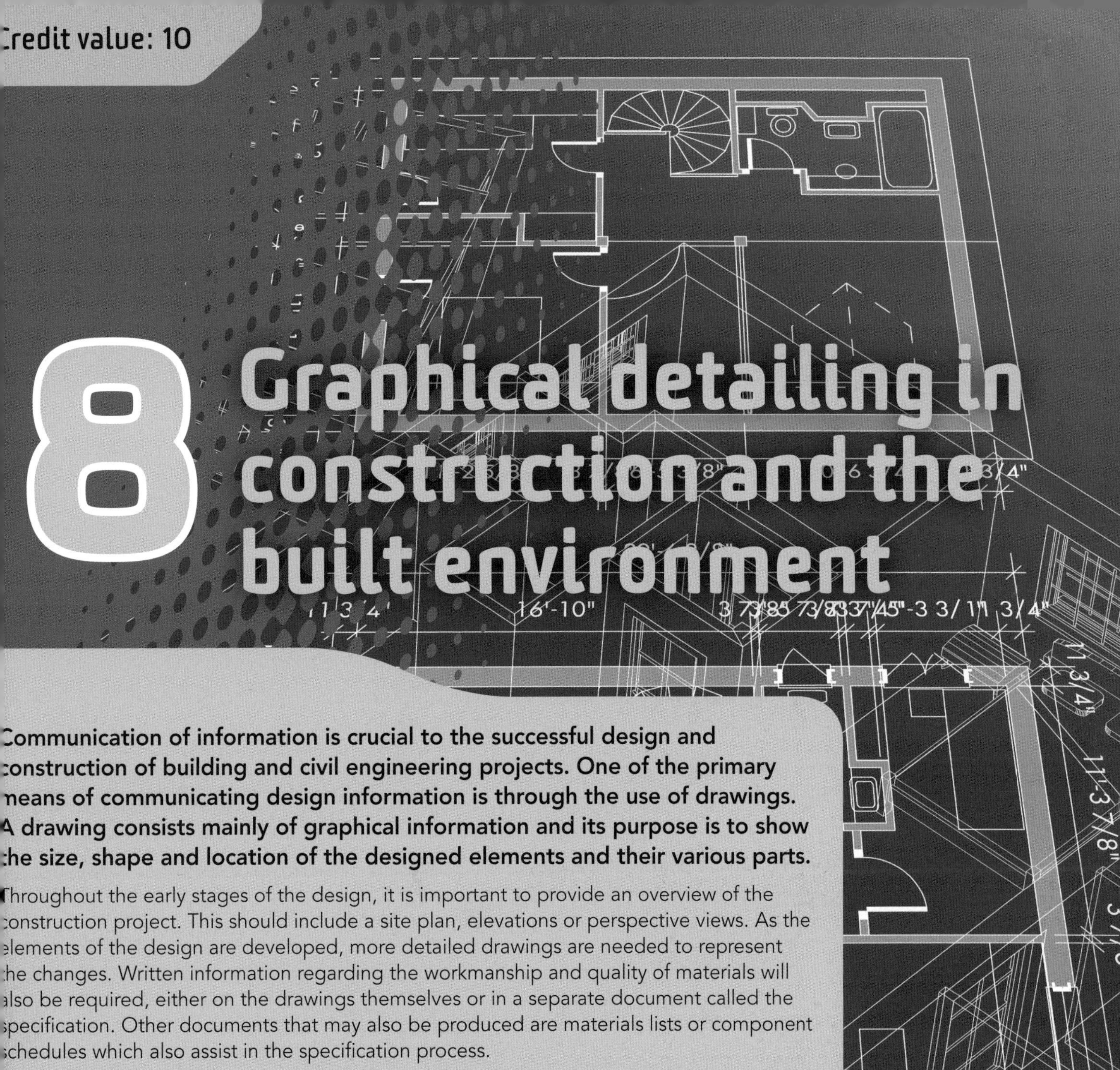

8 Graphical detailing in construction and the built environment

Communication of information is crucial to the successful design and construction of building and civil engineering projects. One of the primary means of communicating design information is through the use of drawings. A drawing consists mainly of graphical information and its purpose is to show the size, shape and location of the designed elements and their various parts.

Throughout the early stages of the design, it is important to provide an overview of the construction project. This should include a site plan, elevations or perspective views. As the elements of the design are developed, more detailed drawings are needed to represent the changes. Written information regarding the workmanship and quality of materials will also be required, either on the drawings themselves or in a separate document called the specification. Other documents that may also be produced are materials lists or component schedules which also assist in the specification process.

Drawings and specifications form a part of the contract documents. It is therefore important that this drawn information is clear and accurate. Poorly produced or presented information may result in mistakes or errors being made on site and the original designer who produced the information may be held liable for any delays or additional cost to the project!

Learning outcomes

After completing this unit you should:

1 know the main equipment, media and techniques used in the production of manual graphical information

2 understand the use of CAD and its benefits in the production and management of graphical information

3 be able to interpret graphical drawings, details, schedules and specifications

4 be able to produce graphical drawings, details, schedules and specifications using manual drafting techniques.

Assessment and grading criteria

This table shows you what you must do in order to achieve a pass, merit or distinction grade, and where you can find activities in this book to help you.

To achieve a **pass** grade the evidence must show that you are able to:	To achieve a **merit** grade the evidence must show that, in addition to the pass criteria, you are able to:	To achieve a **distinction** grade the evidence must show that, in addition to the pass and merit criteria, you are able to:
P1 identify the use of equipment and media used to produce manual graphical information **See Assessment activity 8.1, page 278**	**M1** compare the use of manual and CAD techniques in the production and presentation of graphical information **See Assessment activity 8.2, page 286**	
P2 describe correct drawing standards and conventions **See Assessment activity 8.1, page 278**		
P3 describe manual presentation techniques **See Assessment activity 8.1, page 278**		
P4 explain techniques and uses of different types of CAD information **See Assessment activity 8.2, page 286**		
P5 describe the benefits of using CAD for the production and management of graphical information **See Assessment activity 8.2, page 286**		
P6 interpret graphical drawings, details, schedules and specifications **See Assessment activity 8.3, page 296**	**M2** extract and report clear, accurate and valid information from graphical sources, details and schedules **See Assessment activity 8.3, page 296**	**D1** evaluate how the quality of graphical information relates to the quality of the final constructed project **See Assessment activity 8.3, page 296**
P7 produce 2D and 3D graphical drawings using manual drafting techniques **See Assessment activity 8.4, page 303**	**M3** apply manual techniques and resources to produce complex graphical information **See Assessment activity 8.4, page 303**	**D2** produce manual graphical information to a high level technical skill **See Assessment activity 8.4, page 303**
P8 produce graphical information in the form of simple specifications and schedules **See Assessment activity 8.4, page 303**		

How you will be assessed

The evidence requirements for pass, merit and distinction grades are shown in the grading criteria grid. Evidence for this unit may be gathered from a variety of sources, including well-planned investigative assignments, practical work or reports of practical assignments. You will be given written assessments briefs to complete for the assessment. These will contain a number of assessment criteria from pass, merit and distinction.

This unit will be assessed by the use of three assignments:

- Assignment one will cover P1, P2, P3, P4, P5 and M1
- Assignment two will cover P6, M2 and D1
- Assignment three will cover P7, P8, M3 and D2.

Joe

I always used to think that I wasn't very good at technical drawing but this unit has showed me that by learning a few basic rules, using the right equipment and having a bit of patience, I can produce some good results!

I also learned that graphical detailing is at the core of good design work and follows a gradual process of designing, reviewing and revising drawings and written information.

This unit has inspired me to go on and be a CAD Technician: the person who is the link between the designer in the office and the person on site building the project.

Over to you

- How much do you already know about technical drawings?
- Would you be interested in being a CAD technician?
- What are you looking forward to learning about in this unit?

1 Know the main equipment, media and techniques used in the production of manual graphical information

Build up

House layout

If you are to be successful in your construction career, no matter what discipline you choose to follow you must be able to read, understand and produce graphical information. It is crucial that all professionals in the industry can sketch and draw. The importance of learning manual detailing skills cannot be overestimated. Think about the following questions:

- Would you be able to describe the layout and organisation of spaces in your own home using just written words and no sketches?
- Can you produce a two-point perspective sketch of a simple building?

Equipment for manual detailing

Parallel motion drawing boards

When drawing technical drawings a firm, smooth surface is necessary. There are a number of board types available, but the most common is the standard parallel motion type. This board is free to tilt and is usually on a free-standing frame or can be fitted with a ratchet mechanism to sit on a desk. Some students still opt for the more traditional T-square and flat board, but these are not used in commercial design offices.

Media

Media is the name given to the material that is drawn on. In traditional forms of drafting there are four basic types:

- Paper – the quality of paper is given by its weight in grams per square metre (g/m^2). Very light layout paper is around 60 g/m^2, while ordinary, photocopy paper is about 80 g/m^2. A thicker letter quality paper is around the 90–100 g/m^2 range, and thicker cartridge paper is around 120–150 g/m^2. Paper drawings can be reproduced in black and white or colour by electrostatic photocopying machines, called plain paper copiers, which can now handle larger paper sizes – see Figure 8.1.
- Tracing paper – this is semi-transparent and comes in various grades for draft work; from around 80 g/m^2 up to 110 g/m^2 for master copies. Generally, pencil construction lines are drawn on tracing paper and inked in later with drawing pens. In the past, tracing paper was copied using a two-stage process ammonia copier or dyeline copier. Dyeline copies were excellent at copying pencil shading or any other subtle toning methods but faded in sunlight. Dyeline copiers can still be found in some design offices although plain paper photocopiers are now more popular.

Typical equipment used in manual detailing

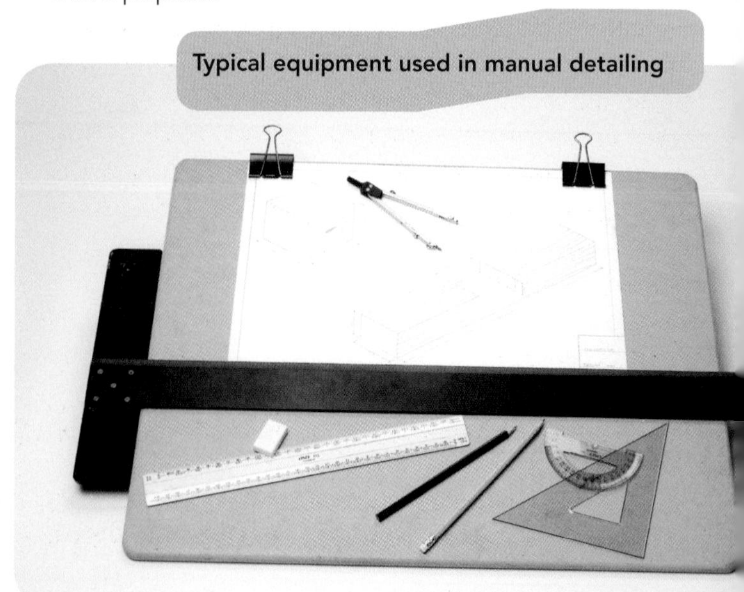

Remember!

Tracing paper is susceptible to changes in moisture content and can stretch unevenly if left taped down to a drawing board overnight, so beware.

- Drafting film – similar to tracing paper, but is easily recognisable by its waxy, silky feel, and, as it is made from polyester, it has high static electricity content. It has the advantage of being strong and virtually tear-proof and resistant to moisture. The copies can be reproduced like tracing paper.
- Linen cloth – was used historically when the existing plans were often drawn in ink on a lacquered linen cloth.

Remember!

Drafting film needs special ink because ordinary ink will take much longer to dry and be non-permanent. Also, ordinary pen nibs will wear out very quickly on polyester paper and so tungsten-tipped nibs are required.

All drawing media are supplied using the international paper size standard, ISO 216, which is based on the metric system. The sizes are based on the A0 sheet which has an area of 1 m² with the sides of the rectangle in the ratio 1:√2. The range of paper sizes is shown in Figure 8.1.

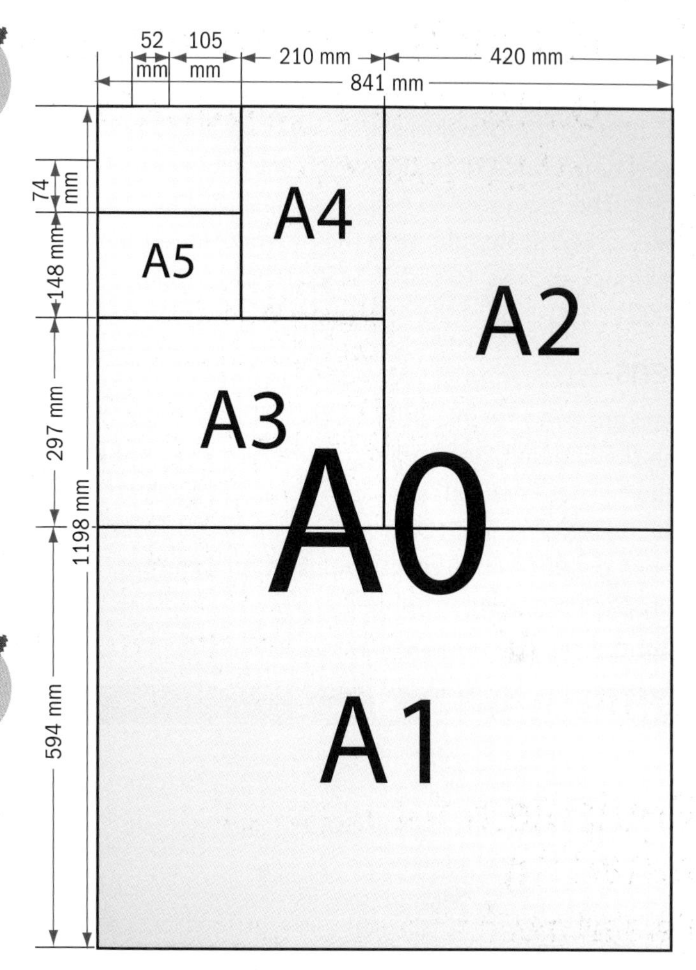

Figure 8.1: Paper sizes – from A0 to A5

Pencils, erasers and erasing shields

Pencils are the main working tool of the draftsperson and come in a range of different hardnesses. Table 8.1 shows the five most used for technical purposes, with B being the softest and 2H the hardest. Pencil can be used on cartridge paper or tracing paper. For plastic film, high polymer lead pencil can be used. These are usually available for use in clutch pencils with leads ranging from 0.2 mm to 0.9 mm.

For the best drawings, pencils must be kept sharp at all times and have regular, smooth points. This is aided by gently turning the pencil in your fingers as you draw the line. For rubbing out pencil lines, a soft rubber eraser is used. It may be necessary to mask off parts of the drawing by using a thin metal or plastic erasing shield.

Remember!

Always use the back of your hand to brush away erased shavings. If you use the palm of your hand, it may be sweaty and you will smudge your work!

Table 8.1: Most used pencils in technical drafting

Darkest				Lightest
B Shading and texturing	HB Rough sketching	F Printing and general line work	H Dimension lines and hatching	2H Construction lines

Scale rules

Drawings are generally produced to a scale which will fit onto the drawing sheet. You very seldom will need to produce drawings which are actual size, that is 1:1. The main tool to help draw scaled drawings is a pre-marked scale rule, with the most common scales included, such as 1:2500, 1:1250, 1:1000, 1:500, 1:250, 1:200, 1:100, 1:50, 1:20, 1:10 and 1:5.

Pens

Pens are used for ink drawing and allow a range of line thicknesses, depending on the purpose of the line drawn. There are four standard common pen widths which are 0.25 mm, 0.35 mm, 0.5 mm and 0.7 mm. These produce lines which are used for different purposes as shown in Table 8.2.

Table 8.2: Pens for different purposes

Pen thickness	Type of line
0.25 mm	Dimension line Hatch line
0.35 mm	General printing and linework details Hidden details
0.5 mm	Section lines Titles
0.7 mm	Titleblock and drawing borders

Ink lines can only be drawn on non-paper media such as tracing paper or plastic film; to correct a mistake, the ink line has to be gently scraped away with a safety blade and resealed with a rubber eraser.

Adjustable set squares

Adjustable set squares are used mainly to draw vertical lines at 90° to the parallel motion arm, but they can also be used to produce a line at any angle on the paper by adjusting the protractor scale. Because of their adjustable nature, these set squares are far more versatile than the traditional '30°/60°/90°' and '45°/45°/90°' fixed set squares.

Compasses

Compasses are used to draw circles and arcs. There are two main types: the traditional spring bow and the longer horizontal beam compass which is used for plotting traditional linear land surveys.

Templates, stencils and flexible curves

Templates, stencils and flexible curves are all useful aids to the draftsperson. Templates provide common outlines of objects in a range of typical scales, such as toilet cisterns and pans. Ink stencils help provide a guide for producing standard text on drawings and the flexible curves, French curves or 'Flexi-curve', can help draw smooth curves.

Drafting tape

Drafting tape is an important aid. It is used to secure the medium to the board. The drafting tape is applied at its four corners; but, you must remember to remove the tape every night, otherwise the medium will not be able to expand and contract as the temperature changes.

Drawings standards

In the past companies used to adopt their own drawings styles which became 'house standards'. However, if every company was using different standards, there would be confusion as to what physical object was being symbolically shown. There have been various British Standards developed over the past thirty years to help the industry develop a consistent set of drawing norms and procedures.

The current British & European drawing standard for all types of technical drawings are provided in the various parts of 'BS EN ISO 128 – Technical Drawings General principles of Presentation'. The parts most relevant to manual drafting are:

- Part 20:2001: Basic conventions for lines
- Part 22:1999: Basic conventions and applications for leader lines and reference lines
- Part 23:1999: Lines on construction drawings
- Part 24:1999: Lines on mechanical engineering drawings
- Part 30:2001: Basic conventions for views
- Part 40:2001: Basic conventions for cuts and sections.

There are also specific drawing standards for specialist construction, surveying and civil engineering applications. These include:

- BS EN ISO 8560:1999 'Construction drawings – representation of modular sizes, lines and grids', which sets down rules for the representation of modular sizes, lines and grids on construction drawings

- BS EN ISO 3766:2003: 'Construction drawings – Simplified representation of concrete reinforcement', which establishes a system for the scheduling of reinforced bars covering dimensions and shapes
- BS EN ISO 7518:1999: 'Construction drawings – simplified representation of demolition and rebuilding' which specifies rules for symbols, markings and simplified representations of demolition and rebuilding on general arrangement drawings and assembly drawings
- BS EN ISO 9431:1999: 'Construction drawings – spaces for drawing and for text, and title blocks on drawing sheets' which specifies requirements concerning the placing, layout and contents of spaces for drawing and for text, and title blocks on construction drawings
- BS EN ISO 11091:1999: 'Landscape drawing practice' which establishes general rules and specifies graphical symbols and simplified representations for landscape drawing practice
- BS 6750:1986: 'Specification for modular coordination in building' which are the rules for modular reference systems for use in the design of buildings; for the position of key reference planes; and for the sizing of buildings, their components and materials.

General techniques used for manual detailing

When deciding the best way to represent the dimensional and constructional information onto paper, the draftsperson needs to consider what the drawing's main purpose is and who is going to be reading it.

A drawing is used to provide information about the size and layout of buildings and their elements. It can also provide detailed information about the specification of the quality of materials and the workmanship required. They form part of the contract agreement, and at various stages throughout the design and construction process different graphical forms of information are used. Above all else, the drawing must have the following qualities whether produced manually or by CAD:

- enough information to be useful for its purpose, but not too busy
- clear and unambiguous construction information
- individual details on the drawing clearly referenced and labelled, such as section marks and individual titles
- a clear system of cross referencing between drawings and other relevant documents.

In any design office there should be a consistent 'house style' that meets all these requirements and is applied to all drawings and documents. There may also be a need to set up a series of standard templates for drawings, word documents and spreadsheets which can be flexible enough to apply to large and small jobs alike.

Projection methods

Drawings are essentially a two-dimensional (2D) representation of a three-dimensional (3D) element of a building; this could be a wall, a staircase or a whole ground floor layout. There are a number of geometrical techniques for showing 3D objects on 2D media, flat paper. The following are the most popular projection methods and they will all be illustrated using the same house:

Orthographic projection

Orthographic projection is where individual views and plans are drawn in flat profile. There are two basic types of orthographic projection: 1st angle or 3rd angle. In construction, we generally use 3rd angle projection to show a building's external elevations and roof plans. It is useful for showing the general arrangement of the features of a house and the relative positions of windows and doors.

In Figure 8.2, the main view is the south elevation with the plan of the roof shown above, that is, a bird's eye view. The side elevations are the east and west elevations and these are drawn next to the south elevation.

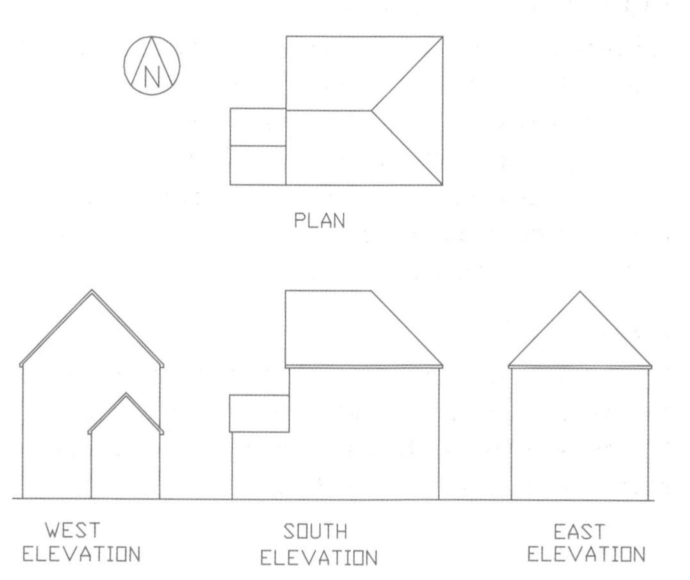

Figure 8.2: Orthographic projection in 3rd angle

Isometric projection

Isometric projection gives a 3D image with all lines drawn at the same scale and at an angle of 30°. All the receding lines are parallel; therefore, it is not a perspective view.

An isometric projection is often used to show the overall massing of construction elements see Figure 8.3.

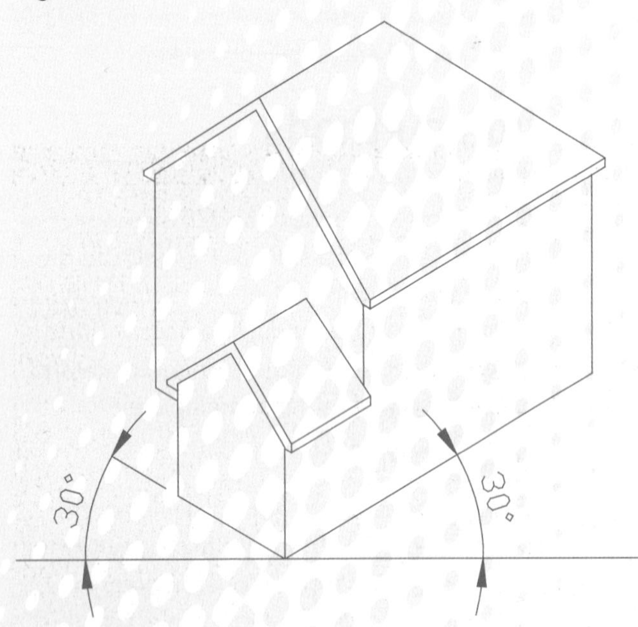

Figure 8.3: Isometric projection

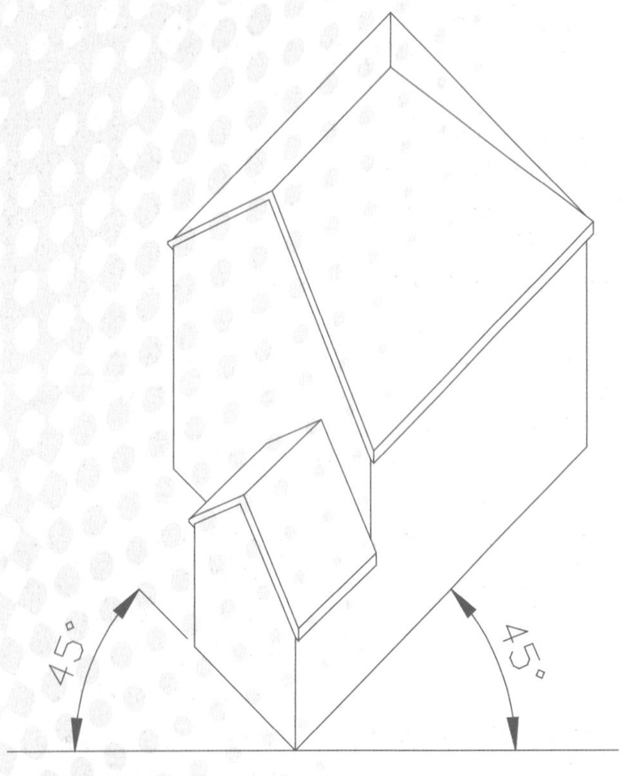

Figure 8.4: Axonometric projection

Axonometric projection

An axonometric projection is similar to an isometric projection, but all lines are drawn at 45°. This method is easier to construct because it uses the true plan view simply rotated by 45°.

The axonometric projection is most suitable for interior and office or kitchen layouts as it appears to give a higher view point than isometric. In other words, it lets you see inside a box.

Perspective

Perspective gives the most realistic view and is particularly good for elevations and external views. For a two-point perspective, all lines converge onto two vanishing points fixed at eye level on the horizon. Perspective drawings can be drawn to scale, however, they are quite complicated to construct. The one shown in Figure 8.5 has been generated from a computer model, Google SketchUp™.

Exploded views

Often in product manufacture and particularly in building services exploded diagrams are provided to show clearly how the various component parts fit together. Exploded views are relatively simple to construct as a 3D model and can be linked to schedules for specification of materials. An example of an exploded diagram of a water pump is shown in Figure 8.6.

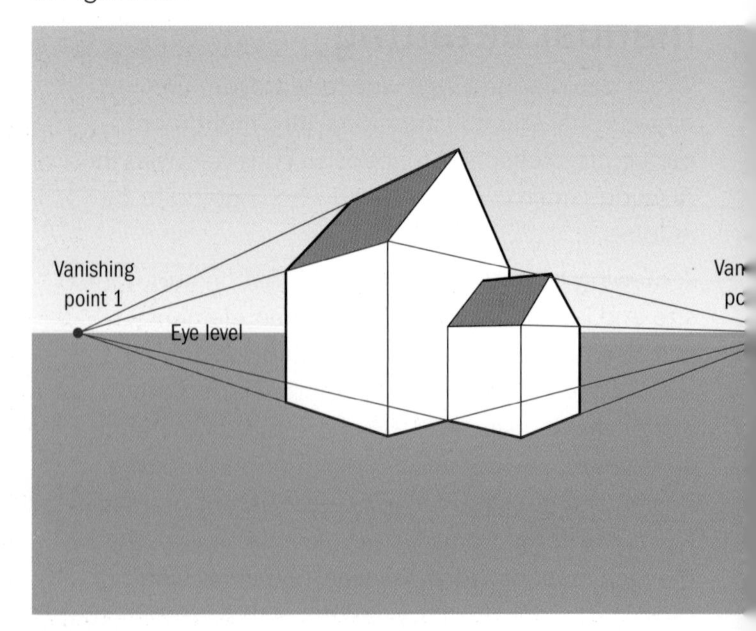

Figure 8.5: Perspective drawing

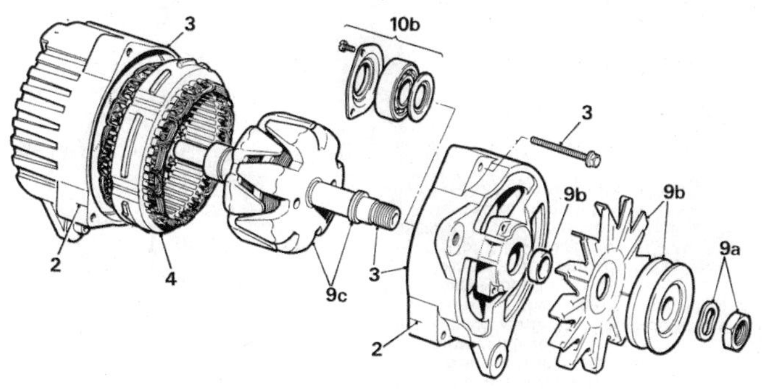

Figure 8.6: Exploded view

Table 8.3: Types of drawing line

Type of line	Function of line	Typical use
thick	Site outline or new building	Site drawings
medium	General details	
thin	Reference grid, dimension lines, leader lines and hatching	
thick	Primary functional elements in horizontal or vertical sections (e.g. load-bearing walls and structural slabs)	General location drawings
medium	Secondary elements and components in horizontal and vertical sections (e.g. non load-bearing partitions, windows, doors); also components etc., in elevation	
thin	Reference grids, dimension lines, leader lines and hatching	
thick	Primary functional elements in horizontal or vertical section (e.g. load-bearing walls, structural slabs)	Assembly drawings
medium	Secondary elements and components in horizontal and vertical section (e.g. non load-bearing partitions, windows, doors); also components etc., in elevation	
thin	Reference grids, dimension lines, leader lines and hatching	
	Medium broken line. The purpose and position of the line should be noted in relation to the plane of section. The line begins with a dash cutting the outline adjoining, and all lines should meet at changes in direction	Work not visible Work to be removed
	Thin line with break in it or if necessary a thin continuous line with a zigzag in it	Breaks in continuity of drawings
	Thick chain lines Medium chain lines	Pipe lines, services, drains
	Thin chain line	Centre and axial lines
	Indicated by a circle at the end of the line	Controlling line, grid line

Techniques used for manual detailing

Text and font styles

General notes should be neatly arranged in panels of regular shape on the drawing and have a clear printed text heading. They should be broken up into paragraphs for ease of reading, and should not be cramped nor become so widely spaced as to become illegible. It is important to avoid over stylised or italicised fonts as these can be difficult to read. A simple, open print text style provides the clearest form of communication. Ideally, text should be in block capitals with plenty of open space inside the characters, for example:

THIS DRAWING SHALL NOT BE SCALED

not ***THIS DRAWING SHALL NOT BE SCALED***

Upper case or lower case is acceptable depending on the house style of the organisation. In each case, the height of the text should range from 3 mm high for general text up to 5 mm high for titles of individual sections and plans.

Types of drawing line

The three most common types of line are continuous, dashed and chain-dotted. Each line has three relative thicknesses: thick, medium and thin. These various conventions are combined to indicate specific construction elements or for controlling information, the functions of which are described in Table 8.3.

Dimensioning

There are a number of ways in which dimension lines can be drawn, two examples of which are shown in Figures 8.7 and 8.8.

It is important that there is a gap between the dimension lines and the detail element to avoid any confusion. For horizontal dimensions, the dimensions should be printed along the dimension line with the ends of the line marked with arrowheads or thick oblique strike lines. For vertical dimensions, the dimension numbers are printed to the left-hand side of the dimension line. The dimensions should always be in millimetres; therefore there is no need to have the abbreviation 'mm' printed after the dimension figures. If the units represent a different unit, for example, metres, then the abbreviation 'm' has to be shown after the dimension figure.

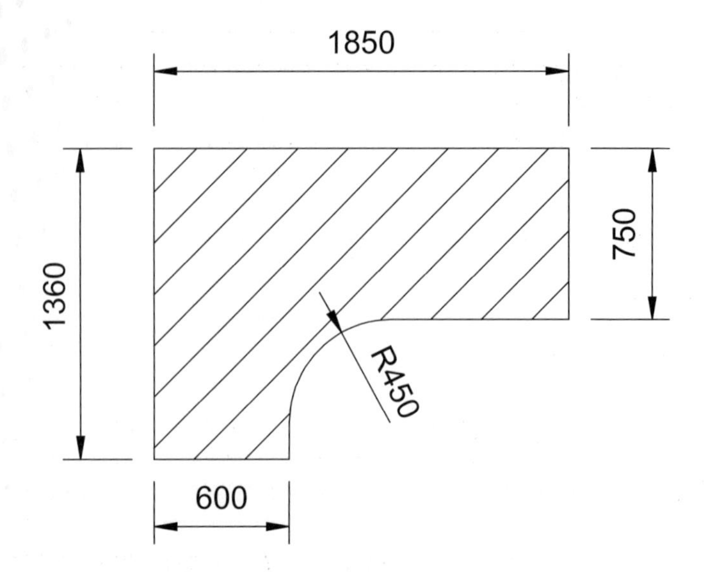

Figure 8.7: Dimension lines with arrowheads

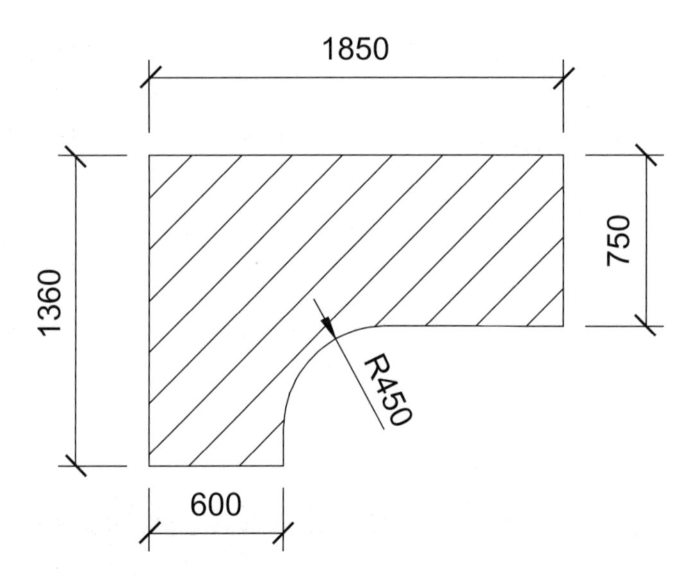

Figure 8.8: Dimension lines with oblique strike lines

Use of standard hatch patterns and symbols

When detail elements are cut through, as in a cross-section in plan or elevation, it is appropriate to show cross hatching to denote the material from which they are made. The most common hatching patterns are brickwork and blockwork. Other typical hatching patterns are given in Figure 8.9.

Another common convention that is used to save time as well as aid clarity is the use of various symbols for common constructional features, such as doors, windows and light switches. Common symbols used are shown in Figure 8.10.

A number of abbreviations are also commonly used on drawings to aid clarity and expression. Some standard ones are shown in Table 8.4.

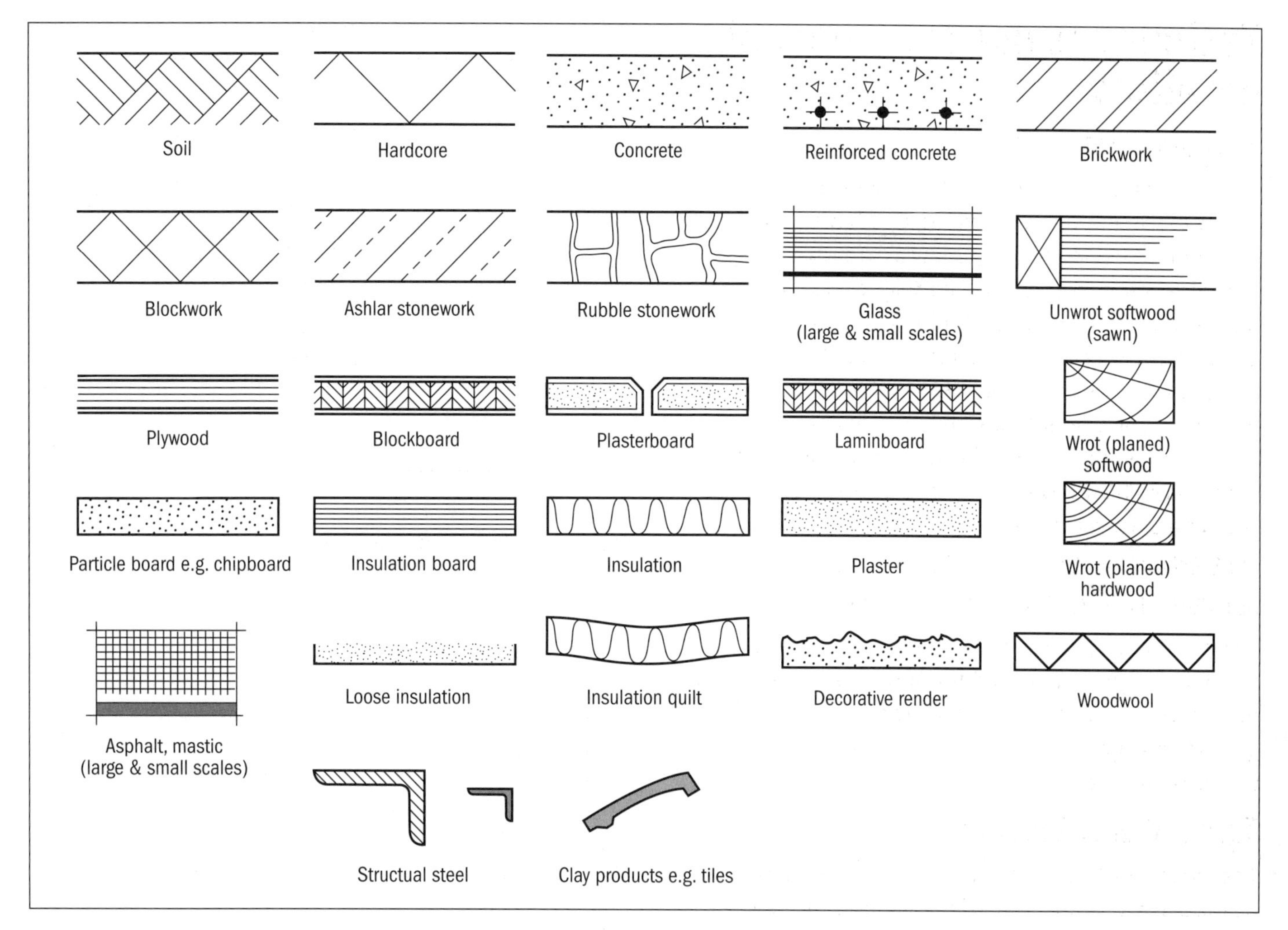

Figure 8.9: Hatching patterns

Table 8.4: Standard abbreviations

Item	Abbreviation	Item	Abbreviation	Item	Abbreviation
Airbrick	AB	Damp-proof course	DPC	Polyvinyl chloride	PVC
Asbestos	asb	Drawing	dwg	Rainwater pipe	rwp
Bitumen	bit	Foundation	fnd	Reinforced concrete	RC
Boarding	bdg	Hardboard	hdbd	Satin chrome	SC
Brickwork	bwk	Hardcore	hc	Satin-anodised aluminium	SAA
Building	bldg	Hardwood	hwd	Softwood	swd
Cast iron	Ci	Insulation	insul	Stainless steel	SS
Cement	ct	Joist	jst	Tongue and groove	T&G
Column	col	Mild steel	MS	Wrought iron	Wi
Concrete	conc	Plasterboard	pbd		
Cupboard	cpd	Polyvinyl acetate	PVA		

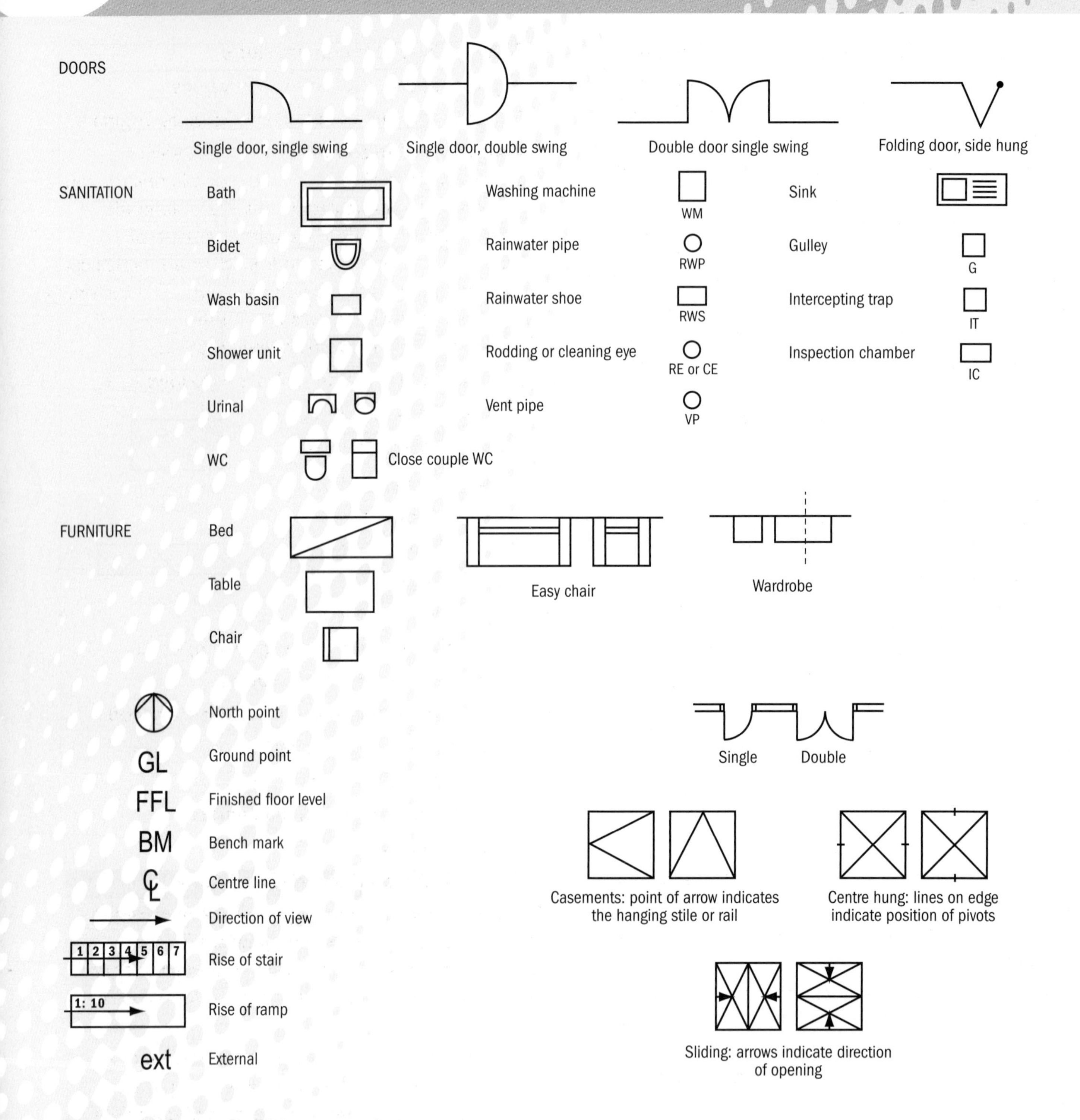

Figure 8.10: Constructional features symbols

Specifying construction methods and materials

A project's **construction specification** communicates in written or tabular form the nature and quality of each construction element required. It is used in addition to the provided visual information about geometry, size, shape and arrangement which is shown on the drawings.

Key term

Construction specification – written information prepared by the design team for use by the construction team, the main purpose of which is to define the products to be used, the quality of work, any performance requirements and the conditions under which the work is to be executed.

On small jobs, such as a 'room-in-the-roof' house extension, all the specifications notes can usually be contained on the drawing; this also allows the work to be priced accurately by the contractor.

On larger jobs, such as a small housing development with a number of drawings, the detailed information, such as how it is to be constructed or the quality of the materials is usually contained within other contract documents like the specification or the **bill of quantities**. It is important to note that for tendering purposes the detailed bill of quantities should provide very clear information as to what the contractor is expected to provide for the proposed project; the drawings and bill of quantities must tie up.

Key terms

Bill of quantities – a list of quantities produced to a standard that is used to price construction tenders.

Standard Method of Measurement of Building Works (SMM7) – a clearly defined method of calculating the materials needed for a project using a systematic and clear measuring procedure based on the dimensions shown in the contract drawings.

Coordinated project information

In order to ensure that information in drawings, specifications and bills of quantities is the same, the construction industry has used a number of specification systems for defining building products and materials in the UK. This has been very important for large projects where there are a large number of drawings produced.

The first coordinated project information system to be widely used was the Swedish CI/SfB – Construction Index system, but, due to problems both with introducing new categories and computerisation, a new all encompassing classification scheme for the construction industry is currently being implemented. It is known as the Uniclass system and has been developed by the National Building Specification (NBS) on behalf of the Construction Project Information Committee (CPIC), which represents the following major sponsoring organisations:

- Construction Confederation
- The Institution of Civil Engineers
- Royal Institute of British Architects
- Royal Institution of Chartered Surveyors
- Chartered Institution of Building Services Engineers.

The Uniclass system comprises 15 tables, each of which represents a different broad area of construction information. Each table can be used as a stand-alone table for the classification of a particular type of information; in addition, terms from different tables can be combined to classify complex subjects. This system is very useful in large projects where there is a lot of information that needs to be communicated.

The full list of classification tables is as follows:

- Table A Form of information
- Table B Subject disciplines
- Table C Management
- Table D Facilites
- Table E Construction entities
- Table F Spaces
- Table G Elements for buildings
- Table H Elements for civil engineering works
- Table I Work sections for buildings
- Table J Work sections for civil engineering works
- Table K Construction products
- Table L Construction aids
- Table M Properties and characteristics
- Table N Materials
- Table O Universal Decimal Classification.

'Table G Elements for buildings' has the most direct relevance to the production of drawings for the construction industry, while 'Table H Elements for civil engineering works' has more relevance for the civil engineering.

Table G is further broken down into a more-detailed classification known as the Common Arrangement of Work Sections (CAWS) from which bills of quantities can be produced using the **Standard Method of Measurement of Building Works, seventh edition (SMM7)**. An example of the way CAWS breaks down the work into sections is
shown below:

Level 1: Group, for example, M Surface finishes

Level 2: Subgroup, for example, M1 Screeds/trowelled flooring

Level 3: Work section, for example. M13 Calcium sulphate based screeds.

In this way, detailed definitions are provided to reduce possible variations and conflicts between documents – even within the same document; in practice it ensures that gaps and overlaps between different work sections are eliminated. Essentially, these systems enable the information on the drawing to complement the information in the written specification and vice versa. This is covered in more detail later in this unit when we look at interpreting and producing drawings.

Theory into practice

Explore the specification of building works in small and large projects by visiting the CPIC website (see www.cpic.org.uk); look at the examples it provides for specifying building works. Copy out a specification clause for a building element. In particular, see 'Documentation for a small project', 'Performance specification' and 'Prescriptive specification' examples.

- What is a performance specification?
- What is a prescriptive specification?

Assessment activity 8.1

1. You have been asked to produce two drawings: the first one, some creative ideas for a proposed landscaping/planting area and patio for a communal garden in medium-sized apartment accommodation. The second one, a draft house floor plan and cross-section – this one would be prior to producing an inked presentation copy.

   Describe the equipment, media and techniques that would be used to produce the graphical detailing information. Clearly describe the use and function of the equipment and the media used. P1
2. Contact a local architectural design practice and ask for some copies of recent projects. Study the drawings and then describe the drawing standards, conventions and presentation techniques used. Base your description on your understanding of British & European drawing standards. Comment on the presentation style of the drawings. P2, P3

Grading tips

For P1 you will need to identify the equipment and media used to produce manual graphical information. Consider the types of paper or media to be used and match that to the most appropriate drawing implements/equipment.

For P2 describe correct drawing standards and conventions. Ensure that you always consider whether the drawing communicates the information clearly.

For P3 you will need to describe manual presentation techniques. In describing the presentation techniques ask yourself if better views could have been chosen or if the layout is confusing.

PLTS

Participating in group research on drawing standards and conventions will help develop your skills as a **team worker**.

Functional skills

In talking to the architects you will have an opportunity to structure a discussion and find out about the type of graphical communication that they use. You will be able to select, read and understand drawings and use them to gather information, ideas and arguments which will develop your **English** speaking and listening skills.

2 Understanding the use of CAD and its benefits in the production and management of graphical information

An overview of CAD techniques

An understanding of computer-aided design (CAD) is an essential requirement for today's design office. There are many advantages over traditional methods when using CAD, which include:

- production of high-quality graphics in a relatively short time
- facility to reuse standard 'house' style details
- ability to easily amend and reissue revised drawings/details
- ease of archiving and saving electronic drawings
- improved management and tracking of work.

Most CAD programs also offer other advantages by combining highly powerful databases to prepare lists of materials, specifications, and cost data.

Typical work station for a CAD operator

Equipment for CAD detailing

Computer hardware

The essential hardware required for CAD work includes:

- a central processing unit (CPU) whose speed is at least 3 Giga-Hertz (GHz), microchip, and a compatible operating system
- an operating disk capacity of 2 Giga-Bytes (GB) RAM and at least 2 GB free disk space on the hard drive
- a visual display unit (VDU) monitor with a resolution of 1280 × 1024 pixels and display adaptor capable of 32-bit colour 128 MB or greater
- a mouse and keyboard, although some professionals prefer using digitisers and digitiser pads or light pens
- a networked plotter capable of minimum A1 size plots in three pen colours/thicknesses.

In addition, the CPU should have capabilities to be networked (wired or wireless) to a local area network (LAN) and also have Internet and email access through a broadband connection.

CAD File Formats

There are many software companies that produce CAD programs. However, currently the three major CAD systems used for construction are Autodesk's AutoCAD, Bentley's Microstation and Graphisoft's ArchiCAD; the CAD drawing files used by these companies come in different formats.

The leading CAD file format is Autodesk's '.DWG' (DraWinG) file system. Microstation uses '.DGN' (DesiGN file) types but can also read and write a variety of other formats including '.DWG'. ArchiCAD uses a slightly different system which stores information in a drawing database depending if it is a 2D or 3D model as '.2DX' or '.3DX', but it can import and read '.DWG' files as well.

Another common file format is the (Drawing Exchange Format (DXF) file, originally developed by Autodesk to enable their competitors to import AutoCAD files. However, as CAD programs have become more powerful, this format has waned, and many commercial CAD software developers have chosen to support DWG as their primary drawing format for CAD data. Autodesk also offers its DWG read/write technology for licence, in a developer toolkit called 'RealDWG'.

The Design Web Format (DWF) format is the latest file format system; it allows electronic drawing information to be accessed by anyone who needs to view, review, or print files of CAD information, without the need for having the specific CAD software. It is particularly beneficial for uploading onto an Internet server where many different organisations need to access information quickly and efficiently.

Intranets and project extranets

There are three typical hardware configurations for a CAD system in a design office and they are dependent on how each configuration is networked and where the data is stored. More information on this is included later in this unit. The three typical hardware configurations are as follows.

- Standalone single user – the CAD software and drawing information is held within a **private domain**. Information is issued by plotting off drawings and sending them by post or sending electronic copies of the drawing files by email.
- Small office-based intranet set up using a local area network (LAN) – the CAD and licensing software, together with drawing and document information, is held within a **project domain** which physically may be located in a separate computer server which itself is linked to the company plotting device. The server may also contain a plot management function to prioritise the work if there are important deadlines to meet. Companies may also use project management software on this system to track drawings, record revisions and control the electronic issuing of information.
- Large multi-location extranet set up for a wide area network (WAN) – drawing information is uploaded or 'published' to a **public domain** on the Internet which is accessible to all the various members of the project team; users can be in different geographical locations. This Internet web domain is provided by subscription from the CAD software company and uses its project management software to monitor and flag up when and what changes are being made. It also allows the user to search for the most up-to-date drawings themselves.

Production of CAD drawings

The development of information and communication technologies (ICT) has revolutionised the process of building design and construction. Computers have greater capacity, work faster, and by means of the world wide web (www) can communicate a vast amount of information securely to all interested parties in any one construction project.

The development of CAD software for personal desktop computers was the impetus for almost universal application in all areas of construction. With 2D in the 1970s, it was initially limited to producing drawings similar to hand-drafted drawings. Advances in solid modelling involving wire frames and surface area treatments in the 1980s allowed more versatile applications of computers in design activities. Autodesk™ was founded in 1982 which led to the 2D system AutoCAD™. Further software developments in the early 1990s led to solid 3D modelling linked to database features. These could capture information about material properties and product specifications of the drawn objects, and not just treat them as a series of lines.

Currently, there are many CAD software products on the market, the most popular ones being Autodesk's AutoCAD™, with its derivative programs based on its DWG file format system, Bentley's Microstation™ suite of programs and Graphisoft's ArchiCAD™ software system. All of these offer capabilities for managing building data to provide secure, clear and timely management of design and construction documents known as Building Information Modelling (BIM) systems.

Today, CAD programs have become highly visual with intuitive interfaces which help make the software easier to use. The keyboard and mouse are still the main input devices, although light pens and digitisers are also used in some design offices. There is also less focus on dialogue or instruction boxes and more emphasis on designing directly in the virtual model space.

Remember!

Originally, CAD stood for computer-aided drafting and was merely a tool to mirror the traditional manual drafting techniques. With increased speed and increased functionality, it became an important design tool in itself and CAD came to represent computer-aided design. You may also come across the term CADD which stands for computer-aided design and drafting!

Basic CAD concepts

There are essentially two methods of producing computer-aided design. The one to use depends on the nature of the construction project.

Small-scale domestic projects such as house extensions and alterations would use 2D CAD systems which, as stated above, mimic the processes involved in creating hand-drafted drawings. To some extent many design practices still use this for the majority of their work. This method uses similar conventions to manual drafting in representing real 3D architectural elements through a series of flat 2D graphical images, typically consisting of plans, sections, and elevations (see Figure 8.11). Often, to provide a more realistic 3D effect traditional geometric techniques are created such as 'two-point perspectives' and isometric views.

The second process for producing CAD drawings or plots utilises powerful hardware systems and complex software programs to produce a set of virtual architectural objects. These objects behave as complete architectural elements and not just a geometric pattern of lines or a wire framework as in the previous 2D method. These objects when combined form a virtual 3D model of the construction project such that they not only contain location and spatial information but also can hold information about the physical specifications of the objects. This is known as a Single Model Environment (SME).

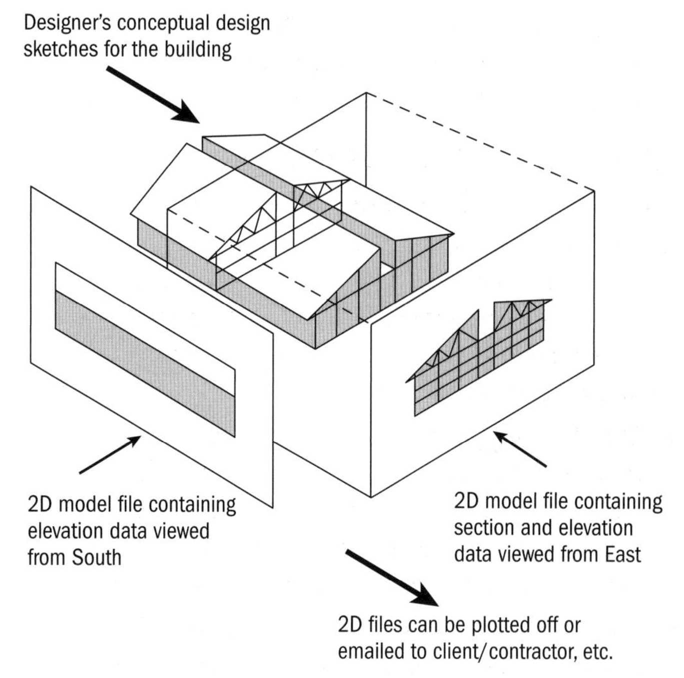

Figure 8.11: A 2D model

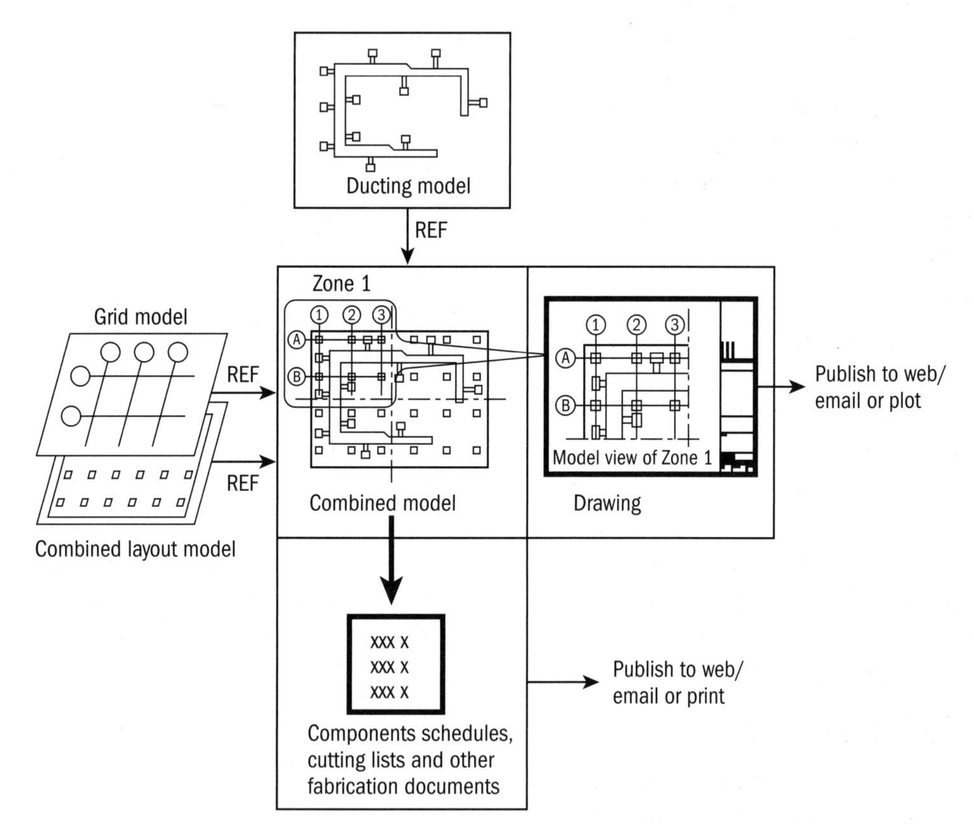

Figure 8.12: A 3D Single Model Environment

Figure 8.12 shows an example of SME. In it, the Column Layout Model specifying the steelwork frame of the building, the Grid Model showing the setting out framework for the building and the Ducting Model showing mechanical services are brought together to form a composite model of the building. This system has a number of powerful advantages over the 2D CAD systems which include:

- flagging up clashes where one object interferes with another, such as where the column positions clash with ducting layout, which is invaluable in checking for errors and speeding up the design process
- improving design productivity and efficiency because objects behave according to the specific properties that these elements have in the real world. For example, a window has a relationship to a wall that contains it; if you move or delete the wall, the window will also be deleted, just as in reality a window could not exist in that location without a wall to support it
- embedding information about the objects in the form of linked schedules which provide details of their material properties, finishes and manufacturer's details which can then be used for costing and procurement purposes. In addition, these intelligent architectural objects can maintain dynamic links. For example, when someone deletes or modifies a door in the model, the door schedule document will be automatically updated
- enabling easier and more efficient methods of specifying components and materials direct from subscription services like the National Building Specification (NBS).

From this 3D model of the construction project, specific cross-sections, elevations and plans can be selected and automatically generated and plotted, and then schedules printed. It combines different objects of pre-existing spatial and technical data to form a new design solution as a Single Model Environment.

Presentation techniques used in CAD

CAD enables projects to be brought to life for impressive client presentations at any stage of the design with integrated rendering and photo-realisation. The streamlined visualisation capabilities make this type of presentation stand out by visually communicating richer design information than mere 2D perspectives drawings or physical 3D architectural models. CAD presentations also help the client to appreciate how the proposals work on a human scale.

As CAD software has become more powerful and computer hardware has evolved greater storage capacity and speed, the virtual representation of construction and civil engineering projects has become increasingly more realistic.

Rendering and photo-realisation

What used to be just wire frames with solid faces have now become realistic structures with natural textures giving the impression of real materials. With the development of powerful surveying software and equipment, the 3D surfaces of the natural world can also be accurately mapped and realistically represented with bright sunlight and shadows and wind-rippled foliage. In addition, some programs offer the capability to map photographs of buildings and people into 3D virtual models. A typical rendered elevation for a major retail refurbishment project which illustrates this type of presentation is shown in Figure 8.13.

Figure 8.13: A 3D virtual model

Walk-throughs

There are times when textures combined with photographic montages in 2D are not enough for some clients who would rather experience a real physical presence of the proposed building; they prefer to move around the spaces that are envisaged. In these cases, animated sequences as seen from a person walking through the property, known as walk-throughs, are ideal.

Activity: Walk-through

On the Internet, search for 'Kaufmann House Falling Water' walk through and watch a video of the famous house designed by the American architect Frank Lloyd-Wright in Pennsylvania.

- As you follow the walk-through, list all the different surface textures that have been created.
- Identify at least three other techniques used to create a very realistic presentation.

Specific CAD techniques

Drawing with a computer requires a different attitude as well as a different set of skills from the traditional approach to drawing. To illustrate this, we will now look at using a 2D CAD package such as AutoCAD LT™.

Setting up the drawing

At the start of the drawing, the 'real' size of the construction project is set up on the screen. This is known as the 'limits' of the drawing. CAD drawings are always drawn real size, that is, at a scale of 1:1. The limits are the x- and y-coordinates of the size of the drawing area.

How you move around the drawing is controlled by the 'limits' of the drawing and the 'zoom' controls. The zoom enables you both to magnify small parts of the drawing and to see an overview of the whole drawing. The screen area could detail the elevation of an electrical plug socket or the complete plan view of a football stadium; it would depend on what limits were set.

Drawing and editing commands

One of the main differences between manual drafting and CAD is the way in which lines and shapes, known as entities, are created. The drawing area is gridded up into x- and y-coordinates which define the exact positions of the drawn entities. This method is known as vector-based drawing; you need to specify the end point relative to its starting point. For example, to draw a horizontal line 3 metres long, the command would be LINE @ 3000,0 where the '@' indicates the chosen start point of the line and the 3000 and 0 the x- and y-coordinates respectively from that starting point.

Once you have created some entities, you can use a variety of different editing commands to copy, rotate and mirror them. Most CAD software comes with its own set of pre-made entities which can be loaded into the drawing as 'blocks', such as items of furniture, people or cars, or you can create your own standard blocks.

Layers and controls

Layers are another useful feature of a CAD drawing. Different constructional elements can be shown on different layers which can be visualised as separate transparent overlay sheets containing grouped or linked objects. For example, one layer would contain all the electrical trunking and duct work and switches, which would overlay the floor plan drawn on another separate layer. Often hatching is shown on a separate layer so as not to slow the computer down. It also makes selection and editing of line work easier.

These separate layers can have their own colour, line thicknesses and line styles all preset, which can be turned off or on as well as plotted separately to aid drawing. Typical layer name conventions using Uniclass elements are shown in Table 8.5. The layer reference is made up of four separate fields, for example, A-25-D-intwall, which means the 'architectural dimensions of the internal walls'.

Plotting methods

As most drawings are larger than A3 size, a plotter is used instead of a printer. These can produce drawing plots up to A0 size, of quality of around 90–100 grams, from a drawing roll which is fed through a series of rollers. The plot is produced by a range of different pens that move against the direction of the fed roll of paper.

The loading of the drawing rolls into the plotter needs careful alignment to ensure that the rolls run smoothly; usually there are guides marked on the feed loading bay to assist in this tricky manoeuvre. The plotter pens will run an automatic check on their operation prior to starting a plot but you will need to ensure that plotter pen reservoirs are full and of the right colour.

Table 8.5: Standard abbreviations for CAD layers

1		2	3	4
Discipline		**Element (from Uniclass Table G or H)**	**Type of presentation information**	**User description**
A	Architect	G2 = Building fabric G13 = Groundworks G21 = Foundations G22 = Floors G23 = Stairs G24 = Roofs G25 = Walls etc.	D – Dimensions G – Graphics H – Hatching T – Text M – Model Graphics P – Page, Plot related	Optional e.g. int-walls = internal walls wall conc = concrete walls roof timb = timber roof, etc.
B	Building Surveyor			
C	Civil Engineer			
D	Drainage, Highways Engineer			
E	Electrical Engineer			
F	Facilities Manager			
G	GIS, Land Surveyor			
H	Heating, Ventilation Designer			
I	Interior Designer			

Many popular CAD software programs use a 'model space' or 'model view' to construct the building, which is then selectively transferred to a flat paper space or paper view layout from which it is plotted. It is only when the drawing is plotted that the virtual drawing model is scaled down from 1:1 to fit onto the plotting paper space layouts. These layouts contain the company's title block, borders and standard notes. In this way, standard details can be reproduced at the same time that the drawing entities are plotted and there is no need for a pre-printed title block/company logo on the drawing paper.

With the increased use of CAD plotting techniques, there is also no longer the need to store hard copies. Drawings can be stored digitally without the need for climate-controlled storage areas that are required for paper- and film-based media.

The Benefits of Using CAD

The Drawing process

CAD programs are quick to use and easy to correct or amend; drawings can be conveniently stored and reproduced. The programs not only allow accurate linework but also incorporate many features to manage the project information, such as linked spreadsheets and databases.

As discussed in earlier in this unit, CAD information is more than flat 2D drawings; instead, they often represent virtual models of the building, such that the building can be built and tested in a virtual reality for example, avoiding clashes between the structural framework of a building and the service ducts which can form a very complex network.

Workflow, tracking and reporting

Until about 10–15 years ago, most design companies would produce drawings as hard copy prints then send them by post. They would arrive a couple of days later, be marked up by hand showing the required amendments, and then posted back. The design company would receive the amended drawings about a week after it initially sent them out. The designer would then revise their original drawing and re-issue it … and the process would start again!

Throughout this process, the designer had to keep on file the drawing issue slips which stated what drawings were sent, their current revision number and who it was sent to.

With the advent of email, postal time was saved at the start, but marked up and amended drawings were still returned by post. This process was probably monitored by an *ad hoc* inspection of the designer's 'Sent Items' to see what they had requested and when they wanted it done by. If there were more than one or two people in the office working on the same project, then error or duplication of work was highly probable.

In recent years, running parallel to the development of the various CAD packages, there has been the development of powerful CAD management software. This has been necessary because of the increasing speed with which drawings can be altered, amended and issued via increasingly powerful drafting/modelling software and Internet communication. For example, a typical information management software application enables all the latest CAD drawings to be uploaded or published in a DWF format onto a third-party Internet server.

With these packages, if a drawing is **red lined**, the coordinator of the project can receive an automatic email providing information of who, when and how data was taken and changed. This can work not only within the design company but also with colleagues working on the project in other companies, such as the structural engineers, contractor or client.

Key term

Red lining – indicates where CAD drawings have been marked up to show proposed or amended details, just as you would annotate manual drawing in red ink.

Real time mark-up and reviews

As stated above, all the major CAD software programs have systems which allow project data to be shared with remote colleagues and clients to review or amend details. Most programs utilise DWF Viewer and DWF Writer programs or similar, which are similar to the commonly available Adobe™ Acrobat for text documents.

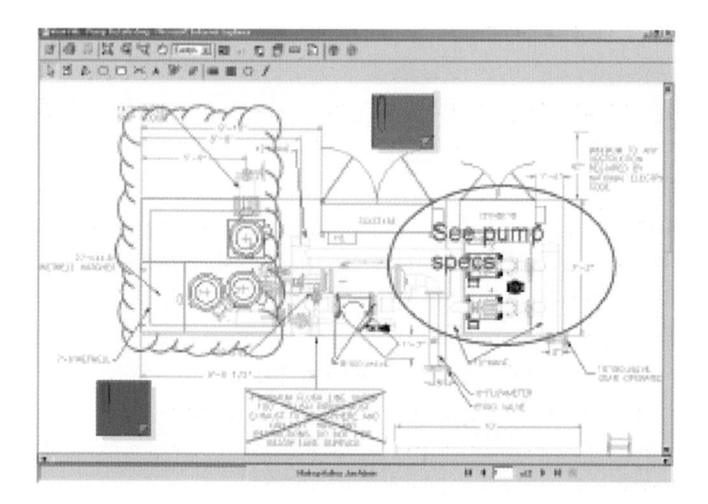

Figure 8.14: A red lined CAD drawing

The main advantages of DWF files is that architects, engineers, project managers, or any of their colleagues can communicate design information without the need to have CAD software installed on their computer network, or even knowing how to use CAD software. Using a DWF writer, anyone working on the project can view, review, comment on or print out the design drawings or information through a variety of standard word-processing, spreadsheet, presentation or web-browsing software.

Some software management systems also enable online design review meetings with a range of people dispersed throughout the country or indeed the world, where electronic drawings are discussed, marked up or red lined and reviewed in real-time.

Sharing/security and back-up issues

Drawing information, like any important data, needs to be kept secure and safe. Some information will be confidential due to its sensitive nature, for example a contentious planning proposal. Access to certain files can be restricted through a technique known as folder-level permissions; sensitive files can be kept in a digital folder which can only be accessed by certain people with the appropriate user permissions and passwords. The system also creates a log of all actions and the name of the persons who carried them out.

The system can also detect when files are checked out and checked back. This importantly prevents two users revising the same document at the same time and protects old revisions overwriting newer revisions; this is similar to 'Read-only' files in Microsoft Office.

Companies can set up a 'project domain' and within this the licensing software is held on the local server. This licensing software is registered and authorised at the time of purchase; thus helping with another security issue regarding the distribution and licences for a given software product as the server is located at the software company's own web-based servers, and it controls the licences directly.

As with all electronic data systems, making regular back-up copies of files is essential. Companies with project domains should back up their local server at least once a week, usually onto a separate removable hard drive which is then kept safe at a separate location. For companies who subscribe to a public domain extranet, then that subscription should include back-up routines as standard and appropriate insurances; it is always wise, however, to undertake a manual back-up by downloading your current projects onto a separate local server at your own offices.

Case study: Site management

The British Airport Authority (BAA) Terminal 5 (T5) at Heathrow was a major architectural and civil engineering project worth more than £4 billion which opened March 2008. The main terminal building, crowned by a distinctive waveform roof, with a span of over 150 metres, is one of the largest single-span structures in the UK. It was conceived by the Richard Rogers Partnership together with a fully integrated design and development team of nearly 30 individual design consultants.

At the start of the project BAA set itself a target to reduce the cost and programme risks by carrying out the design and development of the project on the Single Model Environment. BAA approached a company called Asite to develop a comprehensive CAD management tool. Part of the solution was to provide a platform to create a single graphical, spatial model that covered the full extent of the programme and was divided into model files. All plans, elevations as well as sections, where possible, were spatially located and orientated relative to the global origin and ordnance datum.

This model, with data organised in both 2D and 3D space, enabled design coordination to be carried out between disciplines on and between each floor level. The processes allowed for the data to flow through the design and production/manufacturing processes such that the resultant program model resembled the final virtual construction model.

At one stage, there were 473 CAD document users working on the design SME producing 85,945 CAD drawings, 16,124 2D model files and 12,335 3D model files.

Carry out some Internet research into the BAA T5 Project at Heathrow to find out:

- the names of the key design and contracting companies
- the range of different application software used on the project.
- the estimated cost savings that this Single Model Environment would achieve for the T5 Project. (*Hint*: see Asite's website detailing the collaborative work undertaken with BAA.)

Assessment activity 8.2

BTEC

1. Contact a local design practice that uses CAD as part of its design and production processes. Conduct a structured interview with the company's CAD manager or a senior partner and undertake the following:
   - Identify and describe the CAD systems that they use and find out about the different uses of CAD information that they employ. P4
   - Research into the benefits of using CAD for the production and management of the company's graphical information. P5
   - Reflect on your knowledge of manual detailing and produce a structured report to compare the use of manual and CAD techniques in the production and presentation of graphical information. M1

Grading tips

For P4, in your description make sure that you can demonstrate to your assessor your understanding of the CAD issues.

For P5 think about the local design company you have visited and how CAD improves their productivity and the quality of the work that they do.

For M1 make sure that you provide a balanced view of the use of CAD and manual techniques in terms of speed, convenience, safeguards required, financial outlay, quality of output and storage issues. You need to show that you understand both CAD and manual techniques are suited to different functions but both must be 'fit for purpose'.

PLTS

By undertaking this structured interview you will be demonstrating both that you are an **independent enquirer** and a **reflective learner**.

Functional skills

By making notes during your visit to the designer's office you will be developing your listening and writing skills in **English**.

3 Be able to interpret graphical drawings, details, schedules and specifications

Drawings and details

Planning and surveying drawings

During the early stages of the design process, a great deal of information has to be researched to confirm if the proposed building will be technically, functionally and financially feasible. Existing site drawings as well as historical building drawings are sought, as are reports and records. Measurements are also carried out onsite and a detailed survey of the layout and levels of the site are carried out. A survey drawing, which is a plan of the site showing the physical positions of existing site features, is then produced.

Figure 8.15 shows a typical site survey drawing for a medium size building plot. It shows existing boundaries, trees and buildings. It also shows spot heights as crosses which indicate the slope of the land. The drawing is produced to scale, although overall dimensions are also included.

Design drawings

During the outline and scheme design stages, the building designer uses preliminary sketches to develop design ideas to help the client to understand the proposals and contribute to the design. These drawings are by their very nature sketches and unrefined, but provide sufficient visual and dimensional information for a client to appreciate the proposal. They enable both the designer and client to explore possible elevations and window/door positions, including the massing, symmetry and scale of the building. A typical example of a free-hand design sketch is shown in Figure 8.16.

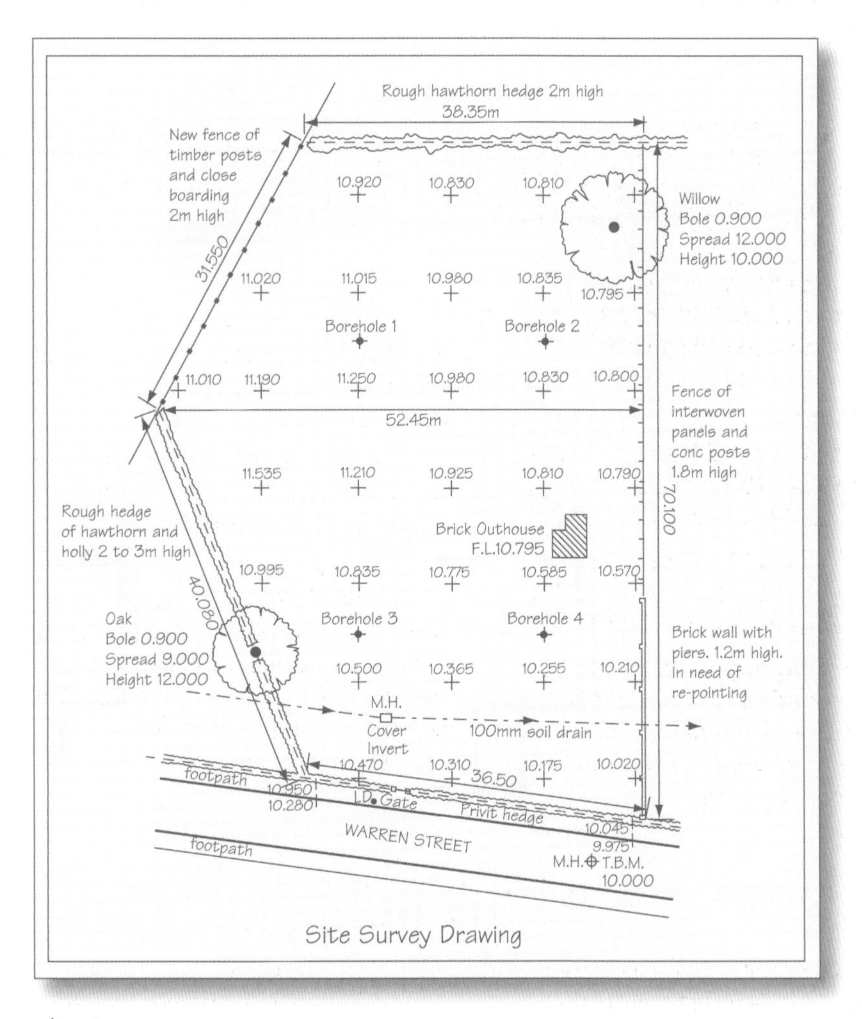

Figure 8.15: A site survey drawing

Figure 8.16: A free-hand design sketch for a bungalow

There are a number of intermediate stages in the design process (see Unit 5 Construction technology and design in construction and civil engineering). It is during these stages that the design is refined and filtered according to the needs and demands of not only the client, who may have a particular requirement for the type of finishes they would like, but also of the other design team specialists, such as structural or building service engineers.

At these early design stages, it is prudent to consult with the local authority planning department to check what its preferences and constraints for the project might be. The local authority has the responsibility to ensure that all the requirements of the various Town and Country Planning Regulations and Building Regulations are being followed (see Unit 6 Building technology in construction). The consultation can be done informally by meeting with the local planner or formally by making an initial outline planning permission application. Before the scheme can proceed it must acquire full approval from the local authority through consultation with both the planning and building control departments.

Typically, a planning permission application needs to include an Ordnance Survey tracing of the area, a site plan and proposed elevations. The information provided should also include the setting out dimensions of the proposed building, surface finishes, positions of external doors/windows and any permanent vehicular access that is required. Figure 8.17 is an example of proposed plans and elevations for a two-storey side extension to accommodate a

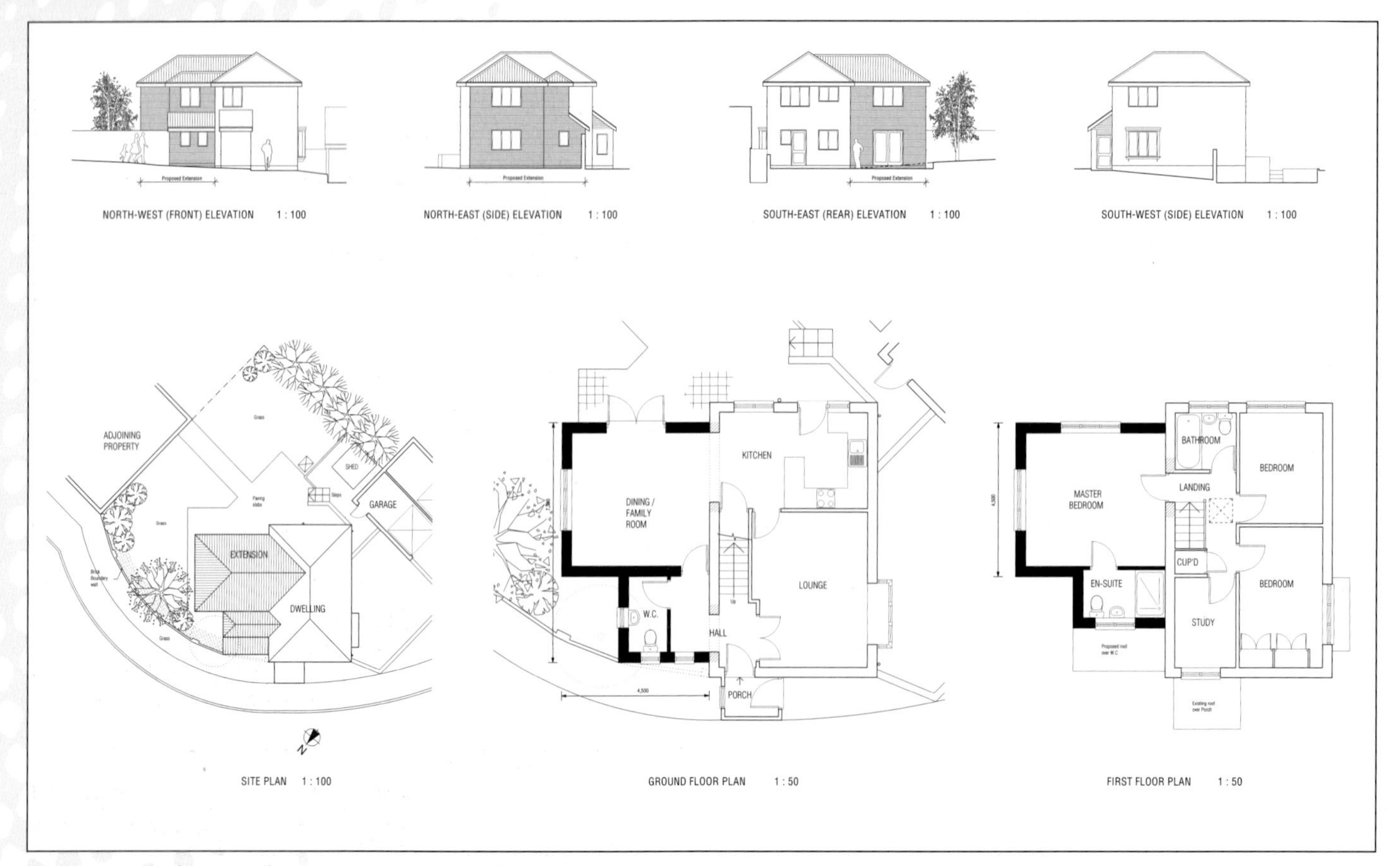

Figure 8.17: Planning drawing *(Source: Munday and Cramer)*

dining room and toilet on the ground floor, and a master bedroom with ensuite on the first floor.

During the final stages of the design process, when the technical issues are being resolved, the drawings are checked for compliance with building regulations by the local building control inspector, who requires far more detailed information than the planners required. For example, sufficient information needs to be provided to show the provision for thermal or sound insulation, the dimensional layout required for disabled access or the minimum headroom requirements for staircases.

Production drawings

Once the final design details are resolved and agreed between the design team, the local authority and the client, the production stage drawings can commence. These drawings will ultimately form the working drawings and contain the detailed information of how the building will be constructed in terms of the quality of materials, the type of fixings and the workmanship requirements. The drawings are structured to work from the 'whole to the part'; the drawings firstly provide an overview of the project (its location, layout, overall size, and general form of construction), then show a detailed breakdown of the project via larger more detailed scale drawings that zoom in on specific areas like the construction of walls, floor and roofs.

The drawings that provide the overview are called location drawings or general arrangement drawings (often abbreviated to GA). These show the relative positions of the proposed works and may also include written specifications. An example of a GA for the construction of a workshop and storage building is shown in Figure 8.18. In this case, because the project is relatively simple, the specification for the materials and construction is supplied on the drawing itself. Figure 8.19 is an example of a more complexed GA drawing. It involves various extension and alterations to a school complex; thus the annotation is kept to a minimum with separate drawings for the sectional elevations and enlarged details.

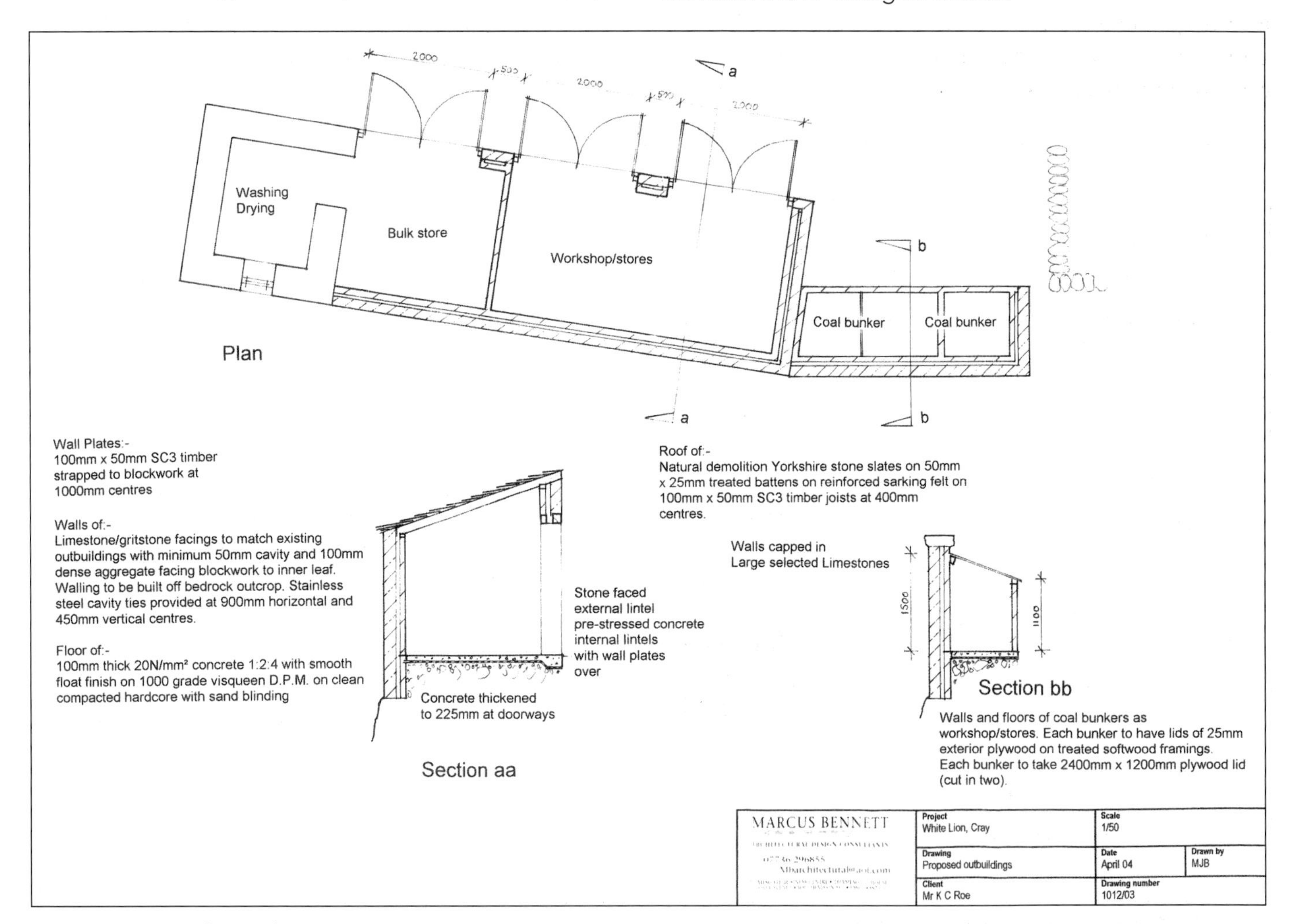

Figure 8.18: Simple GA drawing

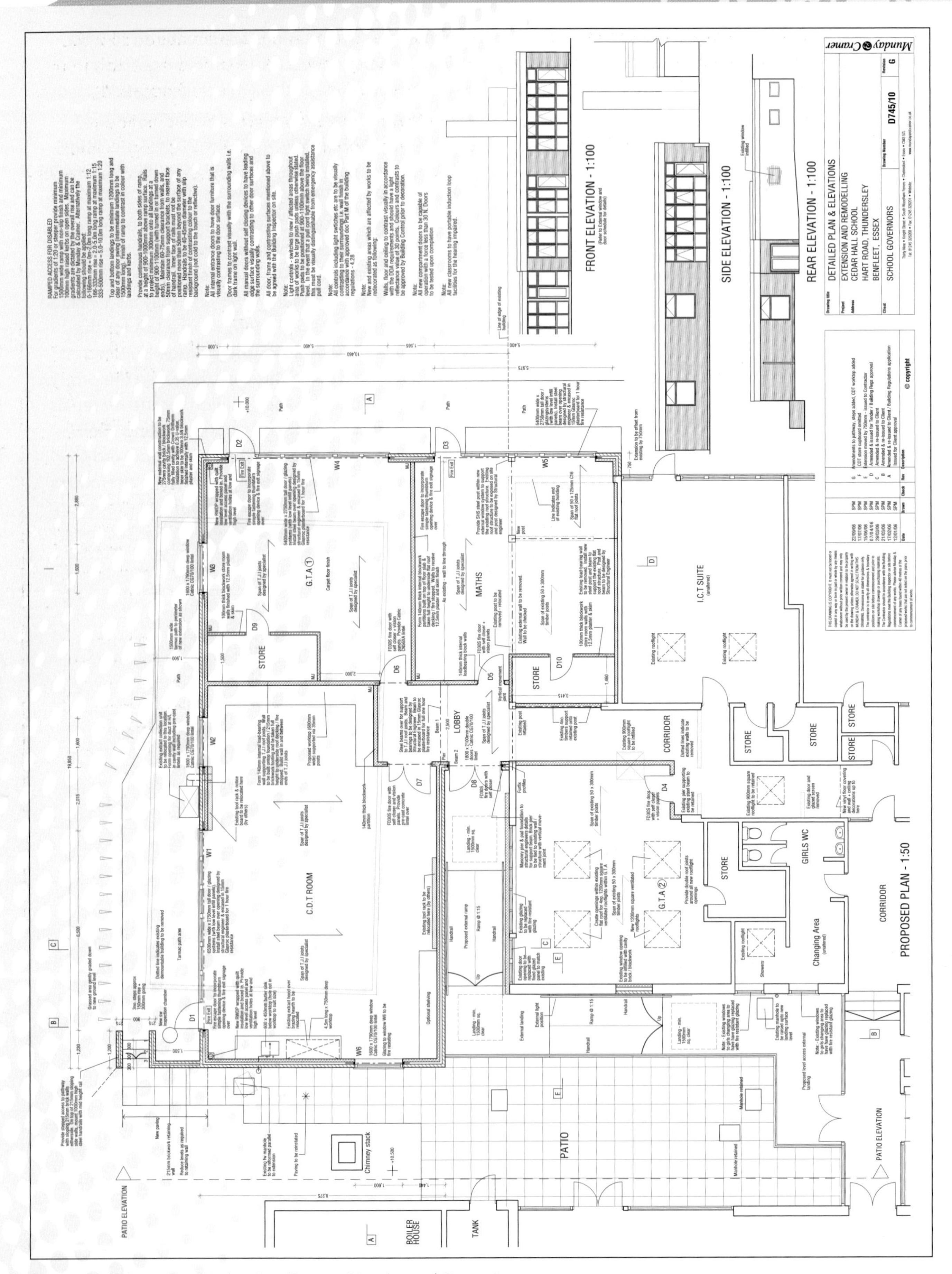

Figure 8.19: Complex GA drawing *(Source: Munday and Cramer)*

The many specialist members of the design team may also produce their own drawings which show their contribution to the project. The structural engineer may include plans and cross-sections of the foundations such as the piling layout or location information about the structural steel framework. While the building services engineer will want to show the positions, fixings and materials needed for the various building services fittings and these too form part of the 'working drawings' (see Figure 8.20).

When there is more detailed information about components and their installation or fixing methods, they are often shown on a larger scale drawing called a detail or assembly drawing.

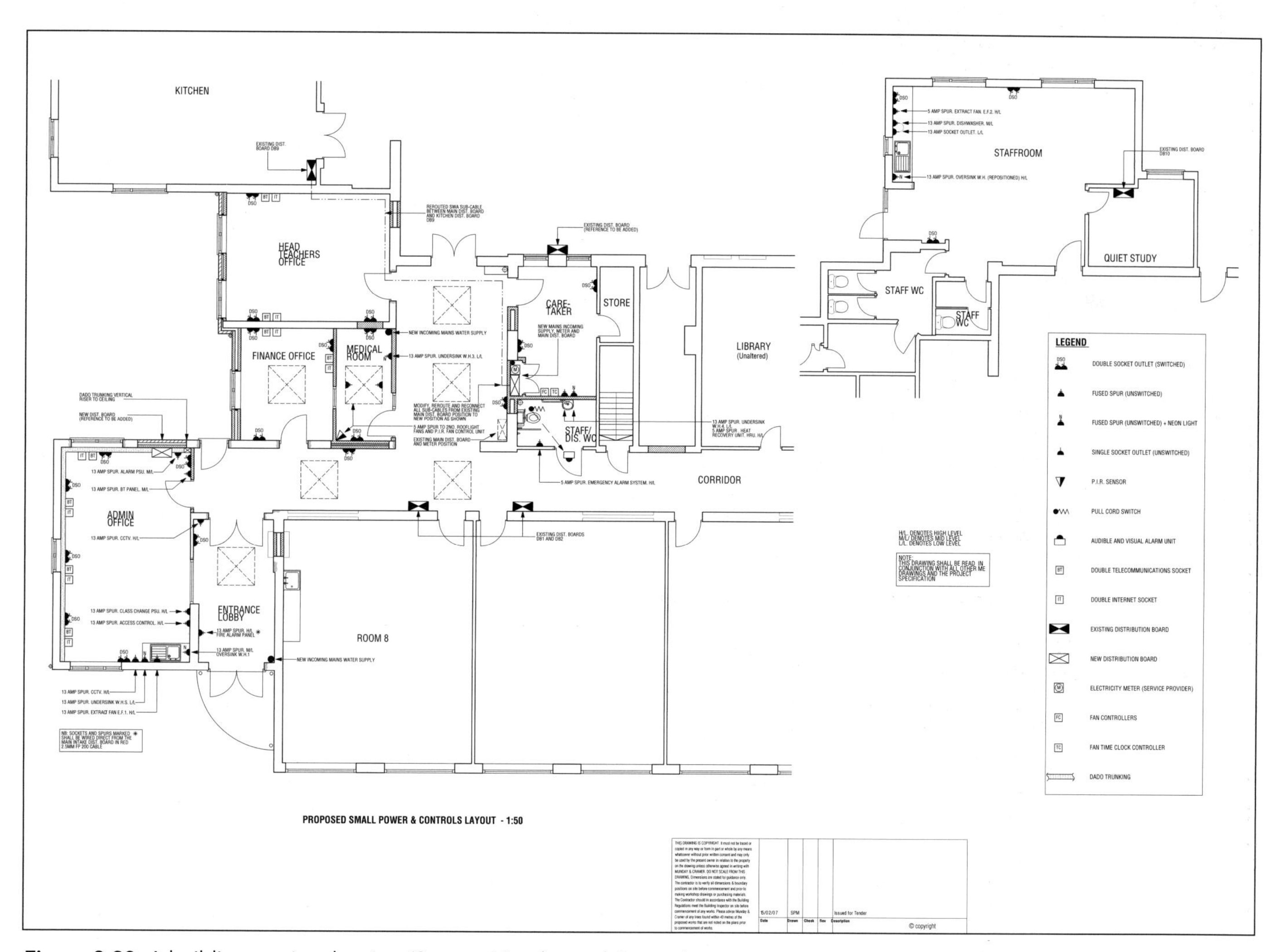

Figure 8.20: A building service drawing *(Source: Munday and Cramer)*

The larger scale facilitates a clearer view of more complicated parts of the project and how they fit together (see Figure 8.21). Ideally, a good detail drawing should clearly show:

- shape and geometry
- position and orientation
- dimensions.

It may also include specification information such as fixings, materials to be used and manufacturer's references; although, these may be included in a separate specification – see later in this unit.

For civil engineering projects, various detail drawings are produced which identify, for example, the construction layout and the position of steel reinforcement. Figure 8.22 shows the general setting

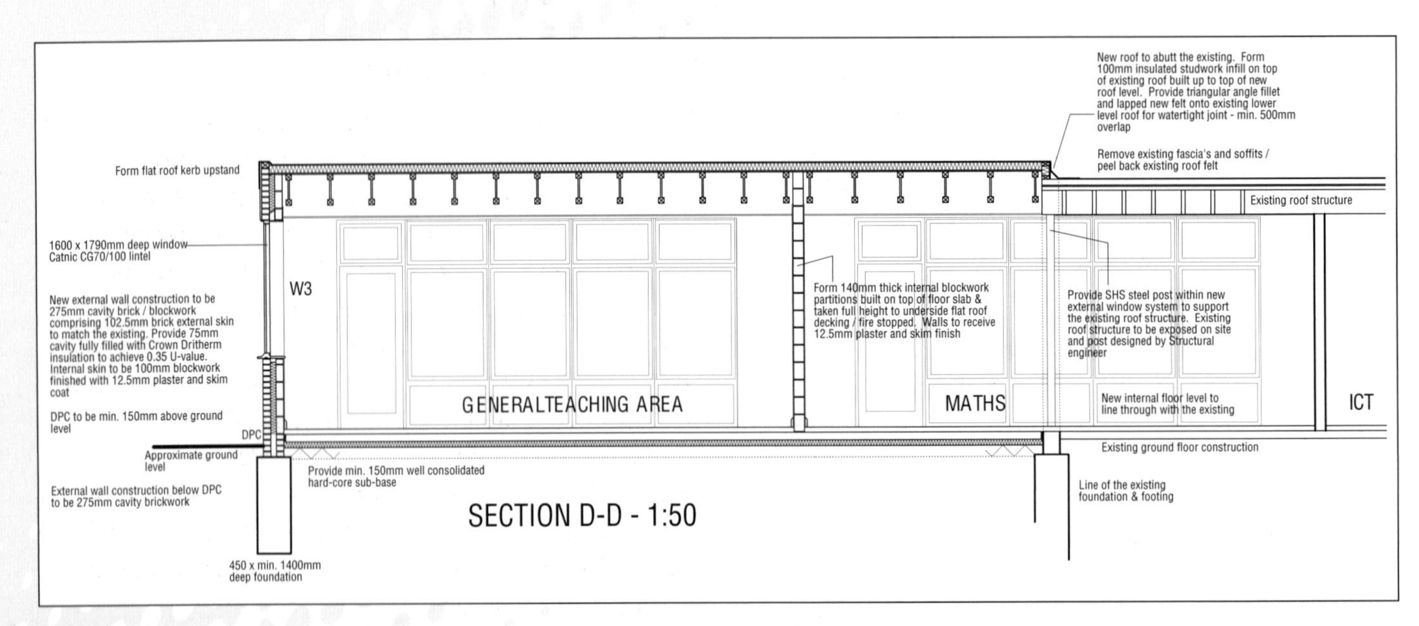

Figure 8.21: A sectional drawing of wall/window details *(Source: Munday and Cramer)*

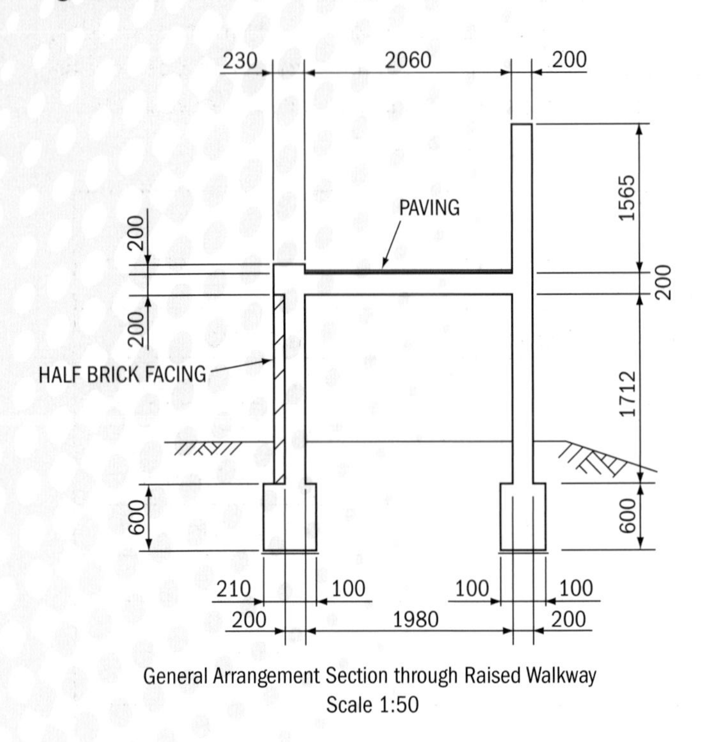

Figure 8.22: General arrangement drawing of a walkway

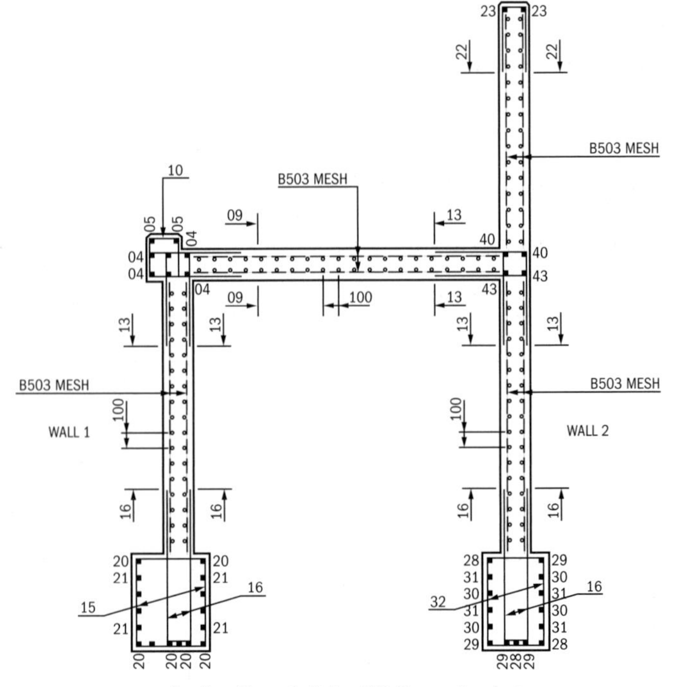

Figure 8.23: Detailed section of a walkway

out details that the formwork contractor would use to construct the overall shape of a raised walkway platform, while Figure 8.23 identifies to a larger scale the exact type of steel reinforcement and where it should be positioned by the steel fixer. Note that numbers shown refer to a coded bar reference which would be used in the steel bar bending schedule.

Component drawings

The component drawing is the largest scale drawing (1:5, 1:2 or 1:1). It enables the fabrication of complicated bespoke off-site or on-site elements such as steel beams, stair flights, sash windows or built-in furniture. These drawings are often used in conjunction with materials lists, bolt lists or cutting list which are essentially tables of exact lengths and sizes of the sections and the jointing/connection treatment required for each section.

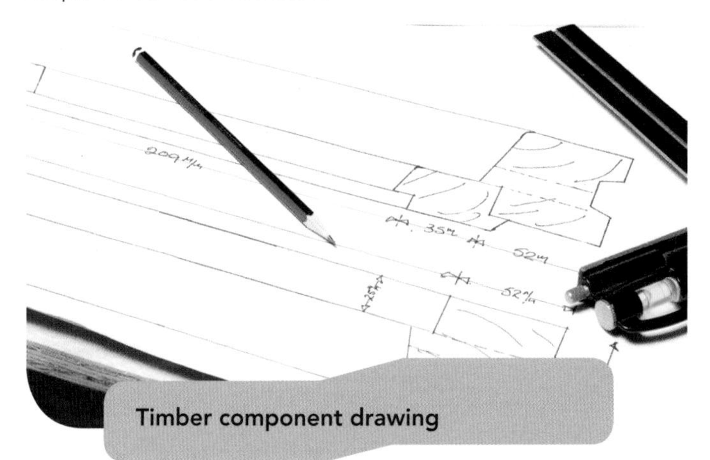

Timber component drawing

The photo above shows an example of a component drawing or 'rod' in the process of being drawn up to fabricate a timber window frame. It is used together with its cutting list which gives the lengths and cross-sectional sizes of timber required to make up the parts of timber window Figure 8.24 shows an example of a timber cutting list for a two-panel door.

Schedules

Component schedules

Component schedules are the most efficient method of providing information on a set of similar components that have minor variations in detail. A good example of the use of component schedules is the specification of doors. These can be specified for internal or for external use, they may be flush or panelled, large or small, made of timber or some other material, pre-finished or they may need to be painted or varnished. Add to this the different glazing possibilities and the various ironmongery variations of locks, hinges and handles and it can be seen that there are many permutations possible. The best way to specify this information is using a door schedule which is a tabular form of collating information.

A typical door schedule is shown in Figure 8.25 where all the minor variations regarding size, finish, and fixing are clearly tabulated. In addition, each component has a unique reference which is also cross-referenced to the appropriate location plan.

Timber cutting list

Job description: Two panel door				**Date:** 8 Sept 2010		
Quantity	**Description**	**Material**	**Length**	**Width**	**Thickness**	**Remarks**
2	Stiles	S wood	1981	95	45	Mortise/groove for panel
1	Mid rail	"	760	195	45	Tenon/groove for panel
1	Btm rail	"	760	195	45	Tenon/groove for panel
1	Top rail	"	760	95	45	Tenon/groove for panel
1	Panel	Plywood	760	590	12	
1	Panel	"	600	590	12	

Figure 8.24: Timber cutting list

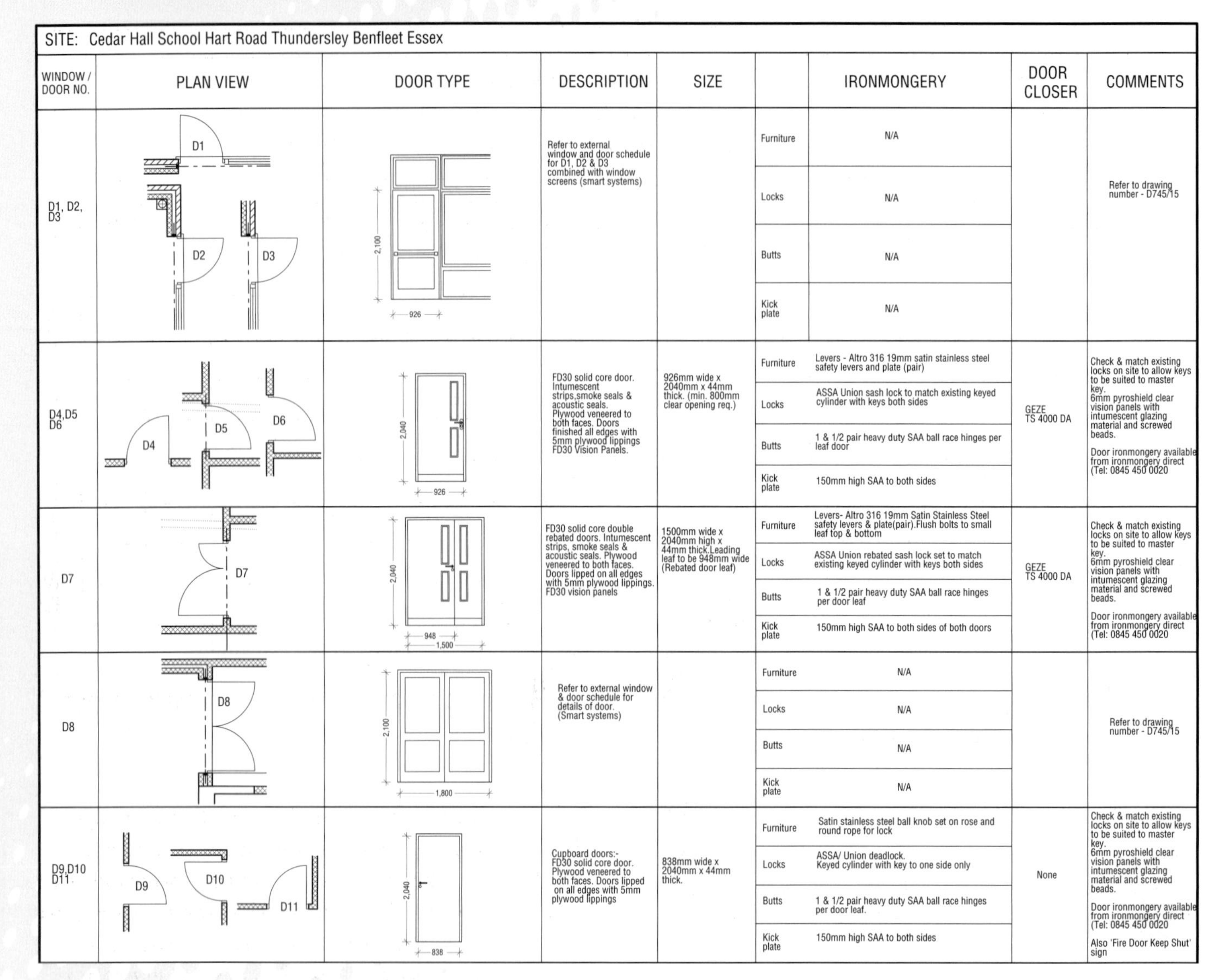

SITE: Cedar Hall School Hart Road Thundersley Benfleet Essex								
WINDOW / DOOR NO.	PLAN VIEW	DOOR TYPE	DESCRIPTION	SIZE		IRONMONGERY	DOOR CLOSER	COMMENTS
D1, D2, D3	D1 D2 D3	2,100 926	Refer to external window and door schedule for D1, D2 & D3 combined with window screens (smart systems)		Furniture	N/A		Refer to drawing number - D745/15
					Locks	N/A		
					Butts	N/A		
					Kick plate	N/A		
D4,D5 D6	D4 D5 D6	2,040 926	FD30 solid core door. Intumescent strips,smoke seals & acoustic seals. Plywood veneered to both faces. Doors finished all edges with 5mm plywood lippings FD30 Vision Panels.	926mm wide x 2040mm x 44mm thick. (min. 800mm clear opening req.)	Furniture	Levers - Altro 316 19mm satin stainless steel safety levers and plate (pair)	GEZE TS 4000 DA	Check & match existing locks on site to allow keys to be suited to master key. 6mm pyroshield clear vision panels with intumescent glazing material and screwed beads. Door ironmongery available from ironmongery direct (Tel: 0845 450 0020
					Locks	ASSA Union sash lock to match existing keyed cylinder with keys both sides		
					Butts	1 & 1/2 pair heavy duty SAA ball race hinges per leaf door		
					Kick plate	150mm high SAA to both sides		
D7	D7	2,040 948 1,500	FD30 solid core double rebated doors. Intumescent strips, smoke seals & acoustic seals. Plywood veneered to both faces. Doors lipped on all edges with 5mm plywood lippings. FD30 vision panels	1500mm wide x 2040mm high x 44mm thick.Leading leaf to be 948mm wide (Rebated door leaf)	Furniture	Levers- Altro 316 19mm Satin Stainless Steel safety levers & plate(pair).Flush bolts to small leaf top & bottom	GEZE TS 4000 DA	Check & match existing locks on site to allow keys to be suited to master key. 6mm pyroshield clear vision panels with intumescent glazing material and screwed beads. Door ironmongery available from ironmongery direct (Tel: 0845 450 0020
					Locks	ASSA Union rebated sash lock set to match existing keyed cylinder with keys both sides		
					Butts	1 & 1/2 pair heavy duty SAA ball race hinges per door leaf		
					Kick plate	150mm high SAA to both sides of both doors		
D8	D8	2,100 1,800	Refer to external window & door schedule for details of door. (Smart systems)		Furniture	N/A		Refer to drawing number - D745/15
					Locks	N/A		
					Butts	N/A		
					Kick plate	N/A		
D9,D10 D11	D9 D10 D11	2,040 838	Cupboard doors:- FD30 solid core door. Plywood veneered to both faces. Doors lipped on all edges with 5mm plywood lippings	838mm wide x 2040mm x 44mm thick.	Furniture	Satin stainless steel ball knob set on rose and round rope for lock	None	Check & match existing locks on site to allow keys to be suited to master key. 6mm pyroshield clear vision panels with intumescent glazing material and screwed beads. Door ironmongery available from ironmongery direct (Tel: 0845 450 0020 Also 'Fire Door Keep Shut' sign
					Locks	ASSA/ Union deadlock. Keyed cylinder with key to one side only		
					Butts	1 & 1/2 pair heavy duty SAA ball race hinges per door leaf.		
					Kick plate	150mm high SAA to both sides		

Figure 8.25: A door schedule *(Source: Munday and Cramer)*

Another example of a component schedule is the bar bending schedules which specify the exact length and shape of individual steel reinforcing bars for reinforced in situ concrete construction (see Figure 8.26). In Figure 8.26 note that there are only seven different types of bar in each beam and that each has its exact diameter and steel grade noted. For example, bar mark 03 is 'H25' which means high yield steel (H) with a diameter of 25 mm. Its length is 2400 mm long and it is a straight bar as '00' is its code. A 'U' shaped bar has a 21 code and an overlapping link is a shape code 51.

Similarly, steel-framed buildings constructed from individual universal beam and column steel sections have their own set of tabulated schedules and component drawings for fabrication purposes, such as steel section 'Materials lists' and steel 'Bolt Lists'.

Take it further!

For more information on the scheduling of reinforcement, refer to BS 8666:2005 'Scheduling, dimensioning, bending and cutting of steel reinforcement for concrete'.

Theory into practice

Go to the National Building Specification (NBS) website and investigate the demo provided that illustrates the content and use of the NBS.

- Briefly list the features of the NBS specification.
- What do you think are the benefits of using this method to create specification notes for a building project?

Company Name:	Bestend Consultants plc	Bar Schedule Reference:	8/ 1298		
Job Reference:	1298/10	Date Prepared:	12/03/10	Date revised:	-
Job Title:	Express Foodmarket, Kings Lynn.	Prepared By:	mjh	Checked By:	jfs

Member	Bar Mark	Type & Size	No. of Members	No. of Bars in Each	Total No.	Length of Each Bar	Shape Code	A	B	C	D	E/R
Ground Beams	01	H20	6	2	12	7600	00					
On Gridlines	02	H20	6	2	12	5400	00					
A to F	03	H25	6	2	12	2400	00					
	04	H25	6	3	18	5600	00					
	05	H20	6	3	18	3125	15	1450				
	06	H20	6	2	12	2875	21	1300	425			
	07	H12	6	37	222	1550	51	450	250			

Figure 8.26: A reinforced concrete bar bending schedule for a ground beam

Specification information

The written specification, together with drawn details or assembly drawings, should define the quality of the construction systems, products, workmanship and finished work such that:

- the designer's and client's detailed requirements will be met
- the contractor, when quoting for the project, can estimate the costs with certainty and accuracy
- the contractor, when managing the project, can plan, execute and supervise the work in an efficient and controlled manner
- the manufacturer's products can be ordered correctly and in good time
- within the design and construction team misunderstandings and unintended variations are minimised.

Therefore, a good specification should be specific to the project, have no irrelevant material and cover every significant aspect of quality to an appropriate level of detail. It should be technically correct and up to date and reflect current good building practice and legal requirements. The specification should also be well coordinated with the drawings and have no conflicts or irregularities with other contract documents. Examples of typical specifications that would be placed on a drawing or in a separate specification document are given in the next section.

The procurement of the contractor and the works onsite

The specifications and drawings together with the bill of quantities form the bulk of the production information documents. Under traditional rules of **procurement**, these are used to select the principal or main contractor. The information, stamped 'Issued for Tender', needs to be sufficiently detailed and specified to enable the contractors to tender fairly for the construction works.

Key term

Procurement – the process of finding and acquiring the expertise, labour, plant and materials resources needed to build a construction project.

Drawings may also be issued to specialist manufacturers, known as nominated suppliers, to supply designed components for the project. This information is in drawing or schedule format to help the nominated suppliers to prepare proposals for the design and installation of a particular element of the project. The information supplied would specify the layout and location required as well as details about the proposed materials or possible fixing points for the particular element. It is important to note, however, that this information may be unsuitable to build from.

Drawn information and written specifications should be complementary and the relationship between the two should be kept simple and clear. All these documents form the basis of the legal agreement between the client and the selected contractor. They are referred to as 'Issued for Contract', particularly where there have been negotiated changes between the tender drawings and the contract drawings.

The drawings and specification provide the contractor with documentation to build from. These are commonly referred to as the 'working details/drawings' but does not necessarily mean that no changes will occur in the future. The designer or the client may want to make changes to the original design, in which case, the drawings are amended and a revision reference appended to the drawing number. It is vitally important that at all times the contractor is working to the latest set of drawings to avoid any possibility for error.

Assessment activity 8.3

P6 M2 D1 BTEC

Get hold of a selection of current graphical information drawings, specifications and schedules that you could use; you can get these from either the design practice you contacted previously or from your college.

1. In groups of two or three discuss the drawings and comment on whether the standard detailing conventions have been met under the following headings:
   - Title block information provided
   - Selection of views and scales used
   - Range of line thickness/hatchings used
   - Clarity of printing and labelling
   - Specification uses
   - Coordinated project references
   - Cross-referencing between other documents

   As part of the written response, take appropriate copies of parts of the drawings studied and clearly label them with your findings. P6
2. Using the drawings and schedules from P6 clearly explain what you understand about the technical issues contained within the drawing and specification. Select one cross-sectional detail and using the National Building Specification or similar provide an alternative specification. Justify your choices. M2
3. Describe how errors in graphical information, and mistakes in how it is managed, can lead to problems during construction. Comment on where responsibilities lie and evaluate how these issues can be resolved. D1

Grading tips

For P6 you are required to interpret graphical drawings, details, schedules and specifications. It is important to show that you are aware that the drawing is fit for purpose in its layout and presentation.

For M2 you will be required to extract and report clear, accurate and valid information from graphical sources, details and schedules.

For D1 you are required to make a judgement on whether the drawing is of a good quality and if it is not what could be done to improve the situation. You also need to include some reasoned case study examples where a poor/inadequate drawing caused construction problems on site.

PLTS

Discussing the drawing standards and techniques used will help develop your skills as an **effective participator**.

Functional skills

By producing the above report you will have the opportunity to enter, develop and format information independently to suit its meaning and purpose which will develop your **ICT** skills.

4 Be able to produce graphical drawings, details, schedules and specifications using manual drafting techniques

Graphical drawings and details

Setting up the drawing

In general, the larger and more complex the project, the more important is the titling and numbering of the production drawings. On smaller projects with up to, say, 15 drawings, this is far less important, particularly if the titles of the drawings are specific and clear and include cross-referencing from general arrangement drawings to details drawings or cross-sections.

The first thing to do when starting a manual drawing is to tape your drawing paper or film onto your parallel motion drawing board. Draw a 10 mm border around the paper using your set square and parallel motion slider. Next, construct the title block and notes section, which is usually drawn on the right-hand side of the drawing (see Figure 8.27). Typical dimensions for the title block contents and layout are given in EN ISO 9431:1999 'Construction drawings – spaces for drawing and for text, and title blocks on drawings sheets' and summarised below.

The information contained within the title block and notes should be as follows.

- Explanations – to provide guidance in reading the drawing, such as abbreviations used, or special symbols or the dimension units used. Conventionally all dimensions should be in millimetres unless stated otherwise.
- Instructions – to provide information such as the drawing should not be scaled. It may also include instructions regarding the inspection of works by the architect, local authority or design engineer prior to them being covered up.
- References – to be made to other relevant drawings, specifications and other documents.
- Location figure – a small 'key' plan to help identify the physical area that the drawing covers in comparison with the whole project. The location figure should be placed so that it remains visible after the drawing has been folded.

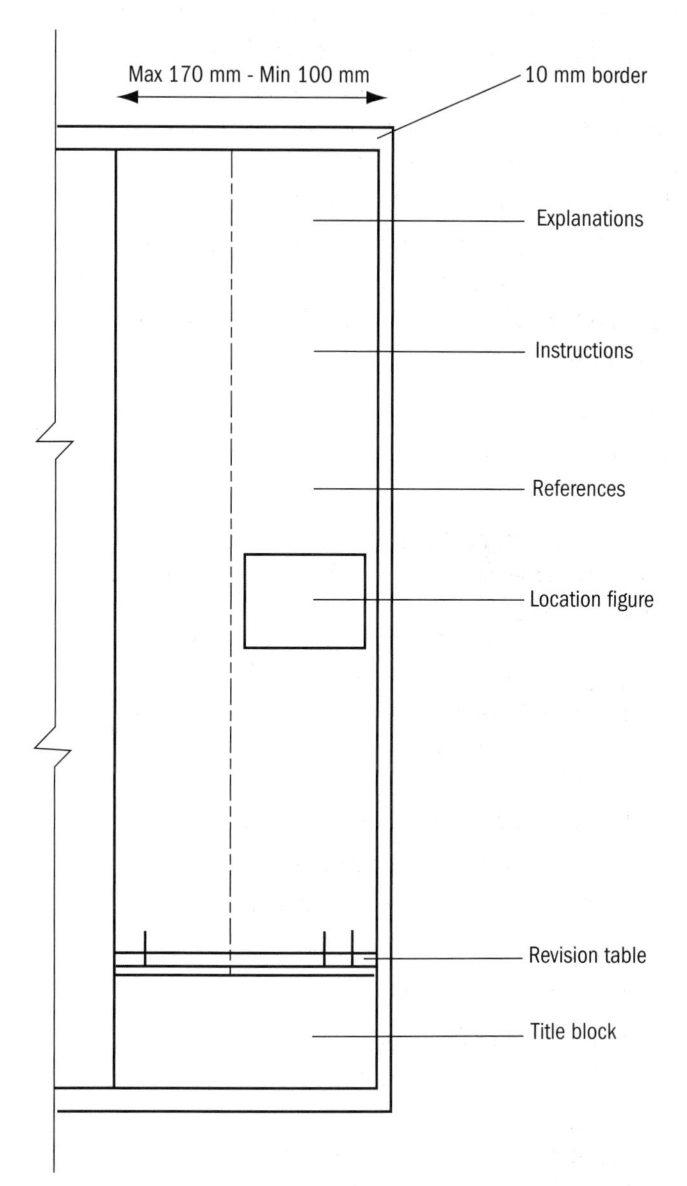

Figure 8.27: Title block layout

- Revision table – to record all revisions such as corrections and/or amendments following the issue of the drawing. The revision should clearly state the date and designation of the revision as indicated by an alpha reference, for example 'A'. It should also describe briefly the details concerning the revision. It should also indicate who made the revision, with suitable initials.

- Title block – this should include separate boxes for the title of the drawing, the address of the project location, the client's name, the company logo panel. Details of the scales used and who drew the drawing should also be indicated.
- The drawing reference number – the drawing should be numbered as simply as possible to facilitate easy filing and retrieval. A typical example is shown in Figure 8.28.

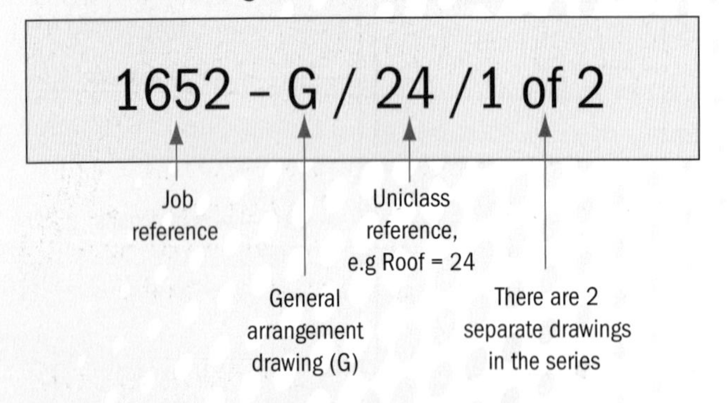

Figure 8.28: The drawing reference number
Note: This is an example of a General arrangement drawing (G). An Assembly Drawing is denoted (A), a Component Drawing is denoted (C).

The details of the drawing are entered on the Register of Drawings which is a record of the drawing title, numbers, issue dates and any revisions. A typical copy is shown in Figure 8.29.

Choosing the correct views and the right drawing scale

Once the drawing border and title block information are in place, the scales of the drawing need to be selected; you need to consider the purpose of the drawing and what information you want to show. The types of view have been covered earlier in this unit (see pages 271–72); the most common views used in construction are orthographic projections and cross-sections.

Next, the type of views on the final drawing need to be considered. This is based on the chosen scale. Typical scales for the various types of drawing are given in Table 8.6. Once the views are decided, the various elements need to be set out on the paper or film. The different views need to sit in the drawing in a well-balanced way and allow for areas of specification text to be included on the drawing.

Co.	DRAWING REGISTER AND ISSUE SHEET							
Origination: **Structural Engineer**		Project No. **02000**			Drawing No. **Drgs 8/G/300 Series**		Sheet No. **1 of 1**	
We enclose copies of the drawings etc listed below		Date of issue						
Project Title: **New Project Substructure**	Day	05	23	30	05	13	04	
	Month	07	07	07	08	08	09	
	Year	06	06	06	06	06	06	
		Date of receipt						
Received by	Day							
	Month							
	Year							
Description	**Ref No.**	Ammendments						
G.A. Foundations - sheet 1 of 2	301	P1	P2	P3	P4	P5	A	
G.A. Foundations - sheet 2 of 2	302		P1	P2	P3	P4	A	
G.A. Ground Floor - sheet 1 of 2	303	P1	P2	P3	P4	P5	A	
G.A. Ground Floor - sheet 2 of 2	304		P1	P2	P3	P4	A	
Foundation & Ground Floor Sections	305		P1		P2	P3	A	

Figure 8.29: The Register of Drawings

Table 8.6: Typical drawing scales

Type of drawing	Purpose	Typical scales used
Site location plans	A map to show the location of the site	1:2500, 1:1250, 1:1000
Layout of block plans	A plan showing the proposal in relation to the boundaries of the site	1:500, 1:200, 1:100
General arrangement (location) drawings	Plans, elevations, perspectives and cross-sections to show the relative position of construction elements	1:100, 1:50
Assembly/Detail drawings	Plans and cross-sections of individual parts of the building showing detailed construction information	1:20, 1:10
Component drawings	Large scale plans and views showing fabrication information of construction components	1:5, 1:2

In order to plan how the views will be set out, draw a rough sketch on your drawing sheet with the proposed views shown as control boxes (see Figure 8.30) The control boxes can be pencilled in on the paper to give you a starting point for each individual view, be it the plan, elevation or section.

If you are going to use a scale of 1:50, then measure the overall width and height of the drawing area to a scale of 1:50. Add up the total width and height area of the control box then subtract this from the overall width and height of the drawing area to give the total amount of free space available. The free space can then be divided by two or three depending on how many equal spaces are required as shown on the sketch in Figure 8.30.

It is important that plan views and cross-sectional views are connected by the correct use of titles and section marks. When drawing a plan view it is conventional that the height at which the horizontal cross-section is taken is at 1.2 m above the finished floor level. In this way, all

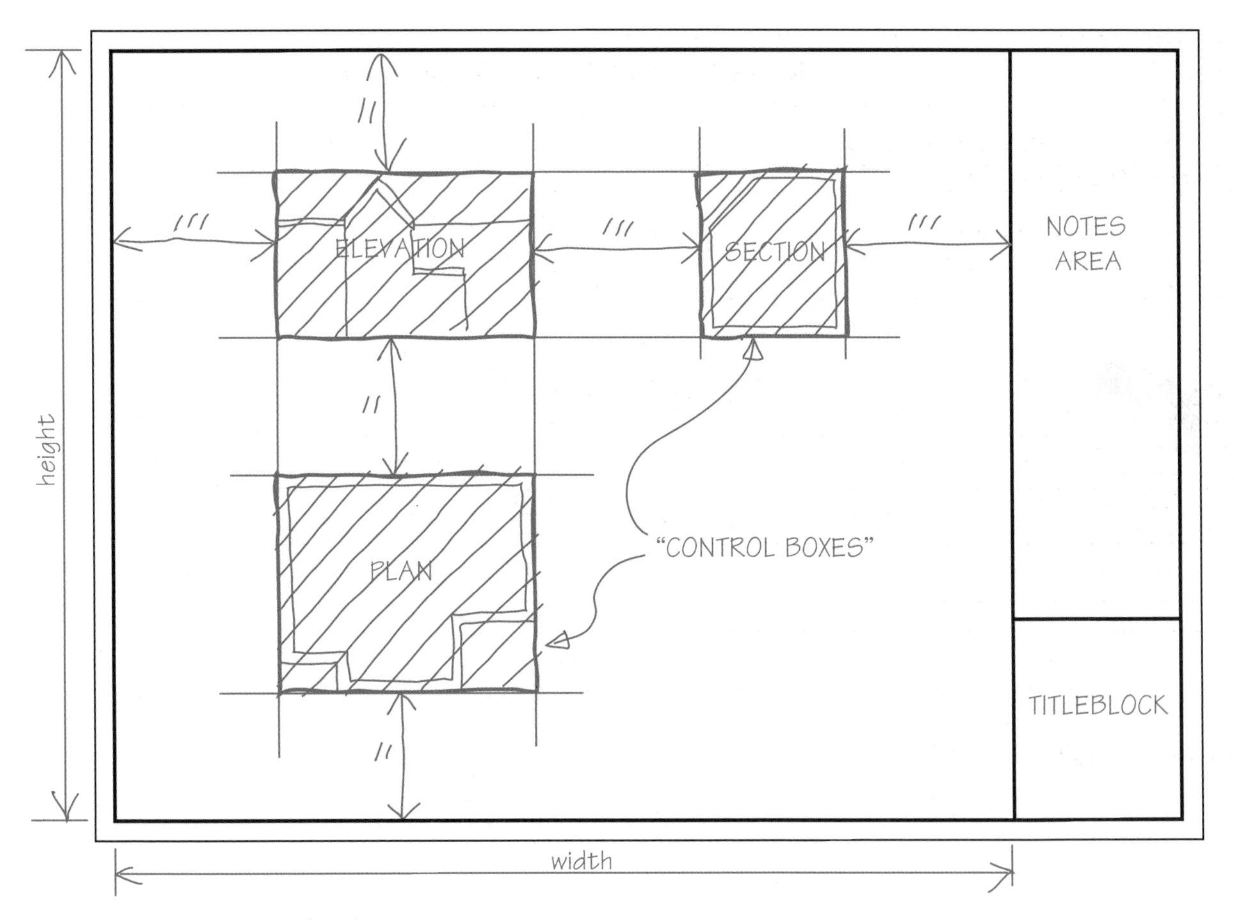

Figure 8.30: Setting out sketch

the main doors and window positions will be picked up together with the positions of staircases and partitions. When selecting where to take a suitable 'cross-sectional view' through a building it is important to select a position that shows the most typical cross-section which applies to the majority of the proposed new work.

Remember!

Always draw to scale as accurately as is possible. Even though contractors should never scale from the drawings, it will give you confidence that you are working accurately and with precision. In this way, you can always double-check as the design is progressing that your calculated dimensions are the same as your scaled dimensions.

What the drawings should contain

All projects will have a high technical content that will rely heavily on your understanding of construction technology and processes. It is therefore essential that you are confident in your knowledge and understanding, as this is ultimately what you are trying to communicate to the reader.

As a guide, here are examples of indicative content of various types of construction drawing. The list is not exhaustive but provides the general scope of detail that you need to know and drawings ought to include the features discussed below.

General

When preparing construction drawings it is good practice to bear in mind the following points.

- Maintain a sequential order of drawings, detail sheets and schedules linked to an itemised register of drawings.
- Create drawing notes and specifications that provide an appropriate level of information tailored to the reader of the drawing.
- Ensure that all work complies with appropriate British Standard, Euro-codes, Building Regulations and Town and Country Planning Regulations.
- Use standard abbreviations, hatching patterns and symbols.
- Ensure that dimension lines and dimension text are clear and there is agreement between separate and overall dimension.
- Title blocks should be complete with correct information and scales shown.
- Doors, windows and other fixtures are referenced to relevant schedules.
- All work is checked by a second person to avoid errors prior to issuing.

Block plan/layout plan

The block plan or layout plan should include the following:

- an Ordnance Survey map view/references provided to locate site
- the North point clearly shown
- vegetation/trees to be removed or retained
- vehicular access and sight splays dimensioned
- setting out dimensions of all new work clearly referenced to identifiable existing features, such as rear kerb line or existing building line
- site datum clearly identified together with all site levels to be achieved including retaining wall positions
- new drainage layout for soil, waste and storm water including location of inspection chambers or soakaways if applicable. Invert levels and cover levels of inspection chambers also should be shown. Location and levels of public sewers
- details of the proposed external works, to identify accurately areas of hard/soft landscaping/planting areas and boundary constructions.

External elevations

The external elevations should clearly show:

- site slope in relation to proposed development
- levels of floors, windows, roofs and other features
- type of external treatment or finish to walls and roof
- position and size of windows and types of openings, position and type of doors
- retaining wall profiles, boundary wall heights and positions of movement joints
- heights of adjacent existing properties/trees.

Building plans

The general arrangements and location plans of the building plans should take the following into consideration.

- Plans should be orientated to block plan with entrances/front of building clearly indicated.

- Grids should be shown on centre lines of structural frames, with numerical/alphabetic grid references.
- Foundation sizes and layout need to be clearly set out about centre lines of superstructural elements/grids.
- Wall construction needs to be correctly hatched with cavities closed at reveals and damp proof membranes/insulation shown.
- Door and window positions must be set out, with door swings shown as arcs.
- Direction of stairs should be indicated by an up or down arrow with numbering of risers. Balustrades/railings need to be identified with the direction of ramps and gradient stated.
- Position of services needs to be noted, including:
  - gas, electric and cold water intakes
  - all service metering devices
  - sanitary fittings to be plumbed
  - electrical consumer unit, luminaries, light switches, power sockets, kitchen appliances and smoke detectors
  - heating appliances such as boilers/flues and radiators, also location of hot and cold water cisterns
  - soil and vent position and associated above ground drainage runs
  - telecommunications/optical cabling sockets.
- Finished and structural floor levels need to be referenced to site datum; this includes direction of floor spans if suspended, positions of lateral restraint straps and strutting (first floor).
- Trimming details around staircase and floor finishes should be specified.
- Roof plans should show direction of falls for flat roofs, position of outfalls, gutters and rainwater downpipes (rwp), direction of span of roof trusses, lateral and diagonal bracing, water cisterns supports, roof lights and loft hatch positions noted, and type of roof covering noted.

Sectional elevations/detail sections

The sectional elevations and detail sections should include:

- construction details of the proposed development with elements clearly cross-hatched to identify materials
- smaller scales used for general arrangement purposes to show interrelation between the different elements; larger scales to show construction details and specifications, for example position of ground floor slab and sub-base in relation to damp proof course (dpc) and insulation materials, etc.
- size and position of stepping of foundations on slopes
- changes in floor levels, staircase/ramp cross-sections, balustrades/railings
- long sections along the length of a proposed sewer and formation and pavement levels of a road gradient, both should use exaggerated vertical scales (see Unit 10 Surveying in construction and civil engineering).

Presentational sketches/details

Presentational sketches can help the client fully appreciate the proposed design. These can be produced free-hand to aid creativity, or in a hard line technical format. The most common projections are:

- isometric – good for showing a 3D aerial type view which looks down onto the building, but because of the rigid box into which the object is placed it appears unnatural as it has no convergent lines of perspective
- axonometric – this is good for showing an aerial view of an object but from a higher viewing point than the isometric projection. Very useful for showing internal fitting out to kitchens and retail areas as the projection is based on a true plan shape
- perspectives – the most complicated to do but often produces the most realistic and natural view with all lines converging to points on the horizon.

Producing good linework and printing

All drawings must be clear to those who will read them. Some drawings may be complicated and so need to be made easy for the design team members, main contractor and subcontractors to read.
Various techniques are used to make views easier to understand; they include the use of different line thicknesses, hatching and shading. With ink pens on tracing paper or film, achieving the correct thickness is relatively easy providing you have a range of nib sizes, usually three from 0.25 up to 0.5, with 0.35 being the most used for text and general linework. However,

with pencil on paper, this effect is more difficult and requires practice.

Thicker lines indicate primary information; for example, a cross-section where members that have been cut through are shown with a thicker outline than those that have not been cut but appear in the same view. Where views are titled, they are printed in thicker lines than the normal specification notes that are printed on the drawing. Section line markers should also be drawn with thicker lines so that they easily stand out from the rest of the view and can be located quickly. Thinner lines are not only used for shading or hatching but also for showing dimension lines as these should not obscure or detract from the main outlines of the physical parts of the constructed element.

Written specification information

Detailed specification notes

On large projects with many drawings produced by many people, inconsistencies may easily occur. Therefore, it is better to have drawings just identifying the elements of construction with brief labels and dimensions together with one definitive specification from which to source all the detailed information about the product/material and how it should be fixed.

For these larger projects a full specification is a separate written document that provides clear and concise requirements and describes exactly the quality of *materials* and *workmanship* to be used in the proposed work. Each sentence in the specification is known as an individual clause, and should be concise and direct; using verbs such as 'excavate', 'pour', 'nail', etc to instruct the contractor/subcontractor. Essentially, each clause should be prefaced by the words 'The Contractor must' and it should state what is to be supplied and then what is to be done with it. The specification should also be supported by the relevant British Standard. For example, a specification for the fixing of natural roof slates:

'The Contractor shall fix natural slates to BS 680-2 with 3.35 mm shank copper clout nails to BS 1202-2 to sawn softwood battens to BS 5534-1. Moisture content at time of fixing of battens to be a maximum 20%. Setting out of battens to be to true lines and regular appearance, with neat edges, junctions and features. Ensure that gutters and pipes are kept free of debris and cleaned out at completion.'

Schedule of work

On small to medium projects, the primary documents are the **schedule of work** and the drawings. These complement each other to give the client an overall appreciation of the project; these documents include notes about the location, materials and dimensions of the construction elements and assembly details.

Key term

Schedule of work – a separate written contents list of the operations to be carried out for the job which includes a brief written specification, often called a reference specification, which describes the product/material to be fixed and how the fixing will be carried out.

For small jobs where only a few drawings are needed to cover the whole project, such as a single house, then the written notes that form the specification may be printed on the drawing itself, provided that it is set out clearly and the notes do not obscure the drawing's individual plans, sections and elevation views.

The design team needs to carefully consider the structure of the information at the earliest stages in the design process. The choice of whether to provide a full and separate specification, a schedule of work and drawings, or just drawings will vary from job to job, but it is important that the amount of information included is consistent and can be revised without causing discrepancies between documents.

Materials schedule

A schedule is a way of presenting information when there are variations in size and specification for a certain element of construction, for example doors schedules, windows schedules, steel bending reinforcement schedules, steel column schedules and drainage schedules. In this way, a great deal of information can be collated in an easy-to-handle table without unnecessary and confusing drawing works.

Typical column headings for some of these schedule formats are shown below:

- Door Schedule: Reference/Size/Description/Frame/Threshold/Hanging/Lock and Latch/Handle/Other
- Steel Bar Bending Schedule: Member/Bar Mark/Type and Size/No. of Members/No. of bars in each/Total No. /Length of each Bar/Shape Code/A–E (individual measurements relating to shape code).

Assessment activity 8.4

BTEC

You are employed in a small, traditional architectural company and have been given the job of producing some of the architectural and construction drawings for a new residential site consisting of three townhouses. The small site is located close to the town centre's popular High Street and has a road frontage of 18 metres and a depth of the plot is 20 metres. The building line is 2 metres from the rear of the front kerb.

The aim of the development is to create three twin-bedroom townhouses with the following features:

- Ground floor – integral garage, WC, hallway, and access stairway to first floor
- First floor – kitchen/diner, living room with small balcony to rear
- Second floor – two bedrooms, bathroom and airing cupboard.

The site is to be built using traditional materials and techniques but still reflect contemporary design in its layout and internal features. There will be space for a small communal garden to the rear of the building.

1. Undertake research into the construction of the development using:
   - examples of similar housing projects in your area
   - construction technology and design notes
   - British Standards/building regulations

   Identify a physical location for this site in your local town and locate it on a large scale Ordnance Survey map of the area.

   Produce simple 2D and 3D graphical drawings using traditional manual drafting techniques for everything listed in Table 1. P7

Table 1

Drg.	Graphical information to be produced	Purpose
1	Preliminary sketches	Free-hand sketches of development to show all floor layouts and overall massing of development and elevations
2	Site location plan	To locate site
3	Block plan	To orientate site and show setting out dimensions, access to rear garden and external works/landscaping
4	General arrangement plan at 1st or 2nd floor for one townhouse	To show layout of walls, doors, windows and fittings
5	Elevations of townhouse development building	To identify the materials and massing of the building
6	Typical cross-section drawing	To show an overview of one of the house's structure and construction details
7	3D isometric view of whole townhouses development	To provide a visual 3D impression of the development
8	Wall details cross-section	To show wall, window and lintel constructions

(Cont.)

Produce in tabular form a window and door schedule for the development. Using the National Building Specification or similar, identify the size, type, opening and ironmongery required for the houses as well as suitable specification for the timber first floor and staircase construction. P8

2. Your manager was impressed with the simple graphical drawings you produced and has now asked that you select and apply manual techniques and resources to produce complex graphical drawings as outlined in Table 2. M3

3. In order to achieve the distinction criteria, you must demonstrate a professional level of competence in the production of the drawings listed in Tables 1 and 2. Your work should show technical skill, neatness and accuracy, and include a full specification for drawings 10 and 11 using the CAWS/Uniclass coordinated project information. D2

Table 2

Drg.	Graphical information to be produced	Purpose
9	3D scaled perspective view of whole townhouse development	To provide a realistic visual 3D impression of the development
10	Roof plan and cross-sectional details at eaves and verge	To show general roof structure, fixings, insulation, drainage and bracing
11	Foundation details	To show foundation and ground floor construction
12	Suitable 3D kitchen layout as an axonometric view	3D view of fitted units in kitchen area

Grading tips

For P7 you will need to produce 2D and 3D graphical drawings using manual drafting techniques. When planning your drawing work always take time to think about the layout of the elevations and plans and how they will be set out over the drawing.

For P8 you are required to produce graphical information in the form of simple specifications and schedules. When writing out specifications or schedules of work always ensure that you are referring to the latest British Standards.

For M3 you are required to apply manual techniques and resources to produce complex graphical information. When dealing with more complex details often two or three additional views are needed to show the true nature of the job. Also in more complex details you need to make full use of different line styles and thicknesses to help differentiate between the various parts of the drawing.

For D2 you will need to produce drawings demonstrating a high level of technical skill that would not look out of place in a professional office; therefore your preparation, planning and execution needs to be thorough and complete. Remember to check your work for errors, including spelling mistakes.

PLTS

Developing your own technical drawings, details and specifications will help develop your skills as a **creative thinker**.

Harry Baxter

Interior design

Harry works for an Interior Design Company working in East London. He is only 22 years old and has been at the company a short time. He started work straight from school as the office junior helping out with the administration and some simple survey and drawing work.

Harry was always interested in graphics and design, so he went to College to study a part time National Certificate in Construction. After passing that course with a double Merit grade profile, Harry went onto study a part-time BA in Interior Design which he is just about to finish.

Harry mostly uses specialist CAD software although on some of the smaller jobs he has to use manual drawings as those are preferred. He works as part of a small team and now takes responsibly for a lot of the small works projects, including the surveying and detailing work. He also is responsible for producing the manual and CAD details on the company's larger projects.

Harry enjoys his job and finds it very varied. There are times when things are quite high pressured, especially when there are client deadlines to meet but he finds the job rewarding.

Think about it!

- Would you be interested in working at a design firm?
- What would be your ideal job in this field?
- Would you prefer to work for a small or large company?

Just checking

1 What is the typical line thickness used for drawing dimension lines?
2 Describe how CAD drawings are reproduced.
3 What type of projection would be used to give a realistic 3D view of a building?
4 Explain why it is important to get the right balance between too much information and not enough information on a drawing.
5 List five reasons why CAD can be more efficient at producing drawings than manual methods.
6 Describe the main security issues that need to be considered when using a CAD system.
7 What are the main purposes of the designer's preliminary sketches done at the early stages of a project?
8 Explain with examples the difference in content between general arrangement drawings and detail drawings.
9 What are the advantages of using a 'schedule' when specifying window components for a small housing development?
10 What is the function of the 'specification' notes?
11 Explain the use of the 'revisions' section in the title block of the drawing.
12 List five construction features that would be shown on a plan view of a new retail shop building.

edexcel

Assignment tips

- When producing a manual drawing always plan out the views and layout of the drawing in rough before starting the drawing. Leave space not only for the view itself, such as an elevation on a house, but also for the dimension lines and the printed notes.
- Try to locate copies of real drawings to see how they are presented. Some can be found on the Internet but you could also approach a local design company who may give you some of their previous project drawings.
- To investigate the use of CAD have a look at the various website of the main CAD companies, such as AutoDesk and Bentley. Some companies even allow students to download free copies of their drawing software.

redit value: 10

9 Measuring, estimating and tendering processes in construction and the built environment

easuring, estimating and tendering are an essential part of a onstruction company's business activities and begin with the estimator. he estimator is the first key person to become involved in pricing the ender for the client and uses measurement to produce quantities, stimating to produce prices and the tendering procedure to submit the ompany's tender.

easurement encompasses the physical act of using a tape measure or scale rule to roduce a value that can be used to create a meaningful quantity. Although it is also sed in surveying, measurement is primarily used in the quantity surveyor's role in the onstruction process.

his unit looks at the processes associated with measurement, the purpose and rocesses of estimating, the production of costs estimates, cost modelling and, finally, e tendering procedure.

Learning outcomes

After completing this unit you should:

1. be able to produce final quantities from dimensions and descriptions of construction work
2. understand the purpose of estimating and the common techniques used to price construction work
3. be able to calculate all-in rates of materials, labour and plant
4. be able to derive approximate quantities and costs to determine the approximate value of building projects
5. understand the process of tendering.

Assessment and grading criteria

This table shows you what you must do in order to achieve a pass, merit or distinction grade, and where you can find activities in this book to help you.

To achieve a **pass** grade the evidence must show that you are able to:	To achieve a **merit** grade the evidence must show that, in addition to the pass criteria, you are able to:	To achieve a **distinction** grade the evidence must show that, in addition to the pass and merit criteria, you are able to:
P1 carry out the measurement of quantities for different applications **See Assessment activity 9.1, page 316**	**M1** apply the Standard Method of Measurement to the production of accurate quantities and descriptions **See Assessment activity 9.1, page 316**	
P2 abstract final quantities from measurements **See Assessment activity 9.1, page 316**	**M2** justify the selection of a new estimating method **See Assessment activity 9.3, page 320**	
P3 explain the purposes of estimating **See Assessment activity 9.2, page 318**	**M3** justify the selection of an appropriate tendering method **See Assessment activity 9.3, page 320 and Assessment activity 9.6, page 340**	**D1** evaluate the limitations of pre-production costing methods **See Assessment activity 9.5, page 333**
P4 explain the uses of different estimating techniques **See Assessment activity 9.3, page 320**		
P5 review the content of a given estimate **See Assessment activity 9.3, page 320**		
P6 determine labour and plant rates, and material costs **See Assessment activity 9.4, page 327**		
P7 produce all-in rates for two classes of construction work **See Assessment activity 9.4, page 327**		
P8 select techniques and processes for use in determining costs **See Assessment activity 9.5, page 333**		
P9 produce approximate quantities and associated cost budgets for two stages of a construction project **See Assessment activity 9.5, page 333**		
P10 explain the common methods of tendering for construction work **See Assessment activity 9.6, page 340**		
P11 discuss the documentation required to support the tendering process **See Assessment activity 9.6, page 340**		
P12 explain the factors that can affect the level of tenders **See Assessment activity 9.6, page 340**		**D2** evaluate the impact of potential variations on the level of tender **See Assessment activity 9.6, page 340**

How you will be assessed

The evidence requirements for pass, merit and distinction grades are shown in the grading criteria grid. Evidence for this unit may be gathered from a variety of sources, including well-planned investigative assignments, practical work or reports of practical assignments. You will be given written assessment to complete for the assessment. These will contain a number of assessment criteria from pass, merit and distinction.

This unit will be assessed by the use of three assignments:

- Assignment one will cover P1, P2 and M1
- Assignment two will cover P3, P4, P5 and M2
- Assignment three will cover P6, P7, P8, P9, P10, P11, P12, M3, D1 and D2.

Jonathan

This unit has given me an understanding of the use of tendering and estimating in obtaining a contractor to undertake the work for a client. It has also helped me realise how measurement can be applied as a tendering tool for obtaining fair prices from contractors.

I've also become aware of how to form a cost budget to approximate a total cost for the client's work.

Over to you

- What does the process of measuring mean to you?
- How would measurements be applied as a tendering tool?
- What are you looking forward to learning about in this unit?

1 Be able to produce final quantities from dimensions and descriptions of construction work

Build up

The use of measurement

Measurement is not just using a tape measure to produce dimensions for drawings, it also has several important uses both pre- and post-contract.

- Why is this important?
- What are the other uses of measurement?
- Do you need a tape measure?

Applications of measurement

Detailed measurement and production of quantities and descriptions for bills of quantities

Detailed measurement is undertaken from the contract drawings and specification and, using dimension paper or a software program, quantities are **taken off** and calculated for each item. The descriptions for the items that are produced for the bill of quantities must follow the rules set out within the **SMM7** rule book. It consists of several chapters that cover aspects of the construction of a project, and the measurement rules that apply to each item. The detailed measurements should contain all the necessary information for the estimator to know what they are pricing.

Key terms

Taken off or taking off – the process of taking dimensions from drawings and producing a quantity.

SMM7 – the Standard Method of Measurement (seventh edition) which is published by the Royal Institution of Chartered Surveyors (RICS).

Take it further!

The SMM7 provides a clear set of rules that can be applied fairly, so that all contractors bid for work on an equal basis that is fair to all, that is, every contractor will be pricing the same set of items. Find out more about the SMM7 and the RICS by visiting their website (go to www.rics.org).

Generally, the rules contained within SMM7 follow the following structure:

- the item's classification, e.g. excavation
- size restrictions, e.g. maximum depth less than 0.25 m
- the unit of measurement, e.g. cubic metres (m^3)
- the measurement rule for that item, e.g. quantities are measured in bulk before excavation
- the definition rule for that item, e.g. site vegetation includes hedges, scrub, trees and stumps
- the coverage rule for that item, e.g. works include removing tree roots
- any supplementary information, e.g. describe filling materials that will be used.

Interim payments

Interim payments occur at regular time intervals during a project's life – usually every 30 days. The client's quantity surveyor will, in agreement with the contractor's quantity surveyor, measure all the work accomplished on site to date. This is called the gross valuation and from this all the previous payments are deducted to give the net valuation, which represents the work achieved that month. The valuation is prepared using the percentage of work done against each item within the **bill of quantities**. Simple multiplication and summing up gives the value of the total amount of work achieved to date.

Final account work

A final account is the summation of all the variations that have occurred on a typical contract and is the final total that the client has to pay the contractor less the previous payments they have received. The final account is adjusted against the original contract sum that was agreed at the commencement of the project.

In compiling the final account, an architect's instruction may require carrying out measurements on site and then valuing against the bills of quantities rates that the contractor entered within their **tender**. All the contract variations are worked through and the final account is then prepared for agreement by the contractor and the client.

Variations

Variations are changes to the construction works on site which can be the result of:

- errors in the design
- errors in the specification
- the expenditure of **provisional and prime cost sums**
- design changes by the client.

Variations may need to be measured in order to ascertain how much the client has to pay the contractor. For example, if a contractor had been asked to increase the length of a brick wall, they would physically measure the length on site or from a revised drawing. This would establish the quantity of wall in square metres (m^2); looking up the bill of quantities' rate for this gives the value of the additional work.

Key terms

Bill of quantities – a list of quantities produced to a standard that is used to price construction tenders.

Tender – to make a formal offer or estimate for a job.

Provisional and prime cost sums – sums of money placed within a tender for unforeseen works or items that cannot as yet be measured. These can also be sums of money for nominated suppliers or subcontractors. They are subsequently omitted and the agreed rate and price is put back when the work is completed.

Claims and disputes

Claims and disputes often arise on construction projects and are principally between the client and the contractor. They can lead to losses and expenses for the contractor and a delay to the project handover for the client. Many factors start disputes such as:

- adverse weather conditions
- late receipt of information from the designer
- a vast number of design changes
- cancellation of some part of the construction work.

Measurement may be needed to substantiate some of the claims and dispute items in order to provide evidence. Any record supplied in evidence will help to determine who is to blame and, ultimately, who will pay for the additional resources used; an adjudicator will then decide who is right in the dispute.

Thinking point

When the Wembley Stadium project ran into delays and additional costs associated with several design changes, the contractor and the Football Association eventually settled out of court.

Processes

Traditional

The traditional process involves the use of dimension and abstraction paper. Dimension paper is used to take off quantities and record their dimensions, size, shape and description, so a bill of quantities item can be produced. Each column has a particular function (See Figure 9.1). Dimension paper is unique as each page contains two pages, as we shall see from the numbering of the following columns.

- Column A – the binding column. No figures or writing go into this column. It is used only to fasten together the other sheets.
- Columns B and F – the 'timesing' columns, used to multiply one particular quantity by several factors above one.
- Columns C and G – the columns where the dimensions are entered. Single dimensions indicate a linear dimension, e.g. metres; two dimensions indicate a squared dimension, e.g. m^2; three dimensions indicate a cubic dimension, e.g. m^3. A single integer in this column indicates a number.
- Columns D and H – the summing up or squaring columns. The final solutions from the dimensions are placed and tallied up.
- Columns E and I – the description columns where the item being measured is described.

The first step in the procedure is to prepare a **take-off list**, which lists all the items to be taken off in a logical sequence. Once the project has been completely taken off, then the process of abstraction follows. Each item of work is recorded as a heading on the top of an abstract sheet. From each dimension sheet, the totals are listed under each heading of the abstract sheet and the final value quantity can be obtained by adding up the column of figures on the abstract sheet, taking across any deductions that occurred. From this, the final figure can be placed into the bill of quantities which is then prepared for the tender process.

Key term

Take-off list – a list prepared by the estimator or quantity surveyor which is used to check that all the items required have been covered. Each item is ticket off as the measurement is worked through.

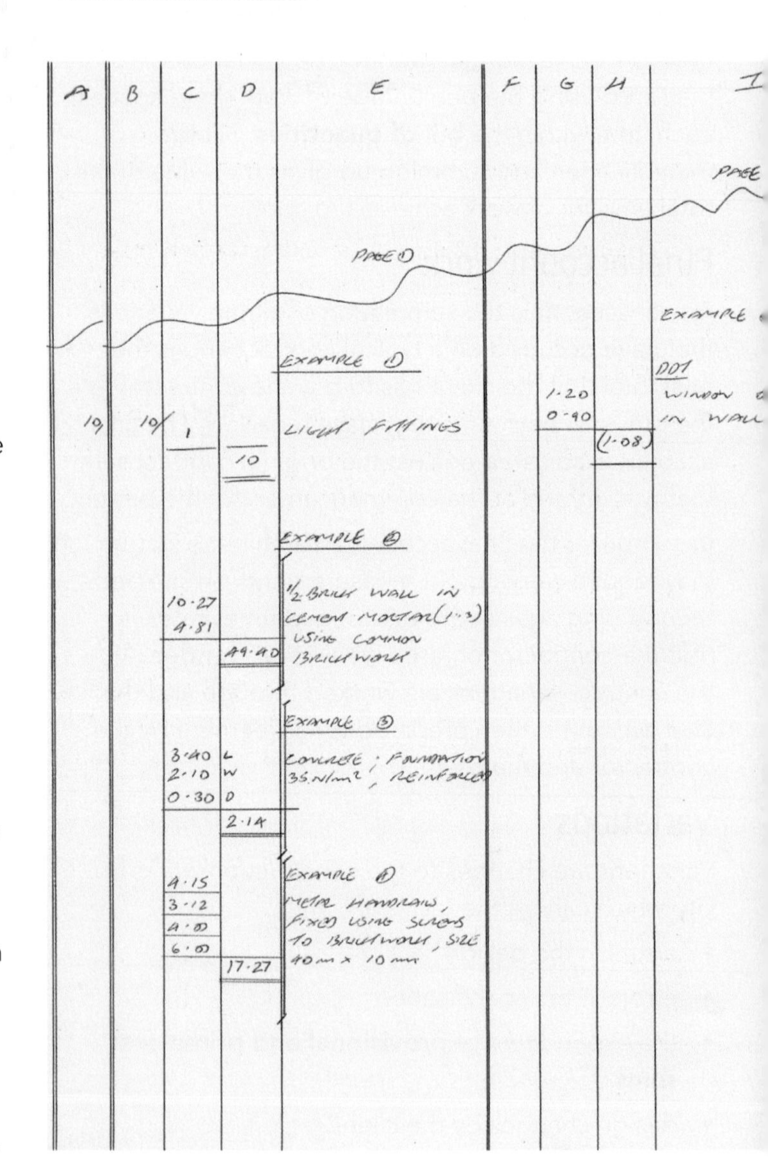

Figure 9.1: Dimension paper

Dimensions are always entered on the dimension paper in the following order: length, width and, finally, depth and are always rounded to two decimal places. Figure 9.1 has some examples of how dimensions are taken off:

- Example 1 is numbered units, for example there are ten light fittings. Always start with one in the dimension column and then times that by ten.
- Example 2 is the dimensions for a metre squared take-off; the SMM7 states that brickwork is measured in square metres with the thickness specified. Always round up to two decimal places in taking off.
- Example 3 is for a cubic dimension, for example, concrete is measured in cubic metres.
- Example 4 uses linear dimensions, for example. handrails to a staircase.
- Example 5 illustrates how we deal with deductions. There are often voids that will need removing from the dimensions, or voids from within shapes such as window openings. This example shows how to deal with a window opening that requires deducting from the brickwork measurement earlier; you often highlight this by putting brackets around it.

Looking at the dimensions illustrated, you will notice a strange s-type shape that is drawn just inside the description column. This is used to group sets of the same dimensions together so they can be added up.

Cut and shuffle

Cut and shuffle is a method that is no longer used; software applications for taking off have made it redundant. This method involved the use of take-off dimension paper that had a perforated centre. Only one item at a time was placed on each sheet. When the take-off had been finished, the sheets for each separate item taken off are all collected together. When all the same sheets were added up together, a total for each item could be established. This avoided the traditional abstraction process but, as you can see, was very complicated to operate.

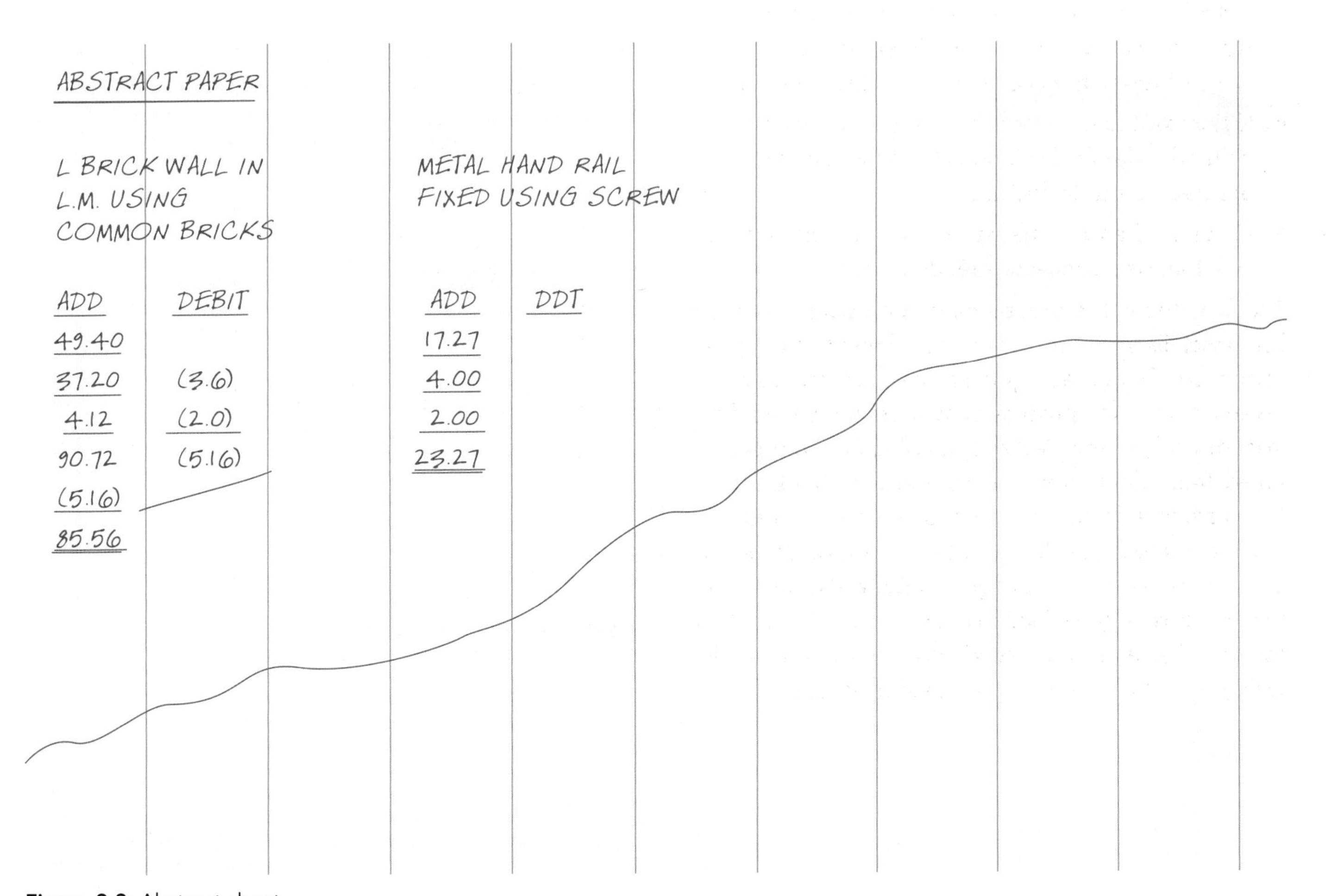

Figure 9.2: Abstract sheet

Production of accurate descriptions and quantities

Compilation of descriptions for works

Descriptions for quantities are often taken from the contract specification and drawings, where the architect has supplied a detailed specification. **NBS** produces specifications that can be used by architects and designers to describe and specify building materials in detail for the contract. Descriptions are compiled from the SMM7 interpretation rules. For example, the classification table gives useful titles to contain within a bill of quantities description.

These general headings must be used to reference the work in the order that SMM7 dictates from the first page to the end, so preliminaries go first in a bill of quantities followed by the other sections. The first column states the type of wall, for example isolated piers; the second column gives descriptions on thickness, whether the work is face work; the third column gives descriptions on shape, for example vertical.

Key terms

NBS – the National Building Specifications publication which can be used in bills of quantities or specifications with drawings.

Mensuration technique – the process of producing quantities that can be used to make calculations simpler and more efficient.

Theory into practice

Find out more about the role of the NBS by looking at their website (go to www.thenbs.com).

Basic mensuration techniques for calculation of accurate quantities for volume, area and length

The centre line calculation method is a basic **mensuration technique** that can be used on closed buildings where the walls rejoin in a closed loop as illustrated in Figure 9.3. This method can be used to find the centre line of any material within the wall's construction, as long as the correct dimensions are

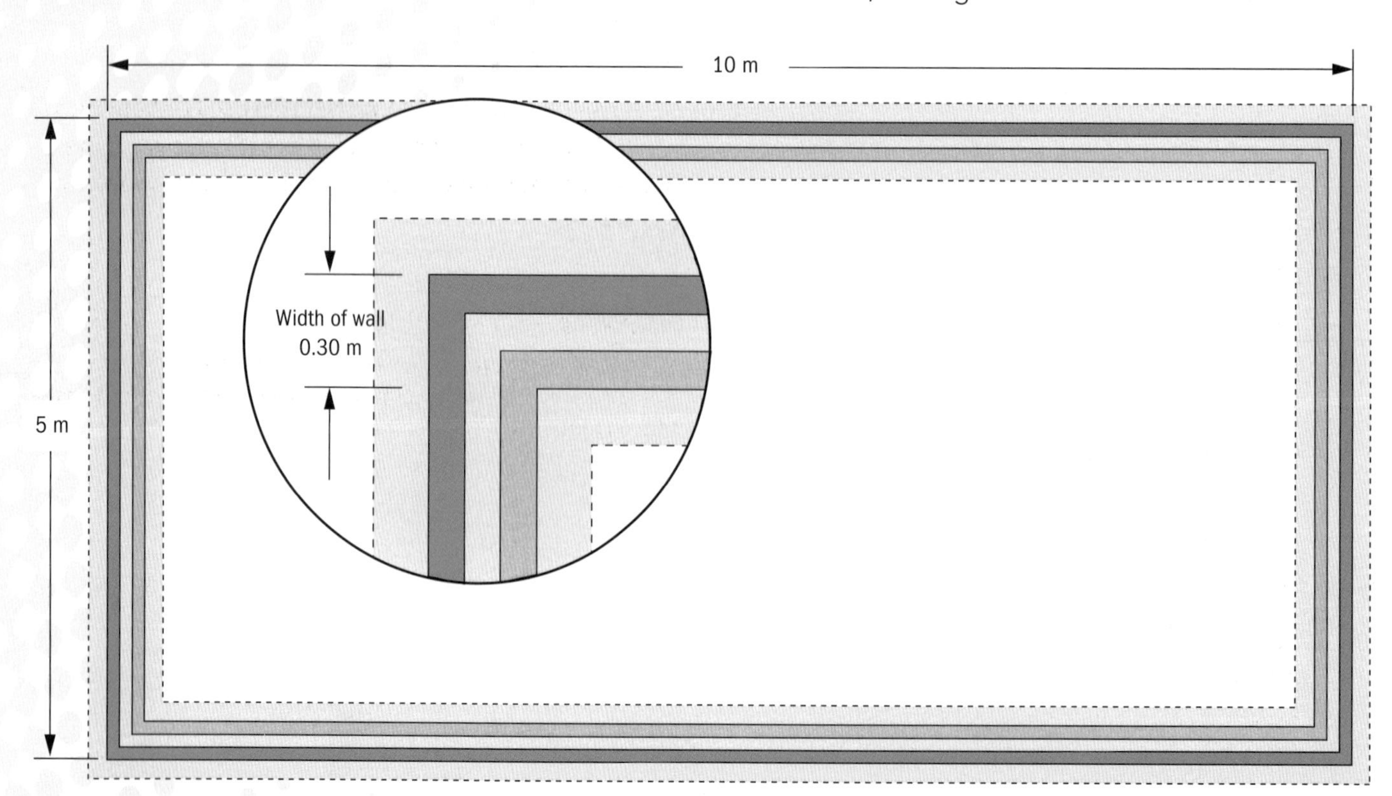

Figure 9.3: Centre line calculation method

taken off or added on. The basic method involves the following:

- Calculate the perimeter from the designer's dimensions. The perimeter is the total length of the outside walls measured on the face from start to back at the point you started.
- Work out the complete wall thickness, in Figure 9.3 it is 0.30 m.
- Take off four full wall thicknesses, in this example 1.20 m (4 × 0.30), from the total perimeter, to arrive at the centre line of the foundation trench, which is the orange line of the diagram.

Application of standard methods of measurement

Standard Method of Measurement for Building Work (SMM7)

As mentioned earlier, SMM7 is the application of a set of rules to the building quantities measurement. It differs from the Civil Engineering Standard Method in that it covers a lot more complex construction works above ground in detail, including services and finishes. The index to the SMM7 contains the following major items:

- preliminary items
- demolition
- groundworks
- concrete
- masonry
- structural carcassing
- claddings
- waterproofing
- linings
- windows/doors/stairs
- finishes
- furniture and equipment
- sundries
- external works
- drainage
- mechanical and electrical.

Civil Engineering Standard Method of Measurement (CESMM3)

CESMM3 is the standard method of measurement also known as the take-off rules for the civil engineering industry. It covers much of the **heavy-side engineering**, such as earthworks and pipework used in reservoir construction. The CESMM3 has a similar purpose to SMM7, however there is very little above-ground building work included within CESMM3. The only other major differences are that the first three columns are called divisions in the CESMM3, whereas in the SMM7, they are called classifications.

The index to the CESMM3 includes the following major items:

- definitions, principles and application
- ground investigations
- demolition and clearance
- earthworks
- concrete
- pipework
- structural steelwork
- piling
- roads
- rail tracks
- tunnels.

You may have noticed that the CESMM3 differs from the building measurement in that it deals with large infrastructure projects such as railways, drainage and roadways.

Key term

Heavy-side engineering – construction work that requires heavy machinery, such as roadways, highways, mass concrete dams and earthworks.

Assessment activity 9.1

Figure 9.4 illustrates a typical modern-day foundation cross-section, which uses a concrete foundation with concrete trench blocks and two skins of engineering brickwork.

Using basic mensuration techniques, calculate and accurately record the following:

- **a** the centre line length from the plan drawings of the foundation
- **b** the volumes of concrete and excavation (ignoring any backfill)
- **c** the area of the earthwork support to the sides of the excavation
- **d** the total square metres of engineering brickwork
- **e** the length of the damp-proof course that will be placed on top of the engineering brickwork. P1

In order to achieve P2 you will need to undertake some further measurements from Figure 9.4, that contain omits and adds, then prepare an abstract of these quantities so a final total can be obtained. Your tutor will direct this assessment.

In undertaking your take-off to the activity P1, ensure that you apply the rules of the Standard Method of Measurement to the production of the accurate quantities and descriptions listed in items **b** to **e**. Ask your tutor for a copy of the relevant sections from SMM7. M1

Grading tips

For P1 you will have to produce some accurate quantities for various applications as listed in the specification.

For P2 you will need to abstract final quantities from measurements.

For M1 the quantities need to be recorded in accordance with the SMM7 rules.

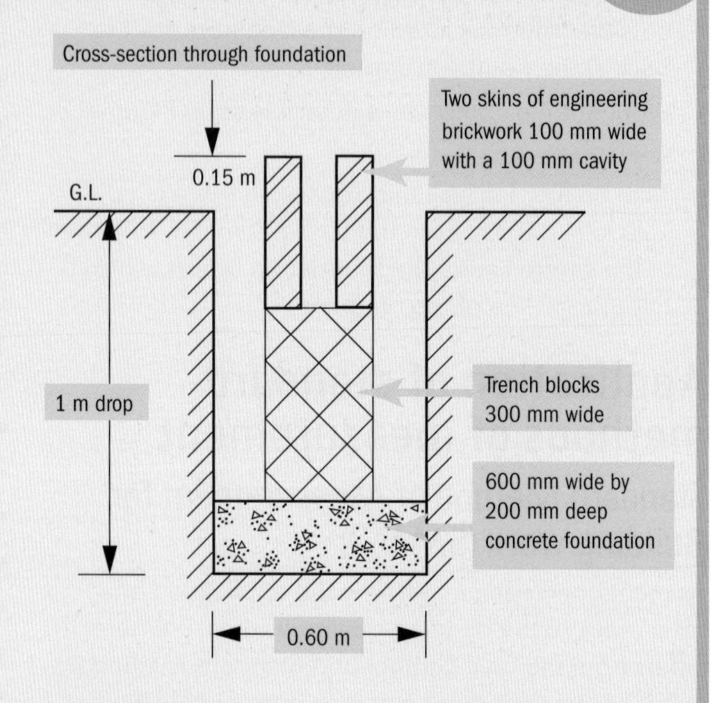

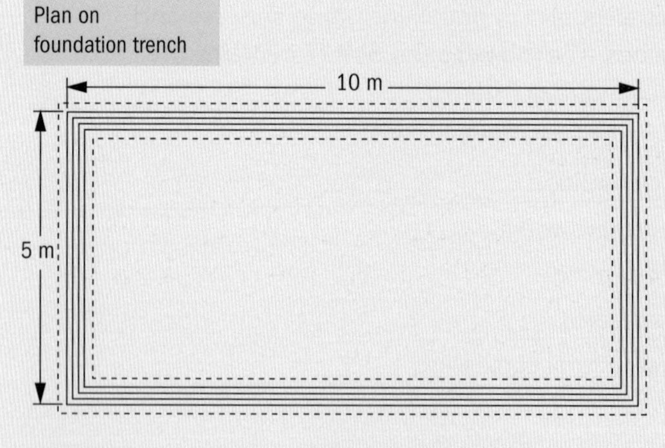

Figure 9.4: A foundation plan and cross-section

PLTS

Analysing and evaluating information while judging its relevance and value in calculating dimensions will help develop your skills as an **independent enquirer**.

Functional skills

Reading and understanding dimensions on drawings to produce some calculations on quantities will help to improve your skills in **Mathematics**.

2 Understand the purpose of estimating and the common techniques used to price construction work

Purposes of estimating

Estimating net cost

The contract documents that the estimator receives may take two forms:

- a set of drawings with a specification
- a set of drawings and a bill of quantities.

From these, the estimator will prepare against each quantity a rate. If using the drawings with a specification, the estimator will have to take off their own quantities. Multiplying the rate by the quantity gives the total price for each taken off item. When all these are summarised, the estimate is said to be the **net costs**. The **gross estimate** contains the items that must be added to the **net estimate**. These items are:

- profit
- overheads
- risk and uncertainty items.

Key terms

Net costs – basic costs of labour, plant, materials, preliminary items and subcontractors.

Gross estimate – the net costs plus overheads and profit and risk items.

Net estimates – net costs.

Pricing of preliminaries

It is a requirement of SMM7 to produce preliminary items and so these need to be priced. However, preliminaries tend to be time-related costs that cannot be easily priced or included within a bill of quantities bill rate. Typical preliminary items would be:

- management and supervision costs
- employers' requirements, such as accommodation
- services and facilities, for example temporary water supply
- mechanical plant
- temporary works
- site accommodation
- transport.

A typical pricing of preliminaries calculation is shown in the worked example below.

Worked example

All the preliminaries are time-related charges so you will need to produce estimated durations on site for these items.

From the following data, calculate the cost of the tower crane which has to be included in the preliminaries.

Data

Cost of delivery of crane = £1200

Number of weeks on site = 20

Cost of erection and dismantling = £2500

Hire charge per week = £5500

Calculation:

Delivery	£1,200
Hire: 20 × £5500	£110,000
Erect/dismantle	£2,500
Total	£113,700

Profit

Construction companies do not consider profit to be measurable and so do not include this within a typical contract. However, businesses have to make money in order to survive, and profit can be included in an estimate by either adding a percentage to each of the rates within the bill of quantities or by adding to each of the items contained within the preliminary section.

The level of profit that is applied to a cost estimate may vary and will depend on the following factors:

- the amount of work that the construction company currently has on its order books
- the level of competition in the location of the work
- the amount of risk associated with the project

- the complexity of the work undertaken
- the nature of the **procurement route**, for example, **partnering agreements**
- the payments terms of the contract.

In essence, there is no set level of profit that can be applied to an estimate. Each tender that is returned should undergo an **adjudication process** where the senior managers of the company discuss the factors listed above and arrive at the level of profit that will be applied to the estimate.

Overheads

A company's overheads are those costs that have to be met in order to run the head office, and include:

- departmental costs
- insurances such as public and employer's liability
- company cars
- IT equipment.

A percentage of the overhead is often recovered from additional costs added into the estimate. To calculate this percentage, the total value of the company's overheads per year needs to be assessed. Then take the turnover for the year (the amount the company takes in receipts) and divide this by the overhead costs × 100 per cent. This gives the percentage that needs to be applied to future estimates as long as the turnover does not drop below this level.

$$\frac{\text{Overheads}}{\text{Turnover}} \times 100\% = \text{Percentage to add to tender}$$

There are a number of ways of **reconciling** the overhead costs and recovering these against tendered works. Overheads can be costed in several ways:

- by not including them, but using an increased profit margin to cover their cost
- by establishing their cost divided by the total turnover and adding this percentage to tender submissions as noted above
- on larger projects, by moving head office functions onto site and recovering these costs through the preliminaries.

Key terms

Procurement route – the method that a client uses to select a contractor to construct the project.

Partnering agreement – a method of procuring a contractor which enables cost sharing between client and contractor on savings.

Adjudication process – a meeting held by senior management to decide what level of risk and profit requires placing on the net tender or estimate.

Reconciling – settling up the values of unknown costs.

Assessment activity 9.2

The estimator has just finished an important tender, and the senior managers of the company are meeting to undertake adjudication before submitting the price. Explain the purpose of estimating. P3

Grading tip

For P3 you need to explain the purpose of estimating.

PLTS

Supporting conclusions while using reasoned arguments and evidence will help develop your skills as an **independent enquirer**.

Functional skills

Explaining what estimating includes to a small group will help develop your speaking skills in **English**.

The effects of quantity and value on the method of estimating chosen

Very small project's costs can be estimated simply by calculating the number of days' labour required plus materials or plant. Much larger, complex projects will require more thorough techniques of measuring and rating items to produce a final cost estimate. This is because they can contain several different installations by specialist subcontractors, each project being a one-off design.

Estimating techniques for labour, plant and materials

A client's budget will need to be accurate when the project costs run into millions of pounds and may have to be financed through loans. There are several techniques that can be employed to achieve this as follows.

The estimation of labour costs is based upon the agreement on pay rates that the company has with the operatives. From this all the additional costs will need calculating, for example employer's National Insurance, holiday credit, sick pay and any other allowances. This will then give a rate that can be used by the estimator to recover all the overheads against the labour in the tender.

Rates per unit of measurement

The **unit** or **number method** estimates cost on a unit basis, for example a seat in a cinema or a bed in a hospital. Very simple calculations can produce a cost for a potential project by using previous contract final accounts and the unit number of occupancy. Obviously, this method may be very inaccurate and does not take into account the complexity of the design.

The **area method** takes previous historical contracts, often by the Royal Institution of Chartered Surveyors, to produce and compile square-metre cost rates and cost estimates. The square-metre cost rate is found by measuring the total floor area of a project and multiplying it by the rate plus or minus any adjustments. This method produces a fairly accurate estimate for the new project.

The **cubic method** is where the volume of a historical project (its length, width and depth calculation) is measured and the original cost divided by this volume. This produces a rate per cubic metre that can then be applied to a new project in order to produce a cost estimate for the client's budget. This method does not provide the most accurate estimate.

The **approximate quantities** method uses the architect's sketch designs to produce some approximate quantities. The estimator's experience needs to be called on here to include the items of work that will be required for the final design. By using current cost rates or **price books**, a realistic estimate can be prepared for the client's budget.

Key term

Price book – a published book that contains current prices and rates for items of work based on the SMM7.

Elemental estimating involves breaking down the proposed design into elements, for example foundations, ground floor construction, first floor construction, structural frame and roof finishes. Cost estimates can then be prepared against each element and a final budget produced.

The **SMM7 quantities** method is undertaken as part of the preparation for tendering and the procurement of a contractor to carry out the contract work. It is the most accurate of the estimate processes compared to the above methods.

The accuracy of all the estimating methods will vary greatly. Preparing area unit or cubic estimates will depend on the accuracy of the drawn information from the architect and the historical cost information used to produce a unit rate, that is, the effects of inflation on prices. Once a detailed design is approved, accurate quantities can be taken off and a better cost estimate refined from the initial studies.

Standard price book rates

There are several price books, such as Spoons, which give a breakdown of a vast number of rates for all aspect of construction work. These can then be used by the estimator as a guide when pricing construction work.

Output tables

Output tables are historical output measurements taken by a company and usually give a realistic output for a unit rate. For example pouring concrete takes $4m^3$ per hour for one person to pour and finish.

Historic rates

Historic rates are measured from actual contracts that have been completed and thus provide an opportunity to deliver rates from the available information. Since the rate is tried and tested it can be used to price future construction work.

Work study

Work study is the method of measuring the actual output of labour. It is done by observing work on site then taking timings and ascertaining the quantities. When these are recorded the unit rate of the actual work done can be worked out. This then can be used to price future work.

Assessment activity 9.3

The estimator has asked you to write a short report on your work experience within their department. Explain the uses of different estimating techniques. 

The chief estimator has asked you to review what is contained within a typical estimate's content. Provide this information. P5

The estimator is considering the following estimating methods:

- using labour, plant and materials as separate items
- using standard price book rates
- using work study rates.

Justify the use and selection of each new method proposed. M2

Grading tips

For P4 describe the techniques used to estimate; look at the list that is contained within the specification page beginning with labour, plant and materials.

For P5 you will need to describe what would be the contents of a typical estimate.

For M2 you will need to justify the selection for each of the methods listed. Include both advantages and disadvantages.

PLTS

Collaborating with others to work towards common goals in any group work will help develop your skills as a **team worker**.

Functional skills

Describing the contents of an estimate in a written document will help improve your skills in written **English**.

Documentation

Code of estimating practice

The code of estimating practice is produced by the CIOB (Chartered Institute of Building) and provides guidance on estimating procedure on the following topics:

- the selection of contractors to undertake the tenders, including methods of selection and compiling the tender list
- the decision to tender, receipt and acknowledgement, including contract conditions, tender documentation and the resources needed to complete the tender
- the management of the estimate, including timetables and workloads of the estimator, site visit, checking the tender documentation and information required
- subcontract enquiries and material enquiries, including how these will be sent out, scheduling material requirement and timetable for receipt
- estimate planning, including the process of **method statements** and the pre-tender programme
- unit rate pricing, including the compilation of the unit rates to be applied to the tender
- provisional sums, including percentage rates and attendances including dayworks
- overheads, including the percentage or calculation of overheads
- completing the estimate, including final calculations
- final review, including the tender adjudication summary
- feedback on whether the tender was won or lost.

Key term

Method statements – documents which identify the methods used to price the work items, the plant and labour required for each activity.

The code sets out a structured management of the tendering and estimating procedure that contractors can follow. It is similar to a quality system and provides checks to ensure no items are missed or mistakes made. It is a very useful guide for the estimator to follow.

3 Be able to calculate all-in rates of materials, labour and plant

Materials costs

Calculation of materials quantities and costs of construction works based on unit costs of materials

Let's consider the calculation involved with a common construction material, as this will explain the process involved in the production of a material unit cost. When calculating the cost of materials, each item should be broken down into common units, so they can all be added together to establish a unit rate for the material delivered to site.

Worked example

The site production of brickwork mortar involves three materials: cement, sand and an additive. Calculate the unit rate for this brickwork mortar using the following data handed to you by the estimator.

Material data:

Mortar is mixed in the proportion specified by the architect or engineer, which is usually a ratio of 1:4 for load-bearing brickwork.

Cement is delivered in a 25 kg bag, which costs £4.25.

Sand is delivered in 20-tonne wagons costing £80 per load.

Additive for workability is £9.00 for 2.5 l; 300 ml covers 100 kg of cement.

Unit rate calculation:

Each 1 m^3 of mortar will weigh approximately 2 tonnes. Therefore, we require:

$$\frac{2000 \text{ kg}}{5 \text{ sets of units per } 1 \text{ m}^3 \text{ mortar}} = 400 \text{ kg per unit}$$

Cement = 1 unit × 400 kg = 400 kg

$$\frac{400}{25} \times £4.25 = £68.00$$

Sand = 4 units × 400 kg = 1600 kg

$$\frac{1{,}600 \text{ kg}}{20{,}000 \text{ kg}} \times £80 = £6.40$$

$$\frac{4 \times 300 \text{ ml}}{2500 \text{ ml}} \times £9.00 = £4.32$$

Therefore, the total cost of 1 m^3 brickwork mortar is £68 + £6.40 + £4.32 = £78.72.

Do you know the correct ratio for mixing mortar?

Labour rates

Calculation of 'all-in' rates for craft workers, skilled, unskilled and gangs

Labour 'all-in' rates need to be established so the estimator can price the tender works and accurately calculate all the costs involved in employing labour directly.

The hourly rate of an operative is not the only cost to the construction company. There are several factors that need to be taken into account when calculating the all-in cost of labour. These are:

- the basic rate of pay per hour that has been agreed between the employer and the worker
- annual holiday entitlement and public holidays
- employer's National Insurance contribution
- the weather (possible loss of production)
- staff sickness
- CITB (Construction Industry Training Board) levy
- travelling time
- bonuses.

Application of labour costs in unit rates

To apply the labour cost per hour, one further piece of information is required: the output rates for the labour, such as the speed a bricklayer can lay bricks. Once this is established, then the unit cost of labour per unit can be calculated and applied to the tender prices. Output rates can be established in two ways:

- timing operatives using **work study**
- using output tables from historical works or price books, which will provide information on unit output rates, how long things took to construct.

Take it further!

The output rate of the labour for a particular item needs to be in the format that the Standard Method of Measurement bill of quantities rate states, for example brickwork is measured in square metres, concrete often in cubic metres.

Do some research on the Internet and find out output rates of labour as stated in the Standard Method of Measurement bill of quantities.

Definition of prime cost of daywork and comparison with 'all-in' rates

Every year, or following price increases, the BCIS (Building Cost Information Service) publishes **daywork** rates which are hourly rates for various specialist trades, for example bricklayers or plumbers. The bill of quantities will have these daywork costs inserted by the client. There may be occasions where a rate for variations cannot be used from the bill of quantities and instead dayworks can be used instead as the agreed method of payment. The contractor inserts a percentage against each of the daywork schedules for labour, plant and materials in the tender to cover for

Key terms

Work study – the timing of work so that a rate can be established. It involves watching the operative work on a known quantity and seeing how long it takes them to finish that work.

Dayworks – unforeseen works that may involve variations. It is an historical arrangement within a contract and records the labour hours and any plant or materials used against items; it is signed by the client's representative.

Worked example

Calculate the annual cost of labour for a general operative using the following data.

Data:

Basic salary = £15,000.00

Holiday pay: 4 weeks at £288 per week = £1152.00

Employer's National Insurance contribution at 12.5% = £1875.00

Lost production time, estimated, 2 weeks at £288 = £576.00

Public holidays: 8 days at £58 = £464.00

Sick pay, estimated, 2 weeks at £288 = £576.00

CITB levy at 2.5% = £375.00

Travelling, estimated, 2 weeks at £288 = £576.00

Bonuses, estimated, £30 week × 48 weeks = £1440.00

Total annual cost = £22,034.00

Unit rate calculation:

In order to calculate a unit rate of labour, we need to establish the total number of productive hours on site. Total hours in one year:

4 days per week × 8 hours + 1 day × 7 hours for Friday × 52 weeks = 2028 hours

Less holidays: 4 weeks × 39 hours = 156 hours

Less public holidays: 7 × 8 + 1 × 7 = 63 hours

Less sickness: 2 weeks × 39 = 78 hours

Total = 1731 hours

The hourly rate of labour is a simple calculation, as follows:

$$\frac{£22{,}034}{1731 \text{ hours}} = £12.73 \text{ per hour}$$

This is the labour rate that an estimator would use to price the works contained within a tender.

profit and overheads. The quantity surveyor can then convert any dayworks into a measured item if there is a rate in the bill of quantities.

In order to present a level playing field, the nationally agreed rates produced by the BCIS are used for the base rate on which the contractors add a percentage to cover all their on-costs, which is added to the tender document.

Plant rates

Calculation of fixed and operating costs

Fixed plant costs are those that are not time-related and include:

- delivery costs
- erection costs
- removal costs.

Information for calculating delivery and removal charges could be obtained from the plant supplier. For example, if delivery was £35 for a dumper, this would be multiplied by two and added onto the total duration time-related costs on site, and then divided by the number of hours to give an hourly rate:

Delivery collection charges = 2 × £35 = £70

Time on site = 150 hours × £9 = £1350

Total cost = $\frac{£1420}{150}$ = £9.47 per hour

Variable or operating costs are those that are time-related to the length of hire on site and would include the coverage of:

- the cost of fuel – to calculate this, the consumption rate of the plant when working will need to be established
- puncture repairs – these will occur randomly and a certain percentage can be added in to recover the costs
- operator costs – the number of hours that an operator is required specifically to use the plant will need to be established.

Calculation of hourly rates

Construction plant can be very expensive to own and operate. There are generally two methods of providing plant for construction sites:

- purchase, service, operate and maintain
- hire in the plant either externally or internally.

The unit rate calculation for an item of construction plant will depend on several factors, including:

- ground conditions
- trained operators
- productivity
- height and reach
- breakdowns and reliability.

There is often a hidden cost associated with large pieces of construction plant which is good to remember – the cost of delivery to site using alternative transport such as a low loader.

Worked example

Calculate an hourly rate for a 1.5-tonne dumper using the data provided.

Data:

Hire rate= £150 per week

Driver costs = £300 per week (50% driving, 50% working on tasks)

Fuel requirements = 150 litres per week at £1.50 per litre

Working week = 39 hours

Unit rate calculation in hours:

Hire = $\frac{£150}{39 \text{ hours}}$ = £3.85 per hour

Driver = £300 × 50% = $\frac{£150}{39 \text{ hours}}$ = £3.85 per hour

Fuel = 150 l × £1.50 = $\frac{£225.00}{39 \text{ hours}}$ = £5.76 per hour

Total cost per hour = £13.46

Application of plant costs in unit rates

The application of plant costs within unit rates for bill of quantity or measured quantity items must be carefully considered. Some plant items are used for several different measured items, activities and operations on site and so cannot easily be included within unit rates. They are therefore often placed within the appropriate section of the preliminaries section of the bill of quantities or specification section (see page 317). Examples of types of plant that would be included within the preliminaries section are:

- rough terrain load-all
- tower cranes
- vans
- skips for rubbish removal
- temporary lighting
- scaffolding.

Examples that would be included within the relevant unit rate would be:

- concrete pumps
- diamond drilling equipment
- compressors and breakers
- screed pumps.

Calculation of unit rates for various classes of construction work

To undertake the calculation of unit rates for use in estimating, you need to know the output rates for two constituent factors – labour and plant. Output rates, as we have seen, can be obtained by actual measurement or from historical work study tables. It is worth looking at the factors that affect the output of labour on a construction site. The following is a list of identified factors that would contribute to a lower output from the workforce:

- poor motivation
- excessive breaks
- wet weather
- dirty conditions
- excessive cold or heat
- dangerous and complicated work
- poor wages
- lack of supervision
- a complicated bonus scheme.

The output of plant and equipment on site is affected by:

- the type of ground conditions
- ground or surface water levels
- the height to be reached
- the distance to move
- the skill of the operator
- the age of the plant
- whether the plant is hired or owned
- the amount of maintenance and servicing
- the correct capacity of plant for the work undertaken.

To calculate a unit rate for construction work, you need to break down the rate into the three main constituent parts of labour, plant and materials and calculate each element before totalling the unit rate. If we use a typical bill of quantities description, it will show the process of compiling the rate that would be applied to each item. Finally, profit can be applied to the rate against each bill of quantities item.

Worked example: Excavation work

Calculate the unit rate for the excavation (see Figure 9.5) using the data provided.

Data:

Excavate to reduced level, maximum depth not exceeding 1 m (unit: m^3)

Excavator with operator = £35 per hour

Output = 6 m^3 per hour, which includes loading the dumper, waiting for it to tip and then starting the cycle again. 4-tonne dumper with driver = £55 per hour. One dumper is required and tips spoil 10 m away.

Unit rate calculation:

$$\text{Cost of excavation} = \frac{£35.00}{6\ \text{m}^3} = £5.83 \text{ per m}^3$$

$$\text{Cost of dumper} = \frac{£55.00}{6\ \text{m}^3} = £9.17 \text{ per m}^3$$

Total cost to excavate and tip = £15.00 per m^3

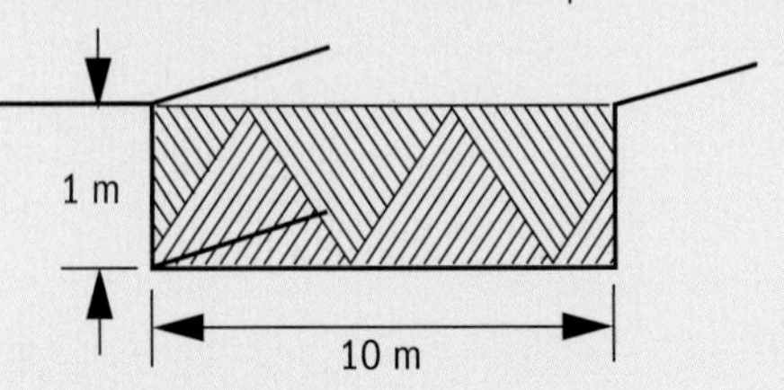

Figure 9.5: Sketch of excavation

Worked example: Brickwork masonry

Calculate the following unit rate for the brickwork masonry using the data provided.

Data:

Facing brickwork one side half brick thick in cement mortar 1:4 built in stretcher bond (unit: m^2)

Brickwork gang (two bricklayers/one labourer):

Bricklayer rate per hour = £12.50

Labourer rate per hour = £9.50

Output per bricklayer = 45 bricks per hour

1 m^2 of brickwork contains 60 bricks

Brick costs = £335 per 1000

Wastage = 5%

Mortar delivered to site = £52.50 per 1 m^3 tub

Mortar use = 0.04 m^3 per m^2 of brickwork

Unit rate calculation:

- Labour (2 + 1 gang cost):

  Bricklayers = 2 × £12.50 = £25.00

  Labourer = £9.50

  Total hourly rate = £34.50

  Output = $\frac{\text{60 bricks per m}^2}{2 \times \text{45 bricks per hour}}$ = 1.5 m^2 per hour for the gang

  Unit rate = $\frac{\pounds 34.50}{1.5}$ = £23.00 per m^2

- Materials:

  Bricks = $\frac{\pounds 335 \times 60}{1000}$ = £20.10

  Add wastage at 5% = £1.01

  Mortar = 0.04 × £52.50 = £2.10

  Total cost = £23.21

  Unit rate = £23.00+£23.21 = £46.21 per m^2

Worked example: Drainage

Calculate the following unit rate for the drainage using the data provided.

Data:

100 mm diameter uPVC pipes in trenches, jointed using collars (unit: linear metres)

PVCu pipes 6 metres long cost £65.78 each

PVCu pipe coupler costs £ 9.95

Silicone lubricant 1 kg per 100 m pipe work costs £25.00

Pipelayer = £15.00 per hour

Output of pipelayer = 20 m per hour

Unit rate calculation:

- Labour:

  Cost per m = $\frac{\pounds 15.00 \text{ per hour}}{\text{20 m per hour}}$ = £0.75 per metre

- Material:

  Pipe = $\frac{\pounds 65.78}{\text{6 m each}}$ = £10.96

  Couplers (one required every 6 m) = $\frac{\pounds 9.95}{6}$

  = £1.66 for each metre run

  Lubricant = $\frac{\pounds 25.00}{\text{100 m}}$ = £0.25

  Total cost per metre = Labour + Materials = £0.75 + £10.96 + £1.66 + £0.25

  = £13.62 per metre

What costs might be involved in laying draining pipes?

Worked example: Concrete works

Calculate the following unit rate for the concrete foundations using the data provided.

Data:

Foundations Gen 1 grade, 10 mm aggregate 75 mm slump, thickness not exceeding 450 mm (unit: m^3)

Concrete will be delivered ready mixed in 6 m^3 trucks

Cost of concrete = £82.50 per m^3

Wastage = 2%

Part load charges = £15.00 per m^3 not delivered in the wagon

Volume required in the foundation = 22 m^3

Two ground workers will be required to place and compact the concrete:

Output of one ground worker = 1.5 m^3 per hour

Cost of ground worker = £12.00 per hour

Compaction will be by hand tamp

Unit rate calculation:

- Labour:

  Ground workers = 2 × £12.00 = £24.00 per hour

  Output = 2 × 1.5 m^3 per hour = 3 m^3 per hour

  Cost per m^3 = $\frac{\text{£24.00 per hour}}{3\ m^3 \text{ per hour}}$ = £8.00 per m^3

- Materials:

  Basic cost of concrete = 22 × £82.50 = £1815.00

  Add wastage × 2% = £36.30

  Part load charge = 2 m^3 × £15 = £30.00

  Total cost = £1881.30

  Unit rate = $\frac{\text{£1881.30}}{22\ m^2}$ = £85.51 per m^3

  Concrete rate = Labour + Materials

  = £8.00 + £85.51

  = £93.51 per m^3

List all the costs involved in laying concrete beds.

Remember!

It is acceptable to calculate the cost of the concrete as in the worked example, but this relies on the fact that when the ground workers have completed their work, they have other work to move on to or their cost will exceed the amount of labour included within the unit rate. Careful site management of resources is therefore essential.

Remember!

To calculate the weekly rate for holidays and sick pay, divide the basic salary by the number of working weeks, add up all your costs for a total, then divide by the number of weeks, then the number of hours per week to find the hourly rate.

Worked example: Timber floors

Calculate the following unit rate for the timber floors using the data provided.

Data:

Floor members, 225 × 38 mm softwood, treated with preservative, built into block walls (unit: linear metres)

Joiner and labourer to fix joists

Joiner = £18 per hour

Labourer = £9.50 per hour

Output = 8 metres per hour

Joists = £3.50 per metre

Wastage = 5%

Unit rate calculation:

- Labour:

  Joiner and labourer = $\frac{£18 + £9.50}{8 \text{ m per hour}}$ = £3.44 per m
- Materials:

  Joists = £3.50 + 5% waste = £3.67 per m

  Total cost = £3.44 + £3.67

  Unit rate per metre = £7.11

Assessment activity 9.4

P6 P7

Calculate the following labour rates, plant rates and material rates from the following data.

Labour data:

Basic salary = £20,000

Working weeks = 42 weeks

Hours per week = 39 hours

Employer's National Insurance contribution = 12%

Holidays = 5 weeks

Sick pay = 2 weeks

CITB levy = 3%

Bonus = £35 per week

Plant data:

Type: compressor and breaker

Rate per week = £85

Fuel at 1.5 litres per hour = £2.50 per litre

Operator = £250 per week

Productivity per week = 65%

Oil = £9 per week

Working week = 39 hrs

Unit rate data:

Brickwork gang (two bricklayers/one labourer):

Bricklayer rate per hour = £14.50

Labourer rate per hour = £10.00

Output per bricklayer = 55 bricks per hour

1 m^2 of brickwork contains 60 bricks

Bricks = £455 per 1000

Mortar delivered to site = £62.50 per 1 m^3 tubs

Mortar use = 0.04 m^3 per m^2 of brickwork.

Provide the following all-in rates:

- 1m^2 of brickwork 
- breaking up 24 m^2 concrete beds if the labourer rate for brickwork is used and 1 person can break up 8 m^2 a day. P7

Grading tips

For P6 you will need to work out all the labour rates, the plant rates, and the material rates.

For P7 you will need to provide an all-in rate for the brickwork and concrete breaking.

PLTS

Analysing and evaluating information, while judging its relevance and value in calculating all-in rates will help develop your skills as an **independent enquirer**.

Functional skills

By calculating routine labour rates you will develop your skills in **Mathematics**.

4 Be able to derive approximate quantities and costs to determine the approximate value of building projects

Traditional cost modelling

Cost per unit

The cost per unit method is used for estimating costs for budget preparation using physical unit, for example a hospital bed, a school pupil place or a workstation. The method relies on the fact that there is some relationship between value and the number of units.

Worked example

A 250-bed hospital costs £350 million to construct. This has been taken from historical information on the cost of building a hospital. Estimate the cost of a proposed 300-bed hospital.

$$\frac{£350{,}000{,}000}{250} = £1{,}400{,}000 \text{ per bed space}$$

Proposed hospital

300-bed space × £1,400,000 = £420,000,000

There can be a vast difference between actual cost and estimated costs and this method does not have a high degree of accuracy. An experienced estimator would need to look at the provisional design against the historical design from which the unit rate was obtained in order to ensure some degree of consistency.

Worked example

A 76,000-seater stadium cost £37 million to construct. Estimate the cost of a proposed 60,000-seater stadium using similar construction methods and design.

$$\frac{£37{,}000{,}000}{76{,}000} = £486.84 \text{ per seat}$$

New stadium = 60,000 × £486.84 per seat

= £29,210,400

Cost per unit area

Calculating the cost per unit area, for example square metres of gross floor area, involves using historical rates, increased in line with inflation, on a similar project in order to produce a realistic rate for a new building project's floor area. This can be illustrated using the following worked example.

Worked example

Historical cost of building A = £300,000

Floor area = 40 m × 20 m = 800 m^2

$$\text{Area rate} = \frac{£300{,}000}{800 \text{ m}^2}$$

= £375 per m^2

Proposed new building B of similar design floor area = 650 m^2

= £375 per m^2 × 650 m^2

= £243,750

Cost of functional element

The cost of functional element method involves breaking down the estimate into functional elements, or small packages of work. Unit rates per square metre, metre or number can then be applied to produce a cost estimate. This method relies on the knowledge and experience of the estimator to apply historical element costs from previous work and upgrade these to the new proposal. It allows costs to be taken from different historical works to provide a budget plan from which further work can be developed. A typical element cost estimate is shown in Figure 9.6.

Approximate quantities

The approximate quantities method provides additional detail compared with the other methods described above. It involves the use of dimensions or sketch proposals. From these, approximate quantities are measured and historical bill of quantity rates applied to produce a reasonably accurate cost estimate. For example, the following single storey building has been taken off to produce approximate quantities and historical rates applied to give a cost estimate, see Figure 9.7 below. (For the purposes of the example, the building is much slimmed down in order to simplify the concept of approximate quantities.)

Design data:

Single storey building 10 m long × 5 m wide × 2.4 m high

Element	Quantity	Element unit rate	Totals
Substructure	150 m²	£50	£7500
Structural frame			£20,000
External walls	2500 m²	£55	£137,500
Roof construction	1500 m²	£45	£67,500
Internal partitions	2000 m²	£28	£56,000
Windows and doors	40	£250	£10,000
Wall finishes	1500 m²	£12	£18,000
Floor finishes	1500 m²	£25	£37,500
Plumbing			£30,000
Electrical			£50,000
External works	2000 m²	£20	£40,000
Drainage	1000 m	£25	£25,000
Total			£499,000

Figure 9.6: A typical element cost estimate

Description	Quantity	Rate	Totals
Concrete foundations	4 m³	£65	£260
Trench blocks	5 m²	£25	£125
Engineering brickwork	2 m²	£23	£46
External solid walls in blockwork	72 m²	£45	£3240
External render	72 m²	£15	£1080
Roof joists	125 m	£9	£1125
Plywood	50 m²	£5	£250
Three-layer roofing felt	50 m²	£22	£1100
Total			£7226

Figure 9.7: Approximate quantities take-off

Application

Feasibility studies

Feasibility studies are undertaken as part of the RIBA Plan of Work; they follow the initial brief taken from the client. At the early stage of a contract, the client will be advised if the contract is feasible and economical to proceed with the detailed stages of the design, or whether it is better to halt the project.

The use of measurement can help determine the feasibility of a project. A brief outline is taken from the client in the form of a sketch and, by using historical cost data, the estimator can produce a reasonably balanced idea of what the projected costs for the project would be. The application of measurement also enables comparison between several project proposals and a decision made in favour of one or a combination of parts to push the feasibility through to detailed design.

Pre-contract cost planning and control

The cost planning of the project is an essential item to link to early stages of the RIBA Plan of Work. Cost planning is undertaken for several reasons, including:

- to ensure that the client's budget is not exceeded
- to produce the right quality for the project
- to ensure that the project will return value for the financial investment.

The pre-contract cost planning and control stage should have a good amount of work put into it to ensure that:

- the tenders received do not exceed the client's budget
- there are no unforeseen or hidden costs
- the amount of financial wastage is reduced and efforts such as **cost value engineering** are put into place when the project overspends the budget.

Measurement only plays a small part in pre-contract cost planning. Other costs must be considered including the cost of finance, consultants' fees and the purchase of the land. It is a good idea to use a lot of resources at this vital stage of the project in order to control the budget of the project.

Key term

Cost value engineering – changing specification, methodology quantities, omitting items or other works to reduce costs.

Processes

Use of historical data

Historical cost data is invaluable in preparing budgets and costing future designs, but requires great care on the part of the estimator undertaking it. A client does not want a completed design which comes in over budget; similarly, the client may not be happy if a completed design is under budget as it may not have realised its full potential. A client will need to be kept well informed; therefore, using historical cost data will have to be discussed with them.

Historical cost data can be obtained from many sources, such as:

- in-house company analysis of tender submissions
- work study methods
- published price books
- BCIS (Building Cost Information Service) and the BMI (Building Maintenance Information).

In-house historical data is established over a period of time by the estimating department and is a very valuable data tool to give a competitive edge to the construction company. It contains an estimating data base from which can be drawn sufficient information from previous projects to assist in pricing future projects; this data includes suppliers' discounts, measured output rates for plant and suitable subcontractors to use. Previous bills of quantities are also a source of in-house historical data that can be used for current information and rates.

Work study data uses the historical physical measurements of the company's labour force to establish output data that is then used to complete current estimates. The data is a realistic output picture on sites and accurately reflects the current situation. For example, if you know how many bricks are contained within a wall and how long it took to build and the costs involved, you can calculate a rate per square metre.

Cost modelling can help determine the approximate value of planned building projects.

Published price books contain the SMM7 rates and are often divided into major and minor works. They provide a rate per item that can be used or adjusted to reflect certain circumstances. However, because they are global figures they tend to be a little generous in the output rates.

BCIS and BMI are compiled and published each quarter by the Royal Institution of Chartered Surveyors. They contain historical cost data which enables the estimator to look closely at a particular product, such as a hospital, school or factory, and examine the complete breakdown of elements, areas, specification and rates. These can then be used to prepare a pre-tender estimate.

Application of tender price indices and location factors

Historical cost data, as its name suggests, is collated over a period of time, which may be over several years. Current costs data will, therefore, require updating in order to take account the following:

- rate of inflation
- local labour rates
- delivery and transport costs
- geographical location of the work.

In order to establish future tender prices, indices are available. These are sets of figures that enable inflation and increased costs to be taken into account when predicting future work costs. Each index has a base rate, usually 100, and prices can then be set above or below the base rate.

Worked example

Cost per m^2 of element A in 2001 was £245.50. Indices for 2001: 171

Proposed project to commence in year 2009. Indices for 2009: 195

Calculate the cost of element A for 2009.

Calculation is as follows for element A:

$$£245.50 \times \frac{195}{171} = £279.96 \text{ per m}^2$$

This can be applied similarly to the different geographical locations around the UK; tender prices in the London area will be greater than in other areas. Cost data used to provide price books are mainly based on London prices so adjustments have to be made.

Principles of wall-to-floor ratios

Design aesthetics and site constraints often play a part in the process and the plan shape may not be designed as economically as possible. Thus, it is often more cost-effective to add something extra to the project at the tender stage rather than in the future when increased costs would make it expensive. Look at the two buildings A and B in Figure 9.8. Their perimeters and floor areas differ in size and shape. The wall-to-floor ratio becomes smaller with the increased floor area, which can be seen by using the small calculations given.

Window-to-floor ratios, plan shape, number of storeys

It is necessary to establish a balance between the amount of daylight and cost savings established. Openings in buildings can be costed against the full wall construction. Building regulations, however, govern the percentage of windows applicable to a building's elevations.

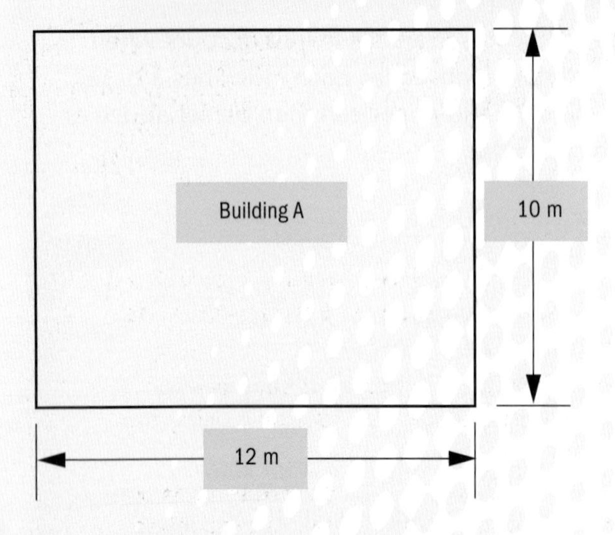

Floor area = 120 m^2
Height 3 m
Perimeter = 44 m
Wall area = 132 m^2

Ratio = $\frac{\text{Wall}}{\text{Floor}}$ = 1.10

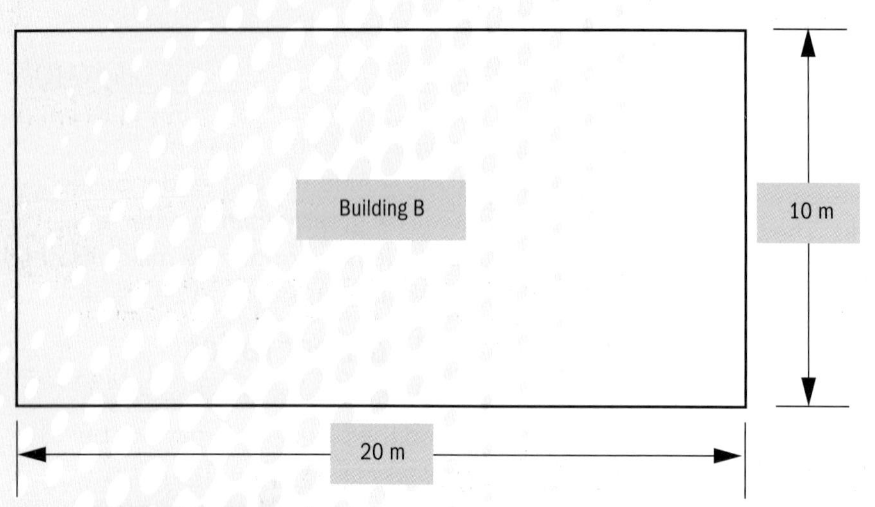

Floor area = 200 m^2
Height 3 m
Perimeter = 60 m
Wall area = 180 m^2

Ratio = $\frac{\text{Wall}}{\text{Floor}}$ = 0.90

Figure 9.8: Wall-to-floor ratios

Cost-savings should be considered when designing the plan shape to maximise the potential within the site boundaries. Square-shaped plans are the most economical as they present the lowest floor-to-wall ratio when comparing buildings of the same height. Using one element for two buildings also cuts costs; savings can be obtained by joining two buildings together thus saving one external wall.

Storey height must be considered when constructing multi-storey buildings. By designing a much smaller storey height, it is possible to increase the gross floor area, which boosts the amount of rental revenue.

For example, look at the comparison of buildings C and D:

Building C:

Plan area = 25 m × 25 m

Height = 59.50 m

Storey height = 3.5 m

Number of floors = $\frac{60}{3.5}$ = 17 floors

Total floor area = 17 × 25 × 25 = 10,625 m^2

Building D:

Plan area = 25 m × 25 m

Height = 59.50 m

Storey height = 2.98 m

Number of floors = $\frac{60}{2.98}$ = 20 floors

Total floor area = 20 × 25 × 25 = 12,500 m^2

Assessment activity 9.5

BTEC

1. A client wishes to know what the cost of a scheme design would be. What methods are available to help determine a cost budget? P8
2. Establish a suitable commercial building locally to where you are now that can be used as a model, and obtain some rough dimensions of the building. Use the table in Figure 9.6 to produce some approximate quantities and a cost budget for the building. P9
3. Following the cost scheme for the client, you have been asked to produce an evaluation on the limitations of pre-production costing methods. D1

Grading tips

For P8 you will need to describe what methods are available that can be used to prepare a cost budget.

For P9 produce approximate quantities and an associated cost plan for a building using the rates from the table.

For D1 you will need to list the advantages and disadvantages of at least three pre-production costing methods.

PLTS

Organising time and resources as well as prioritising actions in completing an assignment will help develop your skills as a **self-manager**.

Functional skills

By reading and summarising information and ideas from a range of sources, you will develop your **English** skills in reading.

5 Understanding the process of tendering

Purpose, aims and objectives of tendering

For main or principal contractors, subcontract and supply packages

The main purpose for the main contractor and, ultimately, the subcontractors and suppliers is to win contract work on a competitive basis and to ensure that the profit margin placed within the tender is maintained or exceeded. If the reverse occurs, and a loss is made, then this can be made up during the trading period of the company with other projects, but large losses cannot be maintained indefinitely and could lead eventually to the closure of the company.

Thinking point

In order to survive, subcontractors and suppliers require construction contractors to operate healthy businesses.

Common methods of tendering

The method of tendering chosen to select a contractor for a project will depend on some of or all the following:

- the size of the project
- the geographical location of the project
- the financial stability of the construction company tendering for the work
- the competency of the contractors with regard to health and safety
- the physical resources of labour, plant and facilities of a company
- the reputation and references of a company.

Construction works vary in many aspects including terms of physical and financial size. Small construction works of less than £10,000 can be run using a local contractor; while larger multi-million-pound projects will require much larger contractors whose organisations may be further away.

The type of project itself also has an effect on the method of tendering. Large civil engineering projects may require a consortium to be put together, in order to pool resources from several contractors so that the work can be completed. In this type of project, a single contractor would be unlikely to be able to sustain the resources required.

Some high-value projects will require the financial accounts of a prospective tenderer to be checked over. This is done in order to ensure that they have the capacity to take on the associated cash flow requirement and financial stability for such a project.

Single stage selective

The single stage selective method involves pre-selecting a number of contractors, usually six, and asking them to submit tenders for the project. The contractors are selected for the list using several factors:

- references
- health and safety record
- experience
- value of the work
- company resources
- type of work.

Local authorities tend to use selective tender lists for different values and categories of work, for example highways and building services.

The benefits of using this method are as follows.

- Contracts are pre-checked and poor performing ones removed from the list.
- Lists can be rotated to provide a constant pool of different contractors.
- The tender procedure is competitive.

Two stage selective

Where time is of primary importance to the client, potential contractors may be invited to initial discussions in order to provide input to the project. The client can canvas ideas, and after these discussions a second stage commences. This stage may include the final selection of a contractor, who will submit a bill of quantities as part of a final tender process using previously agreed rates. This process is complicated to conduct but the client may benefit from the knowledge and experience and ideas gained from the successful contractor.

Open tendering

Open tendering is an open invitation for contractors to tender, normally in a newspaper advert placed by the client. The advert asks for expressions of interest to be placed, and from this a shortlist of prospective candidates is prepared. Such contracts tend to be for services, for example local council road cleaning or waste collection.

The disadvantage of open tendering is that it may attract contractors of which the client has no knowledge regarding quality, costs and reputation. Although, references will normally be requested and checked.

Serial tendering

Serial tendering is done to agree to a series of works set out in the initial offer. This type of tendering could be used for works of a similar nature, for example a series of schools, community projects or community police stations.

There are several benefits to this type of tendering method. The successful contractor will be able to use the experience gained from the first job on the next one. The client will also be ensured a long-term commitment for the series of works.

Target cost

This basis of tendering is very similar to a partnering agreement. The client sets out target costs for the final budge that must not be exceeded. Any savings made under this method are split 50/50 between the contractor and the client, but, any overspend is at the contractor's risk. The benefits of target cost tendering are efficiency savings that are shared, less risk for the client, a less confrontational relationship and a development of trust.

Measured term

Measured term tendering involves the use of a standard specification and bill of quantities or a national measured schedule of rates. The rates are already priced by the client, but the prospective tenderer has to apply a plus or minus to this national schedule, which is then converted into a total tender sum. The winning contractor's work is measured against the agreed rate and payment is made on this basis. This process involves a lot of administrative work in order to control the job, therefore measured term contracts are best used an smaller projects such as pre-paint repairs on housing.

Fee bidding

Fee bidding tenders are sometimes called cost reimbursement or cost plus contracts because they contain a fee element. The contractor agrees to undertake the basic cost of the work plus a fee payable on top of these costs. There are several variations to this method:

- cost plus a percentage as a flat rate which is charged on the construction work
- cost plus a fixed fee amount that is stated in the tender submission
- cost plus a flexible fee which varies under certain constraints.

This method of tendering involves a degree of trust and the awarded contractor has to prove the costs paid out on the contract in order to claim the fee.

TENDER ENQUIRY FORM

Client: __________
Address: __________

Tel: __________
Description of Project __________

Value __________
Timescale __________
Tender return date __________
Decision __________
Decline: __________ Accept: __________
Sign: __________ Dated: __________

Figure 9.9: Enquiry form

Documentation

The enquiry form

The estimating team, contracts manager and managing director use the enquiry form to find out details of the tender proposed by the client. From this, they can then make a decision whether to proceed with the tender or not. Normally, the initial enquiry from the client is a telephone call, asking if the contractor would like to tender for the work. It is good practice for a contractor to make a decision quickly so that the client has an option to add another contractor to the tender list.

The tender information form

The tender information form provides all the data that has been extracted from the tender documents. It is where all the information that is used during the tender adjudication process is placed.

Tender drawings

The set of drawings received with the tender should provide sufficient information for the estimator to price the unit rates within a bill of quantities; if the tender contains drawings and specification only it should provide information for the take-off of quantities.

It is advisable to stamp the drawings with the wording 'tender drawing' and a date. This provides a record reference set of the drawings that accompanied the tender on which the contractor's price was based. These can then be checked against the contract revised drawings during the final account stage of the project in order to highlight changes and any contract variations.

Schedules

Schedules are often prepared for certain material elements and may be issued with the tender documentation. A typical schedule would include:

- windows
- doors
- ironmongery
- joinery
- internal finishes
- concrete reinforcements.

Schedules are an easy way of counting up similar specified items; the one document is easier to look through rather than several drawings as the window schedule in Figure 9.10 illustrates.

Description	W1	W2	W3	W4	W5	W6
1200 x 1200 uPVC window in white with trickle vents to top, including uPVC window board and cill, Pilkington K glass double glazed units with 20 mm air gap.	✓					✓
2400 x 1200 uPVC window in white with trickle vents to top, with two side opening lights 600 mm wide a 1200 mm high, including uPVC window board and cill, Pilkington K glass double glazed units with 20 mm air gap.		✓		✓		
900 x 1200 uPVC window in white with trickle vents to top, including uPVC window board and cill, Pilkington K glass double glazed units with 20 mm air gap.			✓		✓	

Figure 9.10: Window schedule

Specifications

Drawings and specifications are a common form of tender documentation. Often, they do not contain quantities which the estimator will have to prepare in order to price the tender. Specifications often use all-encompassing descriptions against each work activity followed by the unit of 'item' against it. The risk element of taking off accurate quantities is the responsibility of the contractor's estimator and not the client's quantity surveyor, who would normally prepare a bill of quantities.

	WMCC			
Item			£	p
	SUBSTRUCTURE			
A	Excavate to remove hardstandings or turf, and to reduce levels average 250 mm deep, and cart away (approx. 23 m³)	Item		
B	Excavate to reduced level for foundation trench not exceeding 1.00 m deep, 600 mm wide, and cart away surplus spoil, level and ram bottom to receive concrete, part backfill trench with hardcore internally in 150 layers, and with spoil externally to level of existing ground (12 m³)	Item		
C	Provide and lay ground floor complete on prepared ground, comprising 150mm thick consolidated hardcore, well rolled, 25mm sand blinding, 1200 gauge visqueen damp proof membrane turned up wall at edges and lapped into dpc, 100mm thick A 252 mesh reinforced concrete slab, tamped finish Gen 1 concrete – 20 aggregate, 100 thick Kingspan insulation laid butt jointed, floor overlaid with laminate flooring (approx. 45 m²)	Item		
D	Provide and lay concrete Gen 1 foundations poured against face of excavation 250 mm deep (approx 8 m³)	Item		
E	Provide and lay 300 wide dense 7n/mm² concrete foundation blocks, in cement mortar (1:3) (approx. 25 m²)	Item		
F	Provide and lay three course class B engineering bricks 102mm wide, in sand lime cement mortar (1:1:6) (approx. 4m²)	Item		
G	Provide and fix hy-load dpc 100mm wide, including all necessary laps at joints, steps, etc. (approx. 15 m)	Item		
	To Collection	£		
	4.3 Section – Schedule of works			

Figure 9.11: A specification

Bills of quantities

Bills of quantities are included within the tender documents to cover items that have not been measured. They are similar to specifications, but follow the rules set out under the Standard Method of Measurement. The whole project is measured and unit quantities are placed within the bill of quantities. The exceptions are:

- contingencies which are the amounts of money for unforeseen items
- dayworks which are the amounts of money for time-related charges levied by the contractor for variations that cannot be measured
- provisional sums which are the amounts of money included for works not fully designed or specified.

A bill of quantities will have section totals which are carried to collections, which summarise each section. These are then totalled to form a final summary which adds up the tender price.

Subcontractor and supplier enquiries

An enquiry form records the information sent to a supplier and subcontractor which can be used as evidence in the event of a dispute over prices – see Figure 9.12. The enquiry form will be passed to the person who copies the drawings and pages from the specification which are sent to the subcontractor or supplier.

A covering letter on the company's letterhead should be attached to the enquiry packages sent out to the subcontractor and supplier. The letter must include a date by which the subcontractor/supplier will need to respond – this should allow the estimator sufficient time to finish compiling the tender for submission before the due date.

Tender supplier	**Drawing nos**	**Specification pages**	**BoQ pages**
Aggregates Ltd	2007/01/a	P23 item 1	
Tender subcontractor	**Drawing nos**	**Specification pages**	**BoQ pages**
Topliss Electrical	2007/21/a		P47 items 1–8
	2007/22/b		
	2007/22/c		

Figure 9.12: A subcontractor enquiry form

Activity schedules

Activity schedules are programmes that can be used to visually control the tendering and estimating procedure. They enable tenders to be coordinated on a master programme, so that the estimator can control

Project Horizon Laboratories	TENDER TIMETABLE	Simtop Construction

Description	Month	April															May														
	Date	11	12	13	14	15	18	19	20	21	22	25	26	27	28	29	2	3	4	5	6	9	10	11	12	13	16	17	18	19	20
	Latest date	M	T	W	T	F	M	T	W	T	F	M	T	W	T	F	M	T	W	T	F	M	T	W	T	F	M	T	W	T	F
Project appraisal																															
Check documents																															
Tender information sheet																															
Tender timetable																															
Document production (d & b and drg & spec)																															
(Drawing)																															
(Specification)																															
(Bills of quantities)																															
Code bill items and enter computer bill																															
Enquiries																															
Abstract, prepare, despatch - subs	18 Apr																														
Abstract, prepare, despatch - mate	20 Apr																														
Date for receipt of quotations - mate	4 May																														
Date for receipt of quotations - subs	9 May																														
Project appraisal																															
Site visit	19 Apr																														
Tender method statements																															
Tender programme	12 may																														
Pricing																															
Labour and plant																															
Materials																															
Sub-contractors																															
PC and provisional sums																															
Project overheads																															
Reports																															
Checking procedures and summaries																															
Tender																															
Review meeting(s)	17 may																														
Submission documents																															
Submission	19 May																														

Figure 9.13: An activity schedule

the workload. The activity schedule also acts as a programme and can assist the monitoring process by being able to establish if any item lags behind. Where more than one tender is being priced at a time, a master activity schedule may have to be produced to control the whole process within the estimating department.

Codes of procedure for tendering relevant to main and principal contractors, subcontract and supply packages

The CIOB code of estimating practice referred to earlier in the unit provides guidance on the procedures for dealing with estimates and enquiries to subcontractors and suppliers. You should refer to a copy of this document for more information.

The key to getting a response to enquiries is to send as much information as possible with the enquiry. This should contain the following as a minimum:

- tender drawings – relevant to the sections of work the subcontractor/supplier needs to price
- specification pages – the standards that will be required relevant to the subcontractor/supplier's work area
- bill of quantity pages – the pages with the items needing pricing
- preliminary items – any useful information contained with the client's enquiry
- conditions of contract – what type of contract will operate
- return date for enquiry – when the price is needed.
- insurance level required – the amount or indemnity required as a minimum
- any site restrictions – such as restricted access or hours on site
- a provisional programme – to inform the subcontractor/supplier of a start and finish date on site
- a schedule stating what has been sent to the contractor/supplier.

It is useful to maintain a database of subcontractors and suppliers to refer to on future tenders.

Factors affecting the level of tenders

Impact on value, price or level of a tender for main and principal contractors, subcontract and supply packages

Value has a great effect on the level of a tender. Small value tenders usually have a lump-sum overhead and profit element added to them since applying a percentage would mean a small return on investment. Multi-million-pound projects require a large financial commitment by the estimating department. For example, design and build tenders can cost thousands of pounds to complete with no obvious return if the company does not win the contract.

Generally, in a competitive tendering situation a company is competing against five other contractors, so the chances of winning the contract are 1:6. The factors of economic supply and demand also work for construction. A shortage of reliable contractors will have an effect on tender prices. There are several factors that can affect the level of tenders received by the client:

- the economic climate of the country – a buoyant economy also means higher tender prices and enquiries
- the base rate of interest set by the Bank of England – this affects the cost of borrowing required to finance initial stages of contracts and the client's budget
- the geographical location of the work – tender prices within the Greater London area tend to be much higher than the rest of the UK due to the higher cost of living in the capital
- the level and quality of the competition – large organisations may have regional offices to cover the whole of the UK and benefit from **economies of scale**
- specialism – for example, a major contractor specialising in stone work will have more demand for their services due to a shortage of contractors within this field.

Key term

Economies of scale – the ability to buy in bulk, thus receiving greater discounts.

Subcontractors

The tendering methods we have looked at tend to be the common methods adopted in order to secure a main contractor for a project; subcontractors and suppliers also need to be looked at.

Subcontractors can be appointed in two ways:

- as a domestic subcontractor to the main contractor
- as a **nominated subcontractor** engaged by the client in prior negotiations and who is then instructed to be engaged by the main contractor.

Key terms

Nominated subcontractor – a subcontractor who is appointed by the client and has already tendered for the work package for the client. The main contractor is then instructed to appoint this subcontractor and the value is offset against a provisional sum placed within the main tender documents, including profit and **attendances** by the main contractor.

Attendances – the items that will be required by the nominated subcontractor such as a crane, rubbish removal, water and electricity.

Subcontractors are mainly selected by the main contractor who will send out enquiries for the specialist works that they cannot undertake. Sometimes the contractor has to use subcontractors who are approved by the client and, in these cases, a list accompanies the tender.

Subcontractor prices are influenced by:

- how busy they are
- their historical relationship with the main contractor
- the level of main contractor's discount
- how long the main contractor takes to pay the subcontractor
- the location of the work
- how specialised the work is.

Suppliers

Suppliers normally only provide a material or a piece of equipment and do not undertake subcontract works involving installations. Therefore, they are asked to provide material quotations from schedules prepared by the estimator and these materials are purchased from the supplier using a purchase order. Suppliers' prices and rates are affected by the following factors:

- the payment terms negotiated
- nationally agreed rates by the contractors' buying departments
- the level of discount negotiated
- the location of the supplier
- how quickly the materials are required.

Suppliers can also be 'named' by the client; for example a work of art that is ordered and paid for by the client but installed by the main contractor. In this case, the price is already agreed, the main contractor is paid a percentage profit as stated in the tender document plus attendances such as the use of a crane to position the artwork.

Profit element factors

The estimator will prepare a net estimate which contains the basic prices of labour, plant, materials, subcontractors and preliminary items. The overheads and profit items are added to this net value.

These matters are normally discussed at a tender adjudication meeting where senior team members decide on what level of profit to place on the tender. The amount of profit applied to a contract will vary depending on the following factors:

- the level of competition where the proposed work is located
- the current and future workload or capacity of the company tendering
- the current resources of the company
- the level of risk associated with the proposed works
- the complexity of the project
- the personality of a potential client or their representative
- the payment history of a client
- the financial position of a client
- the size of the contract.

Potential variations

Poor-quality tender documents tend to lead to a higher level of contract variations. Similarly, a tender containing a high level of provisional sums has the potential for a high level of variations. However, profit can be generated from these variations as the contract proceeds. Therefore, a lower than expected level of initial profit can be placed with the submitted tender as this will be made up in the long run.

Quality of tender documents

The quality of tender documents is important. Poor-quality tender documents provide an opportunity for an astute contractor's quantity surveyor to exploit any loophole; thus forcing a contract variation from the client along with payment potential to increase the level of profit for the contractor. The types of mistakes that may be made are:

- poor-quality drawings
- incorrect setting out information
- missing bill of quantity items
- poor specification clauses.

Standard form of contract, amended standard form and bespoke contract forms

Standard contracts contain many standard clauses that have been included over several years due to legal litigation. The JCT (Joint Contracts Tribunal) series of contracts are considered fair and reasonable and have been devised by representatives from both the employers' and the main contractors' associations.

Standard forms of contract are produced by the JCT for many types of work. For example, the JCT 98 standard form shares the risk equally between the parties, whereas the JCT standard contractors design and build form sways more risk towards the main contractor. To allow for this, an amended premium may be included in the tender that is submitted.

Clients may choose to use bespoke contracts drawn up by their legal departments. These types of contract are worded in the client's favour. In order to counteract the risk that is placed on them and to allow for onerous contract conditions, the main contractor will normally include a premium in the tender.

Assessment activity 9.6

The client you are working for is concerned about what method to use to tender for the project they are developing. Explain the common methods of tendering for the client. P10

Describe the documentation required for the tendering process. P11

Explain the factors that can have an effect on the level of tender received for the client's project. P12

Justify the selection of these available methods. M3

Evaluate what any variations will have on the level of tenders. D2

Grading tips

For P10 you will need to explain the common methods of tendering for construction work.

For P11 describe the documentation that supports the tendering process.

For P12 explain the factors that can affect the level of tenders.

For M3 you will need to discuss the advantages and disadvantages of each common method of tendering.

For D2 you are required to describe what will happen to the level of tender if several changes are made to the project.

PLTS

Planning and carrying out research on the common methods of tendering will develop your skills as an **independent enquirer**.

Functional skills

Selecting and reading relevant sources of information on tendering methods will help develop your skills in reading **English**.

WorkSpace

Ben Jackson

Quantity surveyor

Ben is employed as a quantity surveyor for a local construction company. He is responsible for the tendering operations and the financial control of all the company's site contracts. He is involved in a project from the tendering of work and estimating through to the valuation and payments for work.

Ben has to organise the pricing of work which arrives as a tender from clients' quantity surveyors. He also has to organise the sending out of work to specialist subcontractors such as electrical and mechanical installation works. Ben enjoys his role and he visits all of the sites as well as working within the company's offices. Ben therefore has a good grasp on what is going on within the business.

Ben's role of the quantity surveyor also includes the use of measurement; for the valuation of work on site for monthly payments, for the measurement of variations and for the completion of the contract final account. Ben therefore has to be good at producing accurate calculations and dimensions and has to be able to understand and interpret drawn information.

Ben often has to price small contract variations by using basic rates produced by assembling labour, plant and material rates. Ben has to gather this information through site records from the site manager who records these against each variation instruction.

Ben has to produce financial reports for the company directors. These are used to ensure that building contracts are monitored to be profitable. Cash flow is also part of Ben's role; he has to ensure that the business monetary outgoings are matched by the incoming cash flow from clients paying for their work.

Think about it!

- What level of mathematics would you need to undertake this role?
- What communication skills would you need?
- Who else would you interact with in this role?

Just checking

1. State three reasons for undertaking measurement.
2. What is the sequence for dimensions in the dimension paper?
3. What is a bill of quantities?
4. What do SMM7 and CESMM3 stand for?
5. What would you find contained within the preliminary section of a bill of quantities?
6. What would be costed within a company overhead?
7. What does daywork mean?
8. What items of plant would you include within the preliminaries section?
9. Name five factors that affect the output rates of labour.
10. What affects the output of construction plant?
11. Name the elements that need to be included within a unit rate.
12. What documents would you send out in a tender enquiry?
13. What factors affect the return level of tenders?
14. What is the difference between net and gross?
15. What is a nominated supplier or contractor?

Assignment tips

- Use some of the quantity surveying textbooks from a library. They will help you with the correct way to lay out and take off quantities using dimension paper.
- The RICS website may contain some useful information on the common methods of tendering and procurement used in the UK.
- When calculating labour, material and plant rates, always remember to include delivery charges and allowances for wastage.

Credit value: 10

10 Surveying in construction and civil engineering

Before a new house can be built, a thorough measured survey of the existing site needs to be undertaken. This maps the existing natural or man-made features and is known as land or geometric surveying. This information can be presented in the form of a two-dimensional plan or three-dimensional CAD model, which is used to help a building designer produce designs for the new building and its situation on the site.

In both land surveying and setting out surveying accuracy is very important. Therefore, all surveyors need to adopt a methodical way of recording their measurement and carrying out suitable checks to make sure no mistakes have been made. Errors must be examined to check if they are within acceptable limits.

This unit introduces you to the practical aspects of going to a site and measuring distances, levels and angles of features, then returning to the office to plot these features on a scaled plan. By the end of the unit you should understand the process of setting out, taking an architect's scaled drawing for a new building and physically marking out the positions on the ground to enable the foundations and walls to be constructed.

Learning outcomes

After completing this unit you should:

1. be able to perform linear surveys to produce drawings
2. be able to perform levelling surveys to produce drawings
3. be able to measure angles and produce results from calculations
4. be able to perform the setting out of small buildings.

Assessment and grading criteria

This table shows you what you must do in order to achieve a pass, merit or distinction grade, and where you can find activities in this book to help you.

To achieve a **pass** grade the evidence must show that you are able to:	To achieve a **merit** grade the evidence must show that, in addition to the pass criteria, you are able to:	To achieve a **distinction** grade the evidence must show that, in addition to the pass and merit criteria, you are able to:
P1 identify linear surveying terminology **See Assessment activity 10.1, page 355**		
P2 carry out linear surveys, using appropriate equipment, to produce accurate drawings **See Assessment activity 10.1, page 355**		
P3 identify levelling surveying terminology **See Assessment activity 10.2, page 368**	**M1** carry out levelling calculations using both height of collimation and rise and fall methods **See Assessment activity 10.2, page 368**	**D1** analyse the methods used for levelling surveys in terms of accuracy **See Assessment activity 10.2, page 368**
P4 carry out levelling surveys, using appropriate equipment, to produce accurate drawings **See Assessment activity 10.2, page 368**		
P5 identify angular terminology **See Assessment activity 10.3, page 377**	**M2** use angular measurements and trigonometry to calculate heights and distances **See Assessment activity 10.3, page 377**	**D2** analyse the methods used to take angular measurements in terms of trigonometric accuracy **See Assessment activity 10.3, page 377**
P6 carry out angular measurements, using appropriate equipment, and calculations **See Assessment activity 10.3, page 377**		
P7 identify setting out terminology **See Assessment activity 10.4, page 385**	**M3** set out and check profiles for a small building **See Assessment activity 10.4, page 385**	**D3** explain the constraints on the positioning of profiles **See Assessment activity 10.4, page 385**
P8 set out and check corner pegs for a small building using appropriate equipment and techniques **See Assessment activity 10.4, page 385**		

How you will be assessed

The evidence requirements for pass, merit and distinction grades are shown in the grading criteria grid. Evidence for this unit may be gathered from a variety of sources, including well-planned investigative assignments, practical work or reports of practical assignments. A variety of activities is included in this unit to help you understand land and setting out surveying. These will comprise both the practical use of various surveying equipment together with linked calculations and drawings or sketches. You will need access to a range of basic surveying equipment as well as appropriate personal protective equipment (PPE); you should seek guidance and support from your tutor in obtaining the appropriate kit.

This unit will be assessed by the use of four assignments:

- Assignment one will cover P1 and P2
- Assignment two will cover P3, P4, M1 and D1
- Assignment three will cover P5, P6, M2 and D2
- Assignment four will cover P7, P8, M3 and D3.

John

I really enjoy the surveying parts of my course because it combines practical tasks and using proper civil engineer surveying instruments with maths, which I am quite good at. I've learned that when you measure things there should always be a way of checking what you've done is correct by applying special calculations.

This unit has taught me that surveying is all about finding the physical position of things, whether in height or in plan coordinates. It has also helped me develop these types of surveying skills which will be very useful when I become a civil engineer!

Surveyors have a lot of responsibility in their job. They need to be numerate, organised and methodical, and be able to produce and understand technical drawings. If they ever get their calculations wrong and set out a structure or building incorrectly, then the whole project will suffer.

Over to you

- Would you be able to check whether two walls were correctly set out at right angles to each other?
- Do you know how to find out the height of an inaccessible point?
- What are you looking forward to learning about in this unit?

1 Be able to perform linear surveys to produce drawings

Build up

Civil engineering projects

Think about some of the major civil engineering projects that are currently being planned or built like the Aquatics Centre for the London Olympics or the Crossrail infrastructure project that will speed rail traffic through the centre of London and beyond. Both these projects required pin-point accuracy in the measurement of the existing terrain/features and the setting out of new works.

In groups, discuss how the surveyors would set out the complicated roof of the Aquatics Centre or the correct alignment for the Crossrail tunnels. You may want to do a little Internet research to find out a bit of information about these projects first.

Methods of measurement

An early method of measuring distances was developed by Edmund Gunter in the seventeenth century; it involved the use of a 22 yards standard-length metal chain with links of a known length. In order to measure **horizontal distances**, these chains would be pulled taut, to reduce sagging and slack; in instances of measuring up a slope, the surveyor might have to 'break' the measurement by raising the rear part of the tape upward, and plumb from where the last measurement ended.

Even in today's fast moving construction industry, with the use of complex electronic measuring devices, these basic principles and methods are still valid. In fact, the principles that are used in modern measuring instruments have not changed since Gunter's time.

Key term

Horizontal distances – the flat distance that would be plotted on a two-dimensional drawing plan.

Linear measurements

In order to understand how linear measurements can be accurately taken, let's have a look at the two basic methods of linear measurement: offset measurement and trilateration.

The best way to understand how each of these methods operates is to consider an example, see Figure 10.1 (page 347). We can fix the position of a brick wall XY in relation to an aerial mast at Z by measuring distances only, as shown.

Measurement by offset relies on being able to locate the point P which is at right angles to the aerial mast. Once this point is found, then the linear distances along the wall can be recorded together with the offset distance PZ. To find point P, place the measuring tape's 'zero' on Z. Hold the tape taut, then swing it in an arc; where it just touches the wall is the shortest distance, and will be the point P. PZ is also known as the perpendicular offset.

The three distances 5100 mm, 1900 mm and 6300 mm can be plotted to a suitable scale to represent the wall and mast positions, with the 6300 mm dimension drawn at 90° to the wall. This is a traditional method for measuring details in linear surveys.

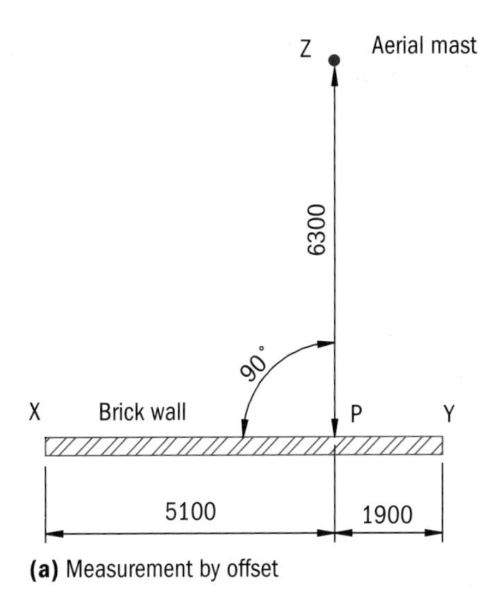

(a) Measurement by offset

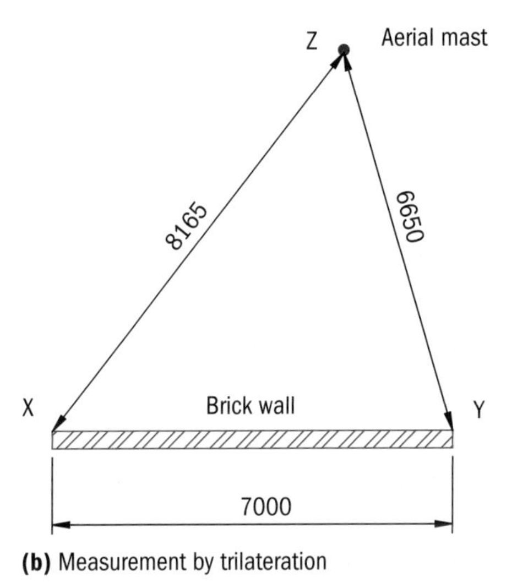

(b) Measurement by trilateration

Figure 10.1: Measurements by offset and trilateration

Measurement by trilateration relies on measuring the three sides of a triangle. It is important to note that for accuracy this method should only be used where **well-conditioned triangles** are present. The length of the wall of 7000 mm can be plotted to a suitable scale with two arcs constructed of 8165 mm and 6650 mm radii centred on point X and Y respectively. Where these two arcs cross will be the correct position for the aerial mast at Z. This method of using two **tie lines** is often used in traditional linear surveys where offset would be too long, generally over 10 metres.

Key terms

Well-conditioned triangles – triangles which are roughly equilateral in shape with no small internal angles.

Tie lines – a pair of horizontal plan measurements used to fix the position of a point.

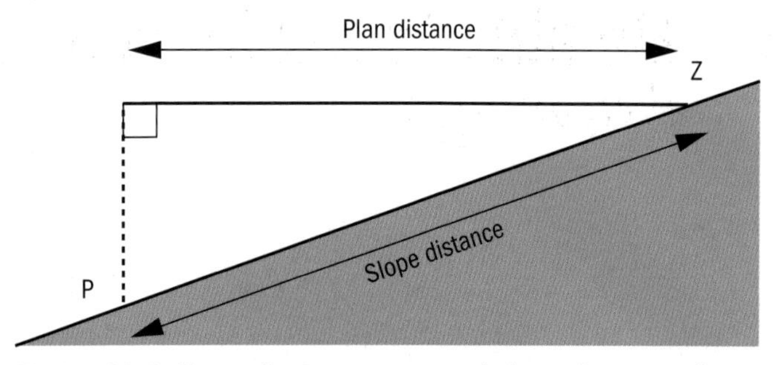

Figure 10.2: Slope distances measured along the ground are longer than the corresponding plan distance!

Remember!

All the dimensions measured must be horizontal plan measurements and not distances measured over sloping ground.

Carrying out a linear survey

The equipment required to carry out a traditional linear survey is shown in Figure 10.3.

Synthetic tape

Synthetic tapes are made from fibreglass or nylon and are coated with PVC. They are a cheaper option than steel tape, but they are only graduated down to the nearest 5 mm and can become permanently stretched if pulled too hard. Synthetic tape has a metal prong on the end to hook it into position, but care should be taken to avoid bending this flat. The tape should be wiped before being re-wound, and care should be taken that it does not become twisted and get jammed into the case.

Steel tape

Steel tapes are durable and more stable than synthetic tapes. They can be enamelled or plastic coated for added protection and are available in 30-metre or 50-metre lengths. Steel tapes are graduated at every millimetre and are calibrated to read true lengths at a standard temperature and pull; both of which are printed on the tape at the start, and are typically 20°C and 50 Newtons force respectively. Steel tapes are used throughout industry for a variety of land surveying and setting-out tasks. As with the synthetic tapes, steel tapes have a metal prong on the end to hook the tape into position. Ideally, the tape should be dried and wiped with an oily rag before being re-wound into its case.

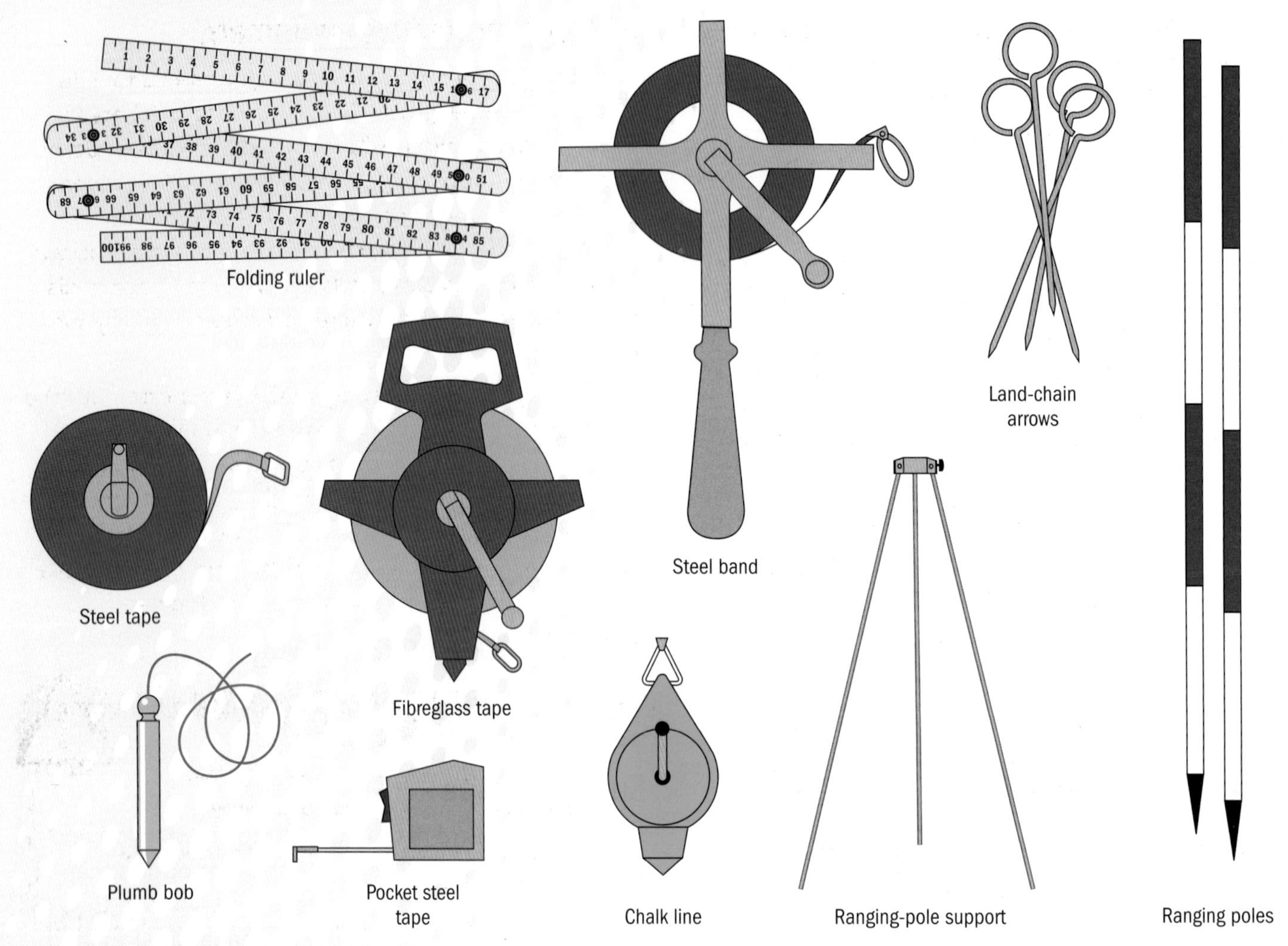

Figure 10.3: Equipment required for a linear survey

Steel band

The steel or surveyor's band is still used today by some surveyors. It functions like a traditional metric chain and is similar to the steel tape but is carried on a four-arm, open-frame winder and has oval handles connected directly to the tape.

25-mm folding wooden rules

There are two main types of 25-mm folding wooden rules: the one-metre long boxwood folding rule, with its characteristic swivel hinge, and the longer two-metre multi-lath rule. Both are graduated to the nearest 1 mm and are used mainly for internal building surveys, but can also be used for short offsets. Care must be taken when opening or closing the hinges to ensure that they are not bent into the wrong position.

Ranging rods or poles

Ranging rods are circular timber poles, usually 2.5 m long, coated with distinctive red and white bands 500 mm in length. They are only used to mark survey stations and for ranging lines. The points are encased in steel shoes which allow them to be pushed into soil. The ranging rods can also be used with lightweight stands for surfaces such as concrete or tarmac.

Arrows

Arrows are 400 mm long steel pins with a point at one end and a ring at the other to aid in carrying. They can be used for marking out points on the ground or for marking survey station positions prior to fixing with a stout wooden peg. A strip of red cloth or plastic tape is attached to the ring so that the arrow can be easily observed. A variation on the normal arrow is the dropping arrow which has a weighted point in the form of a plumb bob; this variation is used in step measurements of slopes.

Chalk-line marker

The chalk-line marker is a useful device which can quickly and easily imprint a straight line on any hard surface. A tight string is impregnated with chalk which is held within the case of the wound string line. This line is pulled out between the two points, e.g. setting out a tile gauge line on a roof, and, as the string line is pinged against the hard surface it leaves a thin chalk line.

Abney level

The Abney level is of the most convenient instruments for quickly finding the angle of elevation or depression of a slope, and hence the horizontal plan distance. It is made of a sighting tube, a spirit level and a scale graduated in degrees With the use of a prism, the instrument can be levelled and the angle of the line of sight recorded.

An Abney level

Electro-distance measuring device

Electro-distance measuring (EDM) devices are a very popular method of measuring and recording distances. The hand-held instruments are quick and accurate and have a range approaching 100 m with an accuracy of + or – 2 mm. Some, like the Leica Disto™, have a continuous reading which allows maximum distances for measuring room diagonals, and minimum distances for finding right-angle distances or offsets distances.

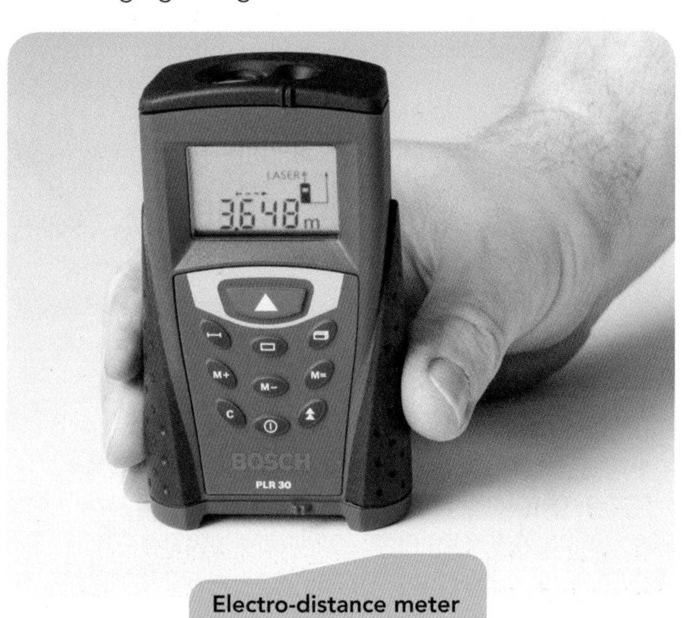

Electro-distance meter

Personal protective equipment

The current minimum statutory requirements for site workers' personal protective equipment (PPE) are to wear a hard hat, high visibility jackets and safety boots or shoes; this applies to site engineers and surveyors. Even when undertaking land surveys prior to any construction work being carried out, you should always wear PPE; it is advisable to keep an appointments diary so office staff can track where you are and know your return time.

When carrying out surveying work it is important for the surveyor to carry out risk assessments on the possible hazards that may be present at the site location. When surveying existing buildings you should always carry out a thorough risk assessment for possible hazards. Typical hazards are the presence of asbestos, unstable floors, trip hazards, un-tethered guard dogs and pigeon droppings!

Safety tip

- Always wear suitable personal protective equipment when carrying out a survey.
- Keep an appointments diary so office staff know where you are.

Procedure for carrying out a linear survey

Reconnaissance

The first stage of any survey involving linear, level or angular measurements is to carry out a full reconnaissance of the area to be surveyed. The basic process involves creating a **survey framework** that covers the whole area, which in itself can then be broken down into smaller parts to record all the necessary details and features of the area.

Key term

Survey framework – a triangular network of measured horizontal lines that fix the end points (survey stations) of those lines.

This process is commonly known as **working from the whole to the part**, taking an overview of the whole extent of the survey area and thinking carefully how it could be split up into manageable parts.

The surveyor should walk over the site to select the most suitable positions for key reference points, which are known as survey stations. These stations should be clearly marked and referenced so that their positions can be found at any time in the future. This process of recording the positions of the stations is called 'witnessing'; the points should be sketched with clear dimensions referenced to at least three existing fixed features such as the corner of an inspection chamber or a steel fence point.

The reconnaissance must also take into account the purpose of the final survey drawing and the accuracy of the final survey plot. In manual plotting the thinnest line width that can be plotted by a sharp pencil or ink pen is approximately 0.25 mm. If the final plot of the survey is to be 1:200, then the smallest dimension that can be measured during the survey and accurately plotted will be 0.25 mm × 200 = 50 mm.

Theory into practice

What is the minimum dimension that can be measured during the survey if the final accuracy requires a plotted scale of 1:50?

The best survey framework for accuracy comprises a network of strong well-conditioned triangles. Also, to ensure high accuracy, there should be one main **baseline** from which all the other triangles are linked. There should also be secondary **check lines** which can be used as further checks on the survey framework.

Key terms

Working from the whole to the part – taking an overview of the survey area and working out how it could be split up into manageable parts.

Baseline – the main survey line or 'back-bone' on which to form all the other survey triangles.

Check lines – an additional measured line to check the accuracy of the survey framework.

Chainage – the cumulative distances measured to specific features as fixed from one end of the survey line.

When deciding the positions of the stations and survey lines that form the surveying framework there should be as few survey lines as possible, but sufficient to form a triangular framework over the site. The following should also be considered:

- the location of the baseline from which to form all the other survey triangles
- the triangles should not have small internal angles
- check lines should be used
- survey lines should pass close to the boundary and any features of the site
- survey lines should be positioned over the more level ground
- stations in any triangle should be intervisible
- obstacles to ranging and chaining should be avoided, such as trees and ponds.

The measured data should be recorded in a special field book, traditionally known as the chain book, which has two thin tramlines drawn down the centre of the page. These represent the tape line laid out between the survey stations and either side of these double lines you should draw the features that are to be measured with offsets or tie lines.

You, as the surveyor, should walk down each survey line and note exactly what physical features are on either side of the line and sketch them. Do not start measuring anything until after you have drawn all the features to be measured first. In this way, you are concentrating first on what and how you are going to measure and then, and only then, you can go on to devote all your energies to measuring and correctly recording your measurements; you must develop this methodical and clear system otherwise you will miss detail or make errors when you do measure.

Remember!

Don't start measuring anything until after you have drawn all the features that you are going to measure.

Measuring and booking the details

After you have drawn all the features that you plan to measure on your booking sheet, you can record the running measurements or **chainage** along the tape from one station to the point on the line where the offset distances (at right angles) occur to selected features;

for example, location of boundary of the site, paths and trees. The offset distances can be measured with a synthetic tape and the way to find the right-angle point is straightforward (see Figure 10.1 on page 347).

The main points to be considered when recording information in the booking sheets are as follows.

- Offset measurements should be as short as possible; they should not normally exceed 10 m. You can estimate a right angle by eye if the distance is less than about 3 m or by swinging the tape in an arc and taking the shortest measurement. The offset dimension lines should not be drawn in the book.
- Where offset distances are greater than 10 m, it is best to use a pair of tie lines as this is more accurate. Both tie line measurements should be shown in the booking sheet. (See the case study below.)
- All measurements should be recorded to two decimal places and an oblique stroke used, not a decimal point, for example 4/25 not 4.25. This avoids blobs of dirt from obscuring the decimal point and making the recorded figures unintelligible.
- Information need not be drawn to scale in the booking sheet, but should be clearly set out so that it can be understood when the survey is plotted in the office. It is best to write the offset distance close to the feature drawn in the booking sheet.
- Several pages of booking may be needed for long or complicated tape lines, and explanatory notes should be added where necessary. It may be very difficult and time-consuming to return to the site to re-do a measurement or check up on information, especially if the site is a long way from the office.
- Straight line measurement may be continued beyond a station to the site boundary, if necessary, and the information recorded. This is so that detail in the corners of the site can be recorded.
- In the booking sheet, circles should be drawn around the stations so they can be seen clearly, and the overall distance of the survey line included within the circle.
- The circumference or girth of a tree should be measured 1 m above the ground and recorded in brackets by the tree. The spread of the tree should also be recorded.

Case study: Eastwood Road

An old factory and storage depot is being surveyed for redevelopment. A reconnaissance has been carried out and the sketch in Figure 10.4 has been produced showing the selected survey stations and survey lines. The longest line AF has been set up as the baseline for the survey and its bearing to magnetic north has also been noted with a magnetic compass. Survey line BC is acting as a check line but will also be used to measure survey details. To ensure that the survey stations can be found at any time, they are pegged out with a stout timber peg if on soil or fixed by a shot-fired nail into asphalt or concrete and marked with surveyor's spray paint.

The booking sheet for GC in Figure 10.5 shows clearly the layout and notation to be used for showing offsets, tie lines and chainage to a standard format.

Baseline AF is approaching 160 m long, which is far longer than any manufactured steel tape. Therefore, a straight line between A and F can be created using a technique called **ranging a line**, in which a ranging pole is set up in a

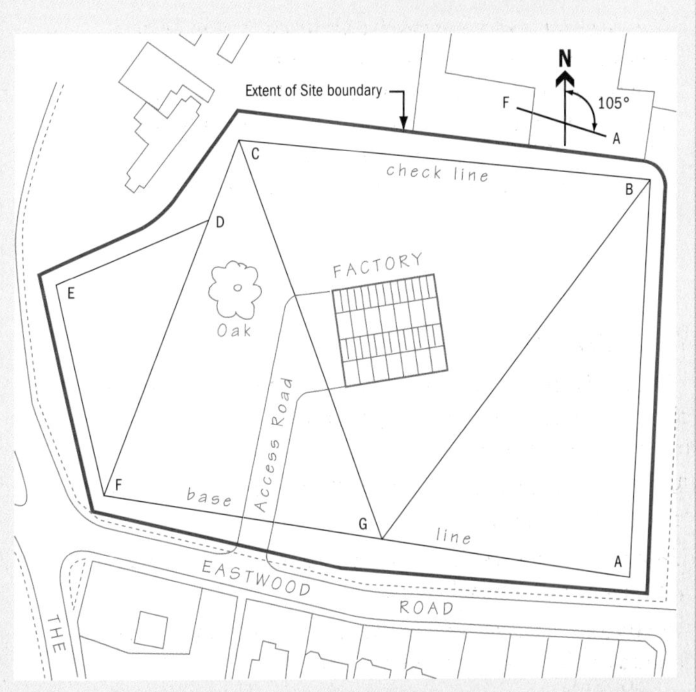

Figure 10.4: Sketch of Eastwood Road

(cont.)

vertical position at stations A and F. One person stands behind the range pole at A; another person holds a range pole in a vertical position at the other end of the tape and roughly in line with stations A and F. The first person sights through A to F and directs the other to move the range pole to the left or right, until it is in line with stations A and F. This new point is marked with a surveying arrow or marked with spray marker. Further ranging and taping may be necessary to position the chain exactly in the direction of the stations A and F; then the end of the tape is marked with an arrow. The length along the tape is measured cumulatively from A right to the end at F.

As you progress along the line, you record the offset distances to the features to be surveyed as before. You must ensure that all the arrows are collected up until you finally reach station F.

- What other survey lines for this redevelopment do you think would need to have a line ranged in?
- The majority of the site is a grassed area. How would the survey stations be physically marked?

Key term

Ranging a line – the process of measuring a straight line between two points which are a long way apart.

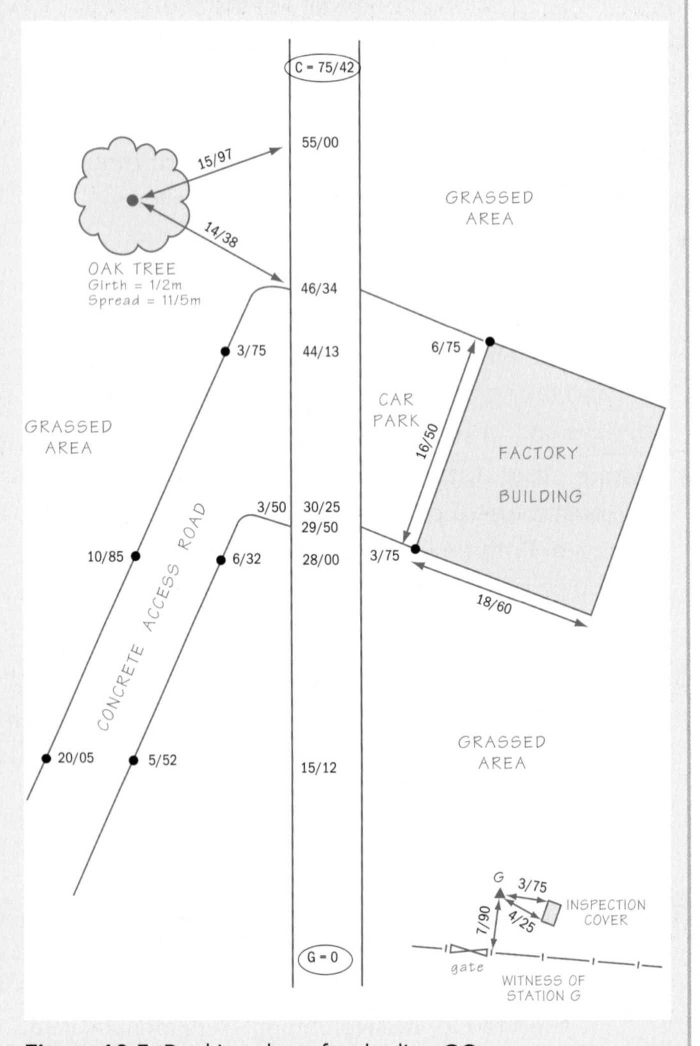

Figure 10.5: Booking sheet for the line GC

Errors in measurement and booking

Measurement of distances will always be prone to inaccuracies due to the continuous nature of the data. As a surveyor, you must be aware of where errors can occur and more importantly how they can be minimised. There are three basic classifications of errors:

- gross or human errors
- systematic errors
- random errors.

Gross or human errors are mistakes made in reading or booking measurements, such as miscounting or wrongly reading the tape, or writing down the wrong measurements in the booking sheet. These errors can be very large and can occur at any stage in the survey. If not picked up early on, the survey may need to be abandoned and redone. The best way to avoid making mistakes is to develop a clear method of reading and booking the results. For example, read the measurement on the tape, write the measurement in the booking sheet, then re-read the tape to check it is the same as what you have just written.

Systematic errors are errors that occur in the **calibration** of the equipment, such as a plastic tape which has become permanently stretched. This inbuilt error will have a gradual cumulative effect on the survey. In the case of the stretched plastic tape, the only way to check this type of error is to compare the tape against 'standard' steel tapes kept for that specific purpose. If the error is found after the fieldwork has been completed, the distance can be corrected by applying a suitable correction to the measured dimensions, so no information is lost.

Key term

Calibration – the process of ensuring that the measuring equipment is correctly adjusted to give true readings.

Random errors are generally small errors that are due to the limitations of the surveyor or the equipment. For example, a person might have a slight visual impediment due to their short-sightedness which affects how they read the tape's graduations. Although these errors tend to be very small and can cancel each other out, they can be avoided by taking suitable check measurements and repeating readings.

Take it further!

You can find full details of how to check the calibration of tapes in BS 7334: Part 2:1990 'Measuring instruments for building construction'.

Accuracy in measurement and booking

In normal practice, linear surveys should try to achieve an accuracy of 1 in 5000, or 6 mm over a distance of 30 m. To do this, you must consider the following as they might affect finding the true horizontal plan distances:

- slope
- sag
- temperature variation
- tension.

Slope – When the measured distance is recorded along the sloping ground and there is a difference of height between the ends of a 30-metre tape of less than 600 mm, then the accuracy above is achievable. Where the drop is greater than 600 mm, use other methods to find the true plan distance such as step measurements or measuring the slope angle (see Figure 10.6 below).

Sag – This occurs when the measured distance is in an arc due to the weight of the tape. To obtain an accuracy of 1 in 5000, the centre of a 30-metre steel tape should not sag by more than 300 mm from the horizontal.

Temperature variation – This occurs when the temperature of the steel tape is warmer than the standard temperature of 20°C, which causes it to expand; in cooler temperatures, the tape may contract. If the temperature does not vary from the standard temperature by a maximum of 18°C, then an accuracy of 1 in 5000 for a 30-metre tape is possible.

Tension – This occurs when a steel tape is pulled to reduce the sag, but as a result is stretched because the steel is 'elastic'. These errors are generally very small and can be ignored.

Step measurements

In this method, while the chain is held in a horizontal position, its end is stepped down to the ground, using a drop arrow. The horizontal distance measured depends on the steepness of the slope of the ground, but generally should not exceed 10 m. The vertical distance should not exceed 1.5 m, for example eyesight level. It is easier to work downhill rather than uphill with this method.

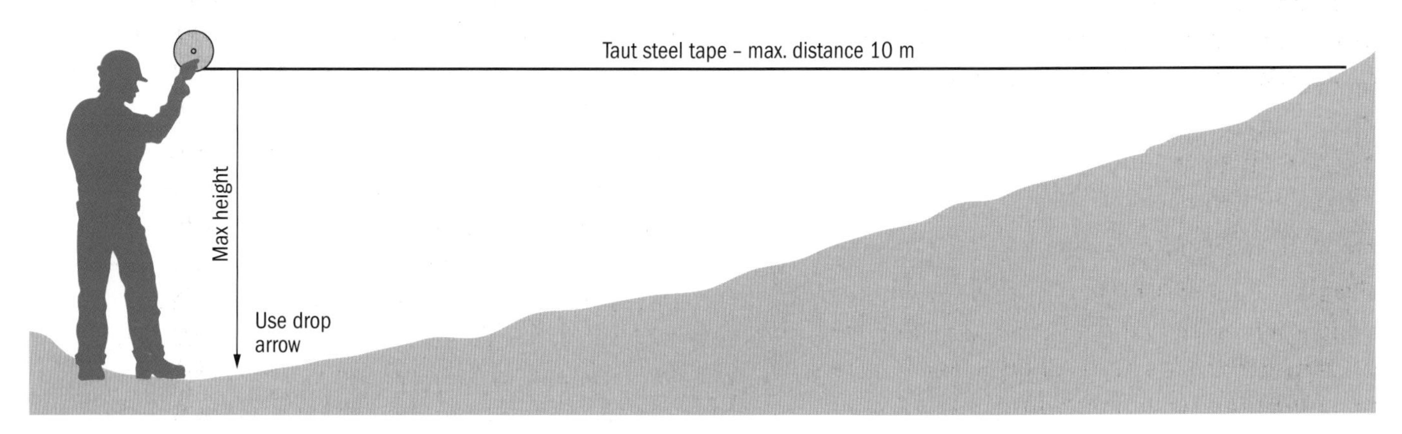

Figure 10.6: Step measurement

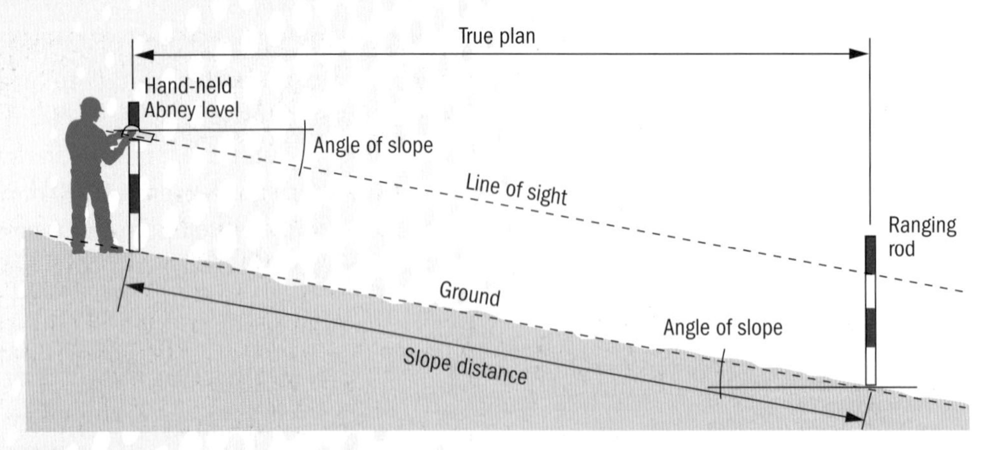

Figure 10.7: Slope measurement

Slope measurement

In this method, the slope distance is measured with the tape, while the angle of slope is measured with an Abney level (see Figure 10.7). The angle is measured by setting up a range pole at each end of the line. The surveyor stands by one range pole and holds the Abney level against it. The surveyor then sights a point at the same height on the other range pole. The angle measured is the angle of slope of the ground. Then the:

True plan distance = Slope distance × Cosine θ
(where θ is the measured angle of the slope)

Plotting the survey

It is very important that your final survey drawings are accurate and neat. A great deal of effort and time will have gone into producing a comprehensive set of field notes and measurement data; this will be wasted if a survey drawing is badly presented. Your client will certainly not want to pay for a poorly presented and inaccurate drawing!

The final survey plot should progress from the 'whole to the part', just as the survey progressed. The lengths of all the survey lines are taken from the booking sheets and the longest line, known as the baseline, is plotted first to the desired scale. By striking arcs with a large radius compass, the other survey stations are established and the network of triangles drawn. Check lines are scaled off and compared with actual distances.

Once the framework has been plotted, the details of all the features of the site can be plotted. Offsets and ties are systematically plotted in the same order in which they were booked; work from the beginning to the end of each line. The right angles for offsets may be set out by set square.

The ranges of available scale are given in Table 10.1.

Table 10.1: Preferred scales

Small-scale maps	Large-scale maps	Site plans	Detail plans
1:1 000 000	1:10 000	1:500	1:20
to	to	to	to
1:20 000	1:1000	1:50	1:1 (full size)

When plotting the survey manually, it is good practice to draft out the survey lines on tracing paper. By overlaying the tracing paper on the paper to be used, the survey may be properly centred on that sheet. The 'North' point should always be shown and preferably pointing towards the top of the sheet.

After the points are plotted, the detail is drawn in, using standardised symbols, some of which are shown in Figure 10.8. The drawing is then inked in, the North point drawn and any necessary lettering, including a suitable title block, scales and annotation, carried out. The conventions for showing landscape features are given in BS EN ISO 11091:1999.

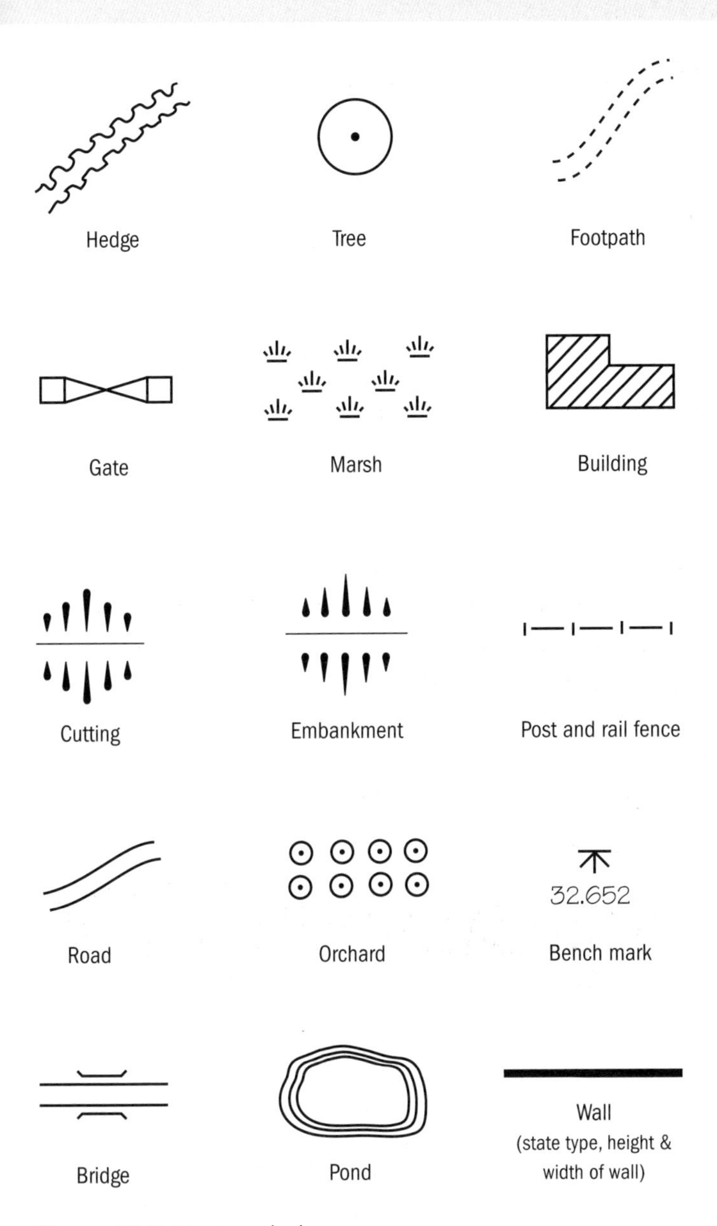

Figure 10.8: Map symbols

Linear surveying calculations

Data and measurements that have been recorded through the linear surveying process can be used to calculate a range of properties including areas and length of perimeters. Details and examples of the various calculation methods can be found in Unit 3: Mathematics in construction and the built environment.

Remember!

- Always consider working out areas or perimeters from lengths that have been measured or calculated horizontally.
- Always use the correct units when working out areas or perimeter lengths.

Assessment activity 10.1

P1 P2

1. Choose a small area of land close to your place of study that you could survey. Select the linear surveying equipment and techniques you would need to perform a linear survey of this site. Working in groups of three, undertake a reconnaissance of the site making appropriate notes and sketches. P1
2. By practical application of tapes or bands, carry out the linear survey of the site and book the survey data using a recognised procedure. Using the survey information draw a sketch to show all the relevant data for calculating the area of the site. Finally from your field notes produce a drawing to a recognised scale showing the main site features. P2

Grading tips

For P1, when planning your practical surveying work think about what activities you are to carry out and the accuracy of the measurements that you are going to take. Write a list of the equipment that you need and match that with the job that you need to carry out. When selecting tapes, for example, think whether you need to read to the nearest 1 mm or will 10 mm be adequate; steel tapes are graduated in mm while fibre tables are in cm. When carrying out linear survey's always show that you are working from the 'whole to the part'.

For P2, in carrying out your linear survey first produce a clear proportional sketch of the survey area and show on it the framework of main survey lines. Then show each line on a separate sketch with the sketched details of everything that you want to survey. Undertake all the measurements and write the values on your sketch. Neatness and a clear layout for your drawings is always necessary. Always use the measured survey dimensions – never attempt to scale off your survey sketches!

PLTS

Undertaking the fieldwork and working collaboratively towards common goals to achieve an accurate final survey will help develop your skills as a **team worker**.

2 Be able to perform levelling surveys to produce drawings

Key terms used in levelling

Levelling is a process that compares heights of points on the Earth's surface. In carrying out a level survey, reference has to be made to a fixed point, or datum, of known height. On large civil engineering projects, such as roads, railways and bridges, levels are linked to the Ordnance Survey Bench Marks (OSBM) system, which is referenced to the height of mean sea level at Newlyn in Cornwall. These heights appear on a range of maps and small-scale plans published by the Ordnance Survey. On smaller local construction projects, it is usually sufficient to relate all heights to an arbitrary fixed point established on site called a Temporary Bench Mark (TBM).

The act of levelling and height control has the following purposes:

- to measure vertical heights of points or stations located on the ground
- to set out profiles on site to give the heights of new construction works, such as depths of foundation relative to the ground slab level or damp-proof course height
- to produce a grid of spot heights to give an indication of the ground surface; this information is often shown on maps and plans as a series of contour lines that join points of equal heights
- to produce a cross-sectional drawing, often called a 'long' section, which helps plan the construction of underground services, such as setting out sewer gradients.

Levelling equipment used in land surveying

Levelling instruments are very versatile and can be used for both setting out points and for recording heights of existing features. They are accurate and, if properly maintained, should provide a long and useful service. All levels comprise two basic parts:

- a high-resolution telescope with magnification of around ×20–×24 which rotates around a vertical axis
- a spirit bubble tube which is set parallel and horizontal to the telescope.

If the level is adjusted correctly, the line of sight through the telescope – the line of collimation – will be parallel to the horizontal level line given by the spirit bubble tube, with both these lines at 90 degrees to the vertical axis of the level. A simple test known as the 'two-peg test' should be undertaken to check that both line of collimation and level line are parallel (see below). The basic set up of these main elements and linked terminology is shown in Figure 10.9.

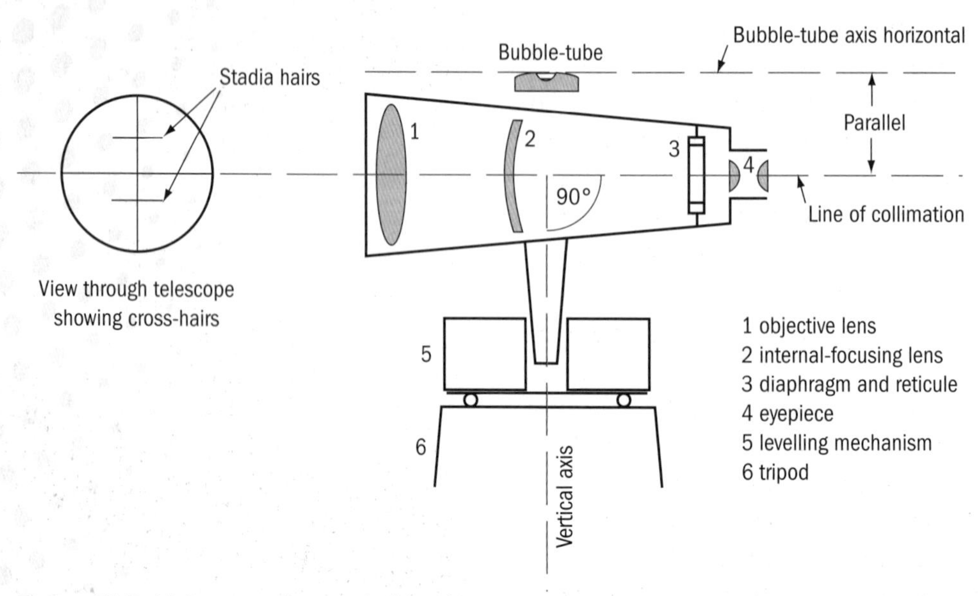

Figure 10.9: Main parts of an optical level

Setting up the level

The following steps should be taken when setting up the level.

- Whenever possible, place the instrument where all stations to be sighted can be viewed.
- Extend the legs on the tripod so that it is at shoulder height, and fix the level to the tripod using the centre screw. Make sure the legs are spread wide to form a stable support to the level.
- Level the instrument according to the type of level and its controls (see below).
- Once the instrument is level, locate the levelling staff on top of the datum, either the TBM or, if available, the OSBM. You may have to adjust the main telescope focus knob on the side of the level to see the staff.
- Once you have located the staff, adjust the eyepiece focus ring and make sure that the cross- and stadia hairs are sharp. This corrects for the phenomenon known as parallax where the cross-hairs move relative to the image of the staff.
- The level is now ready to use. Another person will be needed to hold the staff as upright as possible to ensure an accurate reading can be taken.

Reading the surveying staff

Special staffs known as the 'E' staff are used with the level. These are lightweight, aluminium, telescopic staffs available in 4-metre and 5-metre lengths and graduated in distinctive 'E' shaped black, red and white blocks. Each horizontal bar is equal to 10 mm, making the height of the 'E' as 50mm. Staffs are generally used to measure the distance up to the level line of sight. However, they can just as easily be turned upside down to record the heights of bridge soffits or gable walls by measuring the distance down to the level line of sight. These readings are known as inverted staff readings and are booked as negative values in a level booking form.

To help the person holding the staff to keep it vertical, there is usually a circular spirit bubble on the back face of the staff. However, with older staffs, this can often get damaged or knocked, so may not be all that reliable. An alternative is to gently sway the staff backwards and forwards in line with the instrument, and the surveyor observing the staff reads the minimum distance observed on the staff. With good eyesight, the surveyor should be able to estimate the staff reading to the nearest 1 mm, which is one tenth of one of the 10 mm coloured blocks.

Metric 'E' staff

Automatic level with dumpy levelling mechanism

The dumpy level

The dumpy level is the traditional optical level that is levelled by three foot-screws and a single plate, or tube, spirit bubble. It is a common device used extensively in theodolites and total stations (see page 370). The dumpy level can give an accurate reading up to a tolerance of 3 mm over a distance of 150 m.

The following steps should be taken when setting up the dumpy level.

- Ensure the instrument is stable, the legs are spread and it is at the right height for viewing.
- Ensure the top of the tripod is reasonably level to receive the instrument; the screw adjustment has only limited mobility.
- Ensure the screws of the instrument are approximately in the middle of the thread.
- Position the barrel of the telescope, and therefore the levelling bubble, parallel to any two screws – 2 and 3 in Figure 10.12 (a).
- Centre the bubble by adjusting screws 2 and 3 at the same time, either both away from or both towards each other at the same rate. The bubble will move in the direction your *left thumb* is moving.
- Now turn the barrel through 90 degrees so that it lies over the third screw 1, as shown in Figure 10.12 (b).
- Adjust that screw *only* to centre the bubble.
- Repeat the process on the two original screws to ensure fine accuracy.
- Finally, rotate the level so it is 180 degrees from its starting position as shown in Figure 10.12 (c).
- If the bubble remains in the centre, then the instrument will be level in all directions; if not, then the bubble tube will need to be reset according to the manufacturer's instructions.

The tilting level

Tilting levels are most often used for high accuracy work but can be used in all types of level surveys. Tilting levels can also be used to assist in setting out slopes and sewer gradients with the aid of the tilting screw. The main difference between tilting levels and the dumpy level is the levelling mechanism. The rough levelling of the tilting level is sometimes done by a circular spirit bubble. Every time the instrument is pointed in a new direction; it is accurately levelled by centring a plate bubble with a fine vertical adjusting screw. This does not affect the line of collimation.

The automatic level

The automatic level has a mechanism which allows the instrument to be levelled automatically, provided that it is roughly levelled by using a small circular spirit bubble located on the body of the instrument. The method by which this bubble is levelled varies from instrument to instrument. Some favour a central knuckle joint while others have a pair of sliding circular wedges; some also use three foot-screws such as used in a dumpy level, as shown in Figure 10.12.

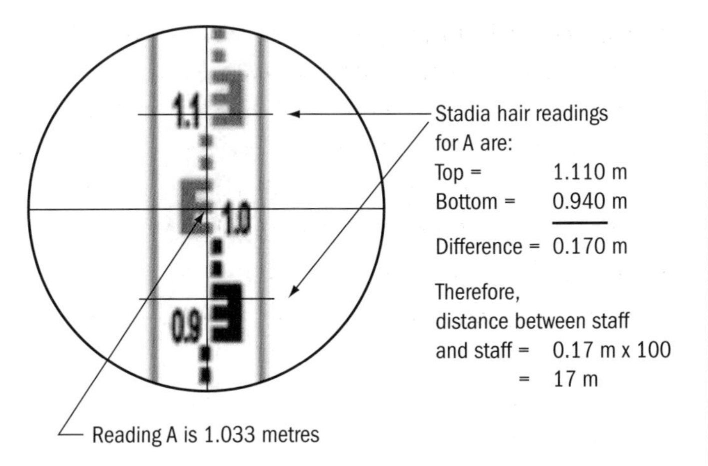

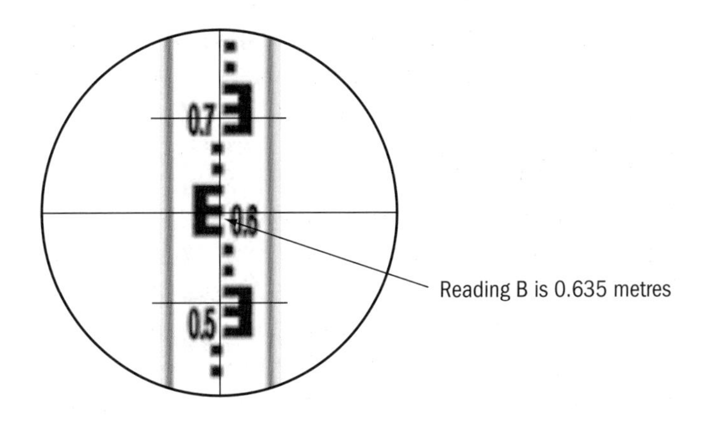

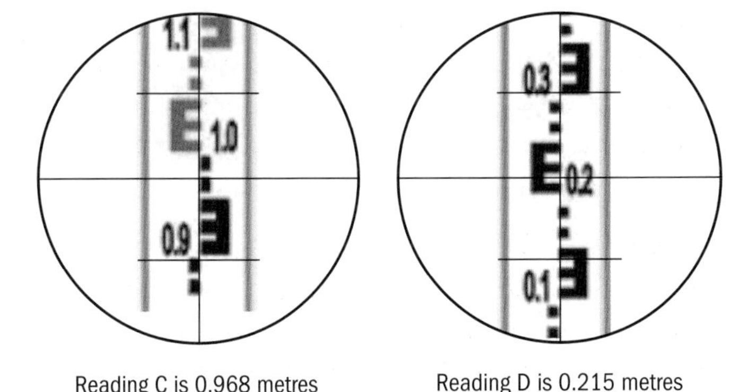

Figure 10.10: Reading a levelling staff

Take it further!

You can also use a level to measure plan distances. This method is known as stadia tachemometry, where the:

Plan distance = 100 × Distance recorded on the levelling staff between the top and bottom stadia hairs.

For example, if the top stadia hair reads 2.255 m and the bottom hair reads 2.103 m, then the direct plan distance between the instrument and the staff positions would be:

$$\text{Plan distance} = 100 \times (2.255\text{ m} - 2.103\text{ m})$$
$$= 100 \times 0.152\text{ m}$$
$$= 15.2\text{ m}$$

For optical level staff readings, calculate the distances for B, C and D.

A circular bubble is centred in the same way as a tube type. A bubble will always move in the same direction as the surveyor's left thumb

Figure 10.11: Automatic levelling

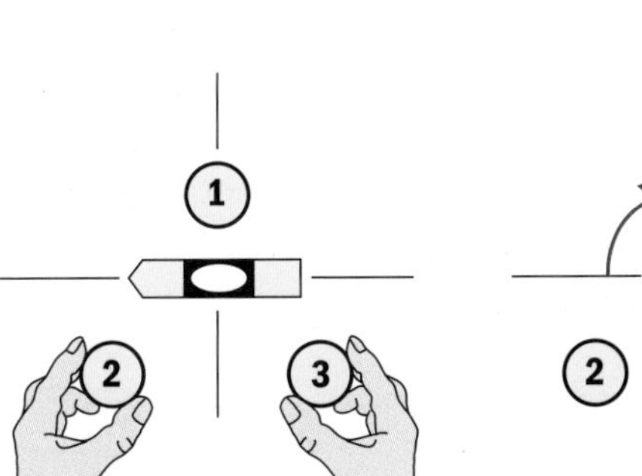

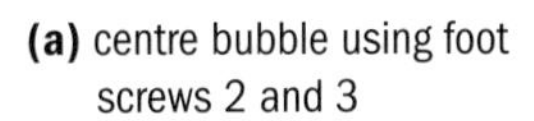

(a) centre bubble using foot screws 2 and 3

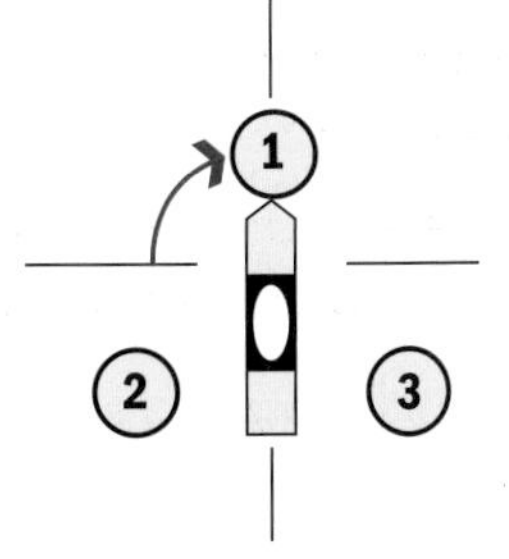

(b) centre bubble using 1 after turning through 90°

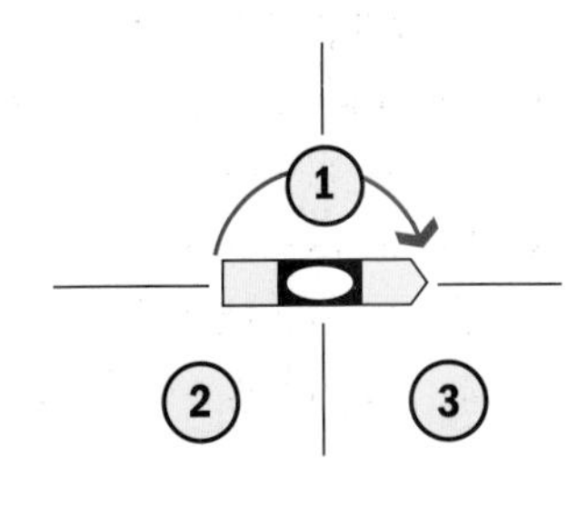

(c) turn bubble through 180° in plan and if it stays central it will do so for all positions

Figure 10.12: Setting up the dumpy level

The digital level and bar-coded staff

Digital levels are automatic levels that read the bar-coded staffs electronically. This is a relatively new development and although more expensive than normal optical automatic levels, they are quicker to read and, when used properly, are less prone to human errors. They can also be set for tracking continuous measurement, for use in setting out height.

After the digital level is aimed and focused on the bar-coded staff using optical sights, the reading is taken electronically and then displayed. They also incorporate facilities to store and then download levelling measurements to a variety of computer software programs for working out cross-sectional areas and volumes.

(a) A bar-coded staff in use

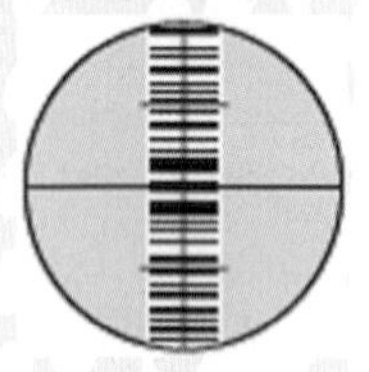

(b) View through telescope

(c) Electronic display of difference in height and reduced level

Figure 10.13: Surveyor using bar-coded staff

The principle of levelling

The basic method of undertaking levelling with an optical, digital or laser level is with the line of collimation that forms a horizontal level line from which vertical measurements are recorded. If a level is set up correctly and sighted on to five points on the ground and a TBM of known height, we can work out the **reduced level** of *any* of these points so that all their heights can be compared. For example, as can be seen in Figure 10.14, we can find the reduced level of points A to F:

point A = 15.0 m + 0.9 m – 1.5 m = 14.4 m

point B = 15.0 m + 0.9 m – 1.2 m = 14.7 m

point C = 15.0 m + 0.9 m – 1.7 m = 14.2 m

point D = 15.0 m + 0.9 m – 2.2 m = 13.7 m

point E = 15.0 m + 0.9 m + 0.8 m = 16.7 m

point F = 15.0 m + 0.9 m – 1.4 m = 14.5 m

Note that point E has the level staff turned upside down, which is known as an 'inverted' staff reading; thus the reduced level of the underside of the footbridge is obtained by adding the staff reading to the height of the instrument.

Key term

Reduced level – the height of a point or horizontal line that is relative to a given point of known height (datum).

The two-peg test for optical levels

The main source of error in optical levelling is when the line of collimation of the levelling instrument is not parallel to the horizontal level line. This error can be detected and adjusted using the two-peg test, which should be carried out before any major level survey. The method for carrying out this test is as follows:

- Select two points, X and Y, approximately 30 m apart and set the level up midway between them. Take readings for the staff at point X and at point Y. The readings when subtracted will give the true difference in height between the points as the collimation error will cancel itself out.
- Move the level beyond point X as close as its minimum focusing distance and take readings on points X and Y again. If the level has no collimation error, the difference between the readings of the two points should be identical to that previously measured.
- If the result of this test shows a difference of more than 2.5–3 mm, the level instrument should be returned for a calibration service. (BS 5606: 'Guide to Accuracy in Building' states that the error should be less than 5 mm over 60 m.)

Remember!

Keep the distance of backsights and foresights approximately equal as this will reduce collimation errors. It is easier to estimate the staff reading to the nearest 1 mm if you keep staff readings to less than 30 m distance.

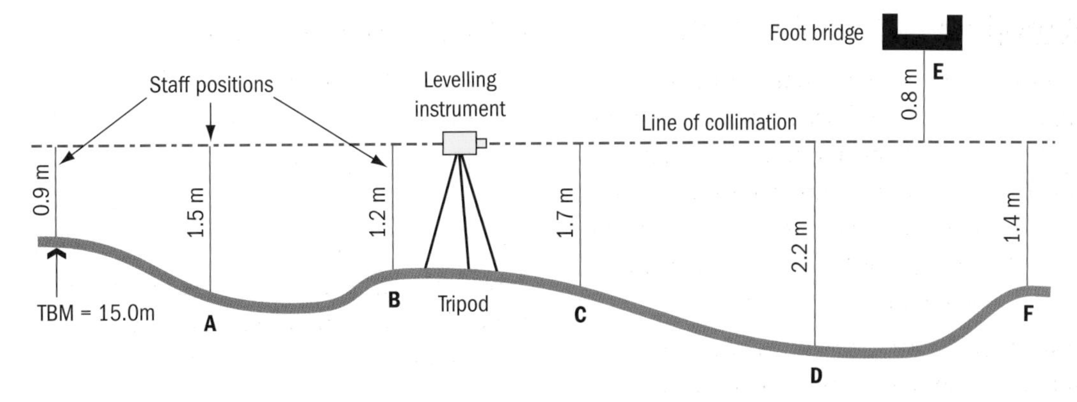

Figure 10.14: The principles of levelling

Traditional methods of booking levels

There are two methods for booking level readings to enable the reduced levels of points to be found:

- the height of the plane collimation (HPC) method
- the rise and fall method.

Each method requires the use of standard booking sheets, which can be purchased in pre-printed bound notebooks; alternatively, they can be drawn up on squared paper.

Both methods use similar terms. The headings of the booking sheets are shown below (Figures 10.15 and 10.16) . Repeated individual staff readings have different names depending on the order in which they are taken. The basic definitions for the different staff readings are shown in Table 10.2.

Table 10.2: Definitions of staff readings

Backsight (BS)	First staff reading of a levelling operation or the first reading after the instrument has been moved
Foresight (FS)	Last staff reading of a levelling operation before the level is moved
Intermediate sight (IS)	Any staff reading taken between the backsight and foresight readings
Reduced level (RL)	Height of any point relative to the datum used for the survey
Change point (CP)	A known point on the ground in which a foresight and then a backsight is taken

Figure 10.15: The HPC method booking form

Job description ______ Surveyor ______
Name of site ______ Job reference ______
Address ______ Date ______

Backsight (BS)	Intermediate sight (IS)	Foresight (FS)	Height of plane of collimation (HPC)	Reduced level (RL)	Distance	Remarks

Figure 10.16: The rise and fall method booking form

Job description ______ Surveyor ______
Name of site ______ Job reference ______
Address ______ Date ______

Backsight (BS)	Intermediate sight (IS)	Foresight (FS)	Rise	Fall	Reduced level (RL)	Distance	Remarks

Often the levelling instrument needs to be moved during the levelling circuit. A point on the ground is selected to be a 'change point' (CP) at which the staff reading is taken. A foresight is read with the staff held on this change point, the staff remains at the change point and the level is moved and set up again. A backsight is then read to the change point and the survey continues. In this way, the reduced levels relative to the original bench mark can be transferred around the circuit.

The staff readings should be recorded with the backsight reading in the left-hand column, the intermediate sight in the second column and the foresight in the third column. It is important to remember that the staff reading for any new point on the ground needs a new line in the booking table, and the only time two staff readings will appear on the same line is at a change point (CP) when the foresight from the previous level position and the backsight from the new level position coincide.

Remember!

When booking the levels in the field, always use a '/' instead of a decimal point '.' because decimal points can become easily obscured by mud or dust.

The levelling operation is undertaken as a circuit of levels between points of known height, usually a bench mark such as a TBM or OSBM. A circuit will start and return to the same point – a closed level survey – or it will go between two different bench marks of known height. Either way, the series of levels can be checked for accuracy such that the closing errors between the first and the final transferred levels can be determined. This survey error is called the 'misclosure' and can be compared against an allowable value given in BS 5606: 'Guide to Accuracy in Building' which is often quoted as:

$\pm$ 20 mm$\sqrt{}$(No. of kilometres travelled around circuit)

or for small surveys less than a kilometre;

$\pm$ 20 mm$\sqrt{}$(No. of instrument positions).

If the misclosure is outside the allowable value, then the survey should be repeated; if it is within the allowable error, the error should be distributed equally between the reduced levels.

Thinking point

The construction and setting out of the Channel Tunnel in 1991 had to achieve an accuracy of 100 mm in level over a distance of 18 km.

Remember!

Keep clear and neat booking entries for all your readings, and always carry out your booking in the field before you return to the office as it is easier and quicker to check and, if necessary, repeat your readings.

In order to understand the process of carrying out a level survey using the HPC and rise and fall methods, let's now consider the following.

The HPC method

The HPC method uses the horizontal line of sight, called the plane of collimation, through the level instrument as a reference for all the individual staff readings. The height of the plane of collimation can be found in this calculation:

$$\text{Height of plane of collimation (HPC)} = \text{Given bench mark height (BM)} + \text{Backsight reading (BS)}$$

Once the HPC is found, the reduced levels (RL) of the ground can be found from:

$$\text{Reduced level of ground (RL)} = \text{Height of plane of collimation (HPC)} - \text{Any staff reading (SR)}$$

This is shown in Figure 10.17.

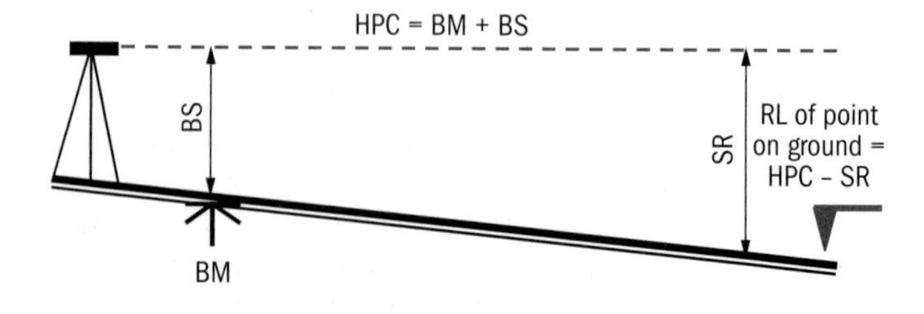

Figure 10.17: The HPC method

Once all the reduced levels have been calculated for the whole circuit of staff readings, a check has to be carried out to make sure that they have been booked correctly and that the arithmetic is right! If the arithmetic is correct, then:

The difference between the sum of the backsights and the sum of the foresights	=	The difference between the first and last reduced level

The rise and fall method

The rise and fall method does not use the height of collimation; instead it compares the difference in height between adjacent staff readings that have been measured.

In the example below, Figure 10.18, the staff reading at the TBM is 1.56 m and the next staff reading at point A on the ground is 1.75 m.

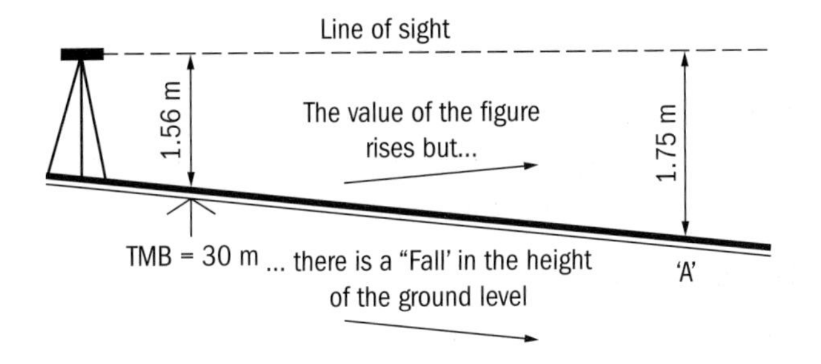

Figure 10.18: The rise and fall method

The difference between these two readings is found by subtracting the second reading from the first reading, i.e. 1.56 m – 1.75 m = –0.19 m. Because it is a negative value, it is a 'fall'.

If the TBM is at a known height of, say, 30 m, then the reduced level of point A is 30 m – 0.19 m = 28.1 m.

This can be seen clearly from looking at the staff readings in comparison to the level line of sight. The figures recorded on the staff 'rise', but the ground levels actually 'fall'.

Remember!

A quick way to remember if the difference is a 'rise' or 'fall' is to say 'If it's a rise in the book it's a fall on the ground'.

The process is repeated for each new staff reading, where each adjacent staff value is compared to see if it is a rise or a fall. The reduced level of these points is then calculated by adding the rises or subtracting the falls from the previous point's reduced level.

As with the HPC method, there are checks that can be carried out to make sure the booking arithmetic is correct. The rise and fall method has an extra check whereby the total rises and total falls are summed up and the difference between the two calculated. This calculated difference should be the same for:

- the difference between the sum of the rises and the sum of the falls
- the difference between the sum of the backsights and the sum of the foresights
- the difference between the first and last reduced level.

If this is true for all three checks, then the levels have been correctly booked.

Worked example

A sewer is to be constructed for a new industrial warehouse on the Premiere Industrial Estate. Part of the initial design work requires the profile of the existing ground level to be found. Therefore, a series of levels is taken along the length of the proposed sewer at regular intervals.

Details of the fieldwork that has been undertaken are shown in Figure 10.19. There were three instrument positions to pick up the ground levels at 15-metre chainage intervals, and the circuit of levels started from an OSBM on an existing building. The individual staff readings that were taken are shown on the sketch.

Book the levels for the Premiere Industrial Estate level survey first using the HPC method then the rise and fall method.

HPC method

The staff level recordings that were carried out as shown in Figure 10.19 are placed in the HPC booking table shown in Figure 10.20. In the table the readings have been colour coded to help you understand how each value was calculated:

Red is the calculated reduced levels of the ground.

Blue is the 1st level instrument position staff readings and the HPC height is found:

HPC = **65.5** + 1.351 = 66.851 m

the RL at 15 m chainage = 66.851 – 1.455 = **66.396 m** etc.

Green is the 2nd level instrument position staff readings and HPC height.

(cont.)

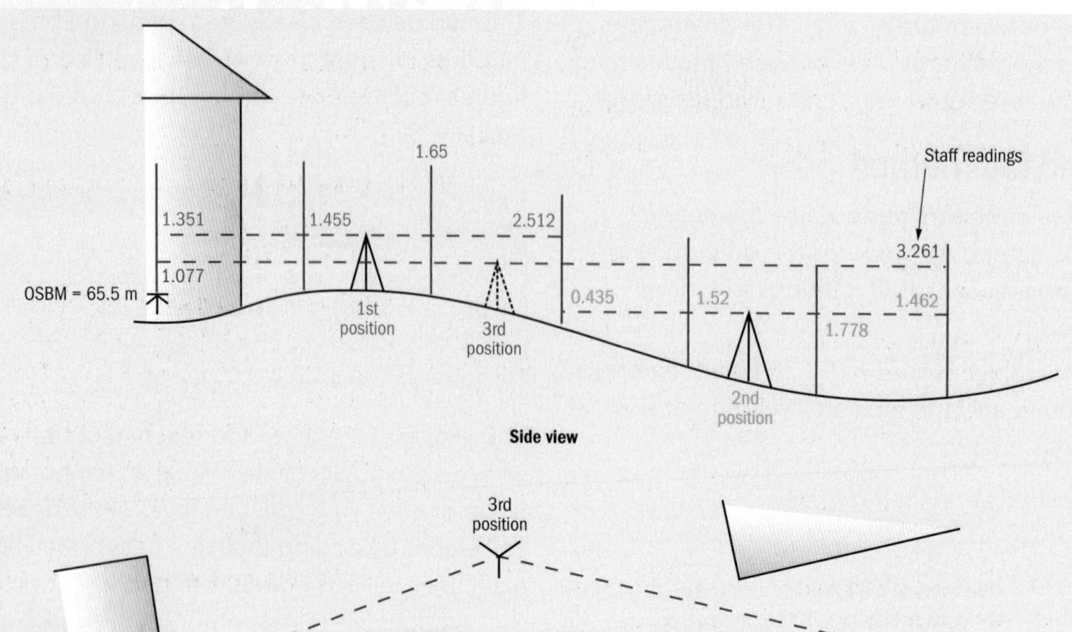

Figure 10.19: Premiere Industrial Estate plan

Job Description...Ground levels along sewer line.. Surveyor...MJH...........
Name of site......Premiere Industrial Estate.. Job Reference...II/19-TT
Address.. Date......15th Sept 10 .....

Backsight BS	Intermediate sight IS	Foresight FS	Height of Plane of Collimation HPC	Reduced Level RL	Distance	Remarks
1/351			66/851	65/500	-	OSBM 65/500
	1/455		"	65/396	15 m	
	1/650		"	65/201	30 m	
0/435		2/512	64/774	64/339	45 m	Change Point (CP)
	1/520		"	63/254	60 m	
	1/778		"	62/996	75 m	
3/261		1/462	66/573	63/312	90 m	Change Point (CP)
		1/077	"	65/496	-	OSBM
5/047		5/051		65/500		
		5/047		65/496		
		0/004		0/004		

Booking Checks
Difference in BS – FS = Difference between 1st & last RL
Therefore Levels have been booked correctly however
there is a misclosure of 4mm

& Allowable Misclosure
The allowable misclosure for this small level survey is
± 20mm √(No. of change points) = ± 20mm x √2
= ± 28 mm > 4mm THEREFORE ACCEPTABLE MISCLOSURE

Figure 10.20: The HPC booking table

(cont.)

New HPC = **64.339** + 0.435 = 66.774 m

and RL at 60 m chainage = 64.774 – 1.52 = **63.254 m** etc.

Purple is the 3rd level instrument position staff readings and HPC height.

New HPC = **63.312** + 3.261 = 66.573 m

and RL at measured back onto the OSBM = 66.573 – 1.077 = **65.504** m etc.

Now we need to check the levels. Totalling up all the backsights and all the foresights and finding their difference gives an answer of 4 mm. Then finding the difference between the first and last reduced levels also gives 4 mm. Therefore, the backsights and foresights have been correctly booked.

Rise and fall method

See Figure 10.20 above. Note that the coloured arrows have been added to show which staff readings are being compared, to give the rise or fall value and the calculated reduced level.

The first readings to be compared are the backsight to the OSBM and the staff reading at the 15 metre chainage:

1.351 m – 1.445 m = – 0.104 m 'fall'

giving a reduced level of the ground at the 15-metre chainage as:

65.5 m – 0.104 = 65.396 m

The next two figures to be compared are the 15-metre and 30-metre chainage:

1.445 m – 1.65 m = – 01.195 m 'fall'

which when added to the reduced level of the 15-metre chainage:

65.396 – 0.195 m = 65.201 m, etc.

At the change point you cannot compare figures written on the same line as these are by definition the same point on the ground. You must compare staff readings written on different consecutive lines, such that at the 45-metre chainage point you have to compare the 0.435 m backsight with the 1.52 m intermediate sight at the 60-metre chainage:

0.435 m – 1.52 m = – 1.085 m 'fall'

This then gives the reduced level of the ground at the 60-metre chainage as:

64.339 m – 1.085 m = 63.254 m

This process is repeated until the last staff reading has been compared which, in this case, is the difference between the backsight at the 90-metre chainage and the final closing foresight onto the original OSBM.

We see from our example that all the differences equal 4 mm.

Job Description...Ground levels along sewer line.. Surveyor...MJH...........
Name of site......Premiere Industrial Estate.. Job Reference...11/19-TT
Address...........Eastwood Road, Southend-On-Sea, Essex......................... Date......15th Sept 06.....

Backsight BS	Intermediate sight IS	Foresight FS	Rise	Fall	Reduced Level RL	Distance	Remarks
1/351					65/500	-	OSBM 65/500
	1/455			0/104	65/396	15 m	
	1/650			0/195	65/201	30 m	
0/435		2/512		0/862	64/339	45 m	Change Point (CP)
	1/520			1/085	63/254	60 m	
	1/778			0/258	62/996	75 m	
3/261		1/462	0/316		63/312	90 m	Change Point (CP)
		1/077	2/184		65/496	-	OSBM
5/047		5/051	2/500	2/504	65/500		
		5/047		2/500	65/496		
		0/004		0/004	0/004		

Difference in BS – FS = Difference in Rise & Fall = Difference between 1st & last RL
Therefore Levels have been booked correctly however
there is a misclosure of 4mm

Figure 10.21: The rise and fall booking table

The rise and fall method v. the HPC method

The HPC method requires less calculation, however, it does not have a built-in check on the intermediate sights, as does the rise and fall method. The rise and fall method is best suited when a level is being transferred over a distance via a number of change points where no intermediate sights are used – this is known as flying levelling. The HPC method is best used when a large number of readings are being taken from the same instrument position, such as plotting or setting out a small levelling grid.

Typical levelling applications

Flying levels

Flying levels is the transferring of levels from one point to another using only backsight and foresight readings. It is a quick and convenient way to establish a new TBM from an existing bench mark. However, the flying levels must be returned to close back at the original bench mark, that is, the levels will need to fly back to close the survey and to carry out all the necessary checks.

Remember!

Make sure that you position the instrument so as not to block access on any footway or make it vulnerable to being knocked. When reading the instrument, do not lean on the tripod or the level, and also be careful when moving around it. Finally, do not leave the instrument unattended – opportunist thieves may steal your equipment, and your prize level may end up at the local boot sale!

Grid levels

Grid levels can be used for contouring an area and also for calculating the volume of earthworks. A rectilinear grid is set out on the ground and marked with pegs or arrows. The levelling instrument is set up in the middle of the points with a backsight to a bench mark. The levels of the ground at each of the grid points are then read and reduced. These grid levels can be manually manipulated to give contour levels at regular vertical intervals or used to calculate volumes of excavation using a variety of formulae and techniques. Alternatively, with the use of digital levels the data can be downloaded directly into a range of software packages for producing virtual three-dimensional (3D) models. These can then be manipulated for contours and volumes as required. A typical example is shown in the NRG Surveys example of a contoured site in Figure 10.22.

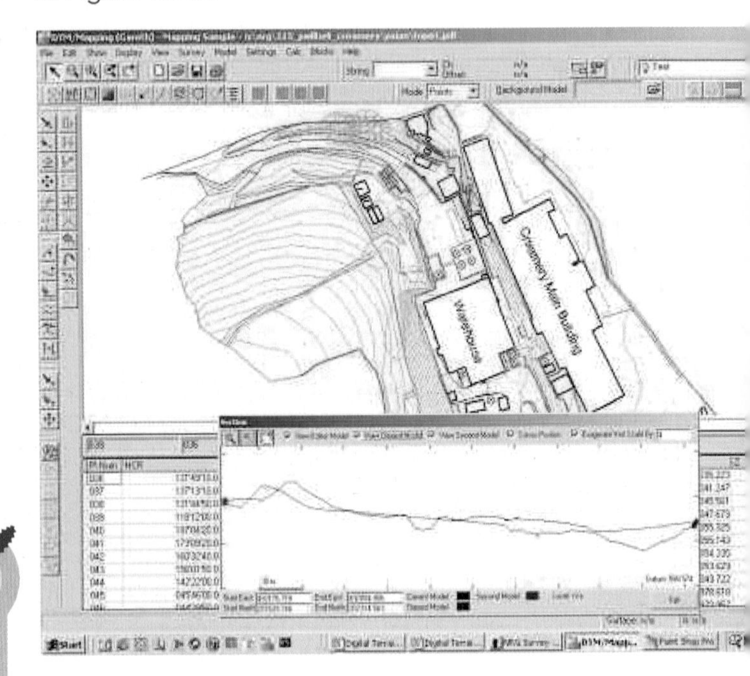

Figure 10.22: Contoured site

Long sections

Long sections are vertical cross-sections taken along the length of proposed sewer or road construction; because these forms of construction are often very long compared to their depth, they are plotted with an exaggerated vertical scale to clearly show the slope and changes in height relative to the length. For example, a new sewer may be 800 m long, but the levels throughout may vary by only 5 or 6 m. Plotting these to the same scale would give no impression of the gradient and how it may change along its length. Therefore, a typical long section will have a vertical scale ten times the horizontal scale, for example 1:500 horizontally and 1:50 vertically.

Using this method, numerical information is tabulated below the long section for the new sewer and will include the cumulative distance or chainage, as well as the reduced levels of the ground, the **invert level** of the sewer, the **cover level**, the **formation level** of sewer excavation and the sight rail heights that aid the setting out. A typical long section is shown in Figure 10.23.

Key terms

Invert level – the reduced level of the lowest part of the internal diameter of the sewer.

Cover level – the reduced level of the top of the inspection chamber cover.

Formation level – the reduced level of the bottom of the excavated sewer trench.

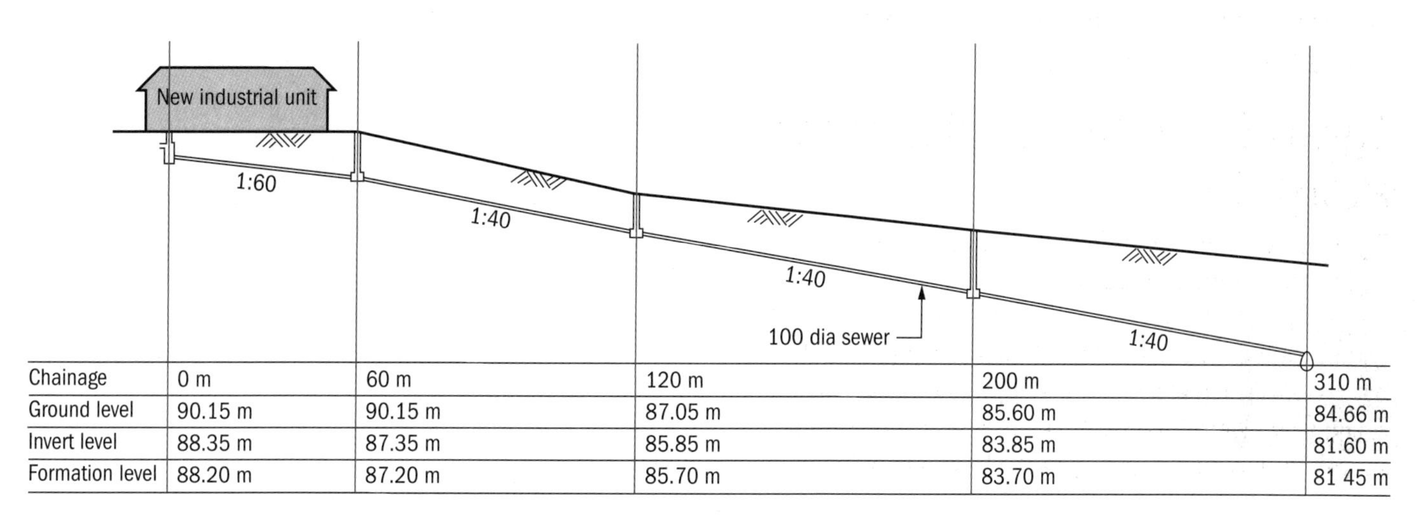

Chainage	0 m	60 m	120 m	200 m	310 m
Ground level	90.15 m	90.15 m	87.05 m	85.60 m	84.66 m
Invert level	88.35 m	87.35 m	85.85 m	83.85 m	81.60 m
Formation level	88.20 m	87.20 m	85.70 m	83.70 m	81 45 m

Long section on sewer

scales - horizontal 1:1250 - vertical 1:200

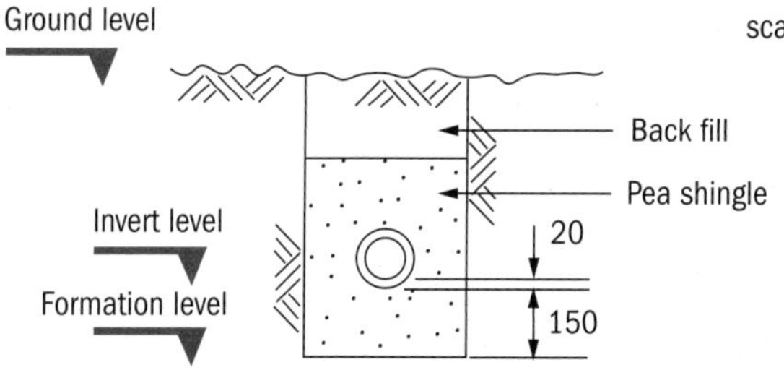

Typical section through sewer

Figure 10.23: Long section of a sewer

Assessment activity 10.2

1. Undertake a practical levelling survey using the levelling equipment provided by your college.

a A levelling survey for a proposed sewer is to be carried out to establish the invert levels at various chainage points for the sewer. Select the equipment needed to perform the levelling survey. Working in groups of three, peg out a straight line a minimum of 40 m long over undulating land. This represents the line of a proposed sewer. Set out pegs or arrows at regular intervals to take account of the changes in the slope of the ground. Establish a suitable point for a TBM and using this, undertake a level survey of the ground marked by the pegs using an optical or digital level. P3

b Carry out the levelling survey in the field and all necessary arithmetic and accuracy calculation checks to determine the 'Reduced Levels' of the points on the ground that have been marked out. Use one of the two available methods for booking levels. Remember in your group to take turns in holding the staff and taking/booking the readings. From your field notes and booking sheet produce a long section in a standard format showing the variation of ground level with chainage. P4

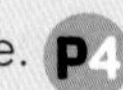

c The new sewer that is to be constructed starts from a point at the end of the line, with a reduced level of 1 m below the existing ground level at that point. Calculate the invert levels of all the chainage points survey along the ground assuming that the sewer falls at gradient of 1 in 40.

Add the reduced levels of the invert of the sewer to the long section produced previously in a standard format. P3 P4

d Undertake further levelling work.

2. Re-survey the line using a **different booking method** from the one used previously and undertake all the necessary booking checks. Comment on the differences between the 'Rise and Fall Method' and the 'Height of Collimation Method' and describe in which situations each would be best used. M1

3. Prepare a report evaluating possible sources of error in your level surveying work, highlighting the practical measures you task to improve accuracy. D1

Grading tips

For P3 you need to demonstrate that you know the right equipment to use for a level survey and what each piece of equipment is called. In your write-up of the practical include labelled sketches or actual photographs of the equipment. On paper set out your calculations in clear easy to follow stages. Your tutor may also like to see you explain each stage. Remember to include all necessary arithmetic checks and comment on the accuracy of your results.

For P4, the most important part of carrying out the level survey is planning ahead. Before starting the survey walk over the site and decide where you will set up your instrument to take all the necessary staff readings. In your assignment write-up, draw a sketch plan of where you sited your instrument in relation to the points that needed levelling and indicate whether you took a backsight, foresight or intermediate site on your sketch to each point.

Your drawings should be large enough to cover the page and should be drawn to scale. For long sections the vertical axis is usually exaggerated by a factor of 10 to clearly show the changes in levels. Remember to show all appropriate units e.g. metres etc.

To achieve M1 you will need to demonstrate that you know the two main level booking systems, how they are booked and checked, and where they are best used. Make sure that you provide some practical examples.

For D1 you need to justify your results by comparing and contrasting your levelling misclosures for both your HPC and Rise and Fall surveys. Describe clearly how you tried to avoid gross, systematic and random errors.

PLTS

Evaluating your results and learning to inform will help develop your skills as a **reflective learner**.

Functional skills

By calculating the invert levels of the sewer you are demonstrating that you can identify and use the **Mathematics** needed to tackle it.

3 Be able to measure angles and produce results from calculations

Angular measurement

So far we have looked at linear and height measurement; the final form of measurement involves angular measurement. In order to understand how angular measurements can be accurately taken, let's now look at the two basic methods of measurement using angular methods:

- polar measurement
- triangulation.

The best way to understand how each of these different methods operates is to consider an example. So, with reference to Figure 10.24, we want to fix the position of a brick wall XY in relation to an aerial mast at Z by using angles, and using the polar measurement method and triangulation method we will show how this can be done.

Polar measurement

Polar measurements measures an angle in relation to a reference direction; in our example it is the angle of 51 degrees relative to the wall. Then, to fix the position of the aerial mast at Z, we need to measure the distance XZ which, in our example, is 8165 mm.

This form of measurement is commonly used in navigation with the reference direction being 'North' and the relative angle known as the 'bearing' angle or '**whole circle bearing**'. This terminology continues to be used in **traverse surveys**.

Key terms

Whole circle bearing – an angle measured clockwise from a reference point, usually North, with a value ranging from 0° to 360°.

Traverse survey – an interlinked polygon of survey stations with known easting and northing coordinates.

The plotting of these measurements is straightforward, with the angle and distances marked out on paper as recorded. Alternatively, the coordinates of Z relative to X can be worked out using simple, right-angle trigonometry and plotted manually or by using computer-aided design (CAD).

Triangulation

Triangulation is ideal to use when the point to be fixed cannot be physically reached. In our example, it may be that the aerial mast and wall are separated by a deep excavation. The only requirement is that there are clear lines of sight between the ends of the wall at points X and Y and the aerial mast at point Z. Only one linear measurement needs to be recorded and that is the length of the wall. In linear surveys this forms the backbone to the survey measurements and is called the baseline. In the drawing office, the length of the wall of 7000 mm is drawn first to a suitable scale, and then the two angles of 51 degrees and 73 degrees are plotted. The lines of sight can be plotted and where they intersect will be the position of the aerial mast Z.

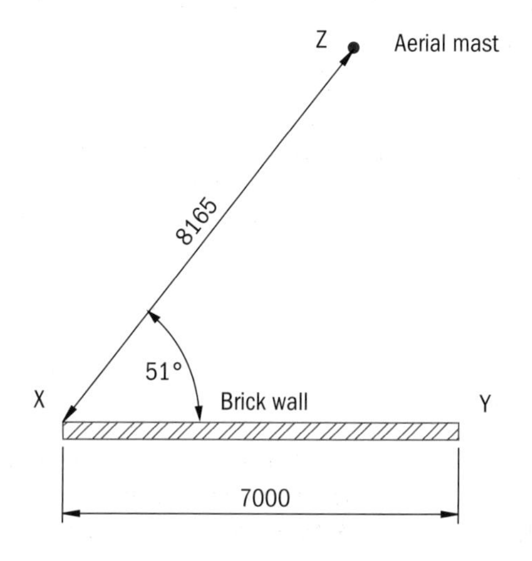

(a) Polar measurement

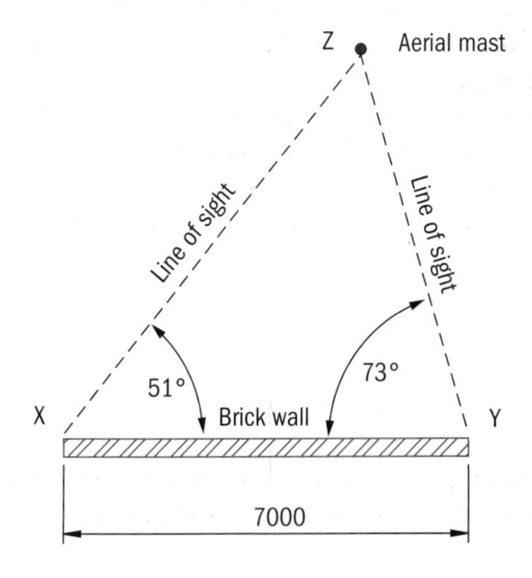

(b) Measurement by triangulation

Figure 10.24: Angular methods of measurement

Using degrees, minutes and seconds

As we have already seen in Unit 3 Mathematics in construction and the built environment, angles in the UK are measured using the sexagesimal system where there are 360 degrees in a whole circle. This system is very common in both land and setting-out surveys where a high degree of accuracy is needed, so it is important to know how to add and subtract in the sexagesimal system, both manually and with a scientific calculator.

When doing manual calculations, you must always remember that you are only using numbers up to 60. For example, 72' is written as 1° and 12' because there are 60 minutes in a degree. Similarly, one minute contains 60 seconds. Let us do some manual calculation examples using the place values of 60.

This one is straightforward with no complications:

192°	30'	20"
62°	20'	10"
130°	10'	10"

The one below, however, is more difficult because we have to subtract 50" from 30". This is done by borrowing 1' from the 28' from the next column and adding that 60" to 30" to make 90". The subtraction can then be carried out:

		+ 60"
	27'	90"
75°	~~28'~~	~~30"~~
26°	20'	50"
49°	7'	40"

Here's another one that needs careful consideration. Subtracting the seconds is fine; however, it is the minutes' column which is not straightforward. We have to borrow 1° from the 273° and add those 60' to give 80' from which 30' can be subtracted.

+ 60'			
272°	80'		
~~273°~~	~~20'~~	40"	
102°	30'	10"	–
170°	50'	30"	

Remember!

All good scientific calculators will be able to deal with these calculations by inputting the degrees, minutes and seconds directly (study the manual carefully to make sure you can do this). Then double-check the result using manual calculations.

The theodolite

Basic principles

The theodolite is a device for measuring horizontal and vertical angles. Its basic components are a telescope and two protractors which measure angles in the vertical and horizontal plane. Old instruments relied on physical protractors etched on brass protractors or 'circles' with a variety of circular scale rules known as vernier reading devices. However, these are no longer used. Today, theodolites are fitted with either an optical or electronic system for reading both horizontal and vertical circles. You will need to familiarise yourself with the theodolites that your college has as these will be the ones that you will use for your practical work, and also seek the guidance of your tutor in the care and the use of this equipment. Figure 10.24 shows the essential parts of a theodolite together with its main controls and moving parts.

The theodolite has two horizontal circular plates that rotate independently of each other, called the upper and lower plates The lower plate of the instrument is calibrated in degrees, minutes and seconds in a clockwise direction and is either free to rotate, or can be clamped to the base (tribrach). The upper plate has an indicator or pointer, thus enabling angles to be read optically or digitally depending on the type of instrument. Control knobs called clamps are provided with slow-motion screws for fine adjustment for both plates. These screws are often referred to as tangent screws. By adjusting these clamps, the theodolite's cross-hairs within the telescope can be set to zero or any given angle and pointed at the required survey target. These same principles apply to the vertical circle also.

Setting up a theodolite

The theodolite is set up correctly when:

- the instrument is *levelled* in all directions, which implies that it is truly vertical
- the instrument is *centred* over the desired survey station point.

This is achieved in two main stages as described below.

Rough levelling and centring – First, set up the tripod of the theodolite over the survey station, which can be a nail head or fine cross made with a site marker pen. Make sure that the tripod is pulled up to at least chest height and that the top of it is roughly level and roughly centred above the nail. Screw the theodolite firmly to the tripod first checking that all the foot screws are at the mid-thread position. Then you need to check if the instrument is centred. Most theodolites have an optical plummet where you can look down the vertical axis of the theodolite. There you can see if the small circular target which represents the vertical axis is centred on the nail. If not, you can adjust the foot screws slightly to get the axis at its base to coincide with the nail, or if there is a large distance, two legs of the tripod can be lifted and moved while still looking through the optical plummet.

Remember!

When taking the theodolite out of its carrying case, always look carefully at how it fits in its cradle within the case. Sometimes the most difficult thing about using a theodolite is getting it snugly back into its case!

Fine levelling and centring – Once the theodolite has been roughly levelled and centred, then the three foot screws are used to finely level the instrument. This operation is exactly the same as levelling a dumpy level (see page 358). Once the theodolite has been levelled on the foot screws, the optical plummet must be checked to see if the instrument's vertical axis has moved off the nail. If so, it can be brought back on centre by loosening the tripod screw and gently sliding the whole instrument across the top of the tripod while looking through the optical plummet. With this operation complete, it will be necessary to recheck the fine levelling by levelling the three foot screws again.

Once the theodolite is centred over the survey nail and levelled, it is ready to use! You will need to practise this procedure if you are to master the use of the theodolite. Make sure you are confident of setting up the theodolite before you undertake a practical assignment.

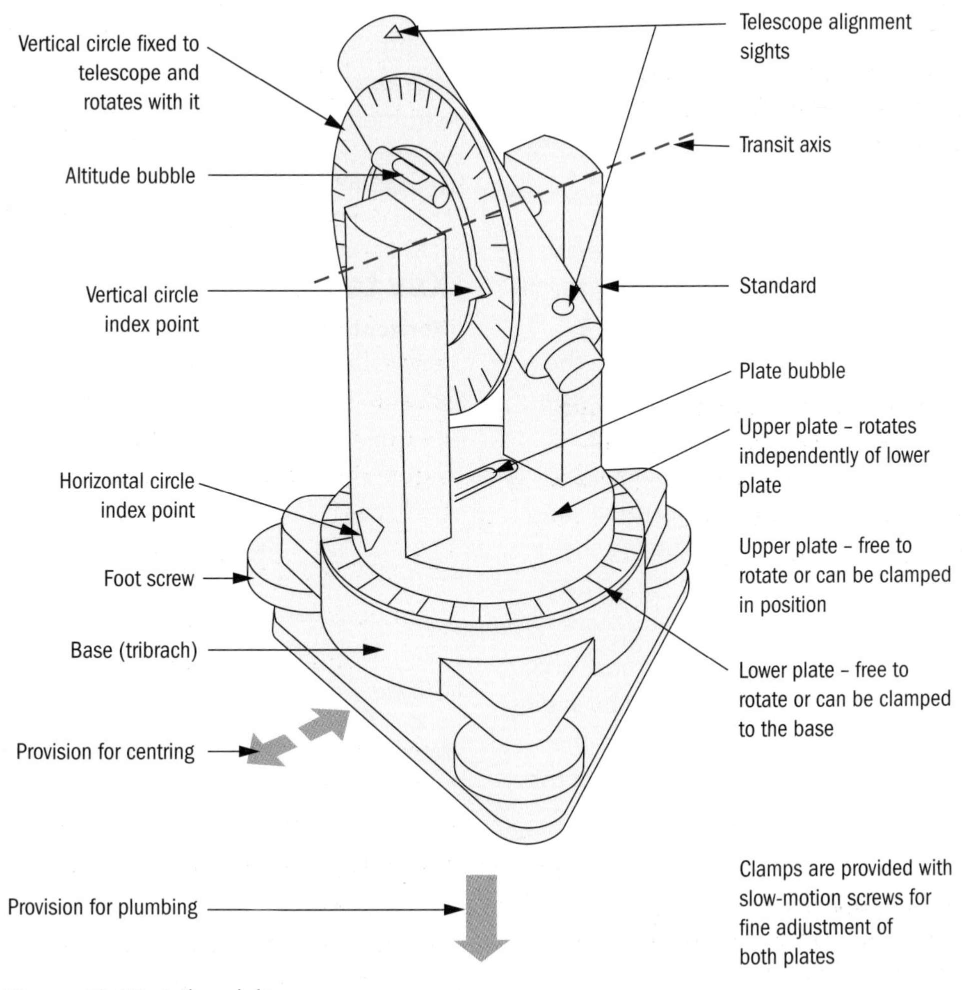

Figure 10.25: A theodolite

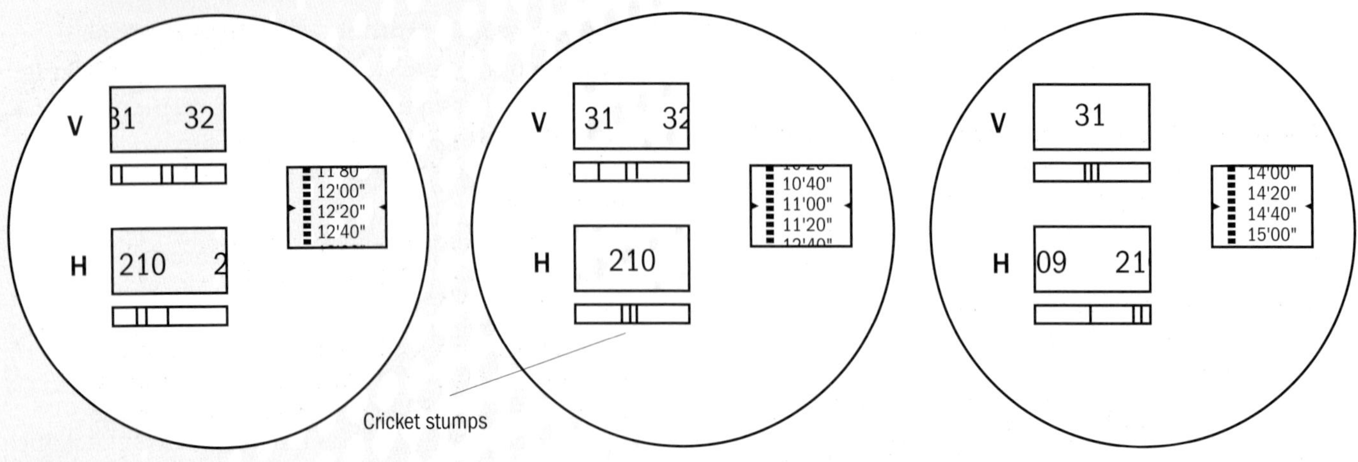

Figure 10.26: Example of optical theodolite reading

Reading an angle on a theodolite

Many modern theodolites have direct reading electronic displays which state clearly the values of angles measured in degrees, minutes and seconds. However, if you are using an optical theodolite, you will need to know how to read its optical micrometer devices. These give the angles in two parts depending on the type of theodolite. You will need to check with your tutor as to the system of optical measurement that relates to your college's optical theodolites.

Different instruments have different displays, but they all work on the same principle. Figure 10.26 shows three sequential views of the optical micrometer reading for a particular sighting. The left-hand view shows that a reading has been taken on the horizontal scale between 210° and 211°. By adjusting the micrometer screw on the side of the instrument, the reading is brought round so that the moving double line coincides with the static single line, forming a 'cricket stump'. This produces a reading of 210° 11' 00", as shown in the middle view. The vertical angle can be read in a similar way by adjusting the micrometer screw for the 'cricket stumps' to give a vertical angle of 31° 14' 30" in the right-hand view.

When using a theodolite, it is necessary to know what 'face' the instrument is on. This can be found by seeing on what side the vertical circle lies in relation to the telescope. When the vertical circle is to the left of the telescope, this is known as 'face left' and when it is on the right, it is known as 'face right'. Moving from face right to face left or vice versa is a simple matter of flipping the telescope over to point in the opposite direction, then turning the instrument 180° to get the telescope to point back to its original sighting. This is called transiting the telescope.

It is common practice in surveying to take at least two readings for a single angle, one on face left and one on face right. In this way, you are using the whole 360° circle and eliminating any centring error that may be present in the instrument. This is explained in more detail below.

Booking procedures for a single horizontal angle

It is important to note in the survey book the number or letter used to denote the survey station over which the instrument is placed and the number or letter of the two stations which are being sighted. It should also be noted whether the instrument is on face left or face right.

Key term

Horizontal angle – an angle measured between three fixed points within a horizontal plane.

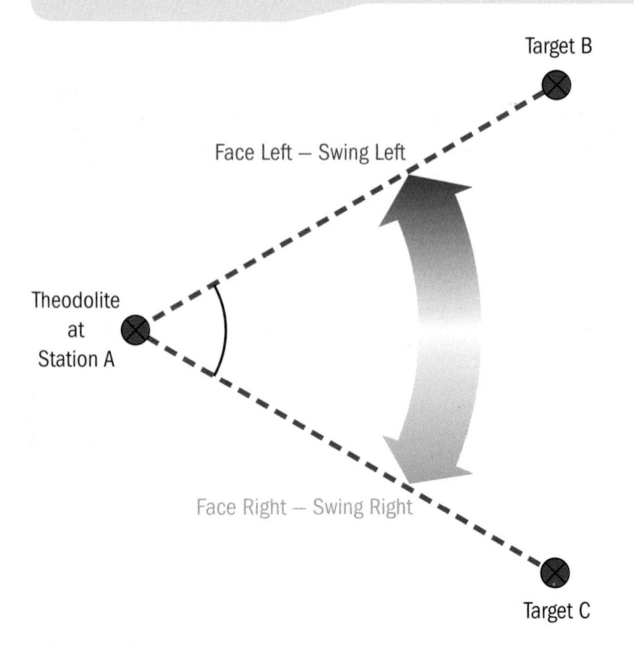

Figure 10.27: Plan on single angle survey measurement

Worked example

The standard booking table has five columns to show where the instrument is located 'At' and at what targets you are sighting 'To'. The next two columns are 'Face right' and 'Face left' and, finally, a place to note down the 'Mean angle' of the two angles.

Here are the readings from Figure 10.26 and an explanation of how field work was done and the readings booked correctly:

At	To	Face right	Face left	Mean angle
A	B	0° 0' 0"	180° 00' 15"	**82° 20' 05"**
	C	82° 20' 15"	262° 20' 10"	
		82° 20' 15"	82° 19' 55"	

Set up instrument levelled and centred over the required station A.

- Using the lower plate clamps and tangent screws, set the index to a low value of angle, preferably 0° 00' 00", and put the instrument in the face right position.
- Sight onto the first station B making sure you are always striking out in a clockwise direction. Using the lower plate clamps and its fine adjacent tangent screw, set the 0° 00' 00" on to station B.
- Release the *upper plate clamp* – not the lower plate clamp – and swing right to the second station at C. Sight accurately onto C using the tangent screws with the upper plate clamps on and, after adjusting the optical micrometer (if it is an optical level), book the reading, which in this case was 82° 20' 15". The instrument is on face right and pointing at station C, so this is where the reading is booked in the booking table.
- Calculate the angle made at A between C and B by subtracting the 'top from the bottom' to give a value of 82° 20' 15".
- Transit the telescope by flipping it over and unclamping the upper plate – *but not the lower plate!* – and turn the telescope to point back to station C.
- Sight on to C using the upper plate clamps and tangent screw, and accurately book the reading, which in this case is 262° 20' 10". The instrument is on face left and pointing towards C, so that is where this reading is booked in the booking table. Also note that, ideally, if there were no errors, this value should be 180° greater than the previous angle.
- Release the upper plate clamp and swing left back onto the original station B and book the reading of 180° 00' 15" in the booking table. This reading is on face left and pointing at B, so this is where it goes in the booking table.
- Calculate the angle made at A between B and C by subtracting the 'top from the bottom' to give a value of 82° 19' 55".
- Where the difference between the angles measured is less than 30", the mean of the pair can be calculated and taken to be the true angle subtended at A measured between points B and C. The mean of the face left and face right angles is 82° 20' 05".

Activity: Finding angles

Determine the mean horizontal angle XYZ given the following angular measurements on face right and face left.

At	To	Face right			Face left		
Y	X	0°	0'	0"	179°	55'	45"
	Z	79°	50'	30"	259°	40'	15"

Booking procedures for a single vertical angle

A similar procedure is used for measuring vertical angles as for horizontal angles, except that the zero is generally fixed in one position, either pointing straight up – known as the zenith – or set at the level horizon. Angles that dip below the horizon are often quoted as angles of depression while those that rise above the horizon are angles of elevation.

It is always good practice to record single vertical angles on both faces, but be aware that once the values go beyond 360°, they revert back to 0°, just as a clock hand moves from 12 o'clock to 1 o'clock.

Accuracy in measuring angles

The theodolite can measure very small angles; in fact, some theodolites can read down to 1 second of a degree, which is equivalent to 1 in 216,000th of a whole circle. An error of just 60" of angle roughly equates to an error of about 15 mm over 50 m. To understand how to reduce errors, you need to know the main axes of the theodolite and these are as follows:

- The vertical axis – at right angles to the horizontal plane and transit axis, which is considered to be truly vertical when the theodolite has been levelled.
- The horizontal circle – at right angles to the vertical axis.
- The transit axis – at right angles to the line of sight and the vertical axis and forms the pivot for rotation of the telescope.

Errors in angular measurement originate from two main sources: instrument or human.

The axes of a theodolite must be in good adjustment with each other to ensure that the instrument is level and rotates perpendicularly about its various axes. It is advisable to have the theodolite serviced at the start of all major contracts and at frequent intervals by a certified calibration and testing company. The rough handling that these instruments may get over the period of a contract can be quite harsh, particularly under site conditions. There are three basic checks for errors:

- Vertical axis check – after levelling the instrument on the three foot screws, simply rotate the instrument 180° and the plate bubble should stay central. If not, there is an error.
- Transit axis check – the transit axis should be perpendicular to the line of collimation; if you book readings on both face right and face left, this error, if any, will be cancelled out when the mean is calculated.
- Spire check – this checks that the transit axis is truly horizontal. It involves sighting a target high up, such as a church spire face, which is then brought down to ground level by dipping the telescope and

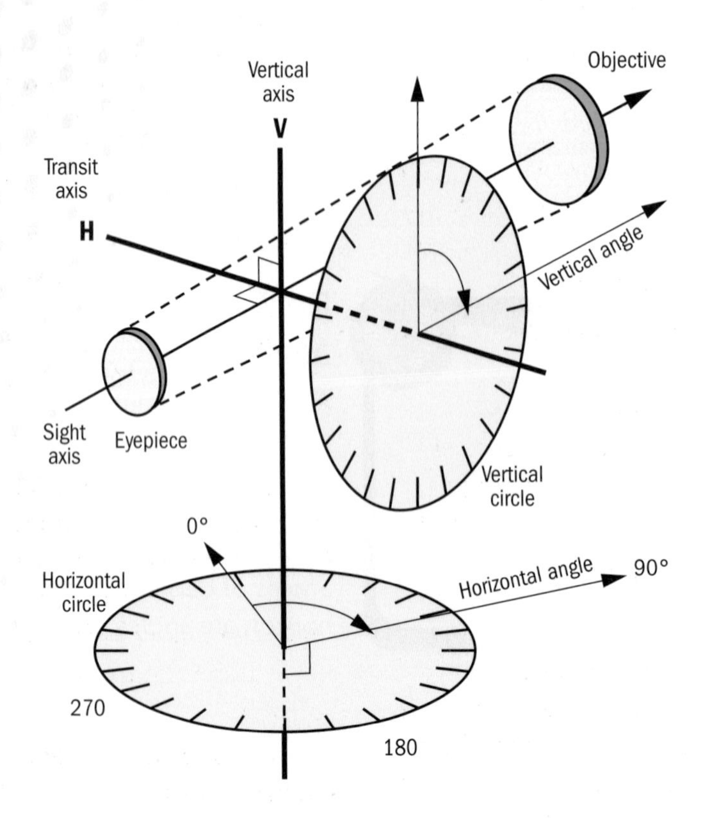

Figure 10.28: Main axes of the theodolite

The procedure is as follows.

- Set up the theodolite on a tripod close to the wall on baseline parallel to wall (point A).
- Set out 90° angle to fix peg at point B.
- Measure the right-angle distance from B to the wall, for example, 23.5 m.
- Set up the theodolite at B and zero circle onto previous station on baseline (at A).
- Draw a sketch of the elevation to be surveyed showing target point to be fixed and plotted. Number these points on the sketch you have made, see Figure 10.30 on p. 376.
- Measure both horizontal angle Hz (θ_1) and vertical angle V (θ_2) to each point and tabulate them in a suitable table:

Target point	Hz (θ_1)	V (θ_2)	x (m)	y (m)
e.g. 3	6° 30' 00"	81° 0' 00"	2.677	3.722

- Calculate the distance x and y and write in the table shown in red from the following:

$x = AB \times \text{Tangent } \theta_1$

$y = \dfrac{AB}{\text{Tangent } \theta_2}$

- Plot x and y as coordinates to fix each point on graph paper or using a suitable CAD program to give the required elevational view of the building.

Assessment activity 10.3

P5 P6 M2 D2 BTEC

Your tutor will set out three pegs on an area of land. Carry out a site survey of the three pegs or stations using a theodolite only.

1. Draw a sketch of the survey pegs and on it indicate the angles to be measured. Identify clearly how you will determine the internal angles in the triangle formed by the three pegs using 'face left' and 'face right' angles at each station, and also indicate how you will check your angular measurements for the 'triangle'. **P5**

   Carry out the angular measurements necessary to fix the position of all three pegs using a standard booking system and method. Carry out a check calculation to show that all the internal angles add up to 180° and hence state your misclosure. If it is an acceptable error, make necessary corrections to spread the closing error to all three angles. **P6**

   Work out the height of an inaccessible object on a slope.

2. In a land survey of a coastal area the reduced level of the top of a cliff needs to be found as part of works to stabilise the cliff face. A baseline 35 m was set up on the foreshore area and the angles of inclination at the ends of the baseline were measured as shown in Figure 10.32. Also, with the theodolite telescope set horizontally at 0°, the heights of collimation of the instrument were found relative to the TBM, which are also shown in the diagram. (*Hint*: Use right-angle trigonometry and Pythagoras' rule to help find the vertical height of point C above point A.) **M2**

   Consider errors in angular measurement.

3. Undertake research to analyse methods used in angular measurement in terms of trigonometrical accuracy. Provide examples which clearly show the effect of increasing distance on angular error and the need for good intersection of lines of sight. **D2**

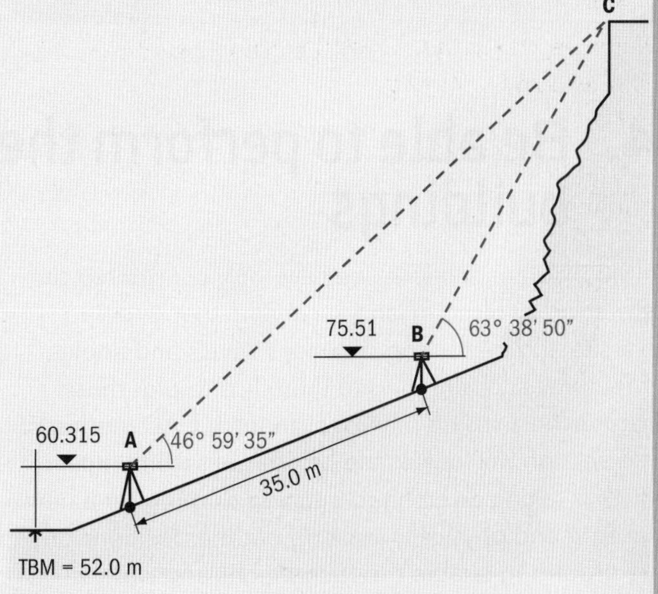

Figure 10.32: Section through cliff face

(cont.)

Grading tips

P5 Your tutor will want to see that you can relate a physical situation to a mathematical model. In the above question you need to show that you know that the internal angles of the triangle are the most important angles to record because you can check that their sum is 180°. You will also need to show that you clearly understand what the terms 'face right', 'face left', 'misclosure' and 'survey station' mean.

P6 This criteria requires you to demonstrate that you can set up a theodolite correctly and use it to measure angles. You tutor will need to witness you carrying out this work so make sure that you let him or her know when to observe you. You will also have to demonstrate the method of booking angles on 'face right' and 'face left'. This is best done by a clear booking table like the one we have used previously in this unit. When doing angular calculations always work in degrees, minutes and seconds as this is how you record angles when using a theodolite. In this calculation you will need to work out the sum of your internally measured angles of the triangle and explain how you would apply any corrections to your results.

M2 In these types of calculations using angular information it is always essential to draw a diagram showing all the information given and also superimposed on it the mathematical geometry, such as the location of the right-angles triangles and 'unknown' distances. This question needs you to solve two simultaneous equations to find the reduced level of point C.

D2 You need to look carefully at how measurements are affected by how survey lines intersect. Your tutor wants to know that you understand the effects it has on the errors that can occur. A series of labelled illustrative diagrams would help to demonstrate your understanding of the important concept of 'well conditioned' triangles.

PLTS

Supporting conclusions using reasoned arguments and evidence will help develop your skills as an **independent enquirer**.

Functional skills

By undertaking the correct calculations above you demonstrating that you can select and apply a range of skills to find the right results using **Mathematics**.

4 Be able to perform the setting out of small buildings

In the previous sections, we have only considered the operation involved in land surveying, that is, going to a site and measuring distances, levels and angles of features and returning to the office to plot these features on a scaled plan by manual or CAD methods. This section will look at the process of setting out, which is taking an architect's scaled drawing for a new building and physically marking out the positions on the ground to enable the foundations and walls to be constructed.

Setting out is an exercise in communication. The requirements of the designer are set out clearly on the drawings and other contract documents. The setting-out engineer or surveyor must interpret these and provide clear dimensional information to the craft operatives so that they can build the structure. This involves creating a clear system of temporary marks and physical profiles so that the layout of the building works can be monitored and controlled in terms of both the horizontal dimensions, referred to as 'line', and the vertical dimensions, referred to as 'level'.

This process starts with a site layout drawing that shows the intended position of roads or buildings (see Figure 10.33).

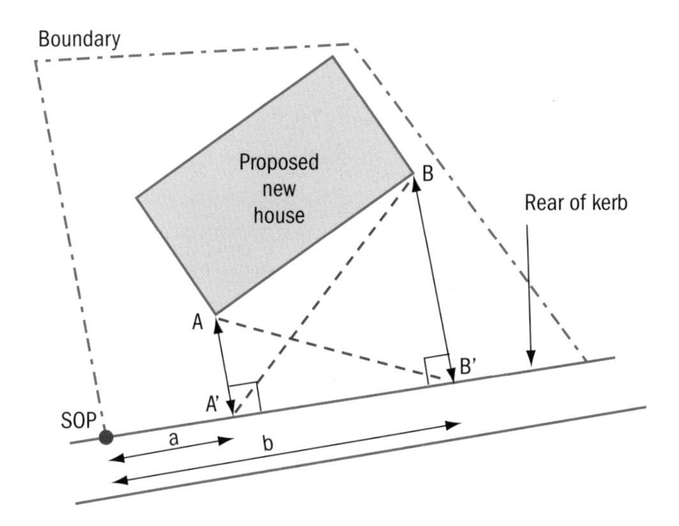

Figure 10.33: Site layout

Reconnaissance and preparation

First, check with the architect or project manger that you have the latest setting-out drawings, usually noted by the date and revision number included in the title block. You should never scale off measurements from drawings or make other assumptions without first checking with the architect or project manager. It is advisable to walk over the whole site before you start in order to familiarise yourself with the boundary positions and all existing features of the site, as well as the slope of the site and visibility issues, for example overhanging trees. Check that the drawing has sufficient dimensions that adequately locate the building on the site relative to two or three existing fixed points, called the **primary setting-out points** (SOPs).

Key term

Primary setting-out points (SOPs) – these are existing identifiable, fixed features to which all new building work is related in terms of their setting out dimensions and angles.

The SOPs could be a pre-existing survey station from a previous control traverse or may just be dimensions from the rear face of the kerb or from the centre-line of the adjacent road. In order to accurately position a building, there must be at least two accurately dimensioned points; a third point is good to act as a check, which should give the horizontal position. There should also be information on the required reduced levels based on an established Temporary Bench Mark on site.

Before you arrive on site, you should have a clear idea of how you are going to set out the given dimensions and do as many calculations and sketches as possible. You may need to discuss with the project manager where they would like the profiles to be placed so that it does not interfere with the movement of vehicles, storage of materials or the like.

Make sure that all your setting-out equipment is correctly adjusted and in order. Typical equipment for a small domestic development includes:

- two measuring tapes (30 m or 50 m), preferably steel
- levelling equipment such as Cowley, optical, laser or digital level as appropriate
- standard 20" theodolite or optical site square for establishing angles on complicated plan layouts
- various hand tools for setting up the corner pegs or constructing the profiles such as lump hammer, claw hammer and hand saw
- selection of round-headed nails and/or cartridge gun
- string to line in wall positions
- spray paint to mark out stations and line in foundation trench positions
- personal protective equipment; hard hat, high visibility jacket and safety boots as a minimum.

Keep your equipment dry and clean at all times. Steel tapes should be lightly oiled and levels or theodolites should be left to dry out before storing way in their boxes. Do not carry instruments attached to a tripod over your shoulder as this will distort the threads and bend the vertical axis of the instrument. Finally, never leave your equipment unattended as it can fall prey to opportunist thieves, particularly when used in public areas.

Setting out the corners of the building

When setting out the corners of the building, it is important that you take your time and do not rush. Pressure from site management and subcontractors can be great, but always remember that accuracy is more important than speed in setting out work. Locate the main SOPs and mark with spray paint. These are your reference datum and must be protected from accidental damage and vandalism with suitable fencing or barriers. Later you may have to re-set out some of the building positions so you will need to locate the SOPs or TBM.

Remember!

When setting out, accuracy is essential! Don't rush – take your time to get the work right.

Step 1: Establish the corners of the front of the building, or a main building line, from the given SOPs. In the example in Figure 10.32, the corners of the building A and B are at right angles from the rear of the kerb to points A' and B'. These points are located from the SOP by given distances a and b.

To establish point A, you need to construct a right angle at point A'. This can be achieved using one of the following:

- a 3:4:5 builder's square, which is a triangle made from three lengths of timber in the ratio of 3:4:5
- a theodolite or site square to swing out an angle of 90° from the rear kerb line
- two tapes and applying Pythagoras' rule to determine the diagonal line X; this is the cheapest and most convenient method as illustrated in Figure 10.33.

Using the third method (two tapes) mark out peg A by pulling out a distance of 6.75 m from peg A' and a distance of 14.15 m from peg B'. Where these two tapes cross will give peg A at right angles to line A'B'. In a similar way, peg B can also be fixed.

$X^2 = Y^2 + Z^2$

$X^2 = 6.750^2 + 12.850^2$

$X^2 = 45.563 + 165.123$

$X^2 = 210.686$

$X = \sqrt{210.686} = 14.5150$ m

Step 2: Establish the other corners of the building by setting out a right angle

The remaining corners of the building can be positioned at right angles to each other by using Pythagoras' rule to calculate the diagonals. As a check measurement, the external perimeter measurements should also be checked. All measurements should be within 5 mm for every 10 m of length (see below).

Step 3: Erect the profile boards

The profile boards will remain in place throughout the job as a permanent reference to fix the position of both the substructure and superstructure. They usually consist of a horizontal timber member fixed to two stout timber posts at a given height (see Setting out the levels on page 382), placed at least 4 m away from the proposed construction works if machine dug.

Nails or saw cuts are used to mark out the face of the wall. You will find these positions by pulling string lines tight between the corner post nails and extending them to the profiles. Once the outside face of the wall is marked on the profile, nails or saw cuts are marked out along the top of the profile showing the centre line of the wall/foundation or the positions of each individual leaf of the cavity wall. They also need to be painted in red and white chevrons so that they cannot be run over or mistakenly pulled up.

Step 4: Excavate the foundations

Using string stretched between the marks as a guide, apply spray paint to the ground. The excavator operator can then use this line to guide the bucket of the machine to pull out the strip footing or ground beam. The corner pegs will, of course, be lost, but with the saw cut marks on the profile boards, they can be re-established to find the position of the following brickwork walls, etc.

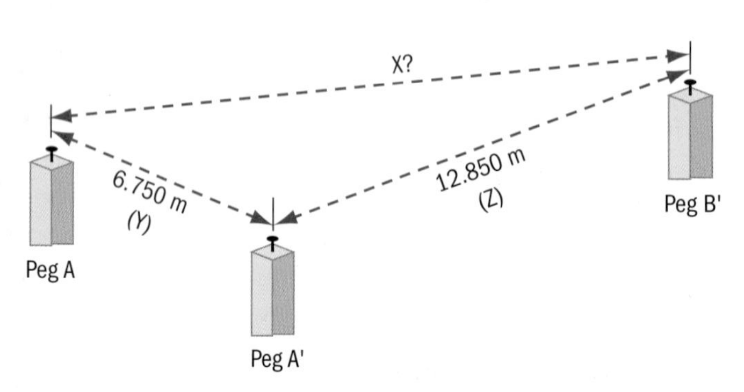

Figure 10.34: Setting out of a right angle by tape: step 1

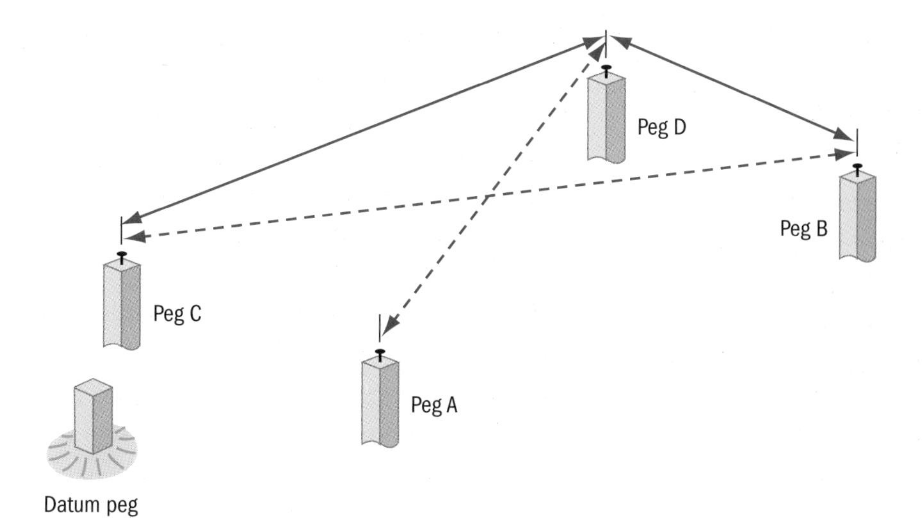

Figure 10.35: Step-by-step setting out of a small building: step 2

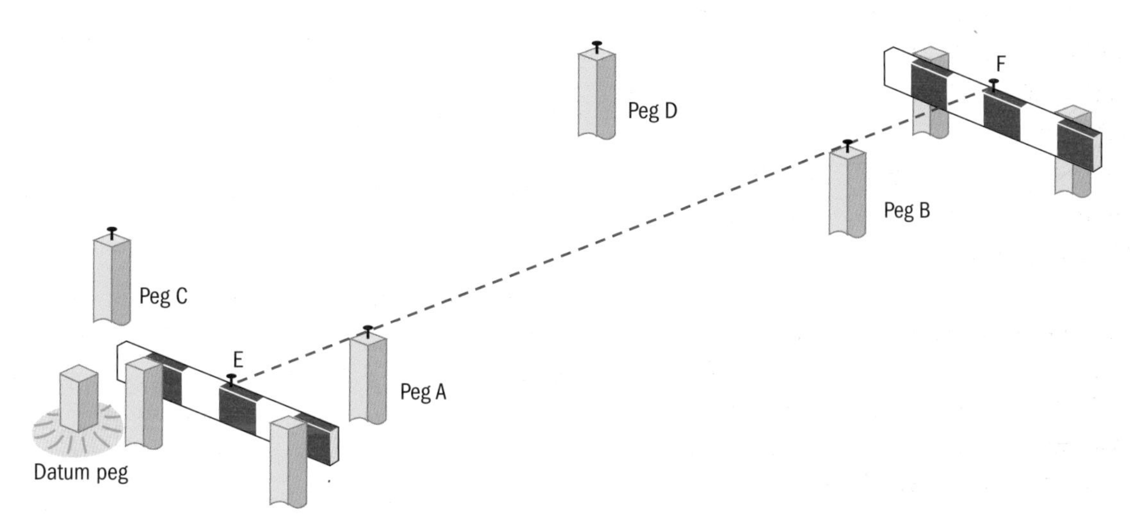

Figure 10.36: Setting out of a small building: step 3

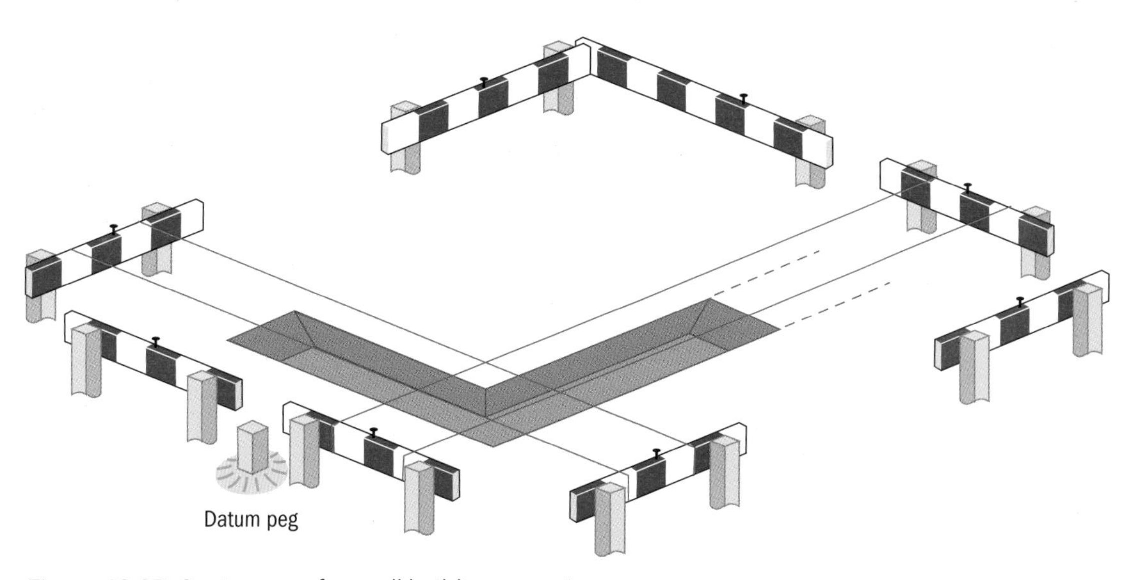

Figure 10.37: Setting out of a small building: step 4

Setting out the levels

The vertical positioning of the building is just as important as the horizontal positioning. The heights of floor levels, window levels and roof ridges, for example, must be constructed exactly how they were planned. There have been cases where levels have been wrong and the height of the building contravened the planning restriction and had to be demolished!

The vertical control is established on site by setting up a Temporary Bench Mark (TBM) to provide a fixed reference point for all levels on site. This is a timber post or steel bolt cast into a concrete block and fenced off with suitably painted barriers or tape. It is given an arbitrary height value, but in some cases may be transferred from a local Ordnance Survey Bench Mark (OSBM). It has to be protected to make sure that it does not get knocked over for the duration of the construction project (Figure 10.38).

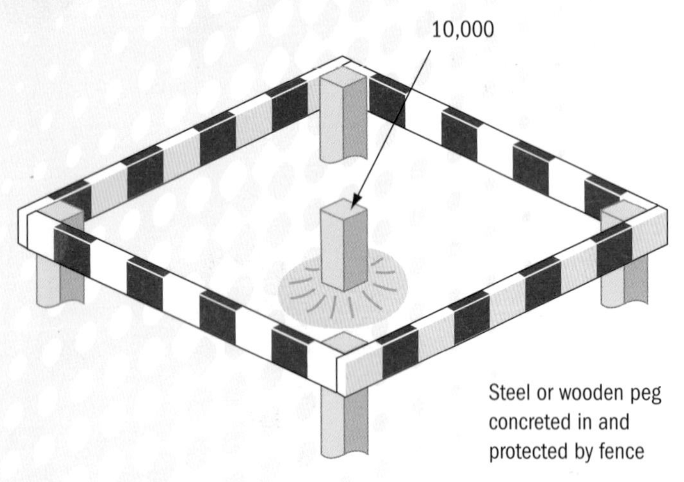

Figure 10.38: Temporary site bench mark (TBM)

Equipment used for setting out levels

Spirit level

The simplest form of transferring a level from one point to another is a spirit level (see Figure 10.38) and straight edge. This is very useful for small-scale craft operations such as setting out pegs for construction of a level area of timber decking. It does have its limitations as it cannot easily be used to transfer levels over longer distances more than about 4.5 m, and is difficult to set out gradients.

Water level

The water level comprises a 15-metre long rubber hose filled with water with Perspex® cylindrical end pieces and screw caps. This is a traditional method of transferring levels, particularly in brick-laying, which relies on water being able to find its own level. This is more useful than using a spirit level as the level can be transferred over much longer distances, and even be taken onto other rooms within the building.

Cowley level

The Cowley level (see Figure 10.39) is even more precise than the water level for transferring levels and can be used accurately to set out slopes and gradients quickly and efficiently. It comprises a system of pivoted mirrors that hang vertically held safely within a metal case (the Cowley level), a lightweight metal tripod and a cross-bar target staff. The level swivels 360 degrees on a pin and is accurate up to about 30 m. It is also a considerably cheaper piece of kit when compared to more complex optical levels. Cowley levels are limited to a range of about 30 m with an accuracy over that distance of +/ –5mm.

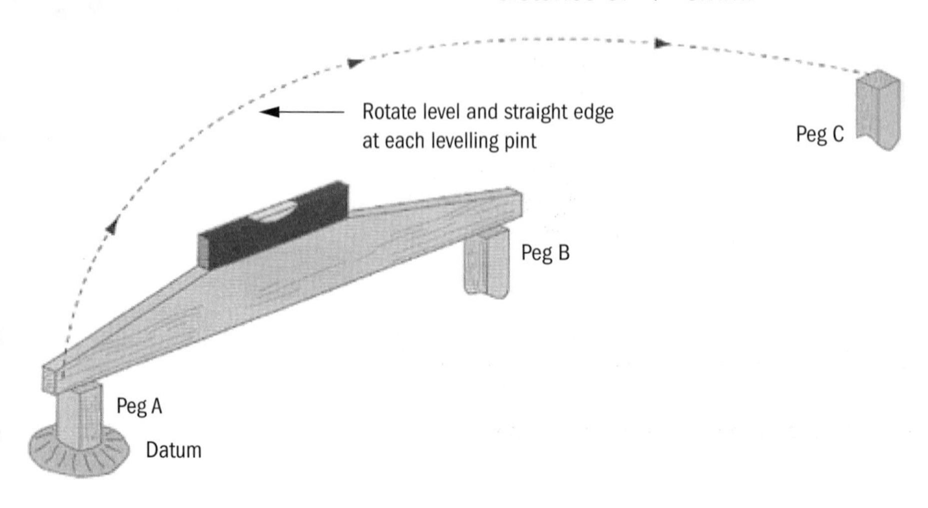

Figure 10.39: Spirit level equipment

The procedure for levelling with a Cowley level is as follows.

- Securely set up tripod and lower level onto pin.
- Position staff on datum and observe through viewfinder.
- Move site rail up or down to obtain image A or D.
- Secure site rail with screw.
- Place staff on peg A and rotate level to observe.
- Adjust height of peg to obtain image A or D.
- Peg A will then be level with datum peg.

Optical level and 'E' staff

The optical level and 'E' staff are some of the most useful pieces of equipment, provided they are handled with care. Their operation and range have been covered earlier on pages 356–57 and are frequently used in setting out all types of levels in construction.

Laser levels

Laser levels are increasingly being used for setting out in preference to the Cowley level. They are far more accurate and have a much wider range – some as much as 300 m. They do not require any visual sightings to be taken and are easy to set up and use. Most instruments also provide an automatic warning signal if the laser gets knocked or is unintentionally moved.

There are many types of laser levels in construction, including pipe laying lasers and rotating lasers. The rotating laser is fixed to a tripod, which when switched on automatically finds its level, then generates a red horizontal line that can be picked up by a suitable sensoring device to give an audible or visual reading of the height. A typical application is in the control of excavation depths and grading level surfaces where the sensor is fitted to the excavating bucket or blade as shown in Figure 10.41.

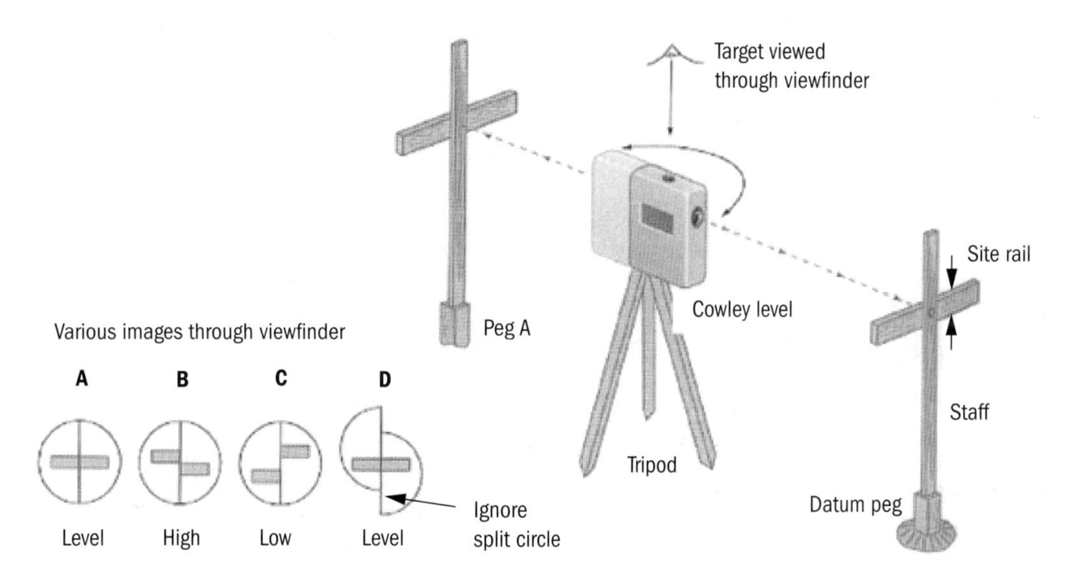

Figure 10.40: A Cowley level

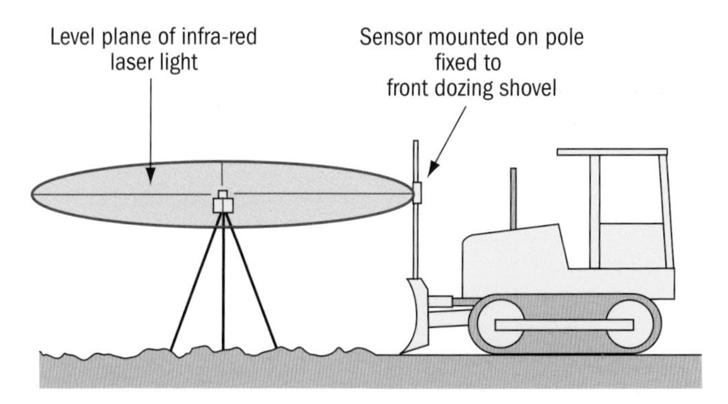

Figure 10.41: Use of the laser level for earthworks

Worked example

The proposed reduced levels of the foundations and floor slab of a house on a site are given on the architect's drawing (see Figure 10.41) as:

Finished floor level (FFL) 10.15 m

Finished ground level (GL) 9.85 m

Formation level of foundations (bottom of foundation) 9.00 m

Top of concrete strip 9.15 m

These are based on a site TBM with a reduced level of 10 m as shown in the cross-section in Figure 10.42.

Studying the reduced level, you can see that the depth of the foundation is:

9.85 m – 9.0 m = 0.85 m

What do you think is the thickness of the concrete strip?

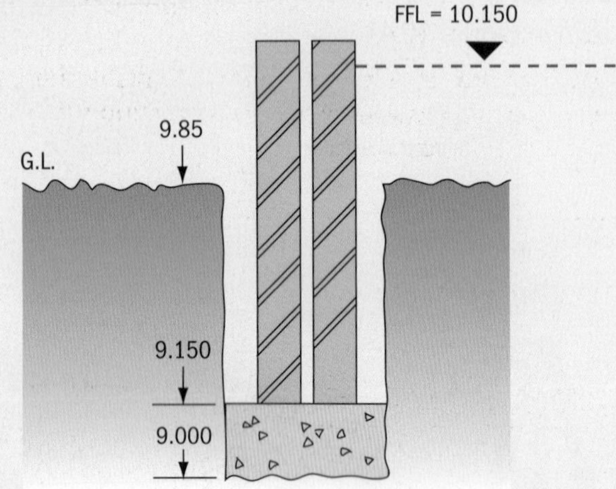

Figure 10.42: Use of reduced levels for setting out house foundations

The profile boards are located close to the building and need to be set at the correct height such that the top of the profile is at the same height as, say, the finished floor level (FFL), in this case 10.15 m. The level instrument is set up, and a backsight taken onto the TBM. This gives the height of collimation (HPC) of the level instrument, i.e. the reduced level of the line of sight. If the backsight reading was 1.355 m, then the HPC will be:

HPC = 10 m + 1.355 m = 11.355 m

Now to set up the profile board at 10.150 m, the value to be read on the staff needs to be:

Staff reading for FFL = 11.355 m – 10.15 m = 1.205 m

The staff is moved up and down by the assistant surveyor until 1.205 m is read on the staff, then the height of the bottom of the staff is marked off on one of the uprights of the profile. This is the finished floor level. The profile board is then nailed to the uprights so that the top of the board coincides with the mark on the upright. The profile can now be used to set out all the required levels by using a spirit level, Cowley level or pulling out a string line between two profile boards set at the same height.

Alternatively, heights can be set out directly by reading the correct value off the staff, for example the staff reading needed to set out the bottom of the foundation is

11.355 m – 9.000 m = 2.355 m

You can use this technique with laser levels that have infrared sensors fitted to the staff. These sensors can be adjusted to the correct position on the staff, so that when the staff is raised or lowered a visual and/or audible signal tells you that you have found the required reduced level.

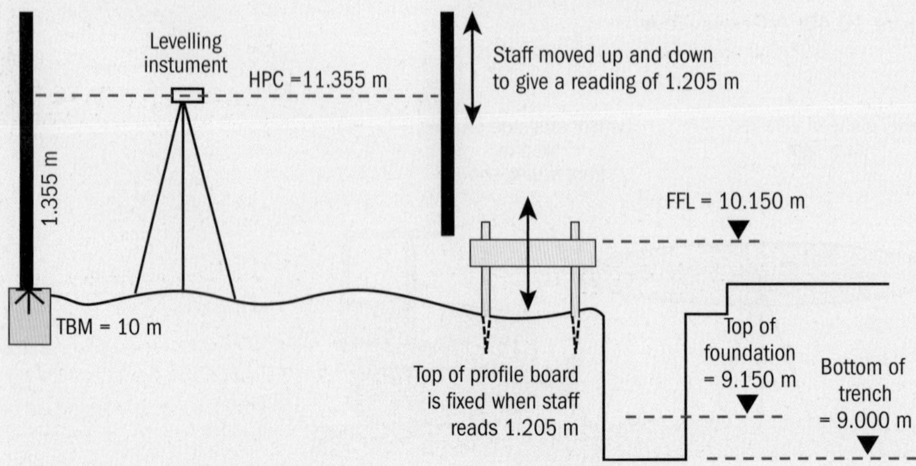

Figure 10.43: Setting out of house using the staff reading method

Assessment activity 10.4

BTEC

1. Study the foundation plan for the small house shown in Figure 10.44. The walls are to be 300 mm cavity walls built centrally on a 600 mm wide strip foundation, and all internal walls are constructed from 100 mm thick concrete blockwork.

   You are required to set out the corner points of this building both in plan and height using traditional methods.

   As part of your preparation:

   a Make a copy of the plan clearly showing all the dimensions you intend to use to set out the corners of the building.

   b Sketch on the plan the positions of all the required profiles.

   c Produce a checklist of all the equipment and materials needed to undertake the setting out, including personal protective equipment. **P7**

2. Select a level piece of land close to your place of study where you are permitted to undertake practical surveying work. In a group of three, set out a baseline approximately 15 m long marked out with ranging rods at each end. This represents the front boundary line of the small house shown in the drawing. One end of the line represents SOP 1 shown on the drawing. Set out the corners of the building using steel tapes, pegs and nails and undertake the necessary diagonal checks. **P8**

3. Establish a suitable TBM close to the building set out above and take its value as 21.90 m. Then, using the corner positions set out previously, erect suitable timber profiles to construct all foundations and ground floor walls in both plan and vertical alignment. The tops of the profiles are to be at the height of the given finished floor level. Check your work for accuracy and take a photographic record of your work. **M3**

   Produce a brief written report that explains the constraints on positioning and protecting the setting-out profiles. **D3**

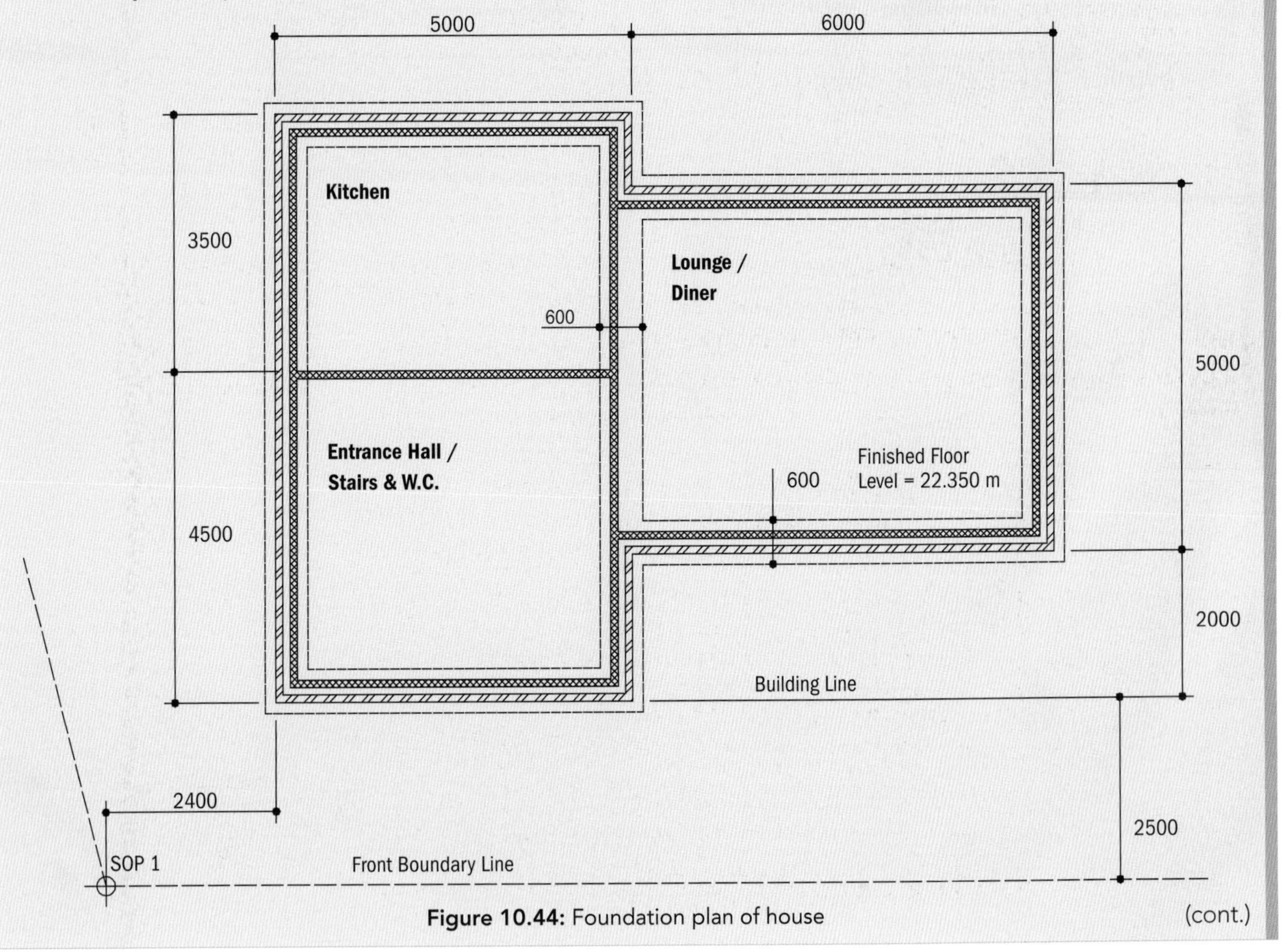

Figure 10.44: Foundation plan of house

(cont.)

Grading tips

P7 In demonstrating that you can identify setting out terminology make sure you include definitions of all the correct setting-out terms such as 'baseline', 'setting out point(s)', 'check lines' for example. It would also be good to identify what you mean by including clear labelled diagrams.

For **M3** you need to clearly show how your site measurements relate to your level height calculations. Provide a labelled sketch of the main features illustrating all the key dimensions and calculated values.

For **D3** you need to outline the practical issues in setting out profiles with reference to buildability, security, health and safety, and accuracy.

P8 In all practical work it is important that you plan the work carefully and carry out all your calculations of diagonal lengths and check distances beforehand. A well annotated sketch including all the calculations helps you to demonstrate your knowledge to the assessor. It is useful if you take a photographic record of your practical work which can supplement any witness statements that your tutor provides.

PLTS

Demonstrating that you can cope with organising your time and resources by prioritising your actions and managing the health and safety risks helps develop your skills as a **self-manager**.

Functional skills

This task gives you the opportunity to demonstrate that you can use appropriate **Mathematics** in your setting-out procedures and can also evaluate their effectiveness at each stage.

Rashdah Jarventre

Highways surveyor

Rashdah works for a local county council as a highways surveyor. Part of her job is to undertake built surveys of existing roads and pedestrian pavements; these are then used to help the highways engineers to plan such things as traffic calming measures in built-up areas, improvements to traffic junctions to help traffic flow and the creation of new user-friendly cycle paths.

The last main project that Rashdah worked on was the survey work used to design a range of speed control measures for a busy suburban road often used as a short cut by rush hour traffic and commercial vehicles. The new measures included a combination of road traffic humps, automatic speed indicator signs and pelican crossings.

Rashdah left school after studying A levels in Geography, Business Studies and Mathematics and went on to study part-time for a BSc (honours) degree in Civil Engineering Surveying. Even though she now uses some of the most up-to-date surveying kit and equipment the basic principles of linear, level and angular measurement that she learned help her to carry out her job effectively.

Rashdah enjoys her job tremendously. She enjoys its variety and likes the interaction with the various members of the build team.

Think about it!

- What basic skills would you need for this role?
- What would be your ideal job in civil engineering?
- Would you be able to work outside almost every day?

Just checking

1 Explain the role site profiles have in the setting out of sewers.

2 What is the most appropriate setting out equipment for setting out the profile heights for a sewer?

3 Select the most appropriate setting out equipment for excavating the road base for a motorway construction.

4 Describe what working from the 'whole to the part' means when carrying out a survey reconnaissance.

5 Define each of the following terms and illustrate with labelled diagrams: (a) baseline, (b) survey line, (c) detail line, (d) check line, (e) offset, (f) tie line.

6 Define the following terms used in levelling: backsight, intermediate sight, foresight, bench mark and reduced level.

7 The standard metric staff uses the 'E' motif to denote the measured height. How high is the letter 'E' and what is the minimum value that can be estimated in good conditions?

8 Briefly explain the main parts of a theodolite and its controls.

9 Describe the three main checks on a theodolite to ensure that it is in good adjustment.

10 State the difference between an angle of depression and an angle of elevation.

Assignment tips

- In your assignment work, always identify the equipment that you use by including photos or diagrams.
- In practical work, always ensure that your tutor has an opportunity to observe you carrying out the work, and take plenty of photographs.
- Your survey reports should emphasise the importance of neat and accurate fieldwork and discuss ways of checking and correcting results where errors have been found.

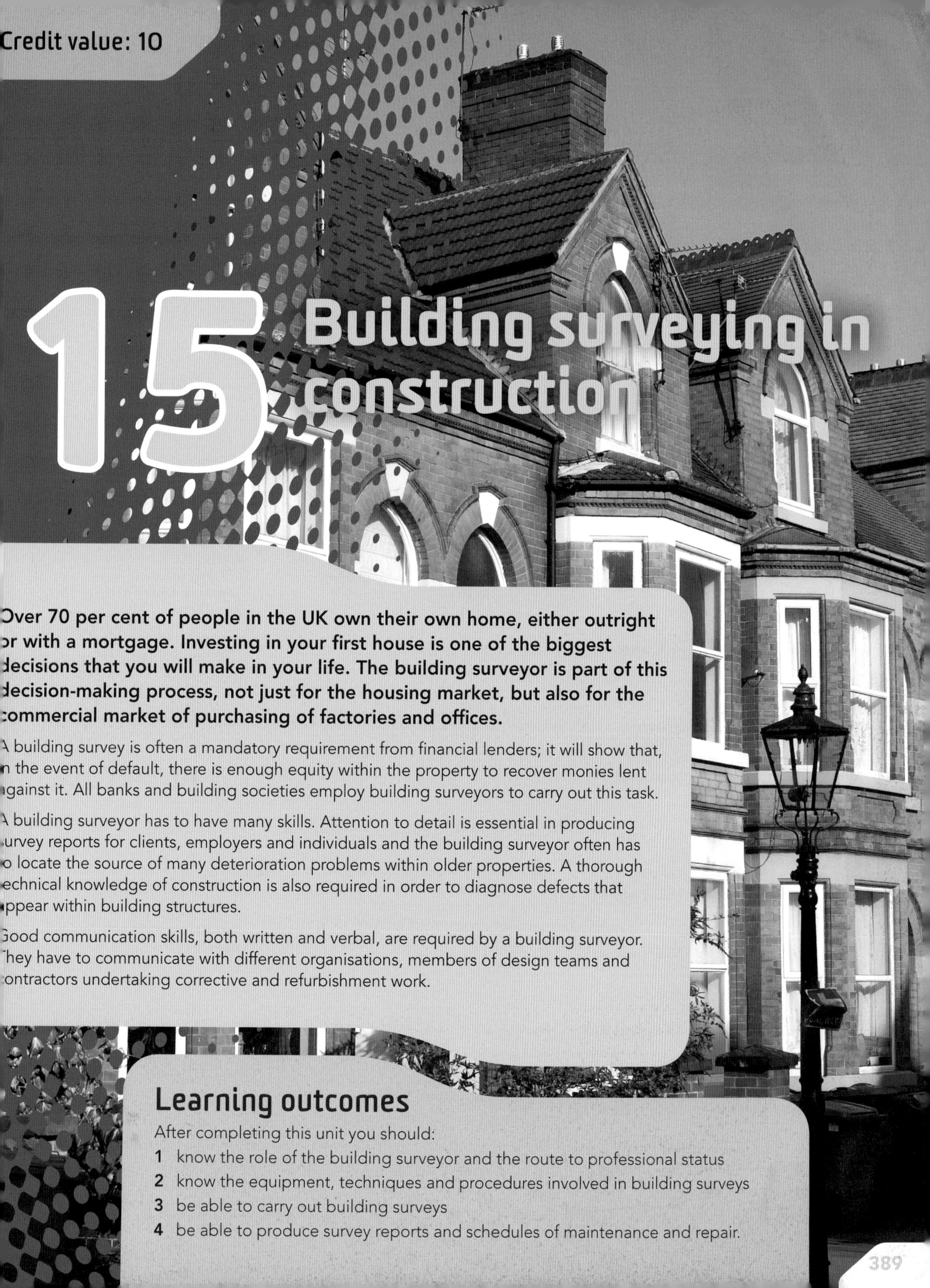

Credit value: 10

15 Building surveying in construction

Over 70 per cent of people in the UK own their own home, either outright or with a mortgage. Investing in your first house is one of the biggest decisions that you will make in your life. The building surveyor is part of this decision-making process, not just for the housing market, but also for the commercial market of purchasing of factories and offices.

A building survey is often a mandatory requirement from financial lenders; it will show that, in the event of default, there is enough equity within the property to recover monies lent against it. All banks and building societies employ building surveyors to carry out this task.

A building surveyor has to have many skills. Attention to detail is essential in producing survey reports for clients, employers and individuals and the building surveyor often has to locate the source of many deterioration problems within older properties. A thorough technical knowledge of construction is also required in order to diagnose defects that appear within building structures.

Good communication skills, both written and verbal, are required by a building surveyor. They have to communicate with different organisations, members of design teams and contractors undertaking corrective and refurbishment work.

Learning outcomes

After completing this unit you should:

1. know the role of the building surveyor and the route to professional status
2. know the equipment, techniques and procedures involved in building surveys
3. be able to carry out building surveys
4. be able to produce survey reports and schedules of maintenance and repair.

Assessment and grading criteria

This table shows you what you must do in order to achieve a pass, merit or distinction grade, and where you can find activities in this book to help you.

To achieve a **pass** grade the evidence must show that you are able to:	To achieve a **merit** grade the evidence must show that, in addition to the pass criteria, you are able to:	To achieve a **distinction** grade the evidence must show that, in addition to the pass and merit criteria, you are able to:
P1 describe the role of the building surveyor **See Assessment activity 15.1, page 396**	**M1** prepare a typical job description for a building surveyor **See Assessment activity 15.1, page 396**	
P2 describe how the building surveyor interacts with other members of the building team **See Assessment activity 15.1, page 396**		
P3 outline the qualification route to professional status **See Assessment activity 15.1, page 396**		
P4 describe the procedures and techniques used in the performance of building surveys **See Assessment activity 15.2, page 401**	**M2** differentiate between the equipment, techniques and processes used in two different kinds of building survey **See Assessment activity 15.2, page 401**	**D1** evaluate the contractual obligations and legal constraints applicable to a building survey **See Assessment activity 15.3, page 408**
P5 identify the equipment used to perform building surveys **See Assessment activity 15.2, page 401**		
P6 identify legislation relevant to building surveys **See Assessment activity 15.3, page 408**		
P7 describe health, safety and welfare issues associated with building surveys **See Assessment activity 15.4, page 409**		
P8 carry out a simple measured survey **See Assessment activity 15.5, page 411**		
P9 carry out a dilapidation survey **See Assessment activity 15.5, page 411**	**M3** relate the nature of the data collected to the type of building survey being carried out **See Assessment activity 15.5, page 411**	**D2** justify the recommendations made in two different types of building survey reports **See Assessment activity 15.5, page 411**
P10 carry out a condition survey **See Assessment activity 15.5, page 411**		
P11 compile data collected during building surveys **See Assessment activity 15.5, page 411**		
P12 record and present survey data in appropriate formats **See Assessment activity 15.5, page 411**		
P13 prepare maintenance and repair schedules **See Assessment activity 15.6, page 412**		

How you will be assessed

The evidence requirements for pass, merit and distinction grades are shown in the grading criteria grid. Evidence for this unit may be gathered from a variety of sources, including well-planned investigative assignments, practical work or reports of practical assignments. You will be given written assessments to complete for the performing element of this unit. These will contain a number of assessment criteria from pass, merit and distinction

This unit will be assessed by the use of three assignments:

- Assignment one will cover P1, P2, P3 and M1
- Assignment two will cover P4, P5, P6, P7, M2 and D1
- Assignment three will cover P9, P10, P11, P12, P13, M3 and D2.

Because of the credit value of the new nationals the third assignment would be better broken down into several hands on class exercises

Joanna

I thought that buying a home was straightforward and that all you did was sign a contract; after reading this unit I now realise that a building survey is often worth its weight in monetary terms when buying a home.

This unit has given me an understanding of different methods available to survey buildings and the equipment needed to undertake these surveys in order to produce a report for a client.

This unit has also made me aware of the financial implications associated with the purchase of a house and how a building survey may affect the final price.

Over to you

- What do you already know about the role of a building surveyor?
- What methods are used to survey buildings?
- What are you looking forward to learning about in this unit?

1 Know the role of the building surveyor and the route to professional status

Build up

The building surveyor

A surveyor has to have a great deal of technical knowledge and they often have to diagnose defects to find the cause of a fault with a building. In groups, think about the following questions.

- What type of qualifications does a surveyor need?
- How long will it take to train as a building surveyor?

Role of the building surveyor

Completion of measured surveys

Measured surveys are, in effect, dimensional surveys of buildings. These can be as a sketch drawing with dimensions laid out on it so that an architectural technician can draw to scale the existing elevations and plans of an existing building.

The building surveyor's role in producing measured surveys is to ensure that they are accurate, can easily be understood by another person and contain sufficient detail and dimensions in order to reproduce a second detailed drawing.

Dilapidation surveys

Dilapidation surveys apply to tenancy agreements as we shall see later. The building surveyor's role in this is one of diplomacy; they are checking a tenant's flat, house or maisonette for damage which may result in the loss of a deposit by the tenant.

The surveyor must be fair and reasonable in their actions, and carefully record the level of wear and tear before a tenant moves in or out so a subjective assessment can be made. This may often result in a disagreement between landlord and tenant so the surveyor must be sure that the facts provide evidence for withholding deposits from tenants.

Condition surveys

Condition surveys may also relate to the condition of housing stock. Once again, the building surveyor may have to enter premises or homes to carry out an inspection with the tenant in occupation. The surveyor has to be accurate in assessing conditions; the whole of the housing stock will require assessment as to how much money is required to be spent. Awareness of financial costs and details are therefore an essential requirement of this role.

Production of survey reports

The building surveyor needs to produce accurate reports for two reasons:

- many clients make large financial investments on properties based on the recommendation of a surveyor as to the properties' value. Monies are being lent to the buyer on the condition of the property. Therefore, the communication within a survey report must be clear, concise and accurate, with no ambiguities
- to enable the surveyor to take out **professional indemnity insurance** cover.

Key term

Professional indemnity insurance – the insurance policy that building surveyors carry so that if they make an error which results in a loss for a client, the surveyor and the company they work for are covered against legal proceedings.

Production of schedules of maintenance and repair

The surveyor's role involves assessing what maintenance is required on a building or structure and evaluating the financial costs involved in making repairs to an acceptable level. Many commercial buildings have a set standard to maintain, such as schools and hospitals, where safety plays an important part.

Interaction with other members of the building team

Client

The client would engage a building surveyor to look at their buildings or housing stock; the surveyor assesses the buildings for the level of repairs, for value and for condition. The client is ultimately the building surveyor's employer. They are engaged for a fee or salary and so must act professionally in all dealings with the client. A building surveyor may interact directly with a client, for example in a house valuation for a potential purchaser, or through an agent such as a mortgage lender. **Approvals** and instructions should be backed up in writing, as would expenditures on a client's budget.

Building owner

Building owners and clients are often one and the same person. A building owner would interact with a building surveyor on many points, including:

- the cost of repair works
- measured surveys in land purchasing
- drainage surveys
- condition surveys
- dilapidation surveys.

This interaction would involve the contractual engagement of the building surveyor to undertake works on the building owner's behalf. The surveyor would also have to obtain any necessary historical information, for example a health and safety file.

Architect

The architect would commission a building surveyor to take all the dimensions of an existing structure so that it could be drawn to scale and proportion. This is done to provide accurate details of the existing building and would be used for planning and building regulation controls. An architect may also engage a building surveyor to advise on areas outside of their expertise, such as **listed building consents**, structural and general building defects, damp surveys and timber surveys. Building surveyors who are employed by English Heritage would liaise with architects on aspects of **listed design approval** to enable the structure to be maintained for the community as a whole.

Key terms

Approvals – the agreement of the cost for the survey and instruction to proceed.

Listed building consent – when a building is graded and listed as a national heritage building and thus cannot be altered or demolished without consent.

Listed design approval – all works on listed buildings must match their original character and materials.

Quantity surveyor

Interaction between the quantity surveyor and the building surveyor is on a financial basis. When a building surveyor acts as a consultant on a refurbishment contract, they liaise with the client's quantity surveyor to ensure that estimates for the work fall within the client's budget. Monthly valuations of the level of works undertaken could also be the responsibility of a building surveyor.

Structural engineer

The building surveyor may liaise with the structural engineer if structural surveys of buildings show signs of structural problems. The building surveyor will call the structural engineer who can then undertake a more detailed structural investigation. The structural engineer will be able to guide the building surveyor on the best possible course of action for the client or building owner.

Clerk of works

A clerk of works may be employed on large construction projects. They would interact with the building surveyor on larger refurbishments run by the building surveyor on issues of quality, instructions required and record keeping. The clerk's duties are to ensure that the work produced is in accordance with the specification and drawings.

Main contractors and subcontractors

If the building surveyor has been engaged by the client to undertake, control and manage a construction project, then a great deal of interaction will occur between the building surveyor and the main contractor. Not much interaction will take place between the building surveyor and the subcontractors.

However; there is no contract between the subcontractors and the client, employer or building owner.

The building surveyor will interact on some common issues with the main contractor, including:

- written instructions
- verbal instructions
- specification queries
- colour schedules
- variations
- drawn information.

Meetings are the most economical way to coordinate these discussions; with all parties face to face and minutes recorded.

Local authorities

Building surveyors interact with several different local authority departments. Examples of these might be:

- highways department – to locate drainage details, runs and depths where drainage surveys are required
- tree and woodlands officer – to discuss **tree preservation orders (TPO)**
- conservation officer – to discuss aesthetical matters of a building within a conservation area
- fire officer – to discuss fire escape provisions
- planning officer – to discuss planning applications
- building control – to discuss aspects of the Building Regulations.

Key term

Tree preservation order (TPO) – an order to protect a tree from being cut down, damaged or altered without permission of the local authority.

Health and Safety Executive

The interaction between the building surveyor and the Health and Safety Executive (HSE) would be on matters of health and safety relating to older structures, including:

- lead – in paints and water pipes
- asbestos – location and potential for harm.

Both lead and asbestos have their own legislation and the building surveyor would have to notify the HSE via a specialist contractor to have these removed from a building before any refurbishment work could commence. The HSE would especially need to know how the contractor plans to deal safely with asbestos, which has a very high long-term risk.

Route to professional status

Secondary education

The qualification route to being a building surveyor starts at secondary education and each of the following stages builds from this foundation (see Figure 15.1). It is important at this stage for students to obtain detailed information on the progression routes that are possible. The flow chart illustrates the development from GCSE to university degree. A good standard of GCSEs, particularly in Maths and English, will be required to gain entry to the Nationals Scheme or A levels. Science and Design Technology would help to reinforce progressive learning. Progression from GCSEs is to A levels either at sixth form college or a further education college and an entry level qualification to a university degree.

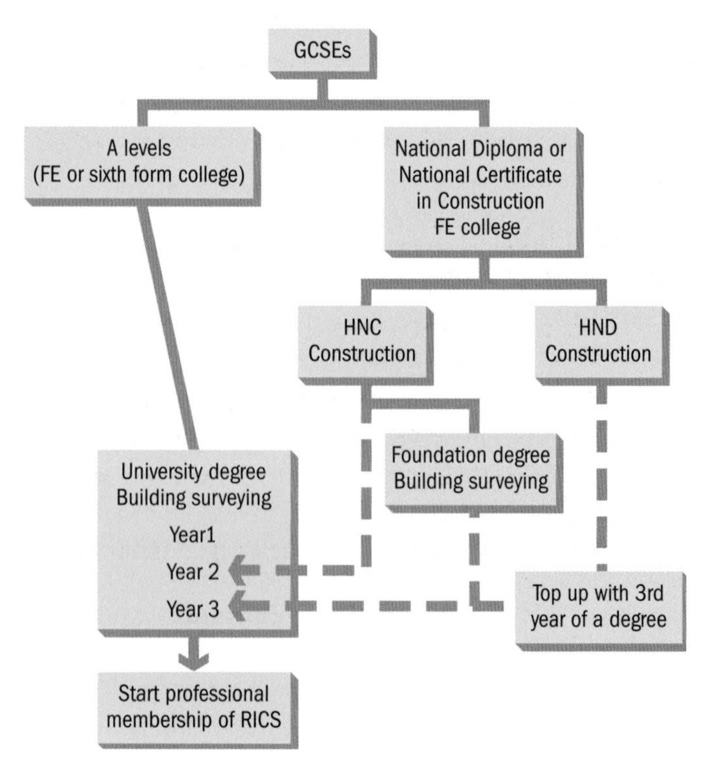

Figure 15.1: Qualification routes

National Certificate/Diploma

The National Certificate or Diploma is an alternative route to A levels. They are vocational qualifications providing detailed construction modules relevant to this career pathway. They also provide an accepted university degree entry route.

The National Certificate is for the part-time learner and the Diploma is for the full-time student. The certificate consists of 12 modules and the diploma 18 modules. Because the qualifications are vocationally specific they give the learners sound background knowledge of construction techniques for both domestic and commercial applications.

Higher National Certificate (HNC)/ Diploma (HND)

The HNC or HND are one progression route from the National Certificate and Diploma in Construction. They can also be achieved via A levels and by mature students with appropriate experience within the construction industry.

The HNC progression can lead into several different areas as follows:

- a second-year degree course
- a foundation degree top-up year
- a final year top-up from HND to full degree.

Both qualifications lead the student into technical pathways with the HNC having 10 units and the HND, 16 units.

Honours degree accredited by professional body

Most professional associations such as the **CIOB** require a higher-level degree in order to attain full membership status and the Honours degree is recognised by the **RICS**, and narrows down the student's career path. Learners should check the status of their proposed degree against the list of accredited degrees that provide exemptions from professional institutions' exams. There is now a great deal of flexibility with degrees from full-time to part-time to distance learning through the Open University.

Key terms

CIOB – Chartered Institute of Building – a professional association for construction personnel.

RICS – Royal Institution of Chartered Surveyors – a professional association for surveyors.

Professional membership of RICS

Membership of the RICS enables surveyors to obtain full indemnity insurance and provides a professional trademark for the chartered surveyor. All members of the RICS follow a code of ethics in the way they conduct themselves and their businesses in dealing with clients, employers and building owners. Membership has several routes:

- a choice of three postgraduate degrees after obtaining a first degree
- an academic route for tutors, lecturers and university teaching staff
- a senior professional route
- a technical route for personnel assisting the chartered surveyor.

The membership routes all involve some experience requirements except for the direct graduate route 1. You should refer to the RICS website for more detailed information on membership (www.rics.org).

Professional membership of CIOB

The CIOB provides a chartered status to its members, who are expected to act under a code of conduct. There are various levels of membership in the CIOB and they also provide qualifications, experience and training.

Assessment activity 15.1

P1 P2 P3 M1 BTEC

1. You are a new recruit to a building surveying practice and have been asked to help with marketing the building surveying posts they are currently trying to fill. You have been tasked with producing a promotional leaflet that needs to cover the following areas.
   - Describe the role of a building surveyor. P1
   - Explain the building surveyor's interaction with other members of the building team. P2
   - Outline the qualification route to becoming a professional surveyor. P3
2. The manager of the practice thinks you have done a fantastic job on the marketing leaflet. He has now asked you to prepare a typical job description for a building surveyor. M1

Grading tips

For P1 you need to describe role of a building surveyor.

For P2 identify who the surveyor may interact with and the nature of these interactions.

For P3 you should describe how to become a full member of the RICS.

For M1 you need to prepare a job description to recruit a building surveyor to a company.

PLTS

Analysing and evaluating information and then judging its relevance and value in becoming a professional surveyor will help develop your skills as an **independent enquirer**.

Functional skills

Discussing who the surveyor will interact with will help develop your verbal skills in **English**.

2 Know the equipment, techniques and procedures involved in building surveys

Types of building survey

Measured surveys

Measured surveys are undertaken for the initial design stages of a refurbishment, remodelling, adaptation or extension of an existing building. Each elevation of the building would be sketched freehand and the dimensions taken in order to reproduce the building to scale, either manually or by using a computer design program (see Figure 15.2). A similar approach is taken with the floor plan, with each floor sketched and the room sizes measured so the plan can be drawn accurately to scale.

The height of a building and different features, such as windows, may be difficult to survey without the use of a ladder. However, if the building is of brickwork, then the courses of bricks could be counted; one metric course is 75 mm, one imperial course is 85 mm. Alternatively, the internal floor to ceiling heights can be measured to establish the eaves' height at guttering level. The roof pitch is often another problem, but this can be overcome by using an adjustable set square and a spirit level.

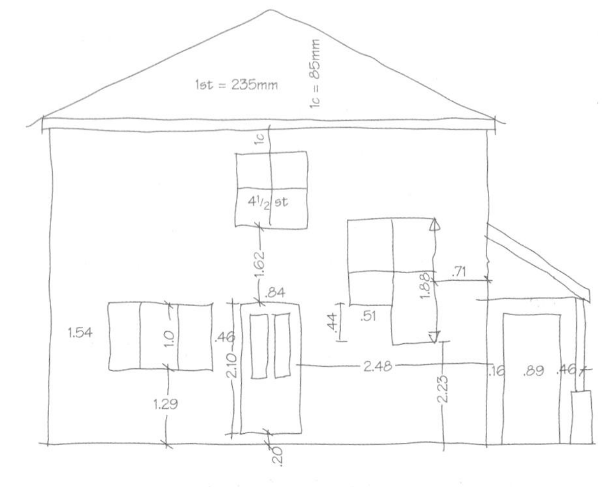

Figure 15.2: Freehand sketch for a measured survey

Access to the loft space will then confirm the roof pitch. It is often a good idea to take along a digital camera which could be used to record all the elevations and details which can then be referred to when redrawing to scale.

Bank or building society surveys

Bank or building society surveys are undertaken by the financial lender to establish whether the valuation of the property being purchased matches its market value. Building surveyors who perform this type of work are almost entirely contracted to the financial institution lending the money to the purchaser, who pays for the valuation as part of the mortgage application process. The building surveyor will assess what the market value is by using recent transactions in the area and by visually inspecting the property for any substantial damage that would reduce the value of the loan to the purchaser.

This type of survey is very simple and often brief; it is only a brief visual inspection of the loft space and possibly some damp readings using a moisture meter, no intrusive work is undertaken. If the valuation report reduces the market value of the property and hence the amount of money that can be borrowed against it, then the prospective purchaser will have to increase the deposit or find the balance in order to proceed with the purchase.

Many lenders require identified works within valuation reports to be undertaken before they will release the whole of the agreed loan. They retain a sum to cover this work, which is released on the works completion. Normally the valuation report is commissioned by the purchaser. The only reason the seller may instruct a valuation is to borrow against the building as an asset which would then require a valuation to confirm value.

House buyer's reports and valuations

The House buyer's report gives a valuation but also includes many other points about the property that the buyer is purchasing. In essence it is a survey that lies between a full structural survey and a valuation and would be instructed by the buyer. The house buyer's report flags any faults or defects that have a high money value. It would contain sections under the following points:

- general description of the property, type and age
- chimney stacks
- general joinery
- services connections for water, gas and electricity
- rainwater goods and pipes
- external decoration
- any evidence of subsidence or movement
- any evidence of woodwork infestation
- any obvious leaks from pipework
- state of electrical wiring
- evidence of damp course work
- evidence of damp to walls and ceilings
- internal decorations
- roof tile check
- loft space check
- outbuildings
- recommendations of further investigation such as damp, sub-floor checks.

This type of report gives the intended purchaser sound building surveying advice on any work that may be required and it can be used as a negotiation tool on the asking price of the seller's property.

If a purchaser is intending buying an older property, then a building surveyor may recommend that a full structural survey report is undertaken to provide evidence of any defects that have occurred over time.

Dilapidation survey

A **dilapidation** survey is used before and after agreements, such as the letting of a house or flat to a tenant or the commencement of construction works on existing sites.

Part of an agreement between a landlord and tenant will often state that any damage as a result of the tenant's occupancy will be paid for by the tenant, usually taken from the deposit held by the landlord. In order to provide evidence of damage, the landlord will undertake a dilapidation survey before a tenant moves in and again when the tenant moves out so the two can be compared for any changes. (See below for the methods involved in carrying out a dilapidation survey.)

Key term

Dilapidation – falling or causing to fall into ruin.

Tenants can also undertake a dilapidation survey to protect themselves against landlords who may falsely charge them for damage that was already there when they took over the tenancy. It is advisable for both parties to agree at the beginning of a lease or rent agreement what state or condition the building is in, and, to some extent normal wear and tear should also be evaluated. This avoids any disputes at the end of the agreement.

Another reason for undertaking a dilapidation survey is to protect other interested parties around a construction site who may suffer damage as a result of the construction works; this is often called a pre-construction survey. An example of when a pre-construction dilapidation survey is done is during demolition works. While undertaking a refurbishment of a building which involves extensive alterations, each neighbour may be asked for access to undertake a dilapidation survey to establish the condition of the shared walls before work starts and then to recheck their condition after the work is completed. If these surveys are undertaken by a professional independent building surveyor, then there should be no disagreement as a result of damage sustained during the construction work. Typical damage could be as a result of:

- excavating close to a neighbour's foundation causing settlement and cracking to internal walls within their property
- demolition works
- piling foundation works
- working on an adjacent chimney causing soot to fall and damage to carpets and furnishings
- vibration causing structural cracks to walls
- excessive dust entering adjacent properties
- damage to water, electric and gas services by excavation
- laying tarmac surfaces to roadways
- the breaking of glass in windows.

If work is being done near something that is very valuable, for example a museum containing many important artefacts, the contractor's insurance company will need to be involved in the dilapidation survey in order to assess the risks associated to the valuable items. This may mean an increase on the insurance premium for the work being undertaken.

Activity: Dilapidation survey

The client you work for owns over 50 flats in the city centre. Part of your work involves landlord/tenant agreements and the release of deposits when each tenant vacates the property and a new tenant moves in. You therefore carry out regular dilapidation surveys.

Outline in detail the benefits of dilapidation surveys for your client.

Condition surveys

The condition survey is a global survey that would, for example, be undertaken by a local authority to check the condition of its housing stock, schools and community buildings. The condition survey tries to predict what future spending levels would be required on upgrades and repairs. Typical sections within a condition survey report would be:

- use of the housing stock: empty or let
- type of dwelling: semi, detached, flat, maisonette
- how many people on average are living in each
- how many dwellings have central heating
- type of fuel used to heat the dwellings
- how many have modern kitchens
- age of housing stock
- repair costs per unit
- fitness standard
- health and safety
- energy efficiency ratings
- renovation grants (Source: headings taken from Dover District Council (2001) 'Private Sector House Condition Survey: Final Report').

Purposes of and processes involved in surveys

Measured survey to produce accurate plans prior to alteration

We have already established the purpose of undertaking a measured survey for initial feasibility design works. The process involves visiting the property in order to draw a freehand outline of each elevation on A4 sheets of plain paper. Horizontal and vertical dimensions are taken to enable a third party to interpret these either manually or digitally. A set of ladders may be needed to accurately establish heights which are out of reach of a 5-metre tape measure. It is also useful to photograph each elevation in order to provide any information that has been missed at the time of the measured survey.

It does not matter where the survey is started as long as the information recorded can be understood by the person who has to reproduce the accurate drawing to scale. Care should be taken to ensure a high degree of accuracy to avoid expensive mistakes. For example, if a surveyor in measuring the building next to the proposed one, records the wrong height, the new building may not marry up with the adjoining properties.

Remember!

Use a pencil to draw freehand sketches. It won't run in wet weather, cannot be washed off easily and does not smudge.

Digital tapes are acceptable, but should only be used on internal dimensions; the accuracy of an infrared or optical beam should not be relied on if the measurement cannot be confirmed from a taped dimension.

Condition survey for prospective lenders and purchasers

The condition survey looks at housing on a smaller scale and takes a snapshot for the prospective buyer of the condition of the house at that moment in time. It is merely an analysis of what condition the building is in, and contains no reasons for defects or any structural or investigation works. This type of report is not as detailed as a house buyer's report.

The basic process involves taking photographs and/or video to illustrate the condition on a digitally dated image. The survey should be divided into external and internal items. Internally, it could be broken down into room-by-room surveys with the condition of each room noted separately. A general section would cover services.

Advice on maintenance, repair and conservation

Clients may wish to know what it will cost to put right the defective parts of a building highlighted within a home buyer's survey report. Advice on maintenance involves how to look after the building's fabric and structure to keep it maintained at a fit-for-use status. Typical maintenance items on a house include:

- ensuring guttering is clean of weeds and runs freely
- external painting is fresh and clean
- weeds removed from gardens, driveways and footpaths
- fences in good condition and painted
- attention to plumbing leaks from overflow pipes causing water staining on external faces of building
- cleaning glazed surfaces and replacing broken glass
- roof tiles and felt in good condition
- internal decorations such as filling cracks and redecorating
- areas of damp treated
- carpets and furnishings clean.

The better presented a property, the more saleable it is likely to be. A building surveyor can identify many of the above 'building' items and provide cost estimates which would enable some negotiation on the asking price of a property.

Repair advice will sometimes involve a specialist. Structural cracks will need a good builder to investigate and repair to a satisfactory level. Often older properties that have settlement may be required to have their foundations **underpinned**.

A good building surveyor can produce a repair schedule for a client with accurate cost estimates so that informed decisions can be made on purchasing a property. Repairs carried out by the property owner can also be cosmetic, and the services of a builder or a construction professional contractor should be sought in order to obtain some form of guarantee. A good building surveyor is most likely to be a member of the Federation of RICS.

Key term

Underpinned – when an existing foundation is settling downwards, which causes structural cracks to the external and internal finishes, and a new, deeper foundation is cast in sections below the old one.

Roof repairs are best left to a roofing contractor who will supply all the specialist equipment required to gain safe access to fix roof tiles or roofing felt problems. While window repairs should be undertaken by a glazier or uPVC window company that can advise on replacements or repair to double-glazed units. As part of the building surveyor's report the quality of the double-glazed units will be checked as will their seals, as there is no guarantee with them.

Conservation

Many local authorities have conservation areas which are designated to have historical significance, natural beauty or important architectural interest and as such have restricted planning permissions. In fact, any tree within a conservation area is automatically protected and must have permission before any work is undertaken on it, for example pruning or felling.

Conservation also covers **listed building** consent, which is a grading given to a building that saves it for our heritage. Any work on a listed building has to be done to the approval of an inspector, which can include having antique paint manufactured! Conservation of materials and the character of a building must often be undertaken by a specialist company; for example replacing a thatched roof where there are few companies left that can undertake this old skill. A building surveyor will be able to recommend specialist companies to undertake the conservation work on old and antique materials.

Key term

Listed building – a building of special architectural or historic interest in the UK. Alterations to these buildings cannot be made without consent or careful consideration.

Assessment of dangerous structures

It is the role and responsibility of the building control officer to be the first person called to a structure to assess its danger. The client or building owner can then hire a building surveyor to take care of the dangerous structure. In some cases of imminent danger the local authority will undertake the work and charge the property owner for the cost of this work.

A building surveyor would be able to offer advice and guidance on dangerous structures such as:

- gable end wall tie failure
- general wall tie failure
- chimney collapse
- bowing of walls at first-floor level
- damage caused by a traffic crash to a structure
- roof tile fixing failure
- retaining walls
- tree root damage.

Structural appraisal

Structural appraisal is the element of a survey that might require the services of a structural engineer or senior building surveyor.

Typical elements of a building that would be investigated during an appraisal are:

- foundation settlement
- roof spread
- wall tie failure
- external cracking
- internal cracking
- removal of any roof timbers
- sagging of roof ridge
- sagging of floors.

Some structures may require investigation works. Sub-floors will require cutting access hatches into timber floors to inspect the floor joists below. Wall tie investigation will involve drilling holes through the outer skin of brickwork and inserting a camera with a mirror to observe and record. To inspect roof timbers, roof access and lighting will be required as will loft access ladders. A theodolite, spirit level and plumb line can be used to check if a structure is leaning. A level can be used to establish horizontal control to see if any part of the building has settled over time and may be a cause of cracking in the structure. Visual checking of door heads will give a clue as to whether a building has moved; as will sticking doors that have been planed to fit uneven openings. Roof spread can be observed by visual lining in of elements from a point at height or by physical measurement within the roof space. Drain problems are best investigated using a specialist company that can drag and steer a mobile camera up a drain to locate a crack or obstruction or break in the drain.

Detailed examination of the building's condition

To help you to understand the processes and equipment involved in surveying, we will now look at different ways in which a building can be examined using more specialised techniques as well as look at the basic surveyor's tool kit.

A **thermal imaging camera** can detect heat and produce a thermal image which can be used for a very specialised survey. The red areas of the thermal image show the hottest points and the blue, the coldest or equal to the background temperature. The image allows any potential cold bridging across cavities and above lintels to be seen; it also shows areas that are subject to excessive heat loss. Thermal images can also demonstrate if a cavity wall contains sufficient insulation to meet current regulations.

A thermal image

A **drainage survey camera** is a mobile, electrically powered camera that can be run up small diameter drains to investigate drainage problems. They can be assisted with smoke bombs and fluid dyes in surveying the direction and runs of drains to establish how the network works. This type of equipment can equally be used to survey chimneys for defects in the flue system.

A **radon gas detector** survey involves the use of a specific radon gas detector to analyse the sub-floor voids to detect this potentially carcinogenic (cancer forming) gas.

Assessment activity 15.2

The potential buyer of a property has approached your building surveying company to seek advice on what type of survey to have undertaken on the property. The details from the commercial sales brochure sound less than convincing:

'This period property is over 150 years old and is in keeping with the conservation area surroundings of the enclosed village. The property has brick walling with timber oak beams and a thatch roof structure. Its character and charm are worthy of the investment that would be required in refurbishing this property to an acceptable standard.'

1. Using the above information describe the procedures and techniques available to undertake a survey on the property. **P4**
2. What equipment would you need to undertake the following surveys on the property:
   - a building survey
   - a damp and timber survey? **P5**
3. Describe in some detail the differences between the above two surveys. **M2**

Grading tips

For **P4** describe what procedures and techniques you would use in undertaking the survey on the property.

For **P5** you need to identify and describe the different equipment needed for performing the two surveys.

For **M2** differentiate between the two types of survey in detail; include information on equipment, techniques and processes used in the two surveys.

PLTS

Supporting conclusions by using reasoned arguments and evidence in differentiation on surveys will help develop your skills as an **independent enquirer**.

Functional skills

Describing the different types of equipment used for building surveys will help to develop your written skills in **English**.

Case study: House surveys

Susan and Nigel had been looking to buy a house close to the city centre and finally found a 1905 property built in the older part of the city. They are now in the process of sorting out a mortgage on the property. This is their first step on the property ladder, and money is tight. The lender has indicated they could have a 90 per cent loan; they will have to find a 10 per cent deposit. The lender has asked Susan and Nigel if they would require any further surveys on this property.

Divide yourselves into teams and discuss the following questions.

- Is it worth spending extra to have these further surveys done?
- What would be the benefits of a structural survey?
- What surveys would you suggest Susan and Nigel have done?

Equipment

Steel and fibre tapes

Steel and fibre tapes come in various sizes, from 30 m to 100 m, and are useful for undertaking a measured survey where long external dimensions are required and where a 5-metre tape is not long enough. Both operate with a wind-up mechanism which rolls the tape into the reel.

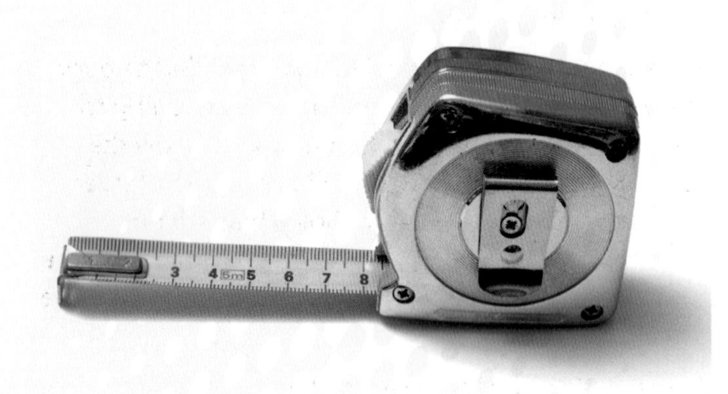

Steel tape

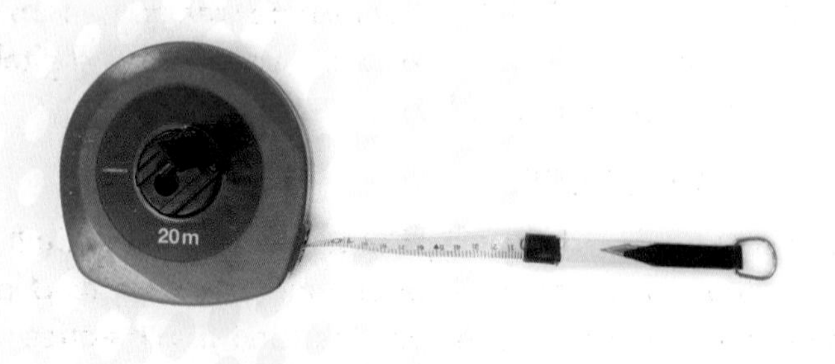

Fibre tape

Remember!

A fibre tape is less accurate than a steel tape because it tends to stretch when pulled and will therefore read short any long dimension. However, steel tape is affected by and expands in temperature.

Folding rule

A folding rule is normally made of timber and graduated in metric. It extends to 2 m but folds down into a convenient length of 300 mm. As it remains rigid until folded back, it is used to take vital dimensions at times when an ordinary tape would bend and collapse. The folding rule can be easily held up to be used for a constant and repeated set of measurements.

Electronic distance measurement device

The electronic distance measurement device uses a laser to measure the distance between two surfaces. The speed of light of a laser is constant and thus the distance between the two walls can be calculated and appears on the readout. The electronic distance measurement device works by pointing the device at the distant surface, for example the wall; the laser bounces from the device to the wall. The data has to then be recorded on the measured survey. Some newer machines have data loggers that enable the measurements to be downloaded back at the office.

Electronic distance measurement devices are good for internal dimensions but, externally, a tape measure will be more effective, as it can be attached by the end clip to the outside angle of the wall.

Electronic distance measure device

Moisture meter

A moisture meter is used to check the moisture content of timber and walls above and below the damp-proof course to check that the damp course is working effectively. The moisture meter is a basic machine that has two metal probes; it measures the electrical resistance across the probe pins. The wetter a material, the less resistance there is across the material and the higher the meter's reading. Several readings should be taken to ensure that an accurate average is obtained. Visual checks, that take into account the age and construction of the building, should also be done to ensure that a damp course is present within a building.

Camera

The improvements to cameras in the last ten years have been phenomenal. Digital imaging has made the old film-based systems nearly obsolete. Time and date recorded photographs of properties and different aspects of the survey can be instantly captured and used as a historical record, especially for dilapidation surveys. A digital camera is very useful when conducting measured surveys as it enables you to instantly take a second look at any aspect that you have missed, for example you can count brick courses from a photograph. The digital technology also enables images to be stored and emailed very easily, enabling a report to be compiled more quickly.

Inspection chamber keys

Inspection chamber keys are used to lift inspection chamber covers from their surrounding frames. There are various sizes of keys and manual handling lifting devices to help reduce the strain on the back. Great care should be taken when lifting a cover due to the weight and length of time that the lid has been down, as dirt and debris can cause the lid to lodge securely within the frame. Larger lids should be lifted by two people. Often, on major roads the health and safety implications are extensive; we shall look at this aspect later.

Safety tip

Never try to lift a large inspection chamber cover on your own. You could damage your back.

Binoculars

Binoculars extend your range of vision, and are especially useful in surveying roof work; small defects and weaknesses within the structure can be spotted more easily with the use of binoculars. Binoculars are especially useful for high-rise buildings where the scope of view of the outer walls is limited from the inside. They will also help to reduce eye strain when surveying exteriors in strong sunlight. A set of 10 × 8 binoculars is all that would be required for a surveyor's toolkit.

Boroscope

The boroscope is a very useful tool for gaining access into a structure with the minimum of disruption and damage to the fabric of the building. It is especially useful for inspecting the cavity of a wall. It is a telescope that has a mirror at the end that diverts the line of sight by 90 degrees. A light within the equipment enables a good visual inspection of inaccessible areas of a structure. Adding a camera to the end of the boroscope enables a record of the inspection to be made.

To use the equipment, a hole must be drilled through the outer part of the structure that you are trying to inspect. The hole must be big enough to allow the camera to slide through easily. Obviously, the larger the inspection area, the more holes that will be needed Once the inspection has been carried out, the boroscope is withdrawn and the hole or holes must be filled in.

Thermal imaging equipment

As we have seen earlier, thermal imaging equipment is very complex and requires specialist training to use. It is often better for smaller building surveying practices to obtain the services of a specialist subcontractor to undertake this work. A thermal camera with some form of recording media will be required along with a digital player to observe the findings.

Sectional ladder

The sectional ladder is a typical surveyor's ladder. Built in sections, it can be easily disassembled and placed in a car boot. Having one, means the surveying practice avoids the need to hire ladders delivered to a survey job or the use of a car roof rack, which would mean a permanent fixture to the car. Sectional ladders are lightweight and have a limited height use.

Spirit level

The spirit level, available in a number of different lengths, measures the horizontal and vertical positions of the surface it is placed on. Care should be taken when measuring vertical walls; a better measurement can be obtained when using a longer level.

Electric torch

An essential item in any surveyor's toolbox is an electric torch. It should be used when entering the following:

- a roof loft space that does not have a mains light installed
- a sub-floor inspection
- a cellar
- a boarded-up property.

A surveyor should always carry spare batteries, or if using a rechargeable torch, ensure that it is fully charged before leaving to undertake the survey. Use a torch of a good size so you can see everything needed on the survey.

Optical levels

An optical or automatic level is used with a staff to establish what the difference in level is across two or more points. It can be used to establish the settlement of a building, to see whether it is even or is dipping in one corner. A known reference point away from the building must be established as a control point to work out if the building levels are moving. Then several levels should be taken over a period of time and used to establish if the building is continuing to settle.

A level and staff

Basic land surveying equipment

Theodolites, ranging poles and optical reflectors are basic equipment that can be used to survey buildings and large areas of land to quickly plot an area to scale. Modern surveying equipment uses satellite global positioning equipment to triangulate a known point. All you have to do is locate a grid coordinate and then print out your survey.

Personal protective equipment

Typical personal protective equipment (PPE) that a surveyor should consider having includes:

- stout steel toecap boots or shoes
- gloves for handling inspection chamber lifting keys and drainage inspections
- a high-visibility jacket
- a safety helmet for using when climbing into and through loft spaces
- a mask to prevent inhalation of glass fibres from loft insulation
- safety glasses for use when cutting any material for testing
- specialist equipment, such as harnesses for abseiling inspections.

PPE should always be used as a last resort; in all cases, a risk assessment should be undertaken first to try to design out the use of PPE.

Safety tip

Always maintain your PPE in a working and clean condition.

Procedures and techniques

Preliminary surveys

The preliminary survey is the first thing that needs to be arranged with the client. This is normally done after the client has initially contacted the surveyor's office. The building surveyor should then contact the client to arrange a mutually convenient time when the survey could be undertaken. Due to the fact that many home owners are employed full-time, this may mean making the arrangements through the selling agent who will have access and keys.

The building surveyor's terms and conditions of engagement must be made clear to the client, and they must also be aware of, and have agreed to, the fee for the survey. Often this is acknowledged by letter with a fee deposit. Advice can be given over the phone regarding the different types of survey that are available with costs so that the client can choose the best survey for their potential purchase.

When a date and time has been established, turn up on time, suitably dressed to undertake the survey, which can take from one to two hours depending on the level of detail required. Photographs can be taken and any measurements established. It is professional not to discuss with a seller or leaser your client's position on the purchase or let.

Once the survey is complete, make sure that the property is left secure and safe and return to the office to have a survey report typed up. This should be checked by a second person before it is sent to a client to ensure accuracy of the information supplied, and to establish that nothing has been missed from the report.

Site location

Site locations will affect how many surveys can physically be undertaken in a working day. Urban locations are easy to undertake surveys in, but care must be taken of the security of vehicles and equipment. Rural locations may require the use of a four-wheel drive vehicle to undertake a survey. Congested locations, where ladders have to be placed across footpaths and roads to gain access to survey, must be carefully considered before proceeding. Great care should also be taken when undertaking survey work within factory site locations; you should ensure that you attend an induction before starting and are aware of all the risks associated with that site. Obviously, the time of year will also have a bearing on the amount of light available for external surveys as in the winter the nights draw in rapidly.

Building location

The building location sometimes has a bearing on the survey being undertaken, for example a third-floor flat is not easily surveyed from the outside. When access to a property is over another owner's land, you should seek permission before crossing that land, otherwise you will be trespassing. If buildings are next to water, railways, harbours or major trunk roads, it should be recorded as such; flooding or vibrations could cause damage to the structure. It should also be recorded if a building is next to an airport, but here the damage is most likely excessive noise levels being a disturbance to the occupiers.

Remember!

Never trespass on adjoining property.

Elemental surveys

It is sometimes easier to commence the survey from the inside out. This is done for two reasons: to prevent dirt and damage from dirty shoes, and to disturb the occupants less and reduce the time they need to secure the building before returning to work or their other business.

External surveys on a detached building should start and finish at the same point, usually a corner. For mid-terrace or semi-detached buildings, it is often worthwhile looking at the defects on adjacent properties to give clues to any potential defect that might occur within the property you are surveying. Typical elements to be inspected externally are:

- guttering and downpipes – blocked or damaged
- chimney stacks – condition of pointing; verticality
- roofs (pitched) – ridge and hip tile bedding; age of tiles
- roofs (flat) – life span of three-layer felt; ponding on roof
- external high level joinery – fascias, soffits and barge boards
- external ventilation vents and airbricks – broken or blocked
- inspection chamber covers and lids – broken
- surface water drainage – blocked
- fencing and property boundaries
- external brickwork – evidence of cracking
- external face of any exposed lintels over openings.

Internal inspections can be divided into two areas: those which are clearly visible and the others which are not, particularly the roof space. Internal inspection of visible areas would include looking at the conditions of the following items:

- internal decorations
- internal joinery – check for evidence of woodworm or rot
- windows and double-glazed units – check if the seals are intact
- doors and frames – check if they are squared
- gas fittings – confirm they have been checked and serviced
- electrical services, fuse board and wiring and fittings – check if they are current or require updating
- radiators, boiler and pipework – check for evidence of leaks
- flooring, solid or timber – check for evidence of bounce, which indicates soft wall plates
- ceilings and plaster finishes – check for evidence of cracking
- chimneys – check leaning or condition of pointing.

Inspection of non-visible areas is limited to those where access can be gained without causing damage to the fabric of the building; for example, loft spaces or roof voids which are easily accessed through hatches. Roof void inspections would normally cover the following items:

- age of tiles and state of fixings
- evidence of underfelt
- state of fixing battens
- general condition of roof timbers
- ventilation to the roof space
- evidence of light passing through roof which may mean slipped tiles
- infestation from wasps, woodworm and birds or bats
- amount of insulation utilised.

An inspection of external works would cover items such as:

- footpaths and driveways – check for cracking and settlement
- external paving – check age and condition; see if there is frost attack
- decking – check age and treatment
- fences – check condition of posts and panels
- gates – check type, metal or timber, and condition
- surface water drainage – check to see if working effectively
- trees – TPOs
- outbuildings – check general condition
- garden – check general condition of plants, bushes and maturity
- brick boundary walls
- retaining walls.

Building services of gas, water, electric and drainage should also be checked. The building surveyor can check the presence of a gas meter, the number of appliances attached to the pipework, isolation valves and visual observation of any pipework exposed. However, gas inspections must be undertaken by a registered heating and plumbing engineer and not an unqualified surveyor. A qualified engineer would undertake two tests: one to detect any leaks from the system and one to check the flues from boilers and gas fires; these need to be functioning correctly as carbon monoxide discharge into occupied areas can kill.

Water services should be checked for **cross bonding** beneath sinks and taps and for general leaks. The existence of old lead pipes should be noted to a potential buyer due to health risks associated with lead poisoning. Leaking overflows will produce external staining of finishes and should be mentioned in the survey report.

Key term

Cross bonding – in the event of an electrical fault, this process prevents any metal service from becoming live with electricity.

Electricity inspections should check the condition of the existing wiring and, if an older property, if it has been rewired. The consumer unit should be inspected to note the presence of **RCDs** or **MCBs** or old fuse wire. A building surveyor can perform a basic plug-in test which can give some diagnostics but if there is any doubt, a certified inspection from a qualified electrician should be sought.

Drainage can be inspected by lifting inspection chamber covers and observing the drainage run. The surveyor should check the condition of the chamber to see if there is any standing water. If there are areas of flooding, these should be inspected to locate a cause. Surface and foul systems that are combined should be noted as against separate systems. Soakaways that allow water from surface drains to percolate into the ground should be noted. If the property is not connected to main drainage, septic tanks should be inspected to see if they have been maintained properly.

Legislative considerations

Contractual obligations

Once an offer is made and accepted, then it becomes a legally binding contract. A building surveyor must ensure that their conditions of engagement are agreed and accepted by a client and the contract contains all the terms of the obligation between the two parties. Once a surveyor has accepted a contract, they are bound to complete their obligations and then receive payment for their services.

Confidentiality and data protection clauses should also be inserted into the agreement to ensure that a report is only used for the purpose for which it was intended. A clause on the concealed areas that cannot be surveyed should be included to limit the extent of the survey and protect the surveyor from legal action by a client.

If a building surveyor is running a contract on behalf of a client, say a refurbishment, then a typical **JCT** contract may be in place which states that the surveyor is the client's or employer's representative. This empowers the surveyor to act and run the contract on their behalf till completion. This role could be likened to that of the architect or designer.

Key terms

RCD – residual current device, which acts as a fuse.

MCB – miniature circuit breaker which switches off power on an earth short.

JCT – Joint Contracts Tribunal.

Legal constraints

Contractually, a building cannot be entered without the owner's permission to do so or else this can technically be called a tort in law, which is **trespass**. A building surveyor could quite easily find that they place themselves in this position and should ensure that they are working with permission and instructions from clients. Gaining permission may be difficult when a house has to be sold in a divorce settlement and is owned by one party; they, or the other party may be obstructive. Again, diplomacy may win through in such cases.

If a building surveyor is commissioned to provide a survey report and accepts remittance for such services, or confirms intention to do so for an agreed fee and then fails to undertake the service, then they may be in **breach of contract**. A client, tenant or building owner can then take the surveyor to court and seek damages for failure to complete the work. Similarly, if a client fails to pay the surveyor, then they can in turn be taken to court to pay the debt due.

Remember!

If the surveyor fails to notice an obvious defect which is not excluded by the report, they could be sued for damages. Reports must be written with care.

Survey reports contain many disclaimers so that a client is left in no doubt as to the legal position of the survey. A survey report must be structured so it is quite clear what terms and conditions it is issued under. If a surveyor does not find a defect which manifests itself within a short period, then the surveyor needs to protect themselves against legal proceedings for damages of negligence. The best way to do this is through professional indemnity insurance which indemnifies the surveyor against any act of omission or negligence on their part which may result in legal claims for damages against them.

Thinking point

Indemnity insurance has often to be paid for the life of the surveyor.

If the client or the surveyor goes bankrupt, then an official receiver would undertake the winding up of the company and establish the liability of the debtors and sums to be paid to creditors.

Assessment activity 15.3

BTEC

1. You have been asked by the survey manager to identify the legislation that is relevant to building surveys. These will be used for insertion into the terms and conditions of operation. Provide a list of relevant legislation. P6
2. Your company has been sued several times over surveys undertaken by an inexperienced surveyor. The managing director is very keen to ensure that all processes and procedures are current and within the legislative framework to avoid any further actions. Evaluate the contractual obligations and legal constraints applicable to surveying. D1

Grading tips

For P6 you are required to produce a list of relevant law applicable to surveying.

For D1 evaluate the contractual obligations and legal constraints applicable to a building survey.

PLTS

Exploring issues, events or problems from different perspectives will help develop your skills as an **independent enquirer**.

Functional skills

Producing a list of relevant legislation from identified websites will help your research skills using appropriate **ICT**.

Health, safety and welfare issues

Health and safety is an important issue in any work the surveyor undertakes. All survey work must be undertaken in accordance with the employer/employees duties under the Health and Safety at Work Act 1974 with regard to the duty of care not only for employees but also the general public and building owners who might be affected by the building surveyor's actions.

The Management of Health and Safety at Work Regulations 1999 must also be considered with regard to the risk assessments that should be undertaken with all survey work. The electrical testing of any electrical survey equipment must conform to the Provision and Use of Work Equipment Regulations 1998 and have an annual PAT test – Portable Appliance Test for electrical safety. Any work at height should be conducted in accordance with the Work at Height Regulations 2005. Any design work must be undertaken in accordance with the Construction (Design and Management) Regulations 2007 with regard to design risk assessments and the duties of the parties involved within the contract. (See Unit 1 Health, safety and welfare in construction and the built environment for more information on legislation.)

Particular care must be taken when surveying very old and derelict properties as the structure can be weakened through timber decay and rot; this can lead to collapse when trying to enter the building or walking over floors and stairs. Buildings that have been empty for some time may contain stale air which will need venting before inspection can take place.

Another aspect of survey work is the stress dealing with tenants. They may be in dispute with a landlord and will try to obstruct a dilapidation survey for fear of losing a deposit on the property. Great diplomacy is required by the building surveyor in this often stressful situation and a great deal of trust can be gained by clearly explaining that you have to undertake this work. A sound professional approach at all times should be maintained.

Pre-visit risk assessments

A pre-visit risk assessment is done to try to assess the risks associated with a generic survey. Such risks could be working at height, working in low light conditions or working within roof voids. Each of these areas will need to be assessed for the required control measures that a surveyor would need to reduce the risk down to an acceptable level.

Potential risks

Potential risks associated with carrying out surveys include:

- electrocution from old, faulty wiring exposed during survey
- diseases caught during drainage inspections; this could be Weil's disease from rats' urine contaminating the water or other diseases associated with human waste
- surveying old buildings frequented by homeless people with drug habits – needle stick injuries, Hepatitis B or C
- standing on unsafe timber floors and structures
- falling down inspection chambers that are unprotected
- falling from height during inspections using ladders
- trips, slips and falls while surveying loft spaces
- injury to the head while accessing loft or sub-floor spaces
- risks associated with surveying near major roads
- inhalation of asbestos which could be present in loose format.

Assessment activity 15.4

P7

BTEC

The survey manager has asked you to review the company's safety policy. Describe the health, safety and welfare issues associated with building surveys. P7

Grading tip

For P7 you are required to describe the issues of health, safety and welfare associated with undertaking building surveys.

PLTS

Reviewing progress and then acting on the outcomes will help develop your skills as a **reflective learner**.

Functional skills

Describing the key health and safety issues in surveying buildings will help your verbal skills in speaking **English**.

3 Be able to carry out building surveys

Building surveys and methods

Measured survey

This section involves undertaking a measured survey of a building and producing the raw data that will need to be processed to form a completed scale drawing.

To undertake the survey, first select a case study building. You will conduct a measured survey on an elevation of the building and produce data in order to redraw the elevation to a scale of 1:100 (see Figure 15.3). You should pick an elevation that contains sufficient detail and proportions to reproduce a good industrial standard drawing. Undertake a measured survey of this building using all or some of the following equipment:

- digital camera
- tape measure
- 30 m metal tape
- A4 paper and pencil.

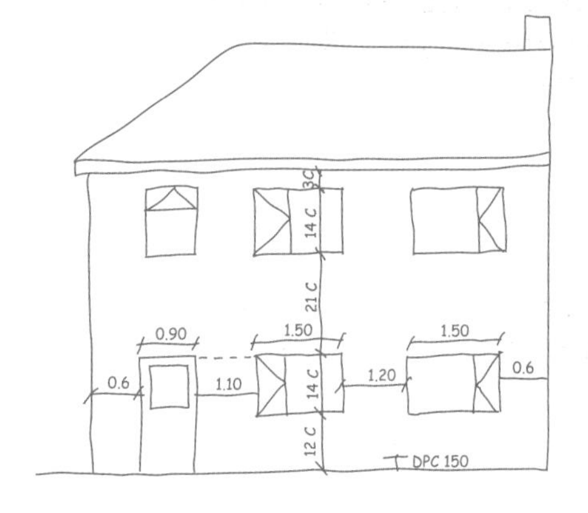

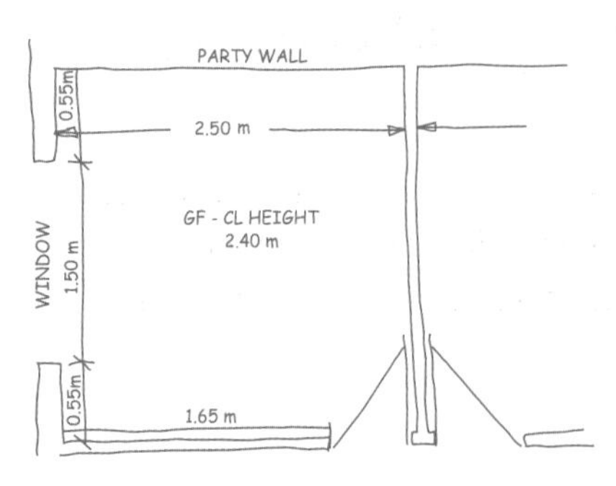

Figure 15.3: A measured survey sketch of a room

Ensure that your measured survey contains sufficient detail and dimensions to enable you to redraw it successfully. With inaccessible heights, use the counting courses of brick method or carry out internal dimension heights to establish the elevation height. Reproduce this elevation on A3 paper to a scale of 1:100 ensuring that it fits onto the paper. If this is not the case, then cut down your elevation. Alternatively, you can reproduce the elevation using a computer-aided design and a plotter.

Dilapidation survey

As a surveyor, you must be able to apply your knowledge and understanding of this type of surveying and actually undertake a dilapidation survey. You must be aware that there is legislation that protects both the tenant and the landlord; therefore, you need to act professionally at all times. The following equipment might be used during a dilapidation survey:

- internal and external checklist
- dictaphone to record notes from survey
- loft ladders
- **PDA** with built in checklist
- digital camera
- previous survey to check against for deterioration.

Key term

PDA – personal digital assistant; a hand-held computer.

Before commencing any dilapidation survey, you should confirm with the tenant or landlord (whichever one has instructed you) the agreed terms of engagement. It is also advisable to obtain copies of all relevant documentation between landlord and tenant including the lease agreement. This will indicate who is responsible for what aspect of the building. For example, the landlord may be responsible for the exterior of the building, and the tenant the internal decorations.

An itemised schedule divided initially into 'External' and 'Internal' should be constructed and subheadings listed under each, for example see Table 15.1.

Once completed, the report should be handed to the person who appointed you for them to discuss and finalise any lease agreement to return deposits.

Table 15.1: An itemised schedule

External	Internal
Guttering	Room 1 decorations
Roof	Room 2 radiators
Walls	

Condition survey

We have already looked at condition surveys on pages 398–99.

Use of checklists for surveys

Standardised checklists

Checklists are a very good idea to standardise in a building surveying practice. They enable a standard methodology to be employed in surveying buildings and can be extended to form the basis of a building surveying report. Checklists can be developed for any particular part or whole of a survey. The main advantage is that they ensure that no single item is missed; this makes the surveyor's job that much easier to undertake. Checklists also allow for the development of hand-held electronic building surveying, especially on large housing stocks where repetition occurs. Figure 15.4 illustrates how a typical checklist can be built up and used; tick the appropriate material used in construction, then check condition writing down any comments.

Element	Description	Tick box	Condition		Comment
			Good	Repair	
Roof tiles	Concrete				
	Clay				
	Slate				
	Other - specify				
Ridge and hip tiles	Concrete				
	Clay				
	Asbestos				
	Other				
Guttering	Cast iron				
	upvc				
	Timber				
	Lead				

Figure 15.4: A typical building surveying checklist

Assessment activity 15.5

P8 P9 P10 P11 P12 M3 D2 BTEC

1. Select a small manageable building, such as your house or a building on your campus. Undertake a measured survey to produce fully dimensioned sketches. Hand these sketches to a colleague and ask them to produce one of the elevations in a scaled drawing. Ensure that you also have one from a colleague to undertake. P8
2. Liaise with a local housing association or tenants' association or the local authority housing office to gain access to an empty rental property. Ensure that a risk assessment is undertaken before entering an unoccupied property, then carry out a dilapidation survey on this property. P9
3. Produce a condition survey on your home, including all internal and external points that should be considered. P10
4. Produce final measured, dilapidation and condition survey reports for the low-rise domestic or commercial buildings surveyed in P8, P9 and P10. Ensure that you collect and compile your survey data using standard techniques and record and present the data in appropriate formats. P11, P12

   In each of the measured, dilapidation and condition surveys you carried out relate the nature of the data collected to the type of survey being carried out. M3
5. Your tutor will hand you two types of survey sample reports. Justify any recommendation made within the reports by analysing their contents. D2

Grading tips

For P8, P9 and P10 you will need to carry out the practical surveys. Don't forget to record and make notes during the surveys.

For P11 and P12 you will need to compile the data then record and present it in appropriate formats.

M3 requires that you relate the nature of the data you collect to the type of building survey being undertaken.

D2 requires that you justify the recommendations contained within the reports.

PLTS

Providing constructive support and feedback to others when surveying in groups will help develop your skills as a **team worker**.

Functional skills

Recording detailed notes during your survey will help with report writing skills in **English**.

4 Be able to produce survey reports and schedules of maintenance and repair

The survey report

The **age of a building** should be recorded on the report. This is often related to the condition of the building; older properties may be subject to long-term settlement. This is because foundation design 100 years ago was not regulated by building control. Also the presence of asbestos in commercial buildings was subject to late legislation.

The **construction method** used can be categorised on the report by type; for example, timber framed, traditional cavity wall construction or solid wall construction. The roof type would be described; for example, mansard or pitched roof. The type of building would also be described; for example, terraced, detached or semi-detached. The **potential defects** recorded as part of the survey would be noted with any recommendations required. These would possibly cover the following areas

- wet rot
- dry rot
- woodworm
- lintel failure
- roof defects
- settlement and subsidence
- ventilation and damp.

Maintenance and repair schedules

Planned maintenance

Following any simple survey, a maintenance schedule should be prepared. It can be given to a landlord or a premises/facilities manager to keep the building in a fit standard and condition. This is done for several reasons, including:

- health and safety
- to avoid devaluation of a building
- to attract tenants.

Maintenance schedule Simtop Ltd

Building..

Date schedule prepared..

By whom..

Organisation..

Item	Interval	Done	Comment
Externals			
Guttering	Annually		Clean out guttering by hand and flush with water
Down-pipes	Annually		Flush with water
Drainage traps	2x year		Remove grids and clear trap flush with water
Inspection chambers	Annually		Remove lid, regrease and flush out benching
Pathways	3x year		Remove weeds and spray with weed killer
uPVC fascias	2x year		Wash down with warm water and mild detergent
External doors	2x year		Oil and lubricate hinges, check closers
uPVC windows	2x year		Spray with silicone lubricant hinges
External vents	3 x year		Remove and clean external ventilation vents
Landscaping	Summer Winter		Weed and mulch beds and water Prune back bushes and trees

Figure 15.5: A maintenance schedule

Emergency maintenance

Emergency maintenance is reactive and has to be done because there is a personal risk of harm or injury. It is worthwhile to assemble a list of spare parts that might be required for an emergency; these can be items that are in constant use; for example, a roller door in a delivery bay.

Repair

A repair schedule lists the repairs that need to be undertaken. A repair schedule for a landlord would contain a full specification on how to undertake the repair along with an associated cost estimate for the repair.

These may be classified as follows:

- urgent and immediate – liable to cause injury to occupiers
- action required within a week
- action required within a month.

Assessment activity 15.6

BTEC

Locate a suitable building that can be easily accessed and that is safe to enter. Undertake a simple survey of the building externally and produce a maintenance schedule and a repair schedule.

The maintenance schedule should instruct the maintenance crew on what will be required over a one-year period.

The repair schedule should be for three items and list the following:

- full specification for the repair
- an estimated cost
- time interval when it should be done by. 

Grading tip

For P13 you will need to prepare maintenance and repair schedules; try to use a table format.

PLTS

Proposing practical ways forward and breaking these down into manageable steps will help develop your skills as an **effective participator**.

Functional skills

Using a spreadsheet package to prepare a schedule will help develop your skills in **ICT**.

WorkSpace

Mohamed Khan

Estate agency building surveyor

Mohamed works at his father's estate agency as a building surveyor. The business is tied to a mortgage company and so can offer professional building surveys.

Mohamed started working on Saturdays in the office giving out details of houses for sale to prospective clients. Mohamed then attended the local technical college and started a National Diploma in Construction, which he successfully completed in two years. He then went on to a part-time degree in Building Surveying at the nearest university in order to continue working with his father to gain as much experience as he could.

After passing his degree, Mohamed gained full membership of the Royal Institute of Chartered Surveyors which gave him the professional qualification that he needed in order to undertake building society valuations.

Mohamed likes his job and he enjoys initially introducing a client to a property, agreeing the sale value, and then seeing the whole process through to completion and handing over of the keys. Often he has to produce further building surveying reports on older properties where there is risk of settlement or subsidence.

Mohamed is recognised for his judgement in giving sound advice on building defects. This is an aspect of building surveying that he is really enjoying and he has now branched out into undertaking refurbishment design and project management work for a local developer.

Think about it!

- Would you enjoy working in an estate agency?
- What aspects of building surveying interest you?
- How long would it take to obtain professional status?

Just checking

1. Name three members of a typical design team with whom a building surveyor would interact.
2. In what way would a member of the local authority interact with the building surveyor?
3. What is a TPO?
4. What do RICS and CIOB stand for?
5. Name six items contained in a valuation report.
6. What does a dilapidation survey look at and whom might it be carried out for?
7. What would you typically look for on a stock condition survey?
8. How would you undertake a measured survey and what equipment would you require?
9. What items would require maintenance on the external envelope of a building?
10. What should a repair schedule contain?
11. Name a typical dangerous structure element.
12. What is radon gas?
13. How does a moisture meter work?
14. What essential PPE would be required for a roof void survey?
15. For what parts of a survey may a building surveyor obtain the services of a specialist contractor?

edexcel

Assignment tips

- A picture tells a thousand words – always take a digital camera with you when you carry out survey work and include photographs in your survey portfolios as they provide valuable reference evidence and might save you having to return for further information.
- Look at some of the example surveys on the Internet. They will give you an idea of how to write a professional survey report.
- Referring to the terms of engagement for a building surveyor will help with the legal and contractual obligations of a building surveyor in the assessment.

:redit value: 10

17 Building Regulations and control in construction

'he introduction of the Building Regulations provided a method of ·nsuring that all domestic and commercial buildings meet a nationally ecognised standard. This helps ensure that structures are safe and ygienic for the occupants that live, work and socialise within them.

uilding Regulations cover many aspects of the construction of buildings from the depth f foundations, the level of insulation, ventilation of rooms, through to the electrical esting of installations. These aspects are known as the set of Approved Documents.

he regulations are essential in order to prevent uncontrolled development which would roduce unsafe structures and potential harm to the occupants. The application of the egulations is enforced through inspection on site by building control officers who check ompliance against the Approved Documents.

Learning outcomes

After completing this unit you should:

1 understand the origins and purpose of building control
2 know how to apply and enforce Building Regulations
3 understand the procedures and documentation involved with Building Regulation approval
4 be able to prepare a submission for Building Regulation approval.

Assessment and grading criteria

This table shows you what you must do in order to achieve a pass, merit or distinction grade, and where you can find activities in this book to help you.

To achieve a **pass** grade the evidence must show that you are able to:	To achieve a **merit** grade the evidence must show that, in addition to the pass criteria, you are able to:	To achieve a **distinction** grade the evidence must show that, in addition to the pass and merit criteria, you are able to:
P1 examine the factors that have influenced the historical development of building control **See Assessment activity 17.1, page 423**	**M1** explain the particular implications of the Building Regulations for low-rise domestic and commercial construction **See Assessment activity 17.1, page 423**	
P2 discuss the legislation and documentation associated with building control, and their application **See Assessment activity 17.1, page 423**		
P3 identify the various Approved Documents that comprise the building regulations **See Assessment activity 17.2, page 427**	**M2** propose answers to two queries related to interpretation of the Building Regulations **See Assessment activity 17.2, page 427**	**D1** justify the proposed solution to two separate Building Regulation issues **See Assessment activity 17.2, page 427**
P4 describe the application and enforcement of the Building Regulations **See Assessment activity 17.2, page 427**		
P5 explain the approval procedures used in building control **See Assessment activity 17.3, page 432**		
P6 evaluate the documentation used to support building control **See Assessment activity 17.3, page 432**		
P7 discuss the powers of local authority building control officers and approved inspectors **See Assessment activity 17.2, page 427**	**M3** explain the procedures used to enforce the Building Regulations **See Assessment activity 17.3, page 432**	**D2** evaluate the specimen building regulation application, with all necessary documentation **See Assessment activity 17.4, page 432**
P8 produce a specimen Building Regulations application, with all necessary documentation **See Assessment activity 17.4, page 432**		

How you will be assessed

The evidence requirements for pass, merit and distinction grades are shown in the grading criteria grid. Evidence for this unit may be gathered from a variety of sources, including well-planned investigative assignments.

You will be given written assessments to complete this unit, which will include the drawing element for the Building Regulation submissions.

This unit will be assessed by the use of three assignments:

- Assignment one will cover P1, P2 and M1
- Assignment two will cover P3, P4, P5, P6, M2 and D1
- Assignment three will cover P7, P8, M3 and D2.

Jack

This unit has made me aware of the regulations that exist in the UK that control how and what you can build. Building work is subject to regulations that cover all health and safety aspects.

I now know what the role of the building control inspector is and the differences in the types of application that can be made. This unit has made me aware of the processes used to submit a full building control application, including the drawings and the specification and documentation that go with it.

Over to you

- What do you already know about Building Regulations?
- What is the role of the building control inspector?
- What are you looking forward to learning about in this unit?

1 Understand the origins and purpose of building control

Build up

Building regulations

The Building Act produced by Parliament establishes regulations upon the construction of buildings covering all aspects. These regulations are contained in what is known as the Approved Documents. Building inspectors use these to check the projects and confirm conformity with the regulations on submitted plans and by site visits to inspect.

- What do the Approved Documents cover?

Development of building standards, hygiene and public health factors

The following issues may help explain why the development and production of buildings in the UK had to be regulated to prevent uncontrolled developments from continuing.

There are two major issues: the Great Fire of London in 1666 and the effects of diseases carried by rats and vermin which was made worse by the inadequate sanitation and waste disposal arrangements. Since there was no drainage or sanitation and no place to bathe or wash under running water, the Great Plague of 1665 killed a sixth of London's inhabitants.

The Industrial Revolution

The Industrial Revolution presented many issues. Before the invention of the steam engine, power was derived from water-driven wheels; buildings that needed power, such as factories, used to be built near water supplies. The invention of the steam engine meant factories, and therefore workers' houses, could be located anywhere in the UK.

The industrial age saw an explosion in construction of factories and workers' accommodation. Some factory owners took great care of living standards for workers, for example, in Saltaire, where the factory owner spent time and effort on making a high-standard purpose-built factory and village. Others did not take this lead and put profit before workers' health; many employees were injured or died because of working conditions.

This led to the development of a rudimentary building control with some standards, but not many.

The Great Fire of London

After the Great Fire of London, which started 2nd September 1666 in a baker's on Pudding Lane, the development of our modern-day Building Regulations began. Buildings at that time were packed so tightly together that the fire swept easily from building to building, moving west driven by winds and the lack of fire-fighting equipment.

After the fire, London needed to be rebuilt. The king favoured a Paris-style layout with wide roads, avenues and linking piazzas, but due to land-holders' rights, this never transpired. However, it was decreed that buildings were to be made from stone and other fireproof materials so that such an event could not occur again. Streets were made wide enough to get fire-fighting equipment down them and to prevent fire from jumping from building to building. Since then, the UK has not had such a fire overcome any major city, apart from the bombing of Coventry during the Second World War.

Ronan Point

Ronan Point was a 23-storey tower block which suffered a catastrophic collapse in 1968. A gas explosion, sparked by a resident lighting her stove, blew out the load-bearing flank walls. This caused the structural support to be lost and a whole corner of the block of flats collapsed killing four people and injuring 17.

The partial collapse of Ronan Point led to many changes; obviously this type of construction had to be regulated to prevent a reoccurrence.

The Great Fire of London made people more aware of the materials buildings were made from.

Part A of the Building Regulations involving structures were amended in 1970 to prevent disproportionate collapses from happening again.

Changes affecting construction and use of buildings

Government policy changes

Recent environmental issues have led to government policy changes that affect construction and the use of buildings. For example, the government has recently signed up to several global targets for the reduction of carbon emissions into the atmosphere. It is hoped that part of this reduction will be achieved by making homes more energy efficient. This has necessitated a change in the regulations under several approved documents, such as Part L on insulation, where levels have been increased and items such as boiler efficiency have been taken into account.

The government drive towards sustainable communities with new eco-village community developments is another government-driven policy aimed at reducing our carbon footprint and greenhouse gas emissions. The government is also looking at the rating of buildings for energy efficiency. Now, there are information signs placed within each public building. These signs are energy performance certificates which give a rating from A to G indicating the energy performance of the building.

Local building by-laws

Local by-laws are laws which relate to specific issues in the local community, and can only be applied in the local area. These by-laws are created by the local authority for the good of the community and may involve things such as footpaths, fencing, playground areas, sports facilities and dog walking. They may, in certain cases, relate to the appearance and construction of buildings local to the area.

However, the Building Regulations take legislative precedence as these are backed by an Act of Parliament. Thus, if a by-law conflicts with a Building Regulation, the by-law will not be applicable.

National standards

There are a set of nationally agreed standards produced by the National House Building Council (NHBC). NHBC is a not-for-profit organisation whose principal aim is to raise quality standards in house construction. They can give advice for all stages, from planning, design and construction up to handover of the project to the client. They offer the full range of services and are equivalent to the local authority building control section.

The NHBC also offers a building control independent inspector service and can pass plans and inspect to NHBC standards. Once inspected, new builds can be awarded a 10-year NHBC defects guarantee on the property.

Purpose of building control

Ensure standards

The aims of the building standards are to produce houses and commercial buildings that are safe and secure for the occupants. Construction is therefore regulated and is controlled closely. Building standards are published, maintained and enforced. Those people who work outside of the regulations may have their property pulled down by the local building control division; they will also be charged for the clearing up of the demolition waste.

The Building Regulations are there for both the safety and the quality provision of buildings where people live, work and socialise. As government policy changes and new technologies emerge the standards are updated through changes in the Approved Documents and revisions to the regulations.

Safety and hygiene

Safety is paramount for the occupants of a building and is achieved through the Approved Documents which cover every safety aspect of a building's construction. Included in these documents is information about:

- structural safety – settlement and collapse
- fire safety – smoke and heat detectors, evacuation routes
- contamination – the care of toxic substances
- ventilation – to provide clean air quality
- hygiene – waste disposal and cleanliness
- drainage and waste disposal
- boilers – regulation of heating appliances.

Employers still have a duty of care within the HASAWA 1974 to protect employees at work which is in addition to these regulations.

Improve energy efficiency

Part L of the Building Regulations is concerned with the conservation of fuel and power. It is split into four parts, which are divided into two subsections:

- New builds
  - domestic
  - commercial
- Refurbishments/extensions
  - domestic
  - commercial.

For new buildings, it is easier to look at the whole house and the insulation levels from the start. With refurbishments and extensions, only part of the building will be looked at, rather than the whole.

Part L covers items such as:

- U-values and calculations
- air permeability
- limiting heat losses
- energy-efficient lighting
- limiting air conditioning
- designing with regard to carbon emissions.

It covers many aspects of improving energy efficiency through the reduction of heat loss from buildings. It also looks at installing methods of supply that are efficient (lighting) and are better for the environment (reducing dependence on air conditioning).

Reduction in water wastage

Part H of the Building Regulations makes provision for separate systems of drainage: one for foul and the other for rainwater or surface run-off. By having separate systems not all of the waste water needs to be treated, which saves energy and water in the process. The split system ensures only the foul water is sent for treatment.

The new planning regulations have a requirement to prevent run-off from hard standings. This requirement reduces the burden of treating surface water with the use of porous materials, such as block paviours, that let the rainwater drain naturally through them.

The new water supply and water fitting regulations provide for the design and installation of systems that prevent misuse, waste, undue consumption and contamination of water. The measures contained within the regulations also prevent water wastage

Table 17.1 The Building Regulations set of Approved Documents

Part	Details
Part A	Approved Document A – Structure (2004 edition)
Part B	Approved Document B (Fire safety) – Volume 1: Dwelling houses (2006 Edition) Approved Document B (Fire safety) – Volume 2: Buildings other than dwelling houses (2006 edition)
Part C	Approved Document C – Site preparation and resistance to contaminates and moisture (2004 edition)
Part D	Approved Document D – Toxic substances (1992 edition)
Part E	Approved Document E – Resistance to the passage of sound (2003 edition)
Part F	Approved Document F – Ventilation (2006 edition)
Part G	Approved Document G – Hygiene (1992 edition)
Part H	Approved Document H – Drainage and waste disposal (2002 edition)
Part J	Approved Document J – Combustion appliances and fuel storage systems (2002 edition)
Part K	Approved Document K – Protection from falling collision and impact (1998 edition)
Part L - Dwellings	Approved Document L1A: Conservation of fuel and power (New dwellings) (2006 edition) Approved Document L1B: Conservation of fuel and power (Existing dwellings) (2006 edition)
Part L - Buildings other than dwellings	Approved Document L2A: Conservation of fuel and power (New buildings other than dwellings) (2006 edition) Approved Document L2B: Conservation of fuel and power (Existing buildings other than dwellings) (2006 edition)
Part M	Approved Document M – Access to and Use of Buildings (2004 edition)
Part N	Approved Document N – Glazing (1998 edition)
Part P	Approved Document P – Electrical safety – Dwellings (2006 edition)
Regulation 7	Approved Document for Regulation 7 (1992 edition)

and loss. In commercial premises you can fit taps that have to be held down to operate and the recent introduction of waterless urinals is a new development that saves valuable water resources.

Thinking point

A dripping tap that produces just 1 drop per second wastes nearly 5,000 litres of water per year!

Legislation and documentation

The Building Acts

The Building Act 1984 and the Sustainable and Secure Buildings Act 2004 are Acts of Parliament that produced the Building Regulations and the Approved Documents. These acts are enforced by law and give power to the building control officer. You can refer to the Office of Public Sector Information (go to www.opsi.gov.uk) for more details on these Acts of Parliament.

Building Regulations

The Building Regulations consist of a set of Approved Documents that cover all aspects of construction work. In 1991, revisions were made to the Building Regulations; various amendments were made to the Approved Documents.

The regulations often change with new technology and government policies. You should always check that you have the most current approved documents. Look at Table 17.1 to see what the Approved Documents cover.

Sustainable Communities Act

The Sustainable Communities Act 2007 was designed to provide some legislation to enforce sustainability. The aim of this act is to reduce energy output and make our environment more sustainable at a local level. It gives a list of matters that local authorities should consider in regard to their planning.

Sustainable construction legislation is slowly creeping into the UK. Both the government and the EU are proposing zero carbon homes. This would mean that the construction and operation of a home would be carbon neutral.

Current developments

Brownfield sites

A brownfield site is a site that is no longer used and can be redeveloped for a new function. This makes maximum use of an existing site and helps protect greenfield sites (areas that have never been built on) from being built on. Brownfield sites are therefore sustainable as their use 'recycles' the land for a second time. This makes the best use of a land resource.

Often the major issue with brownfield sites is soil contamination from previous industrial uses such as power stations, car production or factory units.

Use of contaminated land

Many brownfield sites unfortunately contain contamination within the soil. Therefore, the soil needs to be tested and then a method of neutralising the contamination considered. This may ultimately mean the removal and transportation of the contamination to a registered tip that takes this type of waste. Great care must be taken to ensure that contamination does not cause harm to potential occupants and is risk assessed in its removal and handling procedures.

Energy conservation

Energy conservation developments are at the forefront of many building plans and regulations, both locally and nationally. The use of small wind turbines that can be used on a domestic application have now brought about changes to the planning regulations, making it easier to fix to homes. The use of solar panels has improved with the manufacture of photovoltaic cells that fix together to form a roof tile pattern and so they are also much more aesthetically pleasing than they used to be.

The use of natural insulation products such as sheep's wool is a newer innovation along with NASA technology which uses foil-wrapped insulation to reduce heat loss in lofts. Fuel cell technology is just starting to enter the construction industry from the car industry.

Domestic combined heat and power plants are just starting to enter the market. These use the waste from heating to produce power that can be sold back to the grid; thus they are a more efficient method of producing heating.

Reduction of impact on the natural environment

Town and Country planning regulations have placed restrictions upon developments that eat into the greener areas that surround our towns and cities and turn these into mass conurbations. The development of the 14 green belt areas also helped to reduce the spread of the built environment. Reuse and regeneration of the inner city areas has improved and curtailed the need for greenfield construction and development. The inner city areas of Leeds and Glasgow provide excellent examples where older run down regions have been regenerated with new building work to provide a city lifestyle existence.

Application of building controls

New buildings

Legislation on new buildings is developing. As we have discussed earlier, the Building Regulations Part L have a specific part applicable to new domestic and commercial construction. These set the levels of insulation and the energy efficiency of the new building and even include an air test for air leakage, and hence heat loss, of the new build. The drive towards the new sustainable towns is easier with a new development and allows focusing on one direct policy by the government. New building sustainability provides an 'advertisement' for the rest of the population to take note and perhaps act by wanting the same type of community.

Alterations

It is harder to introduce new legislation regarding alterations as they often occur on older and less energy efficient structures; for example a new 300 mm cavity wall extension added onto an existing 1910 brick wall of solid construction. However the new regulations do enforce the new extension to be built in accordance with the newer energy-efficient parts of the Building Regulations.

Assessment activity 17.1

BTEC

Your client is not sure why they have to apply for building regulation approval. Examine the factors that have influenced the historical development of building control. P1

Discuss the legislation and documentation associated with building control and their application. P2

Explain the implications of the Building Regulations for low-rise commercial and domestic construction.

Grading tips

For P1 you will need to fully explain the development of the Building Regulations historically.

For P2 discuss the various building acts and the various building control documentation.

For M1 you need to examine and discuss what the implications are in applying the regulations to the stated buildings.

PLTS

Planning and carrying out research, while appreciating the consequences of applying regulations will help develop your skills as an **independent enquirer**.

Functional skills

Undertaking research using the Internet on the development of the Building Regulations will help develop your skills in **ICT**.

2 Know how to apply and enforce Building Regulations

Approved Documents

Part A

Part A deals with the structural elements which form the building envelope. These elements include the foundations, walls, floors and roof. Foundations need to be of adequate size and construction to hold the weight they are expected to carry. Floors also need to be built to safely carry the loads through to the walls; and the walls, down to the foundations

Any structural frame, beams or columns must be designed to safely support the loads placed upon it. Part A also covers the expected wind loadings and the height of the buildings that can take these.

Part B

Part B deals with fire safety and is divided into several topic areas:

- means of escape
- use of smoke detectors
- fire doors
- holding back internal spread of fire
- fire resistance
- construction materials
- methods used to stop external spread of fire
- fire brigade access.

The means of escape in the event of fire covers having features such as fire exits and stairs and fire exit corridors, such that the occupants can escape without any external help. In domestic application smoke detectors must be installed and the provision made for escape windows from occupied rooms above ground level. Where loft conversions are carried out on buildings of more than one storey, additional measures, such as the provision of self-closing doors which open onto the staircase and fire protection between the loft rooms and the staircase, are essential.

Smoke alarms give people more time to get out of a burning building safely by alerting them to the fire.

The surface spread of flame on walls needs to be controlled so that the fire can not easily spread. Things such as paper on notice boards can cause a fire hazard. This requirement is critical on escape routes which need to remain clear of smoke.

The resistance of the structure to fire is considered and includes the effects of smoke and the time period that the building should resist fire. In dwellings, traditional construction materials such as concrete, brickwork and plaster usually provide sufficient fire resistance.

The methods used to stop the external spread of fire need to be considered such that fire is prevented from travelling from one house to another. Such methods include the use of fire stops in roof voids in semi-detached properties. Access by the fire brigade needs to be considered, so a turn table ladder can get close enough to fight a fire.

Part C

Contaminants, weather and ground moisture must not be allowed to get into a building as this could cause harm to the occupants. Part C deals with the resistance to both moisture and contaminants. Modern construction incorporates dpcs, dpms and flashings to prevent water penetrating a structure and causing damp and damage to finishes. These regulations, however, do not consider the recent flooding incidents we had in 2008. Often gas membranes have to be installed in areas subject to radon gas.

Part D

Part D deals with toxic substances. It covers materials, such as cavity wall insulation foams, where toxic fumes from installation must be stopped from permeating any occupied building.

Part E

Part E covers the resistance to noise. Walls, floors and ceilings must resist the passage of sound so that the noise does not disturb neighbours.

Part F

The occupants of a building should be provided with sufficient ventilation by natural or mechanical means. Information about this is covered in Part F. Ventilation can be provided through trickle ventilation and extraction fans in bathrooms and kitchen areas.

Part G

The building should provide sufficient hot and cold water, sanitary and washing facilities for the occupants. Part G covers information on these and other hygiene issues.

Part H

Part H deals with drainage and waste disposal. The waste from WCs, bathrooms and washing machines which is foul water should discharge to a suitable drain.

Thinking point

The introduction of the flushing WC reduced death rates by 15 per cent in London.

Part J

Fires, boilers and incinerators burning any fuel must be provided with sufficient air for their safe operation. Part J covers this information and also deals with a satisfactory means of disposing of combustion gases or fumes via a flue or chimney.

Part K

Protection from falling, collision and impact are covered in Part K. Stairs, fixed ladders and ramps should be designed and constructed so that people can move safely in or about a building. All stairways and stairwells should have fixed guardrails designed around them to prevent falling.

Part L

The efficient use of heat and the restriction of its loss are covered Part L. This section deals with the conservation of fuel and power, including the use of energy-efficient artificial lighting.

Part M

Part M covers access and facilities for disabled people. It looks at the provisions that are required for the disabled, such as ramps and lifts.

Part N

The safety of glazing in relation to impact, opening and cleaning is dealt with in Part N. The use of toughened glass is considered in here to prevent accidents from falling through or against glazed areas.

Part P

Part P makes provision for the design, installation, inspection and testing of electrical installations in order to protect people from fire or injury.

Application of the Building Regulations

Design and construction of dwellings

As mentioned above, Part K of the Building Regulations deals with the main staircases in dwellings; it covers topics such as:

- the pitch of the staircase
- the width of the treads
- the height of the risers
- the guardrails around a stairwell
- the clearance height from the stair nosing to the ceiling above
- the use of landings on stairways.

The means of escape, covered in Part B, and disabled access provisions, covered in Part M, also need to be carefully considered in any staircase design, for example the underside of the staircase will need fire protection.

As we have seen, fire precautions are covered quite extensively in Part B. Fire and smoke inhalation are a major cause of building damage and fatalities in domestic homes. The main areas of precaution are:

- means of escape in the event of a fire
- smoke detection
- self-closers to doors
- protected staircases
- surface spread of flame
- fire-resistant structures
- prevention of flame spread to adjacent properties.

Insulation is covered by the amended Part L of the Approved Documents. This covers factors such as:

- air leakage from new buildings
- insulation levels to meet required U-value standards
- boiler efficiency
- energy-efficient lighting.

Drainage should provide a system that can carry waste water from cooking, toilets, washing and bathing to a sewer or cesspool or settlement tank.

If a cesspool or tank is used then it must be impermeable to liquids and have adequate ventilation. It must also have means of access for a contractor to empty the tank.

Rainwater must be guided away from a building to the sewer by an adequate system such as by the use of guttering. Finally, there must be a designated place for a waste bin or dustbin so it can be collected for waste removal by the local authority.

Amendments to Building Regulations

Often there are amendments to the Building Regulations. If you look at the list of Approved Documents (See Table 17.1), you will see that each has a date against it. This was when the regulations were amended and issued as a revised approved document.

Legal obligations to comply

The Building Act was passed by Parliament and as such is a legal and enforceable Act of Parliament. The Building Act gives local authorities, through building control inspectors, the authority to inspect and action work which does not meet the provisions of the Approved Documents. Builders and contractors have a legal obligation to comply when undertaking any construction work that qualifies for Building Regulation control.

Take it further!

Do some research on the Internet and find out what penalties can be enforced or action taken by local authorities if a building work does not follow Building Regulations.

Enforcement of the Building Regulations

Approved inspectors

An approved inspector is a person or organisation that is approved by the Secretary for State as having the authority to control building work in accordance with the Approved Documents. The NHBC, for example, can act in this role.

Building control officers

A building control officer (BCO) is an appointed local authority employee who has the authority under the Building Act to undertake inspections or plan approvals. They first see the proposed work in either a building notice or as a full plan. The council cannot reject any work that has been undertaken in accordance with approved plans if they have been passed by the BCO as 'fit for purpose'.

Both the building notice and the full plans process involve a site inspection at regular intervals in order to check that either:

- the work conforms to the Approved Documents
- or the work meets the approved plans and specification.

The BCO has the authority to enter the premises at any reasonable time for the purposes of inspecting any building work that is being undertaken. But you should ask for evidence of authority and identity before allowing entry to your house as a precaution against theft.

The BCO also has the authority to undertake tests of any material or work that appears not to conform to the regulations. Notices are required to be given to the BCO so any work that may be covered up, for example foundation excavation before concreting, can be inspected and passed. Should this not occur then the BCO has the authority to ask for the work to be opened up.

Any enforcement action taken is also recorded formally within the Land Registry documents so any potential building owner is aware of the enforcement action taken.

Approvals

A Building Notice just informs the BCO of your intentions to undertake some building work; it is not an approval to build. You must ensure that the work undertaken meets the details contained within the Approved Documents or you may have to take down and rebuild any work that does not comply.

Full plans submission is the best way of obtaining Building Regulation approval. It gives the local authority time to review your drawings and the specification to ensure that the plans comply with the Approved Documents. They, then issue an approval notice so you can proceed. You must of course construct the building work in accordance with these Approved Documents.

Building inspection

A building inspector will want to inspect the construction works on site at regular intervals:

- commencement
- foundation excavation
- foundation constructed
- ground floor oversite (before concrete)
- damp-proof course
- steel beams/structural alterations
- drains (before backfilling)
- drains (testing)
- completion.

It is best to contact the BCO and agree the procedures you will adopt with regard to inspection before you commence work. When you commence work it is best to confirm in writing the inspection time; this holds true for the completion stage inspection. Other stage inspections can be booked via telephone.

Assessment activity 17.2

1. Identify the various approved documents that comprise the Building Regulations. P3

   Describe how the regulations are applied and enforced. P4

   Discuss the powers of the local authority building control officers and approved inspector. P7

2. The following are proposed for a buildings construction. Propose solutions to the two issues:
   - a house foundation 600 mm wide x 100 mm deep is found to be defective
   - a lintel with 50 mm end bearing is found to be failing at the padstone. M2

   Justify the solutions that you proposed in M1. D1

Grading tips

For P3 you will need to identify all of the Approved Documents that form the Building Regulations.

For P4 describe how the regulations are applied and how they can be enforced.

For P7 you need to discuss what powers local authority building control officers have to enforce the regulations.

For M2 you will need to propose solutions to the two defective specifications by looking at the regulations.

For D1 you are required to justify your two proposed solutions.

PLTS

Analysing and evaluating information while judging its relevance and value will help develop your skills as an **independent enquirer**.

Functional skills

Reading summaries on the Approved Documents will help improve your reading skills in **English**.

3 Understand the procedures and documentation involved with Building Regulation approval

Approval procedures

The Building Notice

The Building Notice is a form which states that you are going to start building works and informs the building control section of the following:

- where the work is occurring
- when you are starting – it must give at least 48 hours notice prior to commencing on site.

There is a standard Building Notice form available on every local authority building control website. Often this form can be sent as an attachment to an e-mail. The only other documentation required to accompany the notice is a site location plan, which is usually the 1:1250 Ordinance Survey plan, and the appropriate fee for the service. There is often no difference between the fee for a Building Notice and that for full plans submission. The Building Notice form cannot be used if it concerns a workplace or involves the construction or formation of flats.

Advantages of using a Building Notice:

- plans do not need to be drawn up by a designer
- purchase and preparation of a site plan is all that is needed.

Disadvantages of using a Building Notice:

- work may need to be corrected once it is inspected
- all charges, plans and the inspection fee, are paid at the time of deposit
- no approval notice is given
- no Completion Certificate is issued
- plans and calculations may sometimes be required.

Full plans submission

In a full plans submission, fully detailed drawings are submitted along with a specification and any calculations required in order to enable a plan checker and the building control officer to decide if your submission meets the Approved Documents.

You are required to provide the following with your application:

- two copies of drawings. Drawings should include elevations sections and plan views and be at a scale of 1:100 or 1:50
- a specification for the materials and proposed construction methods
- any supporting calculations to justify any structural members or thermal calculations
- a site location plan to a scale of 1:1250
- your completed full plans application form
- appropriate plan charge
- if the work to a building is designated under the Fire Precautions Act 1971 then two further sets of plans detailing fire safety measures must be supplied so the fire brigade can comment on the proposal.

The building control officer has up to five weeks to decide if the plans are approved or rejected; the period can be extended up to 2 months if you agree. Once plans have been approved they are valid for a period of three years, even if the regulations change in this period. A Completion Certificate is issued when all the work has been finished and signed off by the inspecting building control officer.

Advantages of full plans submission:

- plans are approved and a confirmation notice is sent
- plans are valid for 3 years
- the inspection fee is not payable at time of submission
- your builder works from approved plans
- the Approvals notice can be handed onto any future buyer
- on completion of the work a Completion Certificate is issued.

Disadvantages of full plans submission:

- you must wait for plans to be prepared by an architect or designer.

Notice of commencement

Both a building notice and full plans approval require that you inform the local building control officer or approved inspector. You must give 48 hours notice that work will commence on site, naming the site location and giving a location plan (for Building Notice). It is always best if this is done in writing.

Once approval has been gained, inspectors must issue a document called an Initial Notice to the relevant local authority at least 5 days before any controlled building work starts on site.

Stages of notification

Table 17.2 gives some further detail for each stage of notification.

Table 17.2 Stages of notification

Stage	Description
Commencement	You need to give the required notice such that your details are logged at the building control office and a building inspector is allocated to your property
Foundation excavation	Inspect the base of the foundation excavation to ensure that the substructure soil is at a sufficient depth to support the loadings
foundation constructed	Inspect the completed poured concrete, trench blocks and engineering brick foundation before backfilling
Ground floor oversite (before concrete)	Inspect the dpm to ensure that it is lapped sufficiently, is of the correct materials and tucked vertically up and into the dpc
Damp-proof Course	Inspect the dpc for position, correct width and materials
Steel beams / Structural alterations	The padstones for the beams will need inspecting along with the beams
Drains (before backfilling)	Drainage may need inspecting if it passes through a foundation wall or under the floor construction
Drains (testing)	A standing water drain test may be required to check that they hold water
Completion	A final inspection is undertaken, a Completion Certificate is issued

Role of the building control officer or approved inspector

The role of the building control officer or approved inspector may include the following:

- checking of plans for approval
- giving advice on construction
- maintaining an approved standard
- inspections
- dealing with dangerous structures
- demolition notices
- enforcement work.

Documentation

The standard forms for a building regulation approval are normally very similar for each local authority. On pages 430 and 431 you can see some examples.

As you can see, there is very little information that is required. The Building Notice and demolition forms are similar, as seen in the example of a Building Notice form in Figure 17.1.

Local authority or private building control

Local authorities employ building control officers. Private organisations, such as NHBC, will have approved inspectors. In both cases the inspectors and the officers will be certified and able to take appropriate enforcement action.

Approved inspector

An approved inspector is an individual who has been authorised under the Building Act 1984 to carry out building control work as an alternative to the local authority BCO. The Construction Industry Council holds a register of all independent inspectors and is responsible for their approval and registration. Each approved inspector is deemed to be a 'building control body', licensed to approve building work. Approved inspectors are members of the ACAI Association of Consultant Approved Inspectors.

As mentioned earlier, the National House Building Council (NHBC) offers a technical service where they can act as an approved inspector. The NHBC provides a research and investigation facility that publishes standards that improve the quality of home construction.

Take it further!

Visit the CIC website and look for an approved inspector in your area (visit www.cic.org.uk).

Enforcement procedures

A building control officer can enforce any work that does not meet the requirements of the Approved Documents. They have the authority to remove or alter any building work not conforming to the Approved Documents and regulations. This involves serving a 28 day notice on the owner. If the work is not completed then the local authority may do the work and later recover their costs.

An authorised building control officer has the right to enter premises to find out whether or not there is a contravention of the Building Regulations. The section provides for the local authority to obtain a warrant if entry is refused. Legal action could result from failure to gain entry because of obstruction.

The BCO also has the authority to respond to a dangerous structure to avert immediate danger to occupants and public. The building owner is then asked to make safe the work or demolish the structure. Where the BCO is obstructed, the local authority can apply to the courts for an order to carry out the work.

BUILDING REGULATIONS 2000 (As amended)
Please return to:
Building 4, North London Business Park
Oakleigh Road South
London N11 1NP

BUILDING NOTICE FORM

Statement
I/We hereby give notice in accordance with Building Regulation 12(2)(b) of my/our intention to carry out the building work or material change of use as described below.

Applicant's Details (owner)
Mr/Mrs/Ms Initials............ Surname ....
Address ....
Post Code .... Tel: .... Fax: ....

Agent's Details
Mr/Mrs/Ms Initials............ Surname ....
Company Name ....
Address ....
Post Code .... Tel: .... Fax: ....

Builder's Details
Mr/Mrs/Ms Initials............ Surname ....
Company Name ....
Address ....
Post Code .... Tel: .... Fax: ....

Address of building to which work relates
....
Post Code ....

Description of proposed works
....

If the proposed works involves Building Regulation Part P Electrical Safety, Please confirm whether you are intending to use an Authorised Competent Person YES/NO
If no, Please also submit form LBB/BC/Part P

Use of Building
1. If new building or extension please state the use: ....
2. If existing building please state present use: ....

Method of Drainage (e.g. to public sewer, septic tank, cesspit)
1. Foul Water ....
2. Surface Water ....

Building Notice Charge (please refer to separate guidance notes in enclosed booklet)
1. Internal floor area of proposed works .... square metres
2. Estimated 100% cost of the works £...
3. Schedule 1 charge £ ....
4. Schedule 2 charge £ ....
5. Schedule 3 charge £ ....

Total Payable including VAT £ ....

Completion Certificate
On completion of the works I require a completion certificate YES/NO
Please Note:
An additional charge of £25.00 will be made if a completion certificate is requested. Please add this to the Building Notice charge.

Signature
Applicant/Agent/Builder .... Date ....

Date Received:

Please refer to guidance notes overleaf **SM52/63**

Figure 17.1: Building Notice form

Directorate of Environmental Services
Building Control Services
Guildhall 2
Kingston Upon Thames
Surrey KT1 1EU

E-Mail: building.control@rbk.kingston.gov.uk
Tel No. 020-8547-4699 (24 Hour Answering Service)
Fax: 020-8547 4660
www.kingston.gov.uk/buildingcontrol

Plan No.		
Cheque ☐	**Cash** ☐	**None** ☐
Amount: £	**Receipt No.**	
Date:	**(Office Use Only)**	

Royal Kingston

Building Act 1984, The Building Regulations 2000

FULL PLANS SUBMISSION

If this form is unfamiliar please read the notes overleaf or contact the above office for guidance.

1

Applicant (Owner or person on whose behalf the work is to be carried out)

Name: ________ **Tel:** ________

Address: ________ **Fax:** ________

________ **Postcode:** ________ **E-mail:** ________

2

Agent (if applicable - person acting on behalf of applicant and to whom correspondence will be addressed)

Name: ________ **Tel:** ________

Address: ________ **Fax:** ________

________ **Postcode:** ________ **E-mail:** ________

3

Location of building to which work relates

Address: ________

4

Proposed work

Description: ________

If applicable, please confirm the Town Planning Application reference no. to which this work relates: ☐

5

Use of Building

a. Present use ________ b. Proposed use ________

6

Charges (Please read note 2 overleaf and separate Guidance Note on Building Control Charges)

(i) Table 1 - If new houses or flats being built, please state total no. of dwellings: ________
(ii) Table 2 - If domestic extension(s) or domestic detached building(s), please state total floor area: ________ m^2
(iii) Table 3 - If other work (not covered by Tables 1 or 2) please state total estimated cost of work: £________
(iv) Payment enclosed with this submission £ ________ INC. VAT
(Cheques to be made payable to Royal Borough of Kingston Upon Thames)
(v) Person to whom invoice for site inspection charge should be sent (where applicable and if not the applicant):
Name: ________
Address: ________

7

Statements

(i) Is or will the building be a building to which the Regulatory Reform (Fire Safety) Order 2005 applies? (See note 1 overleaf) ***YES/NO**
(ii) Do you agree to a conditional approval if considered appropriate? (See note 3 overleaf) ***YES/NO**
(iii) Do you agree to an extension of the time period for your application to be considered? (See note 4 overleaf) ***YES/NO**
(iv) Does the proposal involve home electrical installation work to which Part P of the Building Regulations applies? ***YES/NO** (If necessary, see separate Part P Guidance Note).
(v) If YES, do you intend to use an electrician who is registered with a Competent Persons Self-Certification Scheme? ***YES/NO** (See Note 5 Overleaf)
(vi) I hereby give notice of intention to carry out the work set out herein. I enclose Full Plans in accordance with Regulation 12(2)(b) of the Building Regulations, 2000 and in case of query would prefer that you contact me by:

Letter ☐ **Telephone** ☐ **E-mail** ☐ **Fax** ☐

Signed: ________ **On behalf of:** ________
(Insert applicant's name if signed by an Agent)

Date: ________

***DELETE AS REQUIRED**

Figure 17.2: Full plans submission form

Assessment activity 17.3

P5 P6 M3 BTEC

1. Explain what the approval procedures are for a client's application to building control. P5
2. Evaluate the documentation used in building control. P6
3. Explain the procedures used to enforce the building regulations in the event of non-compliance. M3

Grading tips

For P5 you will need to explain in terms that a client can understand what the approval procedures are, stage-by-stage.

For P6 you are required to evaluate the documentation used in building control applications.

For M3 you will need to explain the procedures used to enforce the Building Regulations.

PLTS

Adapting ideas as circumstances change will help develop your skills as a **creative thinker**.

Functional skills

Reading and understanding the various types of forms used in building control will help develop your reading skills in **English**.

4 Be able to prepare a submission for Building Regulation approval

The building notice application

You should be able to use the form shown in figure 17.1 to complete a Building Notice for an extension to the home that you live in.

Full plans submission

Your tutor will provide an outline drawing for a semi-detached house. Now provide the following details:

- a completed full plans application form
- a 1:1250 location plan
- plans, elevations and sections of the proposed extension
- a full specification of each element of the extension.

Assessment activity 17.4

P8 D2 BTEC

1. Produce a specimen Building Regulations application with all the associated documentation. P8
2. Evaluate your application. D2

Grading tips

For P8 you will need to produce a full Building Regulations application for a small extension to include all the documentation associated with the application.

For D2 you are required to evaluate your application as fit for purpose.

PLTS

Working towards goals and showing initiative, commitment and perseverance in achieving this criterion will help develop your skills as a **self-manager**.

Michael Rosser

Building Control Officer

Michael works as a building control officer for the local authority. He works within a team of 12 building control officers. Within the team of 12, there are two plan checkers who are office-based and whose job it is to check full plan submissions. The other 10 BCOs are always out of the office on inspections.

Michael has to get into the office early in order to pick up the list of inspection requests from new clients. Michael has to visit about 15 sites a day on his way around his district. He has to organise his day so that he has the maximum efficient use of the time available during the day.

No two days are the same for Michael and he really enjoys this job. He gets to meet a variety of different people and has developed a friendly and approachable rapport with most of his clients. Michael is also the fire expert and often is tasked with looking at commercial buildings and giving advice on what requirements will have to be met for fire and smoke procedures. With this role, he often gets to liaise with the fire brigade officers who are involved in the fire certification scheme for commercial premises.

Recently the team has taken on a trainee. This is the best way to learn about the regulations is from an experienced officer who can pass on their experience to a young apprentice. As the experienced building control officer, Michael is training the new recruit. To do this, Michael takes the trainee out on inspections so he can learn about the regulations and learn to recognise when something is not quite right.

Think about it!

- Would you rather be a plan checker or an inspector?
- How long would the training take?
- What qualifications would you need to become a BCO?

Just checking

1 What is the difference between a building control officer and an approved inspector?
2 What is a Building Notice?
3 How does a Building Notice differ from a full plans submission?
4 What powers does a building control officer have?
5 What does Part L of the Approved Documents cover?
6 What part did the Great Fire of London have in the historical development of the Building Regulations?
7 Why was the significance of the Great Plague?
8 What does the NHBC mean?
9 At what stages would you ask for a site inspection by a BCO?
10 How much notice do you have to give before starting building work?

Assignment tips

- The local authority building control website will often contain a lot of information on the application of the Building Regulations.
- The government planning portal website contains a full set of online versions of the Building Regulations which are free to download.
- All local authority websites will have the forms used for applications free to download and evaluate.

Glossary

A permit to work – a document issued by the person responsible for a particular work area. Such an area is usually one of high and complex activity; the permits detail who is working in that area and lists any precautions that must be taken and any isolation of services that may be required.
Access and egress – entrance and exit.
Adjudication process – a meeting held by senior management to decide what level of risk and profit requires placing on the net tender or estimate.
Air entraining agent – this forms bubbles within the concrete which capture air that is used to insulate the concrete from extreme cold.
Approvals – the agreement of the cost for the survey and instruction to proceed.
Aquifer – an underground storage area created naturally within the Earth's rock strata.
Architect's instructions – written instructions issued on behalf of the client which instruct the contractor to undertake a particular item within the contract.
Attendances – the items that will be required by the nominated subcontractor such as a crane, rubbish removal, water and electricity.
Banksperson – a competent person who supervises the lifting operations of the crane.
Baseline – the main survey line or 'back-bone' on which to form all the other survey triangles.
Bill of quantities – a list of quantities produced to a standard that is used to price construction tenders.
Boreholes – holes sunk into the ground to extract soil samples at differing levels. The information is recorded as the holes are drilled so that the design engineer discovers at what depth each soil is found.
Brief – the client's idea of what they want, which the designer has then to turn into a reality in accordance with current regulations and legislation.
Building control – the local authority department responsible for checking that the building project proposal complies with various legislative controls and regulations.
Building Regulations – these are produced under the Building Act 1974 and control many aspects of construction in order to ensure that energy saving measures are built into new and existing designs.
Calibration – the process of ensuring that the measuring equipment is correctly adjusted to give true readings.
Cames – slender, grooved lead bars used to hold together the panes in stained-glass or latticework windows.
Cash flow – the amount of money flowing into the company from clients and out of the company as payments. Money flowing in should be greater than money flowing out.
Chainage – the cumulative distances measured to specific features as fixed from one end of the survey line.
Check lines – an additional measured line to check the accuracy of the survey framework.
CIOB – Chartered Institute of Building – a professional association for construction personnel.
Cladding – the lightweight material that forms the enclosure of the building. Cladding is usually lightweight because it does not have to carry the structure.
Client – the person who will ultimately own the constructed building or project and who pays for the work.
Closed traverse – an irregular ring of survey stations linked by measured distances and internal angles which can provide 'easting' and 'northing' coordinates to help set out large structures.
Cold bridging – this occurs when the insulation layer within a wall or roof is interrupted by another material or is reduced in thickness. There is greater heat loss through the thinner area of insulation and that part of the wall or roof has a reduced internal surface temperature. When the warm, moist air inside the property comes into contact with the cooler surface, it is chilled and less able to carry moisture. This results in surface condensation or pattern staining of décor.
Collective means of protection – a system that protects the whole workforce and not just the individual. For example, a scaffold with guardrails, handrails, toe boards and netting protects everyone working on or using it.
Compacted – concrete is vibrated either by mechanical means or by hand to remove all the air bubbles.
Compressibility – the ability of soil to compress and withstand a load imposed upon it.
Compressible clay – clay that can be compressed or compacted to increase its strength or load bearing capacity.
Confirmation pads – used to confirm verbal instructions given on site.
Construction specification – written information prepared by the design team for use by the construction team, the main purpose of which is to define the products to be used, the quality of work, any performance requirements and the conditions under which the work is to be executed.
Contract programme – often a simple bar chart, showing activities against time, which offers an effective visual representation of the construction project. The percentage of actual work completed can be plotted against work that was scheduled to be completed.
Contractors – separate companies who work for the main contractor, e.g. a heating engineering company.
Contracts manager – a member of the construction team whose role is to manage several contracts. The contracts manager moves resources around as and when required, and deals with the designer and construction team on each.
Control measure – a method, system or product used to reduce a high risk to an acceptable risk; for example using a fork-lift truck to lift a heavy object rather than trying to lift it by hand.
Cost analysis – the process of providing estimates of what a building project's costs and benefits are likely to be; then comparing these estimates and making a judgement as to whether it is acceptable to proceed to the construction stage.
Cost value engineering – changing specification, methodology quantities, omitting items or other works to reduce costs.
Cover level – the reduced level of the top of the inspection chamber cover.
Cranage – a device used to lift a static tower crane or a mobile crane.
Crane offload – a load is delivered by a crane that is bolted onto the wagon. The crane is operated by hydraulics and lifts solid material off the wagon safely onto the construction site.
Critical path – the link between construction activities crucial for the contract to complete on time. There is no flexibility of time within these activities so any hold up in these activities will delay the final handover date to the client.
Cross bonding – in the event of an electrical fault, this process prevents any metal service from becoming live with electricity.
Damages – financial penalties for every week or part week that the programme has overrun which are usually charged for every day that the programme runs late.
Dayworks – unforeseen works that may involve variations. It is an historical arrangement within a contract and records the labour hours and any plant or materials used against items; it is signed by the client's representative.
Demographics – the characteristics of a population such as gender, location or age.
Desk study – an investigation of information about a piece of ground undertaken by reviewing existing records.
Development of the brief – the development of the client's idea for a design of a building or a concept; this is then extracted and evolved by the designer so it can be taken forward to the feasibility stage.
Dilapidation – falling or causing to fall into ruin.
Dolos units – man-made sea defence blocks, which are piled against quaysides and jetties to protect them from being eroded by wave action.
Down time – when equipment is not being used.
Drawing register – this contains all of the drawings for the project in numbered order with the latest revisions issued.
Duty of care – the duty placed upon everyone by the HASAWA to take care of themselves and others about them.
Economies of scale – the ability to buy in bulk, thus receiving greater discounts.
Embedded energy – the amount of energy that has been used to produce the material. It is often expressed in terms of how much carbon has been released into the atmosphere during its manufacture and transport.
Employees – workers who receive wages for their skills from the employer.
Employer – the person who owns the company which is constructing the building or project; the employer may be represented by a managing director or by a multinational with shareholders and a chief executive officer.
Equilibrium – when all the forces are balanced. For example, upward forces are equal to downward forces and clockwise forces clockwise equal to anticlockwise forces.
F10 – the official document that informs the HSE that a company is undertaking a project.
Feasibility – deciding whether the building is either practicable or will proceed.
Final certificate – this is a certificate written by the designer that releases the contractual obligations of the main contractor with the final payment of monies withheld as retention.
Fit for purpose – a material must be able to be used in the context of the design. For example, glass in windows is required to let light through; any other material would not be appropriate.
Footprint – the shape the outline of the building leaves on the ground when viewed in plan from above.
Formation level – the reduced level of the bottom of the excavated sewer trench.
Ganger – the person in charge of a work gang or sets of gangs which may be trade specific such as a team of ground workers. The ganger acts as the control point so that the site manager or supervisor only needs to talk to one person and not several.
Geotechnical properties – how soil is likely to perform when imposing loads on it or what will happen when water is removed to allow work to take place.
Glulam – a process of gluing timber strips together to form a solid beam within a mould.
Greenfield land – area which is undisturbed by previous construction – in effect, a green field.
Grid lines – an imaginary series of lines running north – south and east – west which allows designers and engineers to plot key positions on-site.
Gross estimate – the net costs plus overheads and profit and risk items.
Health and Safety Executive (HSE) – a body set up by the Health and Safety Commission acting under the Health and Safety at Work Act. The HSE is responsible for inspecting, and enforcing health and safety.
Heavy-side engineering – construction work that requires heavy machinery, such as roadways, highways, mass concrete dams and earthworks.
Helical auger flight – the technical term for the large spiral drill of the auger drill.
Horizontal angle – an angle measured between three fixed points within a horizontal plane.
Horizontal distances – the flat distance that would be plotted on a two-dimensional drawing plan.
Hydroscopic – the ability of timber to easily absorb moisture from the air.
Improvement – identifying what is wrong and how to put it right within a set time.
Inception – an event that is a beginning, a first part or stage of subsequent events.

Interim certificates – Issued by the architect to say that some of the work has been completed. Contractors can then be paid for the work that they have done so far.

International System of Units (SI) – from the French Système International d'unités. The modern form of the metric system developed in 1960 to promote a worldwide measurement system based on the standard properties of metres, kilograms and seconds

Invert level – the reduced level of the lowest part of the internal diameter of the sewer.

JCT – Joint Contracts Tribunal.

Just-in-time deliveries – materials ordered to be delivered just before they are required on site.

Lanyard – an attachment that clips between the safety harness and a secure point to gradually slows the rate of descent.

Legal notices – these are written document requiring a person to do/stop doing something.

Levees – natural banks of silt deposits which are left after a river floods. These are shaped into higher banks to control flood waters. In 2005, some of the levees protecting New Orleans in the USA broke and flooded the city.

Liquidated damages – financial penalties paid by the contractor to the client when the project overruns the agreed completion date through no fault of the client.

Listed building – a building of special architectural or historic interest in the UK. Alterations to these buildings cannot be made without consent or carefull consideration.

Listed building consent – when a building is graded and listed as a national heritage building and thus cannot be altered or demolished without consent.

Listed design approval – all works on listed buildings must match their original character and materials.

Local authority – the elected local council which runs the services within a geographical area.

Local plan – the document that sets out and controls the planning policy for the local authority's area. It sets out where the authority wants industry and housing to grow.

Location plan – a document that is used to guide and identify where the work takes place; a map so that the contractor knows where the site is located.

Made ground – any ground that has been artificially made from material placed from previous works, for example layers of stone compacted and laid to form a level surface ready for construction work.

MCB – miniature circuit breaker which switches off power on an earth short.

Mensuration technique – the process of producing quantities that can be used to make calculations simpler and more efficient.

Method statements – documents which identify the methods used to price the work items, the plant and labour required for each activity.

Mock-up – a sample panel or erection created by contractors and subcontractors as a sample of what the work will look like when it is finished.

Modular coordination – the grid lines on which the designer lays out a building. For example, a window opening must be worked into brick courses to avoid excessive cutting of the material.

National Annex – a document supplementing the British Standards specifications that ensures materials meet quality requirements. These documents apply to all members of the EU but may vary slightly from one member state to another.

NBS – the National Building Specifications publication which can be used in bills of quantities or specifications with drawings.

Net costs – basic costs of labour, plant, materials, preliminary items and subcontractors.

Net estimate – net costs.

Nominated subcontractor – a subcontractor who is appointed by the client and has already tendered for the work package for the client. The main contractor is then instructed to appoint this subcontractor and the value is offset against a provisional sum placed within the main tender documents, including profit and attendances by the main contractor.

Oblique cone – cone where the top vertex or point is not aligned above the centre of the base.

Open-span buildings – these are typically built using a skeleton frame, which allows an open-plan floor area that can be divided into smaller spaces. This clear floor space is particularly well-suited to offices where the final floor plan can be changed to suit any client who wishes to move into the building.

Overburden – the layer of material that has to be removed to get at the minerals beneath.

Over-three-day injury – an injury which results in the person being away from work or unable to do the full range of their normal duties for more than three days.

Pad foundation – a mini-raft similar in function to a raft but not connected to other pads that support structural members.

Panel fencing – this is constructed out of mesh squares welded to a framework. The panels sit into feet and are bolted together with clips; they are high enough so they cannot be climbed over.

Partnering agreement – a method of procuring a contractor which enables cost sharing between client and contractor on savings.

PDA – personal digital assistant; a handheld computer.

pH – the measure of acidity or alkalinity of a substance.

Planning permission – this is the formal consent from the local authority allowing you to build. You must not build without planning permission (except in the case of exemptions).

Pore-water pressure – the pressure that water exerts as it moves through the fissures and cracks in the ground.

Portal – a large gateway or doorway.

Power float concrete – a process whereby the surface of the concrete is machined smooth using mechanical equipment.

PPE – personnel protective equipment which is provided for the individual to use to protect themselves against certain hazards where there is no alternative method.

Price book – a published book that contains current prices and rates for items of work based on the SMM7.

Primary setting-out points (SOPs) – these are existing identifiable, fixed features to which all new building work is related in terms of their setting out dimensions and angles.

Prime cost – when a cost is already known and an allowance has been made to meet this cost.

Procurement – the process of finding and acquiring the expertise, labour, plant and materials needed to build a construction project.

Procurement route – the method that a client uses to select a contractor to construct the project.

Professional indemnity insurance – the insurance policy that building surveyors carry so that if they make an error which results in a loss for a client, the surveyor and the company they work for are covered against legal proceedings.

Programme of works – a bar chart which shows the duration and the resources needed to complete the job. It shows how long the work will take and how long the contractor intends to spend on each task.

Prohibition – banning the use of equipment/unsafe practices immediately.

Provisional and prime cost sums – sums of money placed within a tender for unforeseen works or items that cannot as yet be measured. These can also be sums of money for nominated suppliers or subcontractors. They are subsequently omitted and the agreed rate and price is put back when the work is completed.

Provisional cost – an allowance or a provisional sum that is allowed when the actual cost of something is not yet known.

Raft foundation – a slab that supports the building over a large area.

Ranging a line – the process of measuring a straight line between two points which are a long way apart.

RCD – residual current device, which acts as a fuse.

Reasonably practical – a measure put in place to prevent injuries to workers that is both sensible and sound for that particular situation as well as realistic in safety terms.

Reconciling – settling up the values of unknown costs.

Red lining – indicates where CAD drawings have been marked up to show proposed or amended details, just as you would annotate manual drawing in red ink.

Reduced level – the height of a point or horizontal line that is relative to a given point of known height (datum).

Retaining structures – walls and buildings that hold back or support the earth.

RICS – Royal Institution of Chartered Surveyors – a professional association for surveyors.

Right cone – cone where the top vertex or point is vertically above the centre of the base.

Safe loading – the solution calculated from the soil's ability to carry a load plus a factor of safety.

Schedule of work – a separate written contents list of the operations to be carried out for the job which includes a brief written specification, often called a reference specification, which describes the product/material to be fixed and how the fixing will be carried out.

Section drawings – a profile of the ground using the information from the boreholes next to one another – that way engineers can predict what happens to the ground between each borehole.

Settlement – the way that soil reacts to having a load placed upon it – usually the soil 'sinks' a little due to the extra weight placed upon it, although this settlement should not affect the building a great deal.

Site diary – a record of relevant information about the progress of the construction works completed by the contractor.

Sketch plans – an architect or designer develops a sketch plan according to the needs and wants discussed with a client at the consultation stage. This is the time when the designer is creating, refining and rejecting ideas as appropriate.

Slump test – a test of the concrete's workability that shows that the concrete is of the right quality and can be poured with the correct results.

SMM7 – the standard method of measurement (seventh edition) which is published by the Royal Institution of Chartered Surveyors (RICS).

Soil Association – organisation promoting healthy soils via a certification scheme that enables a producer to use the term 'organic' produce.

Soil parameters – how the soil will react to building work and imposing loads. The soil can be expected to carry a certain amount of weight depending on its parameters and its characteristics.

Specification – a written document that expresses information from the drawing in technical terms. It helps the reader to understand what the drawings represent.

Standard Method of Measurement of Building Works (SMM7) – a clearly defined method of calculating the materials needed for a project using a systematic and clear measuring procedure based on the dimensions shown in the contract drawings.

Stress distribution – this is how the foundation distributes the load of the building. A very wide, flat foundation will support more load than a narrow strip foundation.

Survey framework – a triangular network of measured horizontal lines that fix the end points (survey stations) of those lines.

Taken off or taking off – the process of taking dimensions from drawings and producing a quantity.

Take-off list – a list prepared by the estimator or quantity surveyor which is used to check that all the items required have been covered. Each item is ticket off as the measurement is worked through.

Tender – to make a formal offer or estimate for a job.

Tender documents – a document or series of documents that offer to do the work at a price – contractors tender for work by providing an estimate of how much they will charge for the works to be completed.

Tendering process – when an architect may ask several contractors who are interested and are capable of completing this work to provide an estimate of how much it will cost.

Theodolite – a highly accurate instrument for measuring horizontal and vertical angles for undertaking land surveying and setting out surveys.

Thermal performance – the ability of a material to retain heat in the structure.

Tie lines – a pair of horizontal plan measurements used to fix the position of a point.

Toolbox talks – a time when everyone stops work to discuss a safety aspect of a current job.

TRADA – the Timber Research and Development Association.

Traverse survey – an interlinked polygon of survey stations with known easting and northing coordinates.

Tree preservation order (TPO) – an order to protect a tree from being cut down, damaged or altered without permission of the local authority.

Triple point of water – the temperature and pressure at which the three known phases of a substance can exist, ice, liquid and vapour. The triple point of water is the equilibrium point 273.16 K at 610 N/m²which is one of the fixed points of international standard measurements of temperature.

Underpinned – when an existing foundation is settling downwards, which causes structural cracks to the external and internal finishes, and a new, deeper foundation is cast in sections below the old one.

U-value – a measurement of the rate of heat loss through a structure.

Vapour checks – impermeable barriers or membranes used to prevent moisture passing through the structure.

Variations – items that were not in the client's original budget and therefore are additional to the contract; for example obstructions encountered within the ground during excavation work.

Walk-over survey – visiting the site to enable the surveyor to match the information from the desk study to what they see in the field. Experienced surveyors can get a feel for ground conditions by undertaking a walk-over survey.

Water table – the level of water found in the ground during excavations.

Well-conditioned triangles – triangles which are roughly equilateral in shape with no small internal angles.

Whole circle bearing – an angle measured clockwise from a reference point, usually North, with a value ranging from 0° to 360°.

Work study – the timing of work so that a rate can be established. It involves watching the operative work on a known quantity and seeing how long it takes them to finish that work.

Working from the whole to the part – taking an overview of the survey area and working out how it could be split up into manageable parts.

Index